Y0-EEZ-815

CARCINOGENESIS—
A COMPREHENSIVE SURVEY
VOLUME 10

The Role of Chemicals and Radiation in the Etiology of Cancer

Carcinogenesis—A Comprehensive Survey

Vol. 10: The Role of Chemicals and Radiation in the Etiology of Cancer, *edited by E. Huberman and S. H. Barr, 560 pp., 1985*

Vol. 9: Mammalian Cell Transformation: Mechanisms of Carcinogenesis and Assays for Carcinogens, *edited by J. C. Barrett and R. W. Tennant, 1985*

Vol. 8: Cancer of the Respiratory Tract: Predisposing Factors, *edited by M. J. Mass, D. G. Kaufman, J. M. Siegfried, V. E. Steele, and S. Nesnow, 496 pp., 1985*

Vol. 7: Cocarcinogenesis and Biological Effects of Tumor Promotors, *edited by E. Hecker, N. E. Fusenig, W. Kunz, F. Marks, and H. W. Thielmann, 664 pp., 1982*

Vol. 6: The Nitroquinolines, *edited by T. Sugimura, 167 pp., 1981*

Vol. 5: Modifiers of Chemical Carcinogenesis, *edited by T. J. Slaga, 285 pp., 1980*

Vol. 4: Nitrofurans: Chemistry, Metabolism, Mutagenesis, and Carcinogenesis, *edited by G. T. Bryan, 243 pp., 1978*

Vol. 3: Polynuclear Aromatic Hydrocarbons, *edited by P. W. Jones and R. I. Freudenthal, 507 pp., 1978*

Vol. 2: Mechanisms of Tumor Promotion and Cocarcinogenesis, *edited by T. J. Slaga, R. K. Boutwell, and A. Sivak, 605 pp., 1978*

Vol. 1: Polynuclear Aromatic Hydrocarbons: Chemistry, Metabolism, and Carcinogenesis, *edited by R. Freudenthal and P. W. Jones, 465 pp., 1976*

Carcinogenesis—
A Comprehensive Survey
Volume 10

The Role of Chemicals and Radiation in the Etiology of Cancer

Scientific Editor

Eliezer Huberman

Technical Editor

Susan H. Barr

Division of Biological and Medical Research
Argonne National Laboratory
Argonne, Illinois

Raven Press ■ New York

Raven Press, 1140 Avenue of the Americas, New York, New York 10036

Made in the United States of America

Library of Congress Cataloging in Publication Data
Main entry under title:

The Role of chemicals and radiation in the etiology
of cancer.

(Carcinogenesis—a comprehensive survey ; v. 10)
Based on a symposium dedicated to the memory of
Charles Heidelberger, held on Aug. 26–29, 1984 in
Oak Brook, Ill., sponsored by Argonne National
Laboratory, and others.
Includes bibliographies and index.
1. Carcinogenesis—Congresses. 2. Carcinogens—
Congresses. 3. Tumors, Radiation-induced—Congresses.
4. Heidelberger, Charles—Congresses. I. Huberman,
Eliezer. II. Barr, Susan H. III. Heidelberger, Charles.
IV. Argonne National Laboratory. V. Series. [DNLM:
1. Neoplasms—chemically induced—congresses. 2. Neo-
plasms, Radiation-Induced—congresses. W1 CA7624 v.10 /
QZ 202 R7446 1984]
RC268.5.C36 vol. 10 616.99′4071 s 85-11886
ISBN 0-88167-130-4 [616.99′4071]

Papers or parts thereof have been used as camera-ready copy as submitted by the authors whenever possible; when retyped, they have been edited by the editorial staff only to the extent considered necessary for the assistance of an international readership. The views expressed and the general style adopted remain, however, the responsibility of the named authors. Great care has been taken to maintain the accuracy of the information contained in the volume. However, neither Raven Press nor the editors can be held responsible for errors or for any consequences arising from the use of information contained herein.

The use in this book of particular designations of countries or territories does not imply any judgment by the publisher or editors as to the legal status of such countries or territories, of their authorities or institutions, or of the delimitation of their boundaries.

Some of the names of products referred to in this book may be registered trademarks or proprietary names, although specific reference to this fact may not be made: however, the use of a name with designation is not to be construed as a representation by the publisher or editors that it is in the public domain. In addition, the mention of specific companies or of their products or proprietary names does not imply any endorsement or recommendation on the part of the publisher or editors.

Authors were themselves responsible for obtaining the necessary permission to reproduce copyright material from other sources. With respect to the publisher's copyright, material appearing in this book prepared by individuals as part of their official duties as government employees is only covered by this copyright to the extent permitted by the appropriate national regulations.

Dedicated to the memory of Charles Heidelberger

1920—1983

Charles Heidelberger, through vision and leadership, profoundly influenced our understanding of carcinogenic processes. His ideas and research accomplishments provided the inspiration and laid a foundation for much of the research that is being conducted in this field today.

Preface

The Role of Chemicals and Radiation in the Etiology of Cancer represents the most current thought on the cellular and molecular aspects of cancer induced by chemicals or radiation. It is divided into five sections: Overviews and Hypotheses, Carcinogen Metabolism, Chemical Carcinogenesis, Radiation Carcinogenesis, and Mutagenesis and Chromosomal Changes, and it begins with a tribute to Dr. Heidelberger from Van Rensselaer Potter.

The section Overviews and Hypotheses examines the roles played in carcinogenesis by transforming genes, evolution, and epigenetic factors. The origin and reversibility of malignancy is scrutinized, and the multistage and multifactorial nature of cancer is addressed.

The section Carcinogen Metabolism reviews and discusses recent developments in areas such as the detection of the ultimate active carcinogenic forms of chemicals known to induce tumors in humans and animals; cell, tissue, organ, and species specificity in chemical activation; the role of interindividual differences in carcinogen metabolism; and the role of chemical and biochemical dosimetry in the evaluation of biological risk.

The section Chemical Carcinogenesis addresses questions dealing with the cellular and molecular changes involved in the initiation and promotion of malignant cell transformation *in vitro* and *in vivo*. The role of modulators in carcinogenesis and cell differentiation, as well as the functions of specific receptors and membrane signals in tumor promotion, are brought into perspective.

In the section Radiation Carcinogenesis, detailed consideration is given to the roles of specific DNA damage and repair in the carcinogenic process in human and rodent cells, the involvement of oncogenes, and the biological basis for assessing carcinogenic risks of low-level radiation.

The final section, Mutagenesis and Chromosomal Changes, addresses recent research in the potential role of recessive susceptibility genes in the development of certain human cancers, the role of specific chromosomal changes in the development of leukemia, the mechanisms underlying the unique interactions between chemical carcinogens and DNA (and the consequent generation of specific mutations), and the use of *in vitro* tests in evaluating the potential risk of environmental chemicals.

All of the work described in this volume reflects the impact of the original studies conducted by Charles Heidelberger. His contributions, as highlighted by Potter and others throughout the book have significantly influenced the entire field of cancer research. By collecting these chapters and presenting them as a resource for those interested in the prevention of cancer in humans through an understanding of the mechanisms underlying its full development, I, too, acknowledge his influence and pay tribute to Charles Heidelberger, who was my mentor, colleague, and dear friend.

This volume will be of interest to research scientists, clinical specialists, and students, as well as those in governmental administrative or regulatory agencies dealing with environmental health hazards.

Eliezer Huberman

Acknowledgments

To honor the contributions of Charles Heidelberger, many prominent scientists assembled at the Argonne National Laboratory International Symposium "The Role of Chemicals and Radiation in the Etiology of Cancer" held on August 26–29, 1984, in Oakbrook, Illinois. This book records their presentations, as well as those of other colleagues who could not be present but who also wished to acknowledge the work of Dr. Heidelberger.

I would like to give special thanks to the other members of the symposium organizing committee, Carol A. Jones, Thomas E. Fritz, Colin K. Hill, and Carl Peraino. Then I would also like to thank all our co-sponsors: Argonne National Laboratory, The University of Chicago, the U.S. Department of Energy, the U.S. Environmental Protection Agency, Hoffman-LaRoche, Inc., and the Burroughs Wellcome Co. The details of conference organization were ably handled by Miriam Holden, Conference Manager for Argonne, and her staff. Strong encouragement and support were provided by Dr. Alan Schriesheim, Director of Argonne National Laboratory; Dr. Harvey Drucker, Associate Laboratory Director for Biomedical and Environmental Research, Argonne National Laboratory; and Dr. Henry Pitot, Director, McArdle Laboratory for Cancer Research, Madison, Wisconsin. In addition, I would like to acknowledge the warm support and participation of Mrs. Patricia Heidelberger.

For the preparation of this volume I would especially like to thank Ms. Betty L. Sheaks for her help in administrative matters, Ms. H. Leona Berger for the records processing and careful typing of all manuscripts, and the staff members of both the Graphic Arts Department and the Word Processing Center of the Energy and Environmental Sciences Division at Argonne for assistance in final manuscript preparation.

Eliezer Huberman

Contents

1 Years with Charles Heidelberger
V.R. Potter

Overviews and Hypotheses

15 Genetic Mechanisms of Oncogenesis
H.M. Temin

23 Origin and Reversibility of Malignancy
L. Sachs

35 Transforming Genes of Human Malignancies
S.A. Aaronson and S.R. Tronick

51 Evolution and Cancer
M. Calvin

65 Genetics and Epigenetics of Neoplasia: Facts and Theories
H.C. Pitot, L.E. Grosso, and T. Goldsworthy

81 Single-Hit Versus Multi-Hit Concept of Hepatocarcinogenesis
C. Peraino and S.D. Vesselinovitch

Carcinogen Metabolism

93 Sulfuric Acid Esters as Ultimate Electrophilic and Carcinogenic Metabolites of Some Alkenylbenzenes and Aromatic Amines in Mouse Liver
E.C. Miller, J.A. Miller, E.W. Boberg, K.B. Delclos, C.-C. Lai, T.R. Fennell, R.W. Wiseman, and A. Liem

109 Role of Pharmacokinetics and DNA Dosimetry in Relating *In Vitro* to *In Vivo* Actions of N-Nitroso Compounds
P.N. Magee

123 Analogous Patterns of Benzo[a]pyrene Metabolism in Human and Rodent Cells
J.K. Selkirk

135 *In Vitro* Systems to Study Organ and Species Differences in the Metabolic Activation of Chemical Carcinogens
R. Langenbach

147 Interindividual Differences in the Metabolism of Xenobiotics
A.H. Conney and A. Kappas

167 Chemical and Biochemical Dosimetry of Exposure to Genotoxic Chemicals
G.N. Wogan

Chemical Carcinogenesis

177 Cell Culture Studies on the Mechanism of Action of Chemical Carcinogens and Tumor Promoters
I.B. Weinstein

189 Mechanisms Involved in Multistage Skin Tumorigenesis
T.J. Slaga

201 Cellular and Molecular Changes During Chemical Carcinogenesis in Mouse Skin Cells
S.H. Yuspa

211 Mechanisms of Chemically Induced Multistep Neoplastic Transformation in C3H 10T½ Cells
J.R. Landolph

225 Inhibition and Enhancement of Oncogenic Cell Transformation in C3H 10T½ CL8 Cells
S. Nesnow, H. Garland, and G. Curtis

235 Genes and Membrane Signals Involved in Neoplastic Transformation
N.H. Colburn

249 Receptors and Endogenous Analogs for the Phorbol Ester Tumor Promoters
P.M. Blumberg, A.Y. Jeng, B. König, N.A. Sharkey, K.L. Leach, and S. Jaken

263 Control of Cell Differentiation and Cell Transformation *In Vitro* by Phorbol 12-Myristate 13-Acetate and 1α,25-Dihydroxyvitamin D_3
E. Huberman and C.A. Jones

275 Regulation of Cell Differentiation and Tumor Promotion by 1α,25-Dihydroxyvitamin D_3
T. Kuroki, K. Chida, H. Hashiba, J. Hosoi, J. Hosomi, K. Sasaki, E. Abe, and T. Suda

287 HL-60 Variant Reversibly Resistant to Induction of Differentiation by Phorbol Esters
L. Diamond, B. Perussia, R. Businaro, and F.W. Perrella

Radiation Carcinogenesis

303 Oncogenes and Cellular Controls in Radiogenic Transformation of Rodent and Human Cells
C. Borek

317 Repair and Misrepair in Radiation-Induced Neoplastic Transformation
M.M. Elkind, C.K. Hill, and A. Han

337 Mechanisms of Malignant Transformation of Human Diploid Cells
J.B. Little

355 Cancer in Ataxia-Telangiectasia
R.B. Painter

363 Cellular Responses in Chronic Radiation Leukemogenesis
T.M. Seed, L.V. Kaspar, T.E. Fritz, and D.V. Tolle

381 Biological Basis for Assessing Carcinogenic Risks of Low-Level Radiation
A.C. Upton

Mutagenesis and Chromosomal Changes

403 Retinoblastoma Gene: A Human Cancer Recessive (Regulatory?) Susceptibility Gene
W.F. Benedict

409 Chromosome Abnormalities in Human Leukemia as Indicators of Mutagenic Exposure
J.D. Rowley

419 Relationship of Chromosomal Alterations to Gene Expression in Carcinogenesis
N.C. Popescu and J.A. DiPaolo

433 Mammalian Cell Mutation and Polycyclic Hydrocarbon Carcinogenesis
P. Brookes, H.W.S. King, and M.R. Osborne

449 Role of Intercalation in Polycyclic Aromatic Hydrocarbon Carcinogenesis
R.G. Harvey, M.R. Osborne, J.R. Connell, S. Venitt, C. Crofton-Sleigh, P. Brookes, J. Pataki, and J. DiGiovanni

465 Stabilization of Z-DNA Conformation by Chemical Carcinogens
D. Grunberger, R.M. Santella, L.H. Hanau, and B.F. Erlanger

481 *In Vitro* Models of Mutagenesis
B.S. Strauss, K. Larson, D. Sagher, S. Rabkin, R. Shenkar, and J. Sahm

495 Estimation of the Potencies of Chemicals that Produce Genetic Damage
A.R. Peterson and H. Peterson

511 Potential Use of Gradient Denaturing Gel Electrophoresis in Obtaining Mutational Spectra from Human Cells
W.G. Thilly

529 Subject Index

Contributors

S. A. Aaronson
Laboratory of Cellular and Molecular Biology
National Cancer Institute
National Institutes of Health
Building 37, Room 1A07
Bethesda, Maryland 20205

E. Abe
Department of Biochemistry
School of Dentistry
Showa University
Hatanodai, Shinagawa-ku
Tokyo 142, Japan

W. F. Benedict
Clayton Molecular Biology Program
Children's Hospital of Los Angeles
4650 Sunset Boulevard
Los Angeles, California 90027

P. M. Blumberg
Laboratory of Cellular Carcinogenesis and Tumor Promotion
National Cancer Institute
National Institutes of Health
Building 37, Room 3A01
Bethesda, Maryland 20205

E. W. Boberg
McArdle Laboratory for Cancer Research
University of Wisconsin Medical School
450 North Randall Avenue
Madison, Wisconsin 53706

C. Borek
Radiological Research Laboratory, and Department of Pathology
Columbia University College of Physicians and Surgeons
630 West 168th Street
New York, New York 10032

P. Brookes
Chemical Carcinogenesis Section
Institute of Cancer Research
Fulham Road
London SW3 6JB, United Kingdom

R. Businaro
C.N.R. Laboratory of Cell Biology
Rome, Italy 00196

M. Calvin
Department of Chemistry
University of California
Berkeley, California 94720

K. Chida
Department of Cancer Cell Research
Institute of Medical Science
University of Tokyo
4-6-1, Shirokanedai Minato-ku
Tokyo 108, Japan

N. H. Colburn
Cell Biology Section
Laboratory of Viral Carcinogenesis
National Cancer Institute
Frederick, Maryland 21701

J. R. Connell
Institute of Cancer Research
Pollards Wood Research Station
Nightingales Lane
Chalfont St. Giles, Bucks HP8 4SP, United Kingdom

A. H. Conney
Department of Experimental Carcinogenesis and Metabolism
Hoffmann-La Roche Inc.
340 Kingsland Street, Building 86
Nutley, New Jersey 07110

C. Crofton-Sleigh
Institute of Cancer Research
Pollards Wood Research Station
Nightingales Lane
Chalfont St.Giles, Bucks HP8 4SP, United Kindom

G. Curtis
Environmental Health Research and Testing
Research Triangle Park, North Carolina 27709

K. B. Delclos
McArdle Laboratory for Cancer Research
University of Wisconsin Medical School
450 North Randall Avenue
Madison, Wisconsin 53706

L. Diamond
The Wistar Institute of Anatomy and Biology
36th and Spruce Streets
Philadelphia, Pennsylvania 19104

J. DiGiovanni
The University of Texas System Cancer Center
Science Park
Smithville, Texas 78957

J. A. DiPaolo
Laboratory of Biology
Division of Cancer Etiology
National Cancer Institute
National Institutes of Health
Bethesda, Maryland 20205

M. M. Elkind
Department of Radiology and Radiation Biology
Colorado State University
Fort Collins, Colorado 80523

B. F. Erlanger
Department of Microbiology
Columbia University Cancer Center/Institute of Cancer Research
New York, New York 10032

T. R. Fennell
McArdle Laboratory for Cancer Research
University of Wisconsin Medical School
450 North Randall Avenue
Madison, Wisconsin 53706

T. E. Fritz
Division of Biological and Medical Research
Argonne National Laboratory
9700 South Cass Avenue
Argonne, Illinois 60439

H. Garland
Carcinogenesis and Metabolism Branch, MD 68
Genetic Toxicology Division
Health Effects Research Laboratory
U.S. Environmental Protection Agency
Research Triangle Park, North Carolina 27711

T. Goldsworthy
McArdle Laboratory for Cancer Research
University of Wisconsin Medical School
450 North Randall Avenue
Madison, Wisconsin 53706

L. E. Grosso
McArdle Laboratory for Cancer Research
University of Wisconsin Medical School
450 North Randall Avenue
Madison, Wisconsin 53706

D. Grunberger
Division of Environmental Sciences
School of Public Health
College of Physicians and Surgeons of Columbia University
Cancer Center/Institute of Cancer Research
701 West 168th Street
New York, New York 10032

[1]A. Han
Cancer Research Laboratory
Southern California Cancer Center
University of Southern California School of Medicine
Los Angeles, California 90015

[1]Deceased, May 7, 1984.

L. H. Hanau
Department of Microbiology
Columbia University Cancer Center/
Institute of Cancer Research
New York, New York 10032

R. G. Harvey
The Ben May Laboratory for Cancer Research
The University of Chicago
5841 South Maryland Avenue
Chicago, Illinois 60637

H. Hashiba
Department of Cancer Cell Research
Institute of Medical Science
University of Tokyo
4-6-1, Shirokanedai Minato-ku
Tokyo 108, Japan

C. K. Hill
Division of Biological and Medical Research
Argonne National Laboratory
9700 South Cass Avenue
Argonne, Illinois 60439

J. Hosoi
Department of Cancer Cell Research
Institute of Medical Science
University of Tokyo
4-6-1, Shirokanedai Minato-ku
Tokyo 108, Japan

J. Hosomi
Department of Cancer Cell Research
Institute of Medical Science
University of Tokyo
4-6-1, Shirokanedai Minato-ku
Tokyo 108, Japan

E. Huberman
Division of Biological and Medical Research
Argonne National Laboratory
9700 South Cass Avenue
Argonne, Illinois 60439

S. Jaken
Laboratory of Cellular Carcinogenesis and Tumor Promotion
National Cancer Institute
National Institutes of Health
Building 37, Room 3A01
Bethesda, Maryland 20205

A. Y. Jeng
Laboratory of Cellular Carcinogenesis and Tumor Promotion
National Cancer Institute
National Institutes of Health
Building 37, Room 3A01
Bethesda, Maryland 20205

C. A. Jones
Division of Biological and Medical Research
Argonne National Laboratory
9700 South Cass Avenue
Argonne, Illinois 60439

A. Kappas
The Rockefeller University Hospital
New York, New York 10021

L. V. Kaspar
Division of Biological and Medical Research
Argonne National Laboratory
9700 South Cass Avenue
Argonne, Illinois 60439

H. W. S. King
Chemical Carcinogenesis Section
Institute of Cancer Research
Fulham Road
London SW3 6JB, United Kingdom

B. König
Laboratory of Cellular Carcinogenesis and Tumor Promotion
National Cancer Institute
National Institutes of Health
Building 37, Room 3A01
Bethesda, Maryland 20205

T. Kuroki
Department of Cancer Cell Research
Institute of Medical Science
University of Tokyo
4-6-1, Shirokanedai Minato-ku
Tokyo 108, Japan

C.-C. Lai
McArdle Laboratory for Cancer Research
University of Wisconsin Medical School
450 North Randall Avenue
Madison, Wisconsin 53706

J. R. Landolph
Norris Cancer Hospital and Research Institute
University of Southern California School of Medicine
1441 Eastlake Avenue
Los Angeles, California 90033

R. Langenbach
Cellular and Genetic Toxicology Branch
National Institute of Environmental Health Sciences
P. O. Box 12233
Research Triangle Park, North Carolina 27709

K. Larson
Department of Molecular Genetics and Cell Biology
The University of Chicago
920 East 58th Street
Chicago, Illinois 60637

K. L. Leach
Laboratory of Cellular Carcinogenesis and Tumor Promotion
National Cancer Institute
National Institutes of Health
Building 37, Room 3A01
Bethesda, Maryland 20205

A. Liem
McArdle Laboratory for Cancer Research
University of Wisconsin Medical School
450 North Randall Avenue
Madison, Wisconsin 53706

J. B. Little
Department of Cancer Biology
Harvard School of Public Health
665 Huntington Avenue
Boston, Massachusetts 02115

P. N. Magee
Fels Research Institute, and
Department of Pathology
Temple University School of Medicine
3420 North Broad Street
Philadelphia, Pennsylvania 19140

E. C. Miller
McArdle Laboratory for Cancer Research
University of Wisconsin Medical School
450 North Randall Avenue
Madison, Wisconsin 53706

J. A. Miller
McArdle Laboratory for Cancer Research
University of Wisconsin Medical School
450 North Randall Avenue
Madison, Wisconsin 53706

S. Nesnow
Carcinogenesis and Metabolism Branch, MD 68
Genetic Toxicology Division
Health Effects Research Laboratory
U.S. Environmental Protection Agency
Research Triangle Park, North Carolina 27711

M. R. Osborne
Chemical Carcinogenesis Section
Institute of Cancer Research
Fulham Road
London SW3 6JB, United Kingdom

R. B. Painter
Laboratory of Radiobiology and Environmental Health
University of California
San Francisco, California 94143

J. Pataki
The Ben May Laboratory for Cancer Research
The University of Chicago
5841 South Maryland Avenue
Chicago, Illinois 60637

C. Peraino
Division of Biological and Medical Research
Argonne National Laboratory
9700 South Cass Avenue
Argonne, Illinois 60439

F. W. Perrella
The Wistar Institute of Anatomy and Biology
36th and Spruce Streets
Philadelphia, Pennsylvania 19104

B. Perussia
The Wistar Institute of Anatomy and Biology
36th and Spruce Streets
Philadelphia, Pennsylvania 19104

A. R. Peterson
University of Southern California Comprehensive Cancer Center
1303 North Mission Road
Los Angeles, California 90033

H. Peterson
University of Southern California Comprehensive Cancer Center
1303 North Mission Road
Los Angeles, California 90033

H. C. Pitot III
McArdle Laboratory for Cancer Research
University of Wisconsin Medical School
450 North Randall Avenue
Madison, Wisconsin 53706

N. C. Popescu
Laboratory of Biology
Division of Cancer Etiology
National Cancer Institute
National Institutes of Health
Bethesda, Maryland 20205

V. R. Potter
McArdle Laboratory for Cancer Research
University of Wisconsin Medical School
450 North Randall Avenue
Madison, Wisconsin 53706

S. Rabkin
Department of Molecular Genetics and Cell Biology
The University of Chicago
920 East 58th Street
Chicago, Illinois 60637

J. D. Rowley
The Franklin McLean Research Institute
The University of Chicago
950 East 59th Street
Chicago, Illinois 60637

L. Sachs
Department of Genetics
Weizmann Institute of Science
Rehovot 76100, Israel

D. Sagher
Department of Molecular Genetics and Cell Biology
The University of Chicago
920 East 58th Street
Chicago, Illinois 60637

J. Sahm
Department of Molecular Genetics and Cell Biology
The University of Chicago
920 East 58th Street
Chicago, Illinois 60637

R. M. Santella
Division of Environmental Sciences
School of Public Health
College of Physicians and Surgeons of Columbia University
Cancer Center/Institute of Cancer Research
701 West 168th Street
New York, New York 10032

K. Sasaki
Department of Cancer Cell Research
Institute of Medical Science
University of Tokyo
4-6-1, Shirokanedai Minato-ku
Tokyo 108, Japan

T. M. Seed
Division of Biological and Medical Research
Argonne National Laboratory
9700 South Cass Avenue
Argonne, Illinois 60439

J. K. Selkirk
Biology Division
Oak Ridge National Laboratory
P. O. Box Y
Oak Ridge, Tennessee 37831

N. A. Sharkey
Laboratory of Cellular Carcinogenesis and Tumor Promotion
National Cancer Institute
National Institutes of Health
Building 37, Room 3A01
Bethesda, Maryland 20205

R. Shenkar
Department of Molecular Genetics and Cell Biology
The University of Chicago
920 East 58th Street
Chicago, Illinois 60637

T. J. Slaga
University of Texas System Cancer Center
Science Park—Research Division
P. O. Box 389
Smithville, Texas 78957

B. S. Strauss
Department of Molecular Genetics and Cell Biology
The University of Chicago
920 East 58th Street
Chicago, Illinois 60637

T. Suda
Department of Biochemistry
School of Dentistry
Showa University, Hatanodai
Shinagawa-ku, Tokyo 142, Japan

H. M. Temin
Department of Oncology
McArdle Laboratory for Cancer Research
University of Wisconsin
450 North Randall Avenue
Madison, Wisconsin 53706

W. G. Thilly
Genetic Toxicology Department
Massachusetts Institute of Technology
Room E18-666, 50 Ames Street
Cambridge, Massachusetts 02139

D. V. Tolle
Division of Biological and Medical Research
Argonne National Laboratory
9700 South Cass Avenue
Argonne, Illinois 60439

S. R. Tronick
Laboratory of Cellular and Molecular Biology
National Cancer Institute
National Institutes of Health
Building 37, Room 1A07
Bethesda, Maryland 20205

A. C. Upton
Department of Environmental Medicine MSB 213
New York University Medical Center
550 First Avenue
New York, New York 10016

S. Venitt
Institute of Cancer Research
Pollards Wood Research Station
Nightingales Lane
Chalfont St. Giles, Bucks HP8 4SP, United Kingdom

S. D. Vesselinovitch
Departments of Radiology and Pathology
The Pritzker School of Medicine
The University of Chicago
Chicago, Illinois 60637

I. B. Weinstein
Division of Environmental Sciences and Cancer Center
Institute of Cancer Research
Columbia University, New York 10032

R. W. Wiseman
McArdle Laboratory for Cancer Research
University of Wisconsin Medical School
450 North Randall Avenue
Madison, Wisconsin 53706

G. N. Wogan
Department of Nutrition and Food Science
Massachusetts Institute of Technology
Building 16, Room 333
Cambridge, Massachusetts 02139

S. H. Yuspa
Laboratory of Cellular Carcinogenesis and Tumor Promotion
National Cancer Institute
National Institutes of Health
Building 37, Room 3A23
Bethesda, Maryland 20205

Carcinogenesis, Vol. 10, edited by E. Huberman and S. H. Barr. Raven Press, New York © 1985.

Years with Charles Heidelberger

V. R. Potter

McArdle Laboratory for Cancer Research, University of Wisconsin Medical School, Madison, Wisconsin 53706

As friends and former colleagues of Dr. Charles Heidelberger, we are saddened by his untimely death, but we can take pride in his humanity and his life, which was full of enthusiasm, productivity, and determination. His contributions to science were so formidable while his health permitted that we could only expect the best had he been permitted to continue for another decade. In eulogizing departed comrades, we reaffirm and idealize the patterns of behavior that we admire and that provide us with models to help us improve our own conduct. In this connection, let us recall a line from Lincoln's Gettysburg address, when he said "It is for us, the living, rather to be dedicated here to the unfinished work which they who fought here have thus far so nobly advanced."

Dr. Heidelberger was born in New York City on December 23, 1920, the only son of a famous father, Michael Heidelberger, who became a member of the National Academy of Science when Charlie was only 22, thus establishing for Charlie a goal that was realized in 1978 when Michael and Charles could enjoy each other's membership in the Academy.

Charlie's life was indelibly affected by the death of his mother while he was still quite young. He told me of his personal rage at the cancer that claimed his mother and at the contemporary scene that led her to conceal her disease from husband and son until the end was near. There is no doubt that the choice of his life's work was strongly influenced by his mother as a cancer victim and a courageous woman and by his father as an outstanding scientist.

Charlie attended Harvard University as an undergraduate, majoring in chemistry, with an S.B. degree in 1942, and as a graduate student in organic chemistry receiving the Ph.D. under the supervision of Louis Fieser in 1946. World War II was in progress and Fieser's group pursued several war-related projects. Part II of Charlie's thesis was entitled: "The Synthesis and Antimalarial Activity of Some Naphthoquinones."

After receiving the Ph.D., Dr. Heidelberger transferred to the laboratory of Professor Melvin Calvin in Berkeley for postdoctoral training, particularly to obtain expertise in the use of radioactive isotopes. His paper with Brewer and Dauben on "The synthesis of 1,2,5,6-Dibenzanthracene Labeled in the 9-position with ^{14}C," appeared in 1947 and caught the eye of Dr. Harold Rusch, Director of the McArdle Laboratory. This paper described the first synthesis of its kind and was the forerunner of a continuing series on the metabolism and action of the carcinogenic hydrocarbons. Dr. Rusch traveled to Berkeley in connection with a scientific meeting and arranged a personal interview. As a result, Dr. Heidelberger accepted a position as Assistant Professor of Oncology at the McArdle Laboratory beginning in the fall of 1948. While in Berkeley he had enjoyed as colleagues Dr. Bert Tolbert, now at the University of Colorado in Boulder, and Dr. Peter Yankwich, now at the University of Illinois in Urbana, as well as the sponsorship of Dr. Melvin Calvin, who remains active in Berkeley. While there, Dr. Heidelberger coauthored the first authoritative book on Isotopic Carbon with Calvin, Reid, Tolbert, and Yankwich.

Dr. Heidelberger was brought to McArdle to take responsibility for setting up radioactive counting facilities for the entire department, to provide expertise in the field of organic synthesis of labeled compounds, and to pursue the cancer problem according to his own inclinations. I can speak of the McArdle years from personal experience, for Charlie and I were kindred spirits from the beginning. We were closely associated for 28 years at McArdle and met frequently after he moved to the University of Southern California. Charlie and I were friends. Some friends play golf together or go fishing together. Charlie and I found common ground in science, although, of course, each of us had many research activities that the other had nothing to do with.

I'd like to give some examples of our interaction. Early in 1949 and soon after he had arrived at McArdle, Charlie came into my lab with a copy of Nature dated Dec. 18, 1948, in which A. G. Ogston of Oxford University argued on purely theoretical grounds "that the asymmetric occurrence of isotope in a product cannot be taken as conclusive evidence against its arising from a symmetrical precursor" since, as he said, "it is possible that an asymmetric enzyme which attacks a symmetrical compound can distinguish between its identical groups" (page 963). Charlie said, "Well, what do you think Van," and I said, "I think he's wrong, but let's get out the models." We modeled citric acid and immediately concluded Ogston was right after all. There were no data reported in the Ogston paper and Charlie was eager to do the experiments, since he knew that I had all the enzymatic methods and that he could handle the ^{14}C. We began at once and on June 7 we sent the manuscript to Sir Hans Krebs with whom I had served as a postdoc in 1939. He sent it to Nature on June 30, 1949, with a note of urgency, and it appeared in Nature on July 30, 1949, under the title "Biosynthesis of 'Asymmetric' Citric Acid, a Substantiation of the Ogston Concept." We had been concerned whether we were infringing on Ogston, but he wrote a letter to assure us we were not and offered his congratulations, as did Krebs, Birgit Vennesland, Earl Evans, and Harland Wood, among others. There were many invitations for lectures, which were all given by Charlie, sometimes with lively discussions. This was his first paper from our department. It undoubtedly helped to quickly identify Charles Heidelberger with the McArdle Laboratory. Indeed, on the other hand, his presence surely enhanced our image. I mention this work as an example of Charlie's initiative and enthusiasm. Without him I would never have started the Ogston project.

In 1947 at the International Cancer Congress in St. Louis I had noted that Mitchell's studies with Neurospora suggested that a path from citric cycle intermediates could lead to nucleic acid pyrimidines via orotic acid, but at that time I lacked the technology to do the indicated experiments. With the success of our first collaboration, I asked Charlie if he would devise the synthesis of ^{14}C-labeled orotic acid, which was not commercially available at that time. He supervised the work of my student Robert Hurlbert, and their description of the synthesis appeared in the Journal of the American Chemical Society in 1950, authored by Heidelberger and Hurlbert. They had just begun work on the synthesis in 1949 when the whole project seemed to be "scooped" by the Swedish workers Arvidson, Eliasson, Hammarsten, Reichard, and Sune Bergstrom, who reported in the Journal of Biological Chemistry in May 1949 that N^{15}-labeled orotic acid was converted to nucleic acid pyrimidines in rat liver. Again Charlie came to my lab with journal in hand, only this time he said we might as well not continue with the project. (I should point out that Charlie was not satisfied with just doing good scientific research. Throughout his career, he was strongly motivated to be first.) I insisted on continuing and this we did, while Charlie's role ended with the completion of the synthetic project. Bob Hurlbert, Hanns Schmitz, Anne Brumm, and I went on to report the isolation of the previously unknown acid-soluble pyrimidine and purine ribotides in 1954. Again I emphasize that this work would not have occurred without the enthusiastic cooperation of Charles Heidelberger at the outset. (I regret to record here that two of the listed coauthors have died of cancer, Dr. Hanns Schmitz with pancreatic cancer in 1967, and Anne Brumm McElya with breast cancer late in 1982 at age 57.)

Although Charlie was not involved in pyrimidine or nucleic acid research at this time, he was well aware of the pathways involved. He was also aware of my studies on the action of fluoroacetate, which was converted to an effective inhibitor, fluorocitrate, thus blocking the citric acid cycle. But again he was several jumps ahead of my thinking when one day he came in and said "What would you think of a 5-fluoro-derivative of uracil as a potential blocker of DNA synthesis?" My mouth was agape and all I could say was "Why didn't I think of that?" But I know why! Stored in Charlie's fertile brain were not only pyrimidine pathways and the structures of fluoro-derivatives of several types including fluoro-acetate and the fluorinated carcinogens studied by the Millers, but he also was familiar with the organic syntheses that might be adapted. So he put it all together and came up with the invention of 5-fluorouracil. What a "Eureka" experience that must have been! The first publication was in March 1957 in Nature. This paper reported the synthesis of both 5-fluoro-uracil and 5-fluoro-orotic acid, which drew on the earlier synthesis of ^{14}C-orotic acid, a scheme that was never entered in my own memory bank. (5-Fluorouracil has been able to produce actual cures for a number of cancer patients).

Meanwhile, Dr. Heidelberger maintained his interest in carcinogen metabolism and action, but now he was firmly committed to the maintenance of a second line of research, the invention of new chemotherapeutic agents.

Throughout his 28 years at the McArdle Laboratory, Charlie Heidelberger was always a friend, always stimulating, and always enthusiastic and energetic. His courage in his final years was evident in his vigorous presentations at the U.C.L.A. symposium in February, 1982, and again at the Gatlinburg symposium in April, 1982, after which Charlie performed on his trumpet for 3 hours during the social evening. His performance brought back many happy memories, but this time my nostalgia was etched against my knowledge of Charlie's predicament and of what he had just undergone in the way of treatment. I will never forget the image that I retain on something like a double-exposure photo--Charlie explaining slides from a podium with a superimposed frame of Charlie blowing a Duke Ellington tune on his trumpet that evening in Gatlinburg. And I'll never forget the earlier time when 28-year-old Charlie came into my office back in 1949 with journal in hand to begin a friendship that lasted 34 years and will never be forgotten. I am reminded of lines from the Funeral Oration of Pericles: "For the whole earth is the sepulchre of famous men; not only are they commemorated by columns and inscriptions in their own country, but in foreign lands there dwells also an unwritten memorial of them, graven not on stone but in the hearts of men."

APPENDIX I

In connection with this presentation, I contacted several people who knew Charlie and asked for a letter that could be reproduced totally or in part. Their letters will appear below, more or less in the order in which they first met Charlie.

From Melvin Calvin: February 7, 1983

Some years ago I received a letter from Charlie Heidelberger in which he reminisced about the early years in Berkeley, when he first came to us from his degree with Louis Fieser at Harvard. At the time Charlie was hired, Professor Fieser had recommended someone else for the position--"more highly qualified"--but as things worked out, I believe we were the lucky ones to have Charlie come to Berkeley at such an auspicious time in the scientific development of radiocarbon and tracer chemistry/biochemistry.

When Charlie came to Berkeley in 1947 the availability of carbon-14 was just opening up many doors in the area of organic synthesis, biochemistry, and metabolism studies, and Charlie took advantage of the juxtaposition of the availability of tracer carbon, the availability of sophisticated (for those days!) counting procedures for carbon-14 and the biochemical/biological facilities at the Donner Laboratory. He was responsible for the synthesis of the first carbon-14-labeled carcinogenic hydrocarbon, using one precious millicurie of barium carbonate. He said in the letter to which I refer above: "technically, the training and experience I gained in your group in Donner made me an 'instant expert' in radiocarbon, which opened the door to many job offers, including the one I accepted at McArdle where I spent 27 happy and productive years....My synthesis of the first ^{14}C carcinogenic hydrocarbon with one precious millicurie initiated my career-long interest in chemical carcinogenesis and I am still working in that field (but not with the same millicurie!)."

Charlie's contributions to the field of chemical carcinogenesis and clinical carcinogenesis do not need any elaboration from me. They are well known to the scientific community. I am personally proud that the beginnings of the long search that he undertook, which had its origin in the Donner group in 1947-1948, led him into the research that has had so much importance for us all.

The original "Four Horsemen" (Heidelberger, Reid, Tolbert, and Yankwich) of the Donner group, which had its origins in 1945 and which produced the first and definitive book on "Isotopic Carbon," have scattered throughout the scientific community of the United States. But all of them have left their mark on science and, more importantly, on the continuing of scientific thought. In fact, one of my more recent graduate students, a physical chemist (Joe Landolph), has been working with Charlie Heidelberger now for several years as a postdoctoral/staff associate, which gives an even closer connection to the original beginnings of Charlie's interests. Thus, the biodynamic connection will continue into the future at the Cancer Center of the University of Southern California.

From Bert M. Tolbert: February 8, 1983

It is a pleasure to recall Charlie Heidelberger's research years at Calvin's Bio-Organic Chemistry group in Berkeley. Charlie came to the lab with a well-developed scientific goal--to prepare carbon-14-labeled carcinogens and use the marker to follow the tissue distribution, excretion, and metabolism of these compounds. At that time the Bio-Organic Chemistry group was developing ways to synthesize and count these labeled compounds and Charlie wanted to use these procedures on his problem. His work went well and he became part of the group of five authors from that lab who wrote the book Isotopic Carbon. I went back to the internal progress reports of the lab, which I was preparing at the time Charlie was there. He was on the group roster for February of 1948, but by mid-year was no longer listed. This list gives Charlie's colleagues, most of whom went on to long and productive scientific careers--even as Charlie did.

The progress reports indicate that Charlie had made significant progress on dibenzanthracene metabolism using his labeled product by the time he left the lab. This work was to be continued in Wisconsin.

I remember Charlie as a person of great drive and energy. He was a careful experimentalist. In those days we were young and had many group/lab parties, to which Charlie contributed with his music and good humor.

From Peter E. Yankwich: February 12, 1983

It has been very difficult for me to put onto paper any "brief remarks" about Charlie Heidelberger that derive from our interaction and otherwise common experiences in the earlier days we were together in Melvin Calvin's "Bio-Organic Group" at Berkeley. There are lots of things I remember about those days, which were the start-up and build-up period of what became a world-famous laboratory. After all, Charlie, Jim Reid, Bert Tolbert, Andy Benson, and I were Melvin's first lieutenants--each with a specific area of expertise we were supposed to exercise. Because we interacted to help each other, but did not overlap each other at the Lab, my strongest memories of that period about Charlie are not about Charlie, the lab, the Bio-Organic Group, or what it got started, did, and finished. My strongest memories are of three then-young married couples: Charlie and Judy Heidelberger, Bill and Carol Dauben, and Betty and myself.

From Harold P. Rusch: February 12, 1983

Dr. Potter has summarized briefly the salient points about Dr. Heidelberger's accomplishments. I will mention two of his many other abilities that not only impressed me but also increased my already high esteem of him.

The first was his musical talent. In addition to playing the violin he was an excellent drummer and trumpeter. On a number of occasions when we attended a meeting in a city that had a good dance band, Charlie would invite me to come along with him to enjoy the music. After listening for awhile, he would introduce himself to the band leader and mention that he would be willing to substitute on the drums or with his trumpet, the latter which he brought with him, and the band leader would often grant his request. His ability to play was recognized by the audience who applauded enthusiastically to hear an outsider step in and perform as well as a professional. In order to play with a band in this extemporaneous manner, even for one or two numbers, it was necessary for Charlie to be a member of the musicians union and he continued to pay his annual dues for many years just to get the opportunity to play occasionally with a real professional band. He also played with friends who shared his hobby and who gathered from time to time for an informal jazz session. He also maintained his interest in the violin, which he played mainly at home. Every year, however, he organized a trio to play music at the annual University Hospital Christmas party. He played the violin and usually two people with recorders accompanied him. The second talent that impressed me was Charlies' ability to dictate his scientific manuscripts. Most of us struggle over the preparation of a manuscript, writing, crossing out, rearranging, and rewriting before we achieve a product that is finally ready to submit for publication. But not so with Charlie. He could dictate a complicated manuscript that was well organized, clear, and concise and required only a modest number of corrections before it was submitted to an editor. I think of the vast amount of time that I could have saved over the years if I had possessed the same ability.

From James A. and Elizabeth C. Miller: February 7, 1983

We are very pleased that you will be participating in the commemorative program for Charlie on February 19th and can thus provide his family and friends a direct report on the very high esteem and regard in which Charlie was held by his long-time colleagues at McArdle and on our sadness at his untimely death. Although Charlie made many contributions to cancer research, biochemistry, and organic chemistry throughout his very productive research career, that productivity would surely have been considerably greater had he been able to continue for another decade. Further, had his health permitted, retirement would not have been a time of rest for Charlie; it would instead have been an opportunity to take on new tasks based on his characteristic insight and the background of much experience.

Charlie came to McArdle at an important time in his professional life (just after completion of his postdoctoral work) and at an important time in the development of McArdle. He came as its eighth staff member, when McArdle was only eight years old. Charlie's interest in the metabolism of the polycyclic aromatic hydrocarbons and his pioneering use of carbon-14-labeled compounds for their study strongly complemented the interests and experiences of those of us already at McArdle. His experience with carbon-14 at Berkeley provided expertise that greatly facilitated its early use at McArdle. Our first work on carbon-14-labeled aminoazo dyes (1952) was collaborative with Charlie and depended strongly on his experience and on his early development of a departmental facility for determination of radioactive isotopes. As time went on he saw the advantages to be gained from the use of tritiated compounds of high specific activity and was instrumental in ensuring that McArdle's facilties for analysis of radioactive isotopes were maintained at the state of the art.

Charlie's training as an organic chemist was evident throughout his research. However, it was his coupling of that ability with his interest in cancer research and biochemistry that resulted in his unique contributions. Thus, he early saw the need to study the metabolism of the polycyclic aromatic hydrocarbons by procedures that were very sensitive and were independent of the variations in sensitivity that resulted from dependence on either ultraviolet absorbance or fluorometric methods. At that time, facility with organic synthesis was necessary to attack such a problem, since commercial isotopically labeled compounds were restricted to a few basic intermediates for synthesis. Similarly, from the organic chemical studies on the strength on the carbon-fluorine bond and our early studies on the high carcinogenic activities of fluorinated aminoazo dyes, he conceived the idea that 5-fluorouracil might be a potent inhibitor of DNA synthesis and thus might have utility in the therapy of cancer. This was not a research project for those with limited horizons or for the fainthearted. Indeed, the idea, which was rooted in his knowledge of biochemistry, was executed first through the application of organic chemistry. Once the synthesis was achieved (and it was not an easy success), tests of the efficiency of 5-fluorouracil depended on biological studies in rodents and then on obtaining cooperation from clinical associates. Finally, an understanding of the mechanism of action of the fluorinated pyrimidines came through studies of nucleic acid biochemistry in vivo and in vitro. Both the studies of the metabolism of the polycyclic aromatic hydrocarbons and of the action of the fluorinated pyrimidines became lifelong professional interests.

Characteristically, Charlie early foresaw the probable importance of organ and cell cultures for obtaining insight into the nature of the malignant transformation. Characteristically, too, he decided that he should acquire the biological background needed to work in that area. Accordingly, he spent a year at the Strangeways Laboratory in England to learn the state of the art of organ culture from Ilse Lasnitzki. Progress in applying the organ and cell culture systems to the problem of transformation came slowly, but he persisted. In time Charlie's C3H 10T1/2 cell line became one of the standard systems for studying the malignant transformation of fibroblasts.

We feel deeply privileged to have had 28 years in which we and Charlie were colleagues. His breadth of interests and expertise, his depth of knowledge, his outgoing nature, his interest in discussing a wide range of topics, his aggressiveness in seeking out whatever was needed, his ability to look ahead with considerable clairvoyance, his unusual ability to distill the important contributions from a meeting and report them clearly, and his loyalty to McArdle and his colleagues were all of major importance in the development of McArdle. McArdle's growth into a cancer research laboratory of stature was greatly facilitated by the strong interaction between the members of the faculty; Charlie made a major contribution in building and strengthening those interactions. His career from 1948-1976 was a part of McArdle's strength.

Please convey to those at the program our great feeling of loss in Charlie's death and our great pleasure in having worked with him and having known him well.

From Howard Temin: February 9, 1983

I first met Charlie on my job visit to McArdle Laboratory in 1959. His enthusiasm about McArdle and the University of Wisconsin, as well as the fact that he was working on nucleic acid metabolism were important factors in my decision to come to McArdle Laboratory.

After my arrival at McArdle, Charlie was helpful to me both scientifically and personally. His enthusiasm, his drive, his deep concern for solving the "cancer problem," and his willingness to try new approaches and techniques all provided stimulation and a model for younger cancer researchers.

His outspokenness and willingness to challenge "sacred cows" were very refreshing.

Although Charlie himself is gone, his compounds, his cells, and the question of the transformation of "all" carcinogen-treated 10T1/2 cells will keep his name in the minds of cancer researchers. In addition, I shall long remember him as a friend and valued colleague.

From Norman Kharasch: March 2, 1983

Thank you for coming to pay special tribute, at the University of Southern California, on February 19th, to Professor Charles Heidelberger--to mourn his passing and to celebrate his life.

Thank you also for inviting me to add this letter, at your discretion, to those you may send along with those of others to Patricia Heidelberger, Michael Heidelberger, and to others at the Academy of Sciences, and the laboratories Charles was associated with closely. We will forever continue to be inspired by his example--as a scientist, as a teacher and administrator, and, most importantly, as a very human person and friend.

So many meaningful memories about Charlie come crowding in as I think of him--memories of the years I knew him best from 1976 to the present. I shall only comment on a few of these in anecdotal form.

In December 1976, as Scientific Director of the Intra-Science Research Foundation, I was closely associated with developing one of the Foundation's most significant and successful international symposia, which was published in Cancer Chemotherapy (1976) and was a landmark summary of the subject at that time. Charlie was invited to come as a key speaker and--unknown to him--he was also a front-running contender for the prestigious Intra-Science Medalist Award, which the Foundation gives each year. The voting was very close and the Award Committee could have leaned either way--as we had two top candidates for the award, and the solicited votes put Professor Sol Spiegelman only one vote ahead of Charlie. A joint award would have been quite appropriate, but the Committee felt that a single medalist was more appropriate, and let the selection stand by the actual count. When Charlie learned the results, it was he who heartily and most sincerely congratulated Sol, gave the luncheon address in the medalist's honor, and participated most effectively in every way in making the symposium a great success. He knew that honors like that are in part a game, "you won some, and you lost some," and you rode right along with the winner when that was the case. Later, Charlie helped and befriended the Intra-Science Foundation in many ways, and his name is indelibly inscribed on its list of "Great Friends and Advisors."

In another matter that related significantly to my own career, Charlie appreciated my special innovative efforts, sought my guidance and expertise (especially in sulfur chemistry when he needed it), and used it effectively, because he asked the right questions and dug the answers out of me. Charlie provided the moral support, trust in our ideas, and access to facilities that were crucial. He recognized the importance of a basic effort in the prostaglandin field, and he was truly delighted to learn of our every progress and never tried to take a part of the action and possible glory. His support was steadfast, right to the very last--even when he already knew his own health predicament.

Charlie and Patricia also applauded my decision to set up the Louis Pasteur Library and Science Foundation. Again, Charlie--and Patricia--could have said "Oh, why?" But they did not. With friendship and understanding, they gave it their "blessing," even in the first idea stage--recognizing its potential importance as a mechanism for accomplishing major goals for the University and the community. I am very pleased to tell you that the collection on "Cancer Chemotherapy and Carcinogenesis," in the Louis Pasteur Library, of great original works in medicine since 1940, will be headed by Charlie's works and will be perpetually named in his honor.

For Mrs. Kharasch (Elenora) and myself, our greatest delight in Charlie was as a close and personal friend whom we will never forget.

APPENDIX II

Finally, I present a brief Curriculum Vitae and Charlie's own list of 12 principal contributions up to 1976.

Charles Heidelberger, Curriculum Vitae:

Born December 23, 1920, in New York City; Died January 18, 1983 in Los Angeles. S.B., Harvard University, 1942; M.S., Harvard University, 1944; Ph.D. (Chemistry), Harvard University, 1946. Instructor in Chemistry, Harvard University, 1946. Research Chemist, Radiation Laboratory, University of California, Berkeley, 1946-48. Assistant Professor of Oncology, McArdle Laboratory, University of Wisconsin, Madison, 1948-1952; Associate Professor, 1952-1958; Professor, 1958-1976; American Cancer Society Professor of Oncology, 1960-; Associate Director, Wisconsin Clinical Cancer Center, 1973-76. Professor of Biochemistry and Pathology, University of Southern California, Los Angeles, 1976-1983; Director for Basic Research, Los Angeles County--University of Southern California Comprehensive Cancer Center, 1976-83. American Association for Cancer Research: Board of Directors, 1959-62, 1965-68, 1975-78. Chairman of the Program Committee, Tenth International Cancer Congress, Houston, Texas, 1970. Member, Council of the International Union Against Cancer, 1970-74. Member Board of Scientific Counselors, Division of Drug Treatment, National Cancer Institute, 1975-78.

Awards: Langer-Teplitz Award for Cancer Research, 1958; Lifetime Professorship, American Cancer Society, 1960; American Cancer Society Wisconsin Section Award, 1965; James Ewing Society, Lucy Wortham James Award, 1969; British Association for Cancer Research, The First Walter I. Hubert Lecturer, 1969; G.H.A. Clowes Memorial Award, 1970; American Cancer Society, Annual National Award, 1974; Lila Gruber Award of the American Academy of Dermatology, 1976; Papanicolaou

Award, 1978; Chemical Industry Institute of Toxicology Founder's Award, 1982; U.S.C. Associates Award for Creativity in Research and Scholarship, 1982; C. Chester Stock Award, Memorial Sloan-Kettering Cancer Center, 1982; American Cancer Society, Man of the Year, 1982; the First Athayde International Cancer Prize, 1982 (which he was too ill to accept in person); and the University of Southern California Presidential Medallion, 1983.

Heidelberger's most famous contribution to cancer research was his synthesis and development of 5-fluorouracil (FU), one of the most successful drugs used in the treatment of patients with advanced cancer, particularly of the breast and gastrointestinal tract. With brilliant intuition, he made use of a previous observation that there was an enhanced utilization of uracil by experimental tumors and conceived the idea that the as-yet-unknown 5-fluorouracil should act as an antimetabolite of uracil and exert tumor inhibitory activity. He predicted its mechanism of action, synthesized it first in his laboratory, worked out its biochemical mechanism of action, and collaborated with the clinicians at the University of Wisconsin in the initial clinical trials. In the first publication in Nature in 1957, the chemistry, biochemical mechanisms, and first clinical results with 5-fluorouracil were all announced. This work has made available, not only clinically useful drugs (FU and FUdR) for cancer, and 5-fluorocytosine, which is clinically effective against yeast and fungal infections, but also widely used tools in cell and molecular biology. At present, FU is being used in combination with other drugs at the time of surgery for breast cancer and is very successfully delaying or preventing metastatic recurrences. Heidelberger also synthesized trifluorothymidine, which is the most active compound known against DNA viral infections when applied locally.

In 1947 Heidelberger synthesized the first ^{14}C-labeled carcinogenic hydrocarbon, and he has made important contributions to our understanding of the mechansim of their action every since.

It is unique to Heidelberger that he continued to make major contributions in two unrelated fields of cancer research: chemical carcinogenesis and cancer chemotherapy. In his Clowes Award Lecture in 1970 he described these parallel threads of research.

Charles Heidelberger, personal list of twelve principal contributions to science:

1947 (with P.W. Brewer and W.G. Dauben), The Synthesis of 1,2,5,6-Dibenzanthracene Labeled in the 9-Position with ^{14}C. J. Am. Chem. Soc. 69, 1389-91.

1957 (with N.K. Chaudhuri, P. Danneberg, D. Mooren, L. Griesbach, R. Duschinsky, R.J. Schnitzer, E. Pleven, and J. Scheiner), Fluorinated Pyrimidines, A New Class of Tumor-Inhibitory Compounds. Nature 179, 663-66.

1964 (with H.E. Kaufman), Therapeutic Antiviral Action of 5-Trifluoromethyl-2'-deoxyuridine in Herpes Simplex Keratitis. Science 145, 585-86.

1970 Chemical Carcinogenesis, Chemotherapy: Cancer's Continuing Core Challenges. G.H.A, Clowes Memorial Lecture, Cancer Res. 30, 1549-69.

1971 (with P.L. Grover, P. Sims, E. Huberman, H. Marquardt, and T. Kuroki), In Vitro Transformation of Rodent Cells by K-Region Derivatives of Polycyclic Hydrocarbons. Proc. Nat. Acad. Sci., USA 68, 1098-1101.

1971 (with E. Huberman, L. Aspiras, P.L. Grover, and P. Sims), Mutagenicity to Mammalian Cells of Epoxides and Other Derivatives of Polycyclic Hydrocarbons. Proc. Nat. Acad. Sci., USA 68, 3195-99.

1972 (with D.L. Dexter, W.H. Wolberg, F.J. Ansfield, and L. Helson), The Clinical Pharmacology of 5-Trifluoromethyl2'-deoxyuridine. Cancer Res. 32, 247-53.

1973 (with C.A. Reznikoff, J.S. Bertram, and D.W Brankow), Quantitative and Qualitative Studies of Chemical Transformation of Cloned C3H Mouse Embryo Cells Sensitive to Postconfluence Inhibition of Cell Division. Cancer Res. 33, 3239-49.

1973 Chemical Oncogenesis in Culture, Advances in Cancer Res. 18, 317-66.

1974 (with P.V. Danenberg and R.J. Langenbach), Structures of Reversible and Irreversible Complexes of Thymidylate Synthetase and Fluorinated Pyrimidine Nucleotides. Biochemistry 13, 926-33.

1975 (with U.R. Rapp, R.C. Nowinski, and C.A. Reznikoff), The Role of Endogenous Oncornaviruses in Chemically Induced Transformation. I. Transformation Independent of Virus Production. Virology 65, 392-409.

1976 (with S. Mondal), Transformation of C3H/10T1/2 Mouse Embryo Fibroblasts by Ultraviolet Irradiation and a Phorbol Ester. Nature 260, 710-11.

Carcinogenesis, Vol. 10, edited by E. Huberman
and S. H. Barr. Raven Press, New York © 1985.

Genetic Mechanisms of Oncogenesis

H. M. Temin

*McArdle Laboratory for Cancer Research, University of Wisconsin Medical School,
Madison, Wisconsin 53706*

Cancer is a multistep disease. The kinetics of the disease process can be seen in experimental oncology, where there is separation of the steps of initiation and promotion and also a long latent period for the development of chemically and some virally induced tumors. The same kinetics can be seen in the development of spontaneous cancers (6,11).

Another way to establish the multistep nature of cancer is to examine the differences between normal and homologous neoplastic cells. At a minimum, the phenotypic differences in the neoplastic cells seem to include altered morphology, altered growth control, immortal life, ability to invade, and ability to metastasize. These differences appear to be the result of separate genetic changes. In addition, in the normal natural history of tumors there is continued evolution of the tumor mass, that is, tumor progression.

In contrast, cancer is induced by highly oncogenic retroviruses in one step; that is, infection of a sensitive cell or animal by a single infectious, highly oncogenic retrovirus particle results in a transformed cell and subsequent growth of the transformed cell to a tumor mass. In this paper, I shall describe the mechanism of carcinogenesis by highly oncogenic retroviruses and explain this apparent difference in kinetics to illustrate the multiplicity of different genetic mechanisms that are involved in carcinogenesis (see 14 for a more detailed discussion and references).

LIFE CYCLE OF RETROVIRUSES

Retroviruses replicate through a DNA intermediate usually integrated with cell DNA. The integrated viral DNA is called the provirus. The provirus has long terminal direct repeats (LTRs) ending in small inverted repeats, the attachment sequences. The incoming virus synthesizes in the cytoplasm a linear DNA copy, which moves to the nucleus, is ligated to form a closed circular molecule, and is integrated into the cell's DNA (10). The cis-acting sequences required for these steps and for viral RNA synthesis are all in or near the LTRs. The coding sequences for the proteins involved in these steps and in virion formation are in between the LTRs (9,16).

Thus, the structure of a replication-competent retrovirus is LTR and other cis-acting sequences, coding sequences, and again other cis-acting sequences and LTR. In contrast, the structure of replication-defective, highly oncogenic retroviruses differs in that there is a substitution, an addition in Rous sarcoma virus, in the coding sequences (Fig. 1).

This alteration in the coding sequences represents the introduction of an oncogene into the virus. Experiments with cloned viral DNAs have defined these new sequences as the viral oncogene. A viral oncogene is a gene in a highly oncogenic retrovirus that codes for a protein required for neoplastic transformation of sensitive cells.

About 20 different viral oncogenes have been described. They are listed in Table 1, grouped according to whether they have a tyrosine protein kinase activity, a sequence similar to oncogenes with tyrosine protein kinase activity, guanosine triphosphate (GTP) binding activity, a nuclear location, or are not part of any of these groups.

Viral oncogenes originate from normal cell genes. These normal cell genes are called proto-oncogenes and are recognized by their nucleotide sequence homology to viral oncogenes.

However, there is not exact identity between the products of a viral oncogene and the homologous cellular proto-oncogene. For example, in a pair of genes we have been studying, the viral oncogene rel and the proto-oncogene c-rel (turkey), the two products differ in their amino and carboxy termini, several internal amino acids, and in their regulation--as shown by differences in promoters and 5' mRNA sequences and termina-

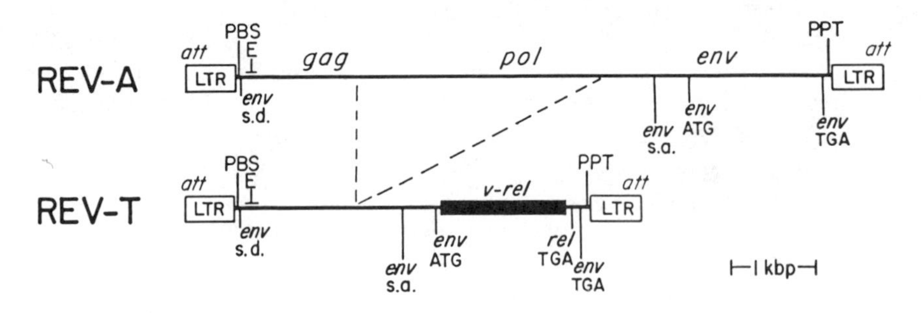

FIG. 1. Genomes of replication-competent helper retrovirus Rev-A and replication-defective highly oncogenic retrovirus Rev-T. Symbols are open box: LTRs; line: other virus sequences; thick bar: substitution of viral oncogene rel; PPT and PBS: sequences needed for viral DNA synthesis; E: encapsidation sequence; att: attachment sequence; s.d. and s.a.: splice donor and splice acceptor; gag, pol, env: viral coding genes.

TABLE 1. Viral oncogenes.

Viral oncogenes	Species of origin	Classification
src	chicken	tyrosine protein
yes	chicken	kinase activity
fps/fes	chicken/cat	
abl	mouse	
ros	chicken	
fgr	cat	
erb-B	chicken	
mos	mouse	sequence related
fms	cat	to src
raf/mil	mouse/chicken	
rel	turkey	?
Ha-ras/bas	rat/mouse	GTP-binding
Ki-ras	rat	activity
myc	chicken/cat	nuclear
myb	chicken	location
fos	mouse	
myb-ets	chicken	
sis	monkey/cat	other
ski	chicken	

tors, poly(A) addition sites, and 3' mRNA sequences (17). In addition, there is a large deletion in the upstream viral sequences in the virus carrying rel.

To establish the role of these differences in neoplastic transformation, we have carried out two types of experiments. In the first, we modified the viral sequences to see what effect these modifications have on the transforming ability of the resultant viruses. In such experiments we manipulate cloned viral DNA in plasmids and then recover virus by transfection of chicken embryo cells under appropriate conditions.

When the deletion was made larger, the recovered virus could only abortively transform cells in culture and could not form tumors. This abortive transformation is the result of the presence of sequences for a small polypeptide before the sequences for the rel protein, resulting in lower expression of the rel protein. Removal of the start codon for this upstream polypeptide restores the full transforming ability of the virus (15).

Alternatively, removing the deletion also abolished transforming activity of the recovered virus. The transforming activity was recovered when the deletion was restored by spontaneous processes (3). (Such deletions occurred at a frequency of about 10^{-4}.)

To look at changes in the oncogene more closely, another oncogene that is easier to manipulate was used. When the Ha-ras gene (from Harvey-murine sarcoma virus) was substituted in one of these avian retrovirus vectors, a virus was recovered that could transform chicken cells but could not transform rat cells. Transcription from the viral LTR is much less efficient in rat than chicken cells, again indicating the importance of quantitative factors in neoplastic transformation (Embretson, personal communication). When the homologous proto-oncogene, c-Ha-ras was substituted for the viral oncogene, the recovered virus could no longer transform (18).

Thus, both qualitative and quantitative factors are important in transformation of sensitive normal cells by highly oncogenic retroviruses.

We can examine more closely the evolution of highly oncogenic retroviruses to determine the nature of the genetic changes needed to activate proto-oncogenes. Integration of retrovirus DNA is very specific as far as which viral sequences are required. The termini of the LTRs make up the att sequence when they are ligated together in the closed circular viral DNA intermediate (10). Four base pairs from the middle of the att sequence are removed during integration and five base pairs of host DNA are duplicated. The sequence of the host DNA is not specified. Interestingly, the same base pairs of the element and the same size duplication of host DNA are found with several cellular movable genetic elements of *Drosophila* and yeast (4).

Other genetic changes occurred upon the insertion of the proto-oncogene sequences into the putative replication-competent retrovirus parent, that is, the loss of intervening sequences from the proto-oncogene and the deletion in the viral sequences. Further genetic changes occurred in the coding sequence of the oncogene, both that part derived from the virus and that part from the proto-oncogene.

Thus, the evolution of highly oncogenic retroviruses involved a great variety of different genetic changes to bring about the formation of the final highly oncogenic virus.

ONCOGENES IN NONVIRAL CANCER AND CANCER INDUCED BY REPLICATION-COMPETENT RETROVIRUSES

If this were all we could say about oncogenes, it would be interesting, but one could ask about its relevance to most cancers. Highly oncogenic retroviruses are not the usual cause of cancer outside laboratories, with the possible exception of feline leukemia. However, a series of important experiments over the last few years has shown that proto-oncogenes are altered in a wide variety of spontaneous and chemically induced tumors of man and rodents. For example, the Ha-ras gene has a base pair mutation in some tumors, the c-myc gene is translocated in others, the N-ras gene is amplified in others, the c-mos gene is activated by insertion of a retrovirus-like element, etc. (5).

Thus, a broader definition of oncogene has been developed. An oncogene is a gene that is altered in a cancer cell and either transforms NIH/3T3 or other cells after transfection, has homology to viral oncogenes, or is transcriptionally activated by retrovirus integration. In no case, however, is it possible yet to prove rigorously that these altered proto-oncogenes are relevant to the cancerous state. In fact, the findings that ras is mutated in 20% of spontaneous tumors and is sometimes mutated already in benign tumors or is only mutated during tumor progression might give one pause (1,13). However, it would be foolish not to pursue the hypothesis that these are true oncogenes until that hypothesis is disproven. (The activated proto-oncogenes could be involved in progression, could mutate frequently, or could be an epiphenomenon of neoplasia.)

This question is on the genetic level similar to the questions that classical oncologists faced first on the biochemical level and then on the cellular level. Because work is now on the DNA level does not make this question less acute.

ANTI-ONCOGENES

However, oncogenes are not the complete story of the etiology of cancer. For a long time it has been known that highly oncogenic retroviruses cannot transform all cells they infect but that they have specific target cell populations. Even Rous sarcoma virus does not cause tumors in embryos, although it does on the chorio-allantoic membrane (7). Increased expression of rel from a particular virus does not transform fibroblasts (15). Thus, there are factors that moderate or suppress the action of oncogenes, even those in highly oncogenic retroviruses (2).

Similar factors are probably responsible for the genetic events in certain childhood cancers, notably retinoblastoma and Wilms' tumors. The distribution of these tumors indicates that two mutational events are required (9). However, molecular analysis of the tumors has shown that the mutations are a requirement for homozygosity or hemizygosity at one locus and the nature of some of the mutations indicates that an inactivation has occurred (8,12). Thus, these mutations are not to activate oncogenes as we have been discussing.

The simplest hypothesis is that the inactivations are of genes whose action is to control or suppress the action of an oncogene. It is an interesting question as to whether or not these genes have evolved to control the oncogene or are genes that control the pre-existing proto-oncogene. My guess is the latter and that in evolution close control of these genes has developed since some mutations can result in cancer.

DEVELOPMENT OF CANCER

Thus, we can see there are several steps in the development of cancer. There must be mutations to give activation of oncogenes, mutations, or differentiation that results in inactivation of anti-oncogenes, then growth of the tumor cell to a tumor mass with further evolution and tumor progression consequent on that growth.

In cancer induction by highly oncogenic retroviruses, a single infection induces the cancer. This induction is only apparently single hit, because the highly oncogenic retrovirus has undergone a multistep evolution from the parental replication-competent retrovirus and proto-oncogene. Because the proto-oncogene was transduced into a retrovirus, all of the genetic changes that made it into an active oncogene were collected in the retrovirus. Thus, a multiply changed proto-oncogene was introduced into a cell by infection.

By contrast, in cells where cancer occurs as a result of mutations, the mutations are selected at the level of the whole cell. Thus, there is an accumulation of mutations in one cell as opposed to in one virus. In cells, mutations will not usually affect an already mutated gene, but will instead affect another one. Thus, in cancer cells we should not expect to see multiply altered proto-oncogenes as we see in highly oncogenic retroviruses. Rather in cancer cells we should expect to see several altered proto-oncogenes, each with only a single alteration. In both cases, a multistage process is required to cause the cancer.

ACKNOWLEDGMENTS

This paper is dedicated to Charles Heidelberger, a valued friend and colleague. The research in my laboratory is supported by Public Health Service Grants CA-22443 and CA-07175 from the National Cancer Institute, U.S. Department of Health and Human Services. I am an American Cancer Society Research Professor.

REFERENCES

1. Balmain, A., Ramsden, M., Bowden, G.T., and Smith, J. (1984): Nature, 307:658-659.
2. Bradley, A., Evans, M., Kaufman, M.H., and Robertson, E. (1984): Nature, 309:255-256.
3. Chen, I.S.Y., and Temin, H.M. (1982): Cell, 31:111-120.

4. Chisholm, G.E., Genbauffe, F.S., and Cooper, T.G. (1984): Proc. Nat. Acad. Sci., USA, 81:2965-2969.
5. Cooper, G.M. (1982): Science, 217:801-806.
6. Dix, D., Cohen, P., and Flannery, J. (1980): J. Theor. Biol., 83:163-173.
7. Dolberg, D.S., and Bissell, M.J. (1984): Nature, 309:552-556.
8. Godbout, R., Dryja, T.P., Squire, J., Gallie, B.L., Phillips, R.A. (1983): Nature, 304:451-453.
9. Knudsen, A.G., Jr. (1971): Proc. Nat. Acad. Sci., USA, 68:820-823.
10. Panganiban, A.T., and Temin, H.M. (1984): Cell, 36:673-679.
11. Peto, R. (1979): Proc. Roy. Soc. Lond. Ser. B, 205:111-120.
12. Solomon, E. (1984): Nature, 309:111-112.
13. Tainsky, M.A., Cooper, C.S., Giovanella, B.C., and Vande Woude, G.F. (1984): Science, 225:643-645.
14. Temin, H.M. (1984): In: Genetic Rearrangements in Cancer, edited by J.M. Goldman and D.G. Harnden. Churchill Livingstone, London.
15. Temin, H.M., and Miller, C.K. (1984): Cancer Surveys, in press.
16. Weiss, R., Teich, N., Varmus, H.E., and Coffin, J., editors (1982): RNA Tumor Viruses. The Molecular Biology of Tumors Viruses, second edition. Cold Spring Harbor Laboratory, Cold Spring Harbor, New York.
17. Wilhelmsen, K.C., Eggleton, K., and Temin, H.M. (1984): J. Virology, 52:172-182.
18. Wilhelmsen, K.C., Tarpley, W.G., and Temin, H.M. (1984): In: Cancer Cells. 2. Oncogenes and Viral Genes, pp. 303-308. Cold Spring Harbor Laboratory, Cold Spring Harbor, New York.

Carcinogenesis, Vol. 10, edited by E. Huberman and S. H. Barr. Raven Press, New York © 1985.

Origin and Reversibility of Malignancy

L. Sachs

Department of Genetics, Weizmann Institute of Science, Rehovot 76100, Israel

Charles Heidelberger was very versatile. He had a wide range of interests from science to jazz music. His death is a tragedy for all of us. He is a loss to the wider scientific community, but even more so for those with whom he was most closely associated. Our meetings will not be the same without him. We often agreed in discussions, and I trust that what I have to say today would have met with his approval.

THE QUESTIONS TO BE ANSWERED

The multiplication and differentiation of normal cells is controlled by different regulatory molecules. These regulators have to interact to achieve the correct balance between cell multiplication and differentiation during embryogenesis and the normal functioning of the adult individual. The origin and evolution of malignancy require genetic changes that uncouple the normal balance between multiplication and differentiation so that there are too many growing cells. This uncoupling can occur in various ways (61-63). Can the change of normal into malignant cells be induced in cell culture by environmental carcinogens? What genetic changes uncouple normal controls so as to produce malignant cells? When cells have become malignant, can malignancy be suppressed so as to revert malignant to nonmalignant cells? Malignant cells can have different abnormalities in the controls for multiplication and differentiation. Do all the abnormalities have to be corrected, or can they be bypassed in order to reverse malignancy? These questions can now be answered.

TRANSFORMATION OF NORMAL INTO MALIGNANT CELLS IN CULTURE BY CHEMICAL CARCINOGENS AND RADIATION

The development of an experimental system for the transformation of normal into malignant cells in culture by chemicals and radiation has made it possible to simplify studies on the identification and mechanism of action of environmental carcinogens. The first such cell culture system was developed with normal diploid hamster fibroblasts, carcinogenic hydrocarbons (2,3), and x-irradiation (8,9). The use of this experimental system has answered basic questions on the action of environmental carcinogens on normal diploid mammalian cells (7,25,58). These answers include the following findings: (i) These agents can directly induce changes in the growth pattern of normal fibroblasts, which grow as a flat layer, to an abnormal pattern in which cells pile up on one another (Fig. 1) (3) with a one-hit dose-response curve (27). (ii) Treatment with these agents also results in other cellular changes, which can arise independently, including an increased cellular life-span (3,8,31) and the ability to grow in soft agar; in addition, another stage is required that results in malignancy _in vivo_ (1,3,8,31). (iii) The changed growth pattern induced by chemicals (3) or radiation (8-10) requires a process associated with cell division to fix the transformed state. (iv) The frequency of cells with a changed growth pattern induced by the active carcinogenic metabolites (28-30,42) is higher than the frequency of single gene mutations for other loci (1,26). This system in which normal diploid fibroblasts are used has now been adopted in various laboratories (1,15,50).

Normal diploid fibroblasts, which have a normal limited life-span in culture, can be changed into fibroblast cell lines whose cells can multiply apparently indefinitely. These cell lines are no longer diploid. However, certain lines can be selected that, although they are heteroploid and have changed in that they no longer have the normal limited life-span, have not

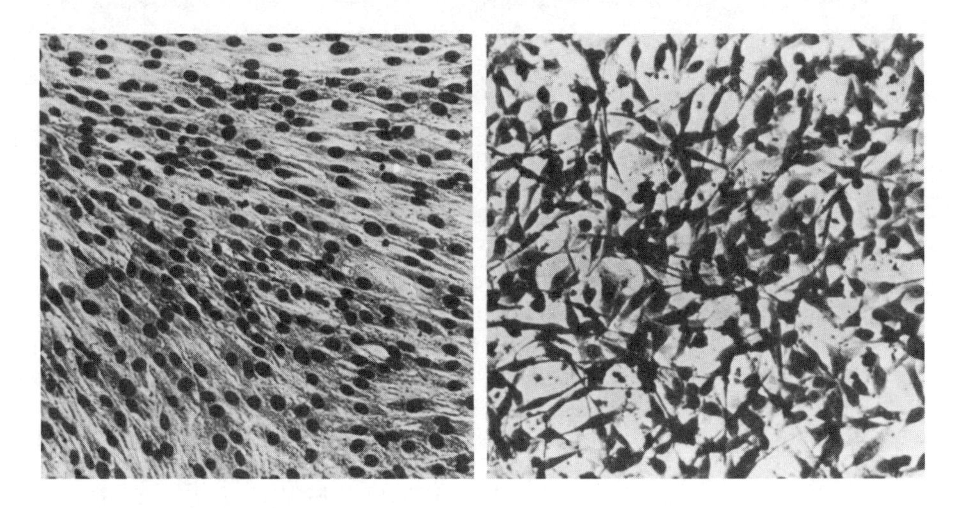

FIG. 1. Cultures of normal hamster fibroblasts (left) and hamster fibroblasts with a changed pattern of growth after treatment of normal cells with the carcinogenic hydrocarbon benzo[a]pyrene (right). From Berwald and Sachs (3).

yet evolved to the stage of being malignant cells that can form tumors in vivo. Such a mouse cell line, C3H 10T1/2 (56), has been established and it has proved to be very useful for studies with environmental carcinogens (11,23,35,55). This cell line has been used in many laboratories. The early events in carcinogenesis can only be studied in experiments starting with normal diploid cells. However, there are many questions that can be answered with 10T1/2 cells (23), and the answers have complemented those obtained from experiments with normal diploid cells. The results with both cell systems have thus elucidated the different early and late cellular changes that are induced by environmental carcinogens. They have formed an essential basis in elucidating how chemical carcinogens and radiation induce malignancy.

GROWTH FACTORS, DIFFERENTIATION FACTORS, AND THE UNCOUPLING OF NORMAL CONTROLS

An understanding of the mechanisms that control multiplication (growth) and differentiation in normal cells would also seem to be an essential requirement to elucidate how carcinogens induce malignancy. The development of appropriate cell culture systems has made it possible to identify the normal regulators of growth (growth factors) for various types of cells, and also, in some cell types, the normal regulators of differentiation (differentiation factors). This approach has been particularly fruitful in identifying the normal growth factors for all the different types of blood cells, first for myeloid cells (22,32,48,49) and then for other cell types, including T lymphocytes (45) and B lymphocytes (44), and in identifying the normal differentiation factors for myeloid cells (19,20, 38,67) and B lymphocytes (44). The normal differentiation factor for myeloid cells is a DNA-binding protein (67). It will be interesting to determine how far this approach applies to identifying normal differentiation factors for other cell types. Experiments with normal myeloid cell precursors have also shown that in these cells the growth factor can induce both growth (cell viability and multiplication) and production of the differentiation factor (37,39,40,61). This mechanism thus ensures the normal coupling of growth and differentiation, a coupling mechanism that may also apply to other cell types. Differences in the time of the switch-on of the differentiation factor would produce differences in the amount of cell multiplication before differentiation.

Identification of these normal growth and differentiation factors, and the cells that produce them, then made it possible to identify the different types of changes in the production of or response to these normal regulators that occur in malignancy. In myeloid leukemic cells, different clones of malignant cells have been identified that have shown all the possible changes that can occur in the normal response to growth and differentiation factors (60-63). There are different leukemic clones that (i) are independent of the normal growth factor for growth; (ii) constitutively produce their own growth factor; (iii) are blocked in the ability of the growth factor to induce production of the differentiation factor; (iv) are changed in their requirement for the normal growth factor, but can still respond normally to the normal differentiation factor (Fig. 2) (19); or (v) are blocked in their ability to respond to the normal differentiation factor. These are thus different ways to uncouple the normal controls for growth and differentiation and different ways for the cells to become malignant (60-63). Growth factors induce cell viability and cell multiplication. Independance from the normal growth factor or constitutive production of their own growth factor can also explain the survival and growth of metastasizing malignant cells in places in the body where the growth factor required for the survival of normal cells is not present.

Studies on protein synthesis in these different types of myeloid leukemic cells have shown that these different ways for the cells to become malignant were associated with changes from an induced to a constitutive expression of certain genes (36,61). These studies indicated that changes from inducible to constitutive gene expression can produce asynchrony in the normal developmental program, uncouple normal controls, and result in malignancy (61). The various types of changes that have been found in myeloid leukemic cells can serve as a model system to identify the different changes in the production and response to normal regulators that can give rise to malignant cells.

NORMAL GENES AND ONCOGENES

The transformation of normal into malignant cells requires a number of genetic changes, and some of the genes involved in this process have been characterized. The genes involved in the expression of malignancy are now referred to as oncogenes and the normal genes from which they are derived as proto-oncogenes (4,13,34). The changes of proto-oncogenes to oncogenes are in all cases associated with changes in the structure or regulation of the normal genes, and the change of normal into malignant fibroblasts requires changes in several proto-oncogenes (34). The normal genes from which the oncogenes are derived, even when they are still normal, are sometimes called cellular oncogenes rather than proto-oncogenes. As in the case of normal genes, not all oncogenes have the same function. Some genes such as myc change in their expression in the normal cell cycle (12,33). It will be interesting to determine further to what extent myc helps to drive the cell cycle or is a passenger that changes in the cell cycle, which is driven by other factors. The sis oncogene is derived from the normal genes for platelet-derived growth factor (16,66), the erb B oncogene from the gene for the receptor for epidermal growth factor (17), and the erb A oncogene from the gene for carbonic anhydrase that is involved in erythroid differentiation (14). The further

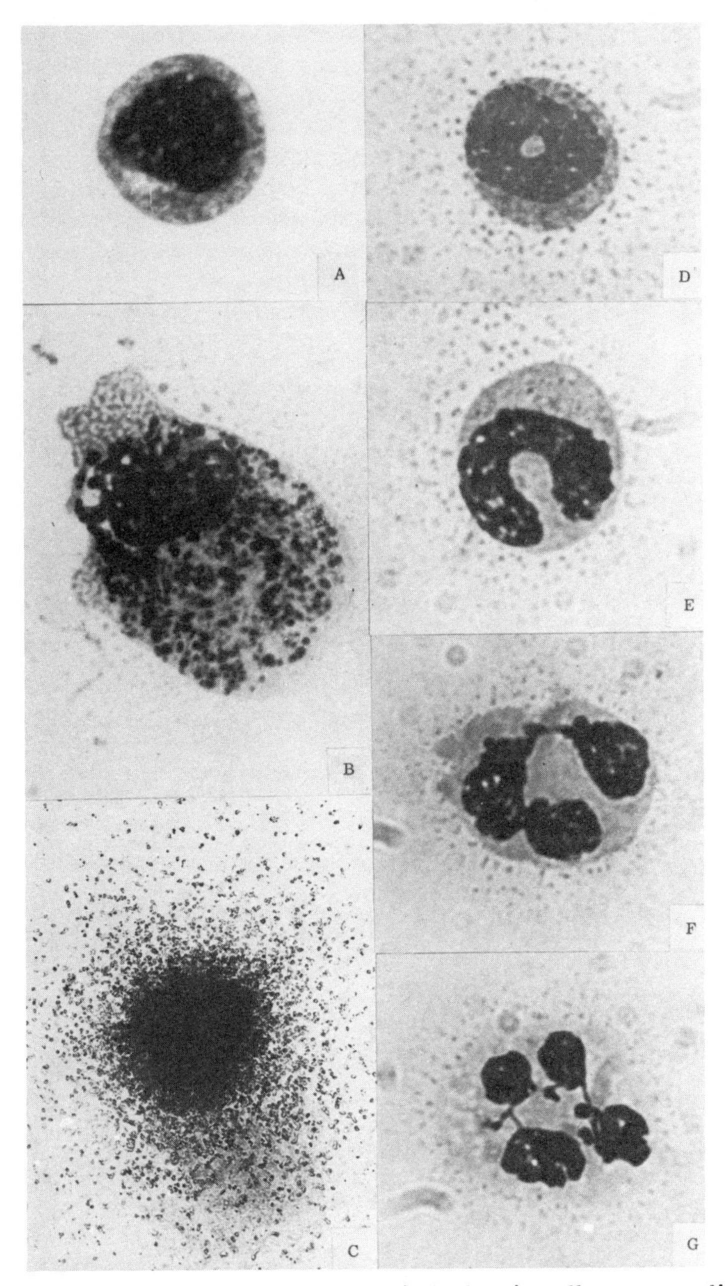

FIG. 2. Differentiation of mouse myeloid leukemic cells to nonmalignant mature macrophages or granulocytes by the normal myeloid differentiation factor. A: leukemic cell; B: macrophage; C: colony of macrophages; D-G: stages in differentiation to granulocytes. From Fibach et al. (20).

characterization of oncogenes, and the normal genes from which they are derived, should be able to define other genetic differences that result in changes in the normal production or response to growth and differentiation factors that occur in malignancy. The origin and evolution of malignancy can involve different genetic changes, including changes in gene dosage (59), gene mutations, deletions, and gene rearrangements (34). The different genetic changes in the structure or regulation of the normal genes that control growth and differentiation can thus produce in different ways the uncoupling of normal controls, which is required for the origin and further evolution of malignancy.

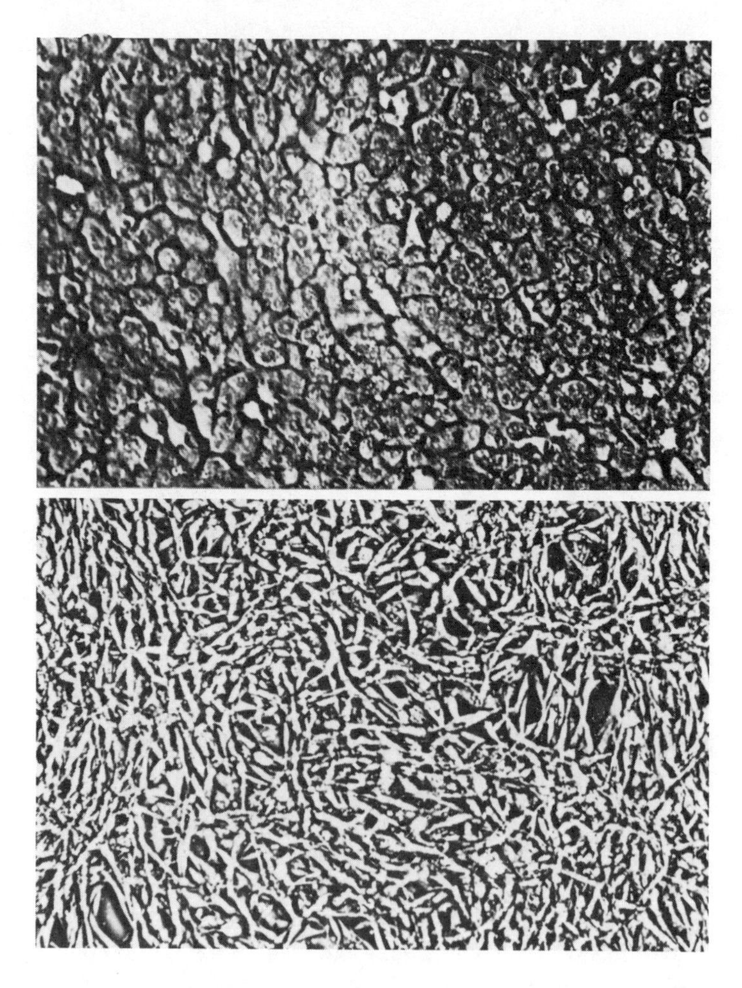

FIG. 3. Cultures of hamster sarcoma cells (bottom) and a nonmalignant revertant (top). From Rabinowitz and Sachs (52).

ONCOGENES AND THE REVERSIBILITY OF MALIGNANCY

The change of normal into malignant cells involves a sequence of genetic changes. Evidence has, however, been obtained with various types of tumors, including sarcomas (59), myeloid leukemias (59,60), and teratocarcinomas (65), that malignant cells have not lost the genes that control normal growth and differentiation. This preservation of these normal genes was first shown in sarcomas by the finding that it was possible to reverse in cultured cells the malignant to a nonmalignant phenotype with a high frequency in cloned sarcoma cells (Fig. 3) (52) whose malignancy had been induced by chemical carcinogens, x-irradiation, or a tumor-inducing virus (52-54,59). In sarcomas induced by chemical carcinogens or x-irradiation, this reversibility of malignancy included reversion to the limited life-span found with normal fibroblasts (54). A comparison of sarcomas with myeloid leukemias then showed that reversion of the malignant phenotype can be achieved by different mechanisms.

The chromosome studies on normal fibroblasts, sarcomas, revertants from sarcomas that had regained a nonmalignant phenotype, and re-revertants have indicated that the difference between malignant and nonmalignant cells is controlled by the balance between genes for expression (E) and suppression (S) of malignancy (5,6, 24,53,59,68). When there is enough S to neutralize E, malignancy is suppresed, and when the amount of S is not sufficient to neutralize E, malignancy is expressed (Fig. 4) (53). Genes for expression of malignancy (E) are now called oncogenes, and these experiments indicated that there are other genes, S genes, that can suppress the action of oncogenes ("soncogenes"). Suppression of the action of the Ki-ras oncogene in revertants (47) is presumably due to such suppressor genes. The balance between E and S genes, oncogenes and their suppressors, seems also to determine malignancy in other tumors including human retinoblastomas (46).

In the mechanism found with sarcomas, reversion was obtained by chromosome segregation, resulting in a change in gene dosage due to a change in the balance of specific chromosomes (Fig. 4). This reversion of malignancy by chromosome segregation, with a return to the gene balance required for expression of the nonmalignant phenotype, occurred without hybridization between different types of cells. The nonmalignant cells were thus derived from the malignant ones by genetic segregation. Reversion of the malignant phenotype associated with chromosome changes has also been found after hybridization between different types of cells (18,57,64).

In addition to this reversion of malignancy by chromosome segregation, another mechanism of reversion was found in myeloid leukemia. In this second mechanism, reversion to a nonmalignant phenotype was also obtained in certain clones with a high frequency but, in contrast to the mechanism found with sarcomas, this reversion was not associated with

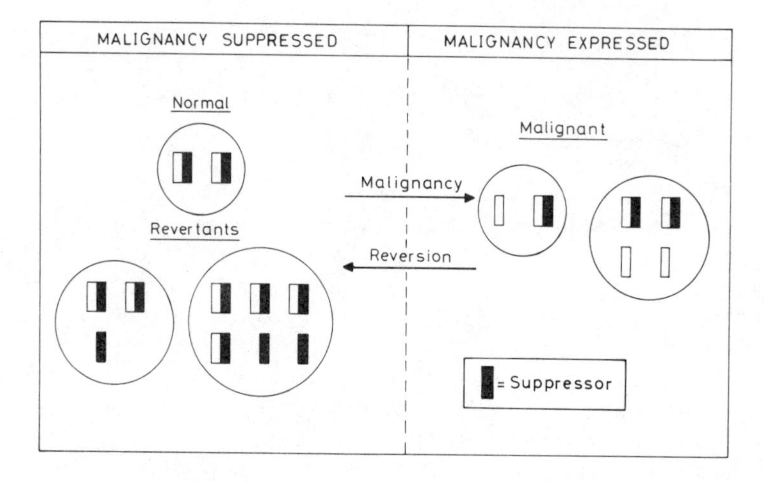

FIG. 4. The expression and suppression of malignancy controlled by the balance between expressor and suppressor genes. From experiments with normal fibroblasts, sarcoma cells, and sarcoma nonmalignant revertants. Genes for □ expression and ■ suppression of malignancy (53). Genes for expression are now called oncogenes and suppressor genes could be called soncogenes.

chromosome segregation. Phenotypic reversion of malignancy in these leukemic cells was obtained by induction of the normal sequence of cell differentiation by the normal differentiation factor (59-62). In this reversion of the malignant phenotype, the stopping of cell multiplication in mature cells by inducing differentiation (Fig. 2) (20) bypasses the genetic changes in the requirement for the normal growth factor that produced the malignant phenotype. Genetic changes that produce blocks in the ability to be induced to differentiate by the normal differentiation factor occur in the evolution of myeloid leukemia. But even these cells can be induced to differentiate by other compounds, either singly or in combination, that can induce the differentiation program by other pathways (59-62). Also in these cases, the stopping of cell multiplication by inducing differentiation by these alternative pathways bypasses the genetic changes that produce blocks in response to the normal differentiation factor. This bypassing of genetic defects is presumably also the mechanism for the reversion of malignancy by inducing differentiation in erythroleukemias (21,43), teratocarcinomas, and neuroblastomas, and it may also be able to induce differentiation in a variety of other tumors including hepatomas (51). There can, of course, also be cases of reversion in which all the oncogenes are lost, or in which changes of normal genes into oncogenes are actually reversed.

It can, therefore, be concluded that the changes of normal genes into oncogenes that result in the expression of malignancy do not mean that this expression of malignancy cannot again be suppressed. The results on the reversibility of malignancy have shown that there are different ways of reverting malignancy, that reversion does not have to restore all the normal controls, and that genetic abnormalities in normal controls can be bypassed by inducing differentiation. The results have also shown that reversion of malignancy may be useful as an approach to cancer therapy (41,59, 60,62). The use of appropriate cell culture systems, coupled with experiments in vivo, have thus answered many of the basic questions on the origin and reversibility of malignancy.

REFERENCES

1. Barrett, J.C., and Tso, P.O.P. (1978): Proc. Nat. Acad. Sci., USA, 75:3297-3301.
2. Berwald, Y., and Sachs, L. (1963): Nature, 200:1182-1184.
3. Berwald, Y., and Sachs, L. (1965): J. Nat. Cancer Inst., 35:641-661.
4. Bishop, J.M. (1983): Annu. Rev. Biochem., 52:301-354.
5. Bloch-Shtacher, N., and Sachs, L. (1976): J. Cell. Physiol., 87:89-100.
6. Bloch-Shtacher, N., and Sachs, L. (1977): J. Cell. Physiol., 93:205-212.
7. Borek, C. (1982): Adv. Cancer Res., 37:159-232.
8. Borek, C., and Sachs, L. (1966): Nature, 210:276-278.
9. Borek, C., and Sachs, L. (1967): Proc. Nat. Acad. Sci., USA, 57:1522-1527.
10. Borek, C., and Sachs, L. (1968): Proc. Nat. Acad. Sci., USA, 59:83-85.
11. Brankow, D. W., and Heidelberger, C. (1976): Cancer Res., 36:2254-2260.
12. Campisi, J., Gray, H.E., Pardee, A.B., Dean, M., and Sonensheim, G.I. (1984): Cell, 36:241-247.
13. Cooper, G. M. (1982): Science, 218:801-806.
14. Debuire, B., Henry, C., Benaissa, M., Biserte, G., Claverie, J.M., Saule, S., Martin, P., and Stehelin, D. (1984): Science, 224:1456-1459.
15. Di Paolo, J.A., Donovan, P.J., and Nelson, R.L. (1969): J. Nat. Cancer Inst., 42:867-876.
16. Doolittle, R.F., Hunkapiller, M.W., Hood, L.E., Devare, S.G., Robbins, K.C., Aaronson, S.A., and Antoniades, H.N. (1983): Science, 221:275-277.
17. Downward, J., Yarden, Y., Mayers, E., Scrace, G., Totty, N., Stockwell, P., Ulrich, A., Schlessinger, J., and Waterfield, M.D. (1984): Nature, 307:521-527.
18. Evans, E.P., Burtenshaw, M.D., Brown, B.B., Hennion, R., and Harris, H. (1982): J. Cell. Sci., 56:113-130.
19. Fibach, E., and Sachs, L. (1976): J. Cell. Physiol., 89:259-266.

20. Fibach, E., Landau, T., and Sachs, L. (1972): Nature New Biol., 237:276-278.
21. Friend, C. (1978): Harvey Lectures, 72:253-281. Academic Press, New York.
22. Ginsburg, H., and Sachs, L. (1963): J. Nat. Cancer Inst., 31:1-40.
23. Heidelberger, C., Landolph, J.R., Fournier, R.E.K., Fernandez, A., and Peterson, A. R. (1983): Proc. Nucleic Acid Res., 29:87-98.
24. Hitotsumachi, S., Rabinowitz, Z., and Sachs, L. (1971): Nature, 231:511-514.
25. Huberman, E. (1978): Environ. Pathol. Toxicol., 2:29-42.
26. Huberman, E., Mager, R., and Sachs, L. (1976): Nature, 264:360-361.
27. Huberman, E., and Sachs, L. (1966): Proc. Nat. Acad. Sci., USA, 56:1123-1129.
28. Huberman, E., and Sachs, L. (1974): Int. J. Cancer, 13:326-333.
29. Huberman, E., and Sachs, L. (1976): Proc. Nat. Acad. Sci., USA, 73:188-192.
30. Huberman, E., Sachs, L., Yang, S.K., and Gelboin, H.V. (1976): Proc. Nat. Acad. Sci., USA, 73:607-611.
31. Huberman, E., Salzberg, S., and Sachs, L. (1968): Proc. Nat. Acad. Sci., USA, 59:77-82.
32. Ichikawa, Y., Pluznik, D.H., and Sachs, L. (1966): Proc. Nat. Acad. Sci., USA, 56:488-495.
33. Kelly, K., Cochran, B.H., Stiles, C.D., and Leder, P. (1983): Cell, 35:603-610.
34. Land, H., Parada, L.F., and Weinberg, R.A. (1983): Science, 222:771-778.
35. Landolph, J.R., and Heidelberger, C. (1979): Proc. Nat. Acad. Sci., USA, 76:930-934.
36. Liebermann, D., Hoffman-Liebermann, B., and Sachs, L. (1980): Develop. Biol., 79:46-63.
37. Liebermann, D., Hoffman-Liebermann, B., and Sachs, L. (1982): Int. J. Cancer, 29:159-161.
38. Lotem, J., Lipton, J., and Sachs, L. (1980): Int. J. Cancer, 25:763-771.
39. Lotem, J., and Sachs, L. (1982): Proc. Nat. Acad. Sci., USA, 79:4347-4351.
40. Lotem, J., and Sachs, L. (1983): Int. J. Cancer, 32:127-134.
41. Lotem, J., and Sachs, L. (1984): Int. J. Cancer, 33:147-154.
42. Mager, R., Huberman, E., Yang, S.K., Gelboin, H., and Sachs, L. (1977): Int. J. Cancer, 19:814-817.
43. Marks, P., and Rifkind, R.A. (1978): Annu. Rev. Biochem., 47:419-448.
44. Möller, G., editor (1984): Immunol. Rev., No. 78. Munksgaard, Copenhagen.
45. Morgan, D.A.F., Ruscetti, F.W., and Gallo, R.C. (1976): Science, 193:1007-1008.

46. Murphree, A.L., and Benedict, W.F. (1984): Science, 223:1028-1033.
47. Noda, M., Selinger, Z., Scolnick, E.M., and Bassin, R.H. (1983): Proc. Nat. Acad. Sci., USA, 80:5602-5606.
48. Pluznik, D.H., and Sachs, L. (1965): J. Cell. Comp. Physiol., 66:319-324.
49. Pluznik, D.H., and Sachs, L. (1966): Exp. Cell Res., 43:553-563.
50. Pienta, R.J., Poiley, J.A., and Lebherz, W.B. (1977): Int. J. Cancer, 19:642-655.
51. Potter, V.R. (1978): Br. J. Cancer, 38:1-23.
52. Rabinowitz, Z., and Sachs, L. (1968): Nature, 220:1203-1206.
53. Rabinowitz, Z., and Sachs, L. (1970): Nature, 225:136-139.
54. Rabinowitz, Z., and Sachs, L. (1970): Int. J. Cancer, 6:388-398.
55. Reznikoff, C.A., Bertram, J.S., Brankow, D.W., and Heidelberger, C. (1973): Cancer Res., 33:3239-3249.
56. Reznikoff, C.A., Brankow, D.W., and Heidelberger, C. (1973): Cancer Res., 33:3231-3238.
57. Ringertz, N.R., and Savage, R.R. (1976): Cell Hybrids. Academic Press, New York.
58. Sachs, L. (1967): Current Topics Develop. Biol., 2:129-150. Academic Press, New York.
59. Sachs, L. (1974): Harvey Lectures, 68:1-35. Academic Press, New York.
60. Sachs, L. (1978): Nature, 274:535-539.
61. Sachs, L. (1980): Proc. Nat. Acad. Sci., USA, 77:6152-6156.
62. Sachs, L. (1982): Cancer Surveys, 1:321-342.
63. Sachs, L. (1984): Adv. Viral Oncol., 4:307-329.
64. Stanbridge, E.J., Der, C.J., Doersen, C., Nishimi, R.Y., Peehl, D.M., Weissman, B.E., and Wilkinson, J.E. (1982): Science, 215-259.
65. Stewart, T.A., and Mintz, B. (1981): Proc. Nat. Acad. Sci., USA, 78:6314-6318.
66. Waterfield, M.D., Scrace, G.T., Whittle, N., Stroobant, P., Johnsson, A., Wasteson, A., Westermark, B., Heldin, C.H., Huang, J.S., and Deuel, T.F. (1983): Nature, 304:35-39.
67. Weisinger, G., and Sachs, L. (1983): EMBO J., 2:2103-2107.
68. Yamamoto, T., Rabinowitz, Z., and Sachs, L. (1973): Nature New Biol., 243:247-250.

Carcinogenesis, Vol. 10, edited by E. Huberman
and S. H. Barr. Raven Press, New York © 1985.

Transforming Genes of Human Malignancies

S. A. Aaronson and S. R. Tronick

*Laboratory of Cellular and Molecular Biology, National Cancer Institute,
Bethesda, Maryland 20205*

Our understanding of the molecular mechanisms by which normal cells become converted to the malignant state has increased within the past few years because of a series of remarkable discoveries. Many different types of human tumors have been shown to possess discrete genetic sequences capable of inducing the transformation of appropriate assay cells. These genes were subsequently found to be the homologues of the cell-derived oncogene sequences responsible for the transforming activity of a group of well-characterized RNA tumor viruses. Investigations of the specific chromosomal translocations that occur in certain human tumors have also revealed that the movement of a specific cellular gene, again a homologue of a retroviral transforming gene, to a new regulatory environment alters its expression, thus contributing to tumor development.

In the past two years, remarkable progress has also been made in deciphering the normal functions of cellular genes that give rise to oncogenes. In what follows, we summarize the properties of retroviral transforming genes, their products, and their relationships with cellular genes. In addition, we discuss some of the mechanisms by which cellular genes can become activated to become transforming genes of human tumor cells.

VIRAL ONCOGENES, PROTO-ONCOGENES, AND THEIR FUNCTIONS

Proto-oncogenes comprise a small family of cellular genes that have been highly conserved throughout vertebrate evolution. These genes were initially discovered as the normal cellular progenitors of retroviral oncogenes by hybridization of viral oncogenes with normal cellular DNAs of a wide range of vertebrates (4,19,70). Their high degree of evolutionary conservation implies that they must serve very important functions in normal cellular growth processes.

If any of the thousands of genes that exist in our cells could acquire neoplastic properties when captured by one of these viruses, then we would never expect to observe the same sequence incorporated by more than one of the known two dozen or so acute transforming virus isolates. In fact, there is tremendous redundancy such that several virus isolates from the same or even different species have been shown to have captured identical onc sequences (8,70). These findings imply that the number of cellular genes that can acquire transforming properties when incorporated within the retrovirus genome must be rather limited.

There is increasing knowledge of the biochemical functions of some oncogenes, as well as emerging evidence that a number of these genes code for proteins with related structural and functional properties. One class of oncogenes, which at present includes only v-sis, codes for a growth-factor-like molecule (Table 1). This protein is closely related to human platelet-derived growth factor (PDGF) (17,68). We have recently shown that the normal human sequence corresponding to sis/PDGF can be activated as a transforming gene by causing it to be expressed in a cell type responsive to its growth stimulatory effects (27). These findings raise the possibility that genes for other normal growth factors or hormones might also act as oncogenes when inappropriately expressed in a suitable target cell.

Members of another class of oncogenes code for protein kinases, which usually have specificity for tyrosine residues (Table 1) (4). These oncogene products are located in the cytoplasm or plasma membrane. Some members of this class code for proteins that have kinase activity specific for serine and threonine, but these genes still show significant homology to the tyrosine kinases.

Recent studies have shown strong homology between the protein sequence of the EGF receptor and the predicted product of erb B, implying that this oncogene is likely to be an altered form of the normal growth factor receptor (18). Receptors for a number of other growth factors and hormones appear to be tyrosine kinases, suggesting that additional relationships between known oncogenes and such receptors are likely to be found.

Another major class comprises the ras family, which encodes very closely related 21,000 dalton transforming proteins (Table 1) (4). These proteins possess GDP binding and GTPase activity and appear to possess protein kinase activity as well. Ras transforming proteins are also located in the plasma membrane. Recent evidence suggests that these proteins may act in an as yet undefined way to modulate the adenylate cyclase pathway.

A fourth class of viral oncogenes encode nuclear proteins whose functions may be linked with cell cycle regulation and DNA replication. For example, recent experiments have demonstrated a specific cell-cycle dependence for the expression of the proto-oncogenes c-myc and c-fos (32,40). There is also evidence that these genes code for proteins that

TABLE 1. Cellular and viral oncogenes.

Oncogene	Species of origin	Protein Product: Biochemical function	Protein Product: Subcellular localization
Class I			
sis	woolly monkey/ cat	related to PDGF	cytoplasm and plasma membrane
Class II			
src	chicken	tyrosine kinase	plasma membrane
yes	chicken	tyrosine kinase	plasma membrane
fps/fes	chicken/cat	tyrosine kinase	cytoplasm
abl	mouse	tyrosine kinase	plasma membrane
ros	chicken	tyrosine kinase	cytoplasm
fgr	cat	tyrosine kinase	?
erb B	chicken	(related to EGF receptor)	plasma membrane
fms	cat	serine/threonine kinase	cytoplasmic membrane
mos	mouse	serine/threonine kinase	cytoplasm
Class III			
H-ras/bas	rat/mouse	GTP binding	plasma membrane
N-ras	human	GTPase activity	plasma membrane
K-ras	rat	autokinase	plasma membrane
Class IV			
myc	chicken	DNA binding	nuclear matrix
n-myc	human	?	?
myb	chicken	?	nuclear matrix
ski	chicken	?	nucleus
fos	mouse	?	nucleus
Class V			
rel	turkey	?	?
erb A	chicken	?	cytoplasm
met	human	?	?
B-lym	chicken/human	?	?

enhance expression of certain other cellular genes through direct or indirect effects on enhancer elements associated with such genes (31).

The remaining oncogenes are grouped within an additional class (Class V) because, at present, little is known about their function or subcellular location.

ACTIVATION OF ONCOGENES IN HUMAN MALIGNANCIES BY CHROMOSOMAL TRANSLOCATION

Proto-oncogenes related to retroviral oncogenes are well conserved in human DNA. It is possible to map these genes to specific human chromosomes by testing for the presence of human DNA fragments related to a specific oncogene in somatic cell hybrids possessing various numbers of human chromosomes, as well as in segregants of such hybrids. The chromosomal assignments of proto-oncogenes indicate that such genes are distributed through the human genome. The results of chromosome mapping studies by our laboratory and others are summarized in Table 2.

Nonrandom chromosomal aberrations occur in a number of malignant disorders. Most of these disorders affect hematopoietic tissues, and their consistent translocations are of importance because they are often specifically associated with morphologically well-defined subtypes of leukemia and lymphomas. In particular, specific chromosomal translocations have been constantly associated with Burkitt's lymphomas and chronic myelogenous leukemia (CML) (54). The mapping of proto-oncogenes to chromosomes has indicated not only that chromosomes involved in such specific rearrangements harbor onc-related genes (Table 2) but also that some of these proto-oncogenes are located at the breakpoints of these translocations. It has been possible to show that, in the case of Burkitt's lymphoma with the t(8;14) translocation (46,74), c-myc is translocated from chromosome 8 to chromosome 14 into the immunoglobulin μ chain switch region (13,21,67). In the other two Burkitt's lymphoma translocations, t(2;8) and t(8;22) (3), c-myc remains on chromosome 8, while the loci for the constant region of κ chain, located on chromosome 2, and of γ chain, located on chromosome 22, translocate to chromosome 8 into a region 3' to c-myc (12,22).

It has been postulated that, in Burkitt's lymphoma, chromosome translocation could increase c-myc transcription and, as a consequence, induce transformation (11,12,22). It has, in fact, been shown that, whether it is structurally rearranged or not, the c-myc proto-oncogene involved in the translocation is transcriptionally active, while the c-myc proto-oncogene on the normal chromosome 8 is transcriptionally silent. With somatic cell hybrids between Burkitt's lymphoma cells and either human lymphoblastoid cells or mouse plasmacytoma cells used for analyses of the expression of the translocated c-myc oncogene, it has also been shown that the translocated c-myc is transcribed in plasma cells but is suppressed in lym-

TABLE 2. Chromosomal assignments of the human proto-oncogenes.

Chromosome number	Proto-oncogene	Chromosomal aberration	Tumor[a]
1	N-ras	del 1p	Neuroblastoma
3	raf-1	t(3;14); t(3p); del 3p	Renal & lung CA, parotid tumors
4	raf-2	-	-
5	fms	del 5q	AML
6	K-ras-1; myb	$6q^-$; t(6:14)	ALL, ovarian CA
8	mos; myc	t(8;14); t(8;22); t(2;8)	Burkitt's lymphoma
		t(8;21)	AML
9	abl	t(9;22)	CML
11	H-ras-1	$11p^-$	Wilms' tumor
12	K-ras-2	12+	CLL
15	fes	t(15;17)	APL
20	src	-	-
22	sis	t(9,22)	CML

[a]CA - carcinoma; AML - acute myelogenous leukemia; CML - chronic myelogenous leukemia; APL - acute promyelocytic leukemia; CLL - chronic lymphocytic leukemia; ALL - acute lymphocytic leukemia.

phoblastoid cells (11). It has been postulated that the translocated c-myc comes under the transcriptional control of enhancer elements associated with the immunoglobulin locus. Studies to date suggest that the activity of such an enhancer may depend on its interaction with trans-acting factors active in plasma cells and Burkitt's lymphoma cells but not present or active in lymphoblastoid cells (11).

Another specific chromosomal translocation has been associated with CML where chromosome 9 is translocated to chromosome 22, the Philadelphia chromosome (47,53). The human c-abl gene was initially mapped on the long arm of chromosome 9 (38). It has subsequently been shown that in CML patients, the c-abl gene is translocated from chromosome 9 to chromosome 22 with the breakpoint occurring near the 5' end of c-abl gene (15,34). More recent studies have shown an aberrant 8-kp c-abl transcript in leukemia cells of 5 of 6 patients who have CML and the Philadelphia chromosome (5,26). Moreover, an altered human c-abl protein (p210) has been detected in K562 leukemia cells, established from a CML patient (41). This altered p210 c-abl protein was found to be phosphorylated on tyrosine in vivo and was also phosphorylated on tyrosine during in vitro kinase reactions. Thus, this translocation appears to alter the structure and function of the c-abl product.

The high degree of specificity of c-myc and c-abl rearrangements in Burkitt's lymphoma and CMLs, respectively, and the subsequent alteration of their biologic characteristics support the concept that these genes play a role in the etiology of such hematopoietic tumors. It seems likely that highly specific translocations associated with other hematopoietic malignancies may also activate as yet undiscovered oncogenes.

RAS PROTO-ONCOGENES OF HUMAN CELLS

Studies on the Kirsten and Harvey strains of murine sarcoma viruses (Kirsten-MSV, Harvey-MSV) led eventually to the identification of a small group of highly conserved cellular oncogenes, designated as the ras family. Early research on Harvey-MSV and Kirsten-MSV both isolated from rats infected with either Moloney or Kirsten leukemia viruses, demonstrated that each was genetically related in nonhelper-virus-derived regions of their genomes (59). With the advent of molecular cloning techniques, it was possible to demonstrate that most of the nonhelper-derived sequences were contributed by endogenous rat retroviral information (20). However, an additional set of sequences in each viral genome, designated ras, was shown to be derived from the cellular genome and was not homologous to any known viral sequences. The ras sequences in Kirsten-MSV and Harvey-MSV were found to be only partially related and thus represent different members of a gene family (20).

Subsequent to the isolation of Kirsten-MSV and Harvey-MSV, two other murine sarcoma viruses, BALB-MSV and Rasheed-MSV, were shown to contain ras oncogenes derived from the BALB/c mouse and Fisher rat, respectively (1,29). Another ras family member, designated N-ras, has been identified in mammalian cells but has not been identified to date as a transforming gene of any known retrovirus (24,36,61).

Three ras genes of human cells have been molecularly cloned and characterized in detail. The organization of their coding sequences is similar in that each gene contains four exons; however, the exons are distributed over a region anywhere from 4.8 kbp (H-ras) to 45 kbp (K-ras) in length (8,56,59). Pseudogenes of H-ras and K-ras have also been identified (designated H-ras-2 and K-ras-1) (8). The molecular cloning and nucleotide sequence analysis of ras genes of yeast has demonstrated a remarkable degree of evolutionary conservation of ras gene structure (14,49).

RAS GENES ARE FREQUENTLY DETECTED AS HUMAN TRANSFORMING GENES

The involvement of the ras gene family in naturally occurring malignancies came to light in studies in which investigators asked whether DNAs of animal or human tumor cells possessed the capacity to directly confer the neoplastic phenotype to a susceptible assay cell. Some human tumor DNAs were shown to induce transformed foci in the continuous NIH/3T3 mouse cell line (39), which is highly susceptible to the uptake and stable incorporation of exogenous DNA (10,69).

The first molecularly cloned human transforming gene, whose source was the T24 bladder carcinoma cell line, was demonstrated to be the activated homologue of the normal H-ras gene (28,51,56,58). Subsequent analysis of oncogenes detected by transfection assays has established that the majority belong to the ras family. Thus, K-ras oncogenes have been observed at high frequency in lung and colon carcinomas (50). Carcinomas of the digestive tract, including pancreas, and gall bladder, as well as genitourinary tract tumors and sarcomas have also been shown to contain ras oncogenes. N-ras appears to be the most frequently activated ras transforming gene in human hematopoietic neoplasms (25). These results are summarized in Table 3.

TABLE 3. Detection of ras oncogenes in human tumors.

Tumor source	Percent positive	ras oncogene activated		
		H-ras	K-ras	N-ras
Carcinoma (lung, gastrointestinal, genitourinary)	10-30	4/12	6/12	2/12
Sarcoma (fibrosarcoma, rhabdomyosarcoma)	10	0/2	0/2	2/2
Hematopoietic (AML, CML, ALL, CLL)[a]	10-50	0/9	1/9	8/9

[a]AML - acute myelocytic leukemia; CML - chronic myelogenous leukemia; ALL - acute lymphocytic leukemia; CLL - chronic lymphocytic leukemia.

Not only can a variety of tumor types contain the same activated ras oncogene, but the same tumor type can contain different activated ras oncogenes. Thus, in hematopoietic tumors, we have observed different ras oncogenes (K-ras, N-ras) activated in lymphoid tumors at the same stage of hematopoietic cell differentiation, as well as N-ras genes activated in tumors as diverse in origin as acute and chronic myelogenous leukemia (25) (Table 1). These findings suggest that ras oncogenes detected in the NIH/3T3 transfection assay are not specific to a given stage of cell differentiation or tissue type.

Retroviruses that contain ras-related transforming genes are known to possess a wide spectrum of target cells for transformation in vivo and in vitro. In addition to inducing sarcomas and transforming fibroblasts (35), these viruses are capable of inducing tumors of immature lymphoid cells (48). They also can stimulate the proliferation of erythroblasts (37) and monocyte/macrophages (33), and can even induce alterations in the growth and differentiation of epithelial cells (71). Thus, the wide array of tissue types that can be induced to proliferate abnormally by these onc genes may help to explain the high frequency of detection of their activated human homologues in diverse human tumors.

It should be noted that not all oncogenes detected by NIH/3T3 transfection analysis are related to known retroviral oncogenes. For example, Cooper and coworkers have identified and molecularly cloned an oncogene, B-lym, from a B-cell lymphoma (30). B-lym appears to be activated in a large proportion of tumors at a specific stage of B-cell differentiation (16). Studies to date indicate that this oncogene is relatively small in size (<600 bp), and sequence analysis indicates that it possesses distant homology to transferrin (30). These investigators have also detected oncogenes that appear to be specifically activated in tumors at other stages of lymphoid differentiation (44) or in mammary carcinomas (43). None of these transforming genes appears to possess detectable homology with known retroviral onc sequences. Transforming genes unrelated to the ras family have also been detected in the NIH/3T3 transfection assay by investigators in several other laboratories (9,27,50,57).

MECHANISM OF ACTIVATION OF RAS ONCOGENES

The availability of molecular clones of the normal and activated alleles of human ras proto-oncogenes made it possible to determine the molecular mechanisms responsible for the malignant conversion of these genes. The genetic lesions responsible for activation of a number of ras oncogenes have been localized to single base changes in their p21 coding sequences. In the T24/EJ bladder carcinoma oncogene, a transversion of a G to a T causes a valine residue to be incorporated instead of a glycine into the 12th position of the predicted p21 primary structure (6,52,64,66).

An activated H-ras oncogene of a human lung carcinoma derived cell line (72) was found to possess a single base alteration at codon 61. The change of an A to a T resulted in the replacement of glutamine by leucine in this codon. Thus, a single amino acid substitution was sufficient to confer transforming properties on the product of the Hs242 oncogene. These findings established that the site of activation in the Hs242 oncogene was totally different from that of the T24/EJ oncogene (70).

In subsequent studies, we and others have assessed the generality of point mutations as the basis for acquisition of malignant properties by ras proto-oncogenes by molecularly cloning and analyzing other activated ras oncogenes. Activation of an H-ras transforming gene of the Hs578T human breast carcinosarcoma line has been localized to a point mutation at position 12 changing glycine to aspartic acid in the amino acid sequence (42). Recently Wigler and coworkers (65) reported that the lesion leading to activation of the N-ras oncogene in a neuroblastoma line was due to the alteration of codon 61 from CAA to AAA causing the substitution, in this case, of lysine for glutamine. Another N-ras transforming gene, this one isolated from a human lung carcinoma cell line, SW1271, has been shown to result from a single point mutation of an A to a G at position 61 in the coding sequence, resulting in the substitution of arginine for glutamine (73).

Investigators analyzing K-ras oncogenes (7,62) have achieved strikingly similar results. In two K-ras transforming genes so far analyzed, single point mutations in the twelfth codon have been shown to be responsible for acquisition of malignant properties. Thus, mutations at positions 12 or 61 appear to be the genetic lesions most commonly responsible for activation of ras oncogenes under natural conditions in human tumor cells (Table 4).

The high frequency of detection of ras oncogenes in human tumors, coupled with the knowledge that ras proto-oncogenes can be activated as transforming genes by single point mutations, suggested that these genes might be preferential targets for somatic mutations. As an approach toward validating this assumption, we and others began to analyze animal tumor cells induced by specific carcinogens. Transforming DNAs from several independently derived mouse cell lines transformed by methylcholanthrene (MCA) had been shown to possess identical patterns of susceptibility to restriction endonuclease inactivation (60), suggesting that the same transforming gene was inactivated in each case. Thus, we sought to analyze tumor cells induced in vivo for the presence of activated ras oncogenes (27). Two of four MCA-induced fibrosarcomas were shown to contain activated C-K-ras genes. Our experience and that of other workers in detecting ras oncogenes in chemically induced tumors is summarized in Table 5 (2,23,63). It is striking that these genes are very frequently activated as a result of exposure of animals to chemical carcinogens. In certain cases (63), it has been possible to show that the activation

TABLE 4. Genetic lesions that activate ras oncogenes of human tumors.

ras oncogene	Tumor	Base/amino acid change	Codon no.	Reference
H-ras				
T24	Bladder carcinoma	Gly→Val (GGC→GTC)	12	Tabin et al. (64) Reddy et al. (52) Taparowsky et al. (65) Capon et al. (6)
Hs0578	Mammary carcinoma	Gly→Asp (GGC→GAC)	12	Kraus et al. (42)
Hs 242	Lung carcinoma	Gln→Leu (CAG→CTG)	61	Yuasa et al. (72)
K-ras				
Calu-1	Lung carcinoma	Gly→Cys (GGT→TGT)	12	Shimizu et al. (62) Capon et al. (7)
SW480	Colon carcinoma	Gly→Val (GGT→GTT)	12	Capon et al. (7)
SK-N-SH	Neuroblastoma	Gln→Lys (CAA→AAA)	61	Taparowsky et al. (65)
SW1271	Lung carcinoma	Gln→Arg (GGT←GTT)	61	Yusa et al. (73)

of ras oncogenes by chemical carcinogens is identical to that found to occur in human tumor cell lines. These studies provided model systems for the study of the multistep process of chemical carcinogenesis, and they have also indicated the importance of ras proto-oncogene activation in this process. It may now be possible to determine exactly when in the course of carcinogenesis that activation occurs.

IMPLICATIONS

The large fund of knowledge that has rapidly accumulated regarding the role of oncogenes in the neoplastic process has enabled investigators to

TABLE 5. Ras oncogene activation in carcinogen-induced animal tumors.

Carcinogen	Species	Tumor	Oncogene
MCA[a]	Mouse	Fibrosarcoma	K-*ras* (~50%)
DMBA[b]	Mouse	Skin carcinoma	H-*ras* (~80%)
NMU[c]	Rat	Mammary carcinoma	H-*ras* (~90%)

[a]Methylcholanthrene.
[b]Dimethylbenzanthracene.
[c]Nitrosomethylurea.

conclude that by focusing on the limited set of cellular genes that we have described in this review, it may be possible to define the mechanisms of malignant transformation. This prospect is particularly encouraging when one considers that normal growth and differentiation may involve interaction of many thousands of cellular genes and their products.

Although it has long been known from clinical observations that cancer arises as the result of a multistep process, the number of the steps may be relatively limited, as indicated by recent studies in which certain rodent embryo cells were transformed by the addition of two different oncogenes, whereas either one alone had no activity (45,55). Furthermore, these studies have suggested classes of complementing oncogenes. It is now of critical importance to uncover additional cellular proto-oncogenes and define their normal functions and mechanisms of activation.

The detailed information available regarding the structure and mechanisms of *ras* gene activation has made it possible to devise assays to rapidly distinguish activated from transforming *ras* alleles. Knowledge that some oncogene products act normally as either growth factors or their receptors may allow intervention by agents that interfere with receptor function. Thus, in addition to shedding light on basic mechanisms of normal and malignant growth, studies on cellular oncogenes are starting to provide a molecular armamentarium from which clinicians can select tools useful in the diagnosis, staging, and eventually even in the treatment of cancer.

REFERENCES

1. Anderson, P.R., Devare, S.G., Tronick, S.R., Ellis, R.W., Aaronson, S.A., and Scolnick, E.M. (1981): *Cell*, 26:129-134.
2. Balmain, A., and Pragnell, I.B. (1983): *Nature*, 303:72-74.
3. Bernheim, A., Berger, R., and Lenoir, G. (1979): *Cancer Genet. Cytogenet.*, 1:9-15.

4. Bishop, J.M. (1983): In: Annual Review of Biochemistry, Vol. 52, edited by E.E. Snell, P.D. Boyer, A. Meister, and C.C. Richardson, pp. 301-354. Academic Press, Palo Alto, California.
5. Canaani, E., Gale, R.P., Steiner-Saltz, D., Berrebi, A., Ashai, E., and Januszewicz, E. (1984): Lancet, 5:93-95.
6. Capon, D.J., Chem, E.Y., Levinson, A.D., Seeburg, P.H., and Goeddel, D.V. (1983): Nature, 302:33-37.
7. Capon, D.J., Seeburg, P.H., McGrath, J.P., Hayflick, J.S., Edman, U., Levinson, A.D., and Goeddel, D.V. (1983): Nature, 304:507-512.
8. Chang, E.H., Gonda, M.A., Ellis, R.W., Scolnick, E.M., and Lowy, D.R. (1982): Proc. Nat. Acad. Sci., USA, 79:4848-4852.
9. Cooper, C.S., Park, M., Blair, D.G., Tainsky, M.A., Huebner, K., Croce, C.M., and VandeWoude, G.F. (1984): Nature, 311:29-33.
10. Cooper, G.M. (1982): Science, 217:801-806.
11. Croce, C.M., Erikson, J., Ar-Rushdi, A., Aden, D., and Nishikura, K. (1985): Proc. Nat. Acad. Sci., USA, in press.
12. Croce, C.M., Thierfelder, W., Erikson, J., Nishikura, K., Finan, J., Lenoir, G., and Nowell, P.C. (1983): Proc. Nat. Acad. Sci., USA, 80:6922-6926.
13. Dalla-Favera, R., Bregni, M., Erikson, J., Patterson, D., Gallo, R.C., and Croce, C.M. (1982): Proc. Nat. Acad. Sci., USA, 79:4957-4961.
14. Defeo-Jones, D., Scolnick, E.M., Koller, R., and Dhar, R. (1983): Nature, 306:707-709.
15. DeKlein, A., Van Kessel, A.G., Grosveld, G., Bartram, C.R., Hagemeiger, A., Bootsma, D., Spur, N.R., Heisterkamp, N., Groffen, N., and Stephenson, J.R. (1982): Nature, 300:765-767.
16. Diamond, A., Cooper, G.M., Ritz, J., and Lane, M.A. (1983):Nature, 305:112-116.
17. Doolittle, R.F., Hunkapiller, M.W., Hood, L.E., Devare, S.G., Robbins, K.C., Aaronson, S.A., and Antoniades, H.N. (1983): Science, 221:275-277.
18. Downward, J., Yarden, Y., Mayes, E., Scrace, G., Totty, N., Stockwell, P., Ullrich, A., Schlessinger, G., and Waterfield, J. (1984): Nature, 307:521-527.
19. Duesberg, P.H. (1983): Nature, 304:219-226.
20. Ellis, R.W., DeFeo, D., Shih, T.Y., Gonda, M.A., Young, H.A., Tsuchida, N., Lowy, D.R., and Scolnick, E.M. (1981): Nature, 292:506-511.
21. Erikson, J., Ar-Rushdi, A., Drwinga, L., Nowell, P.C., and Croce, C.M. (1983): Proc. Nat. Acad. Sci., USA, 80:820-824.
22. Erikson, J., Nishikura, K., Ar-Rushdi, A., Finan, J., Emanuel, B., Lenoir, G., Nowell, P.C., and Croce, C.M. (1983): Proc. Nat. Acad. Sci., USA, 80:7581-7585.
23. Eva, A., and Aaronson, S.A. (1983): Science, 220:955-956.

24. Eva, A., and Aaronson, S.A. (1984): In: Genes and Cancer, edited by J.M. Bishop, J.D. Rowley, and M. Greaves, pp. 373-382. Alan R. Liss, Inc., New York.
25. Eva, A., Tronick, S.R., Gol, R.A., Pierce, J.H., Aaronson, S.A. (1983): Proc. Nat. Acad. Sci., USA, 80:383-387.
26. Gale, R.P., and Canaani, E. (1984): Proc. Nat. Acad. Sci., USA, 81:5648-5652.
27. Gazit, A., Igarashi, H., Chiu, I., Srinivasan, A., Yaniv, A., Tronick, S., Robbins, K., and Aaronson, S. (1984): Cell, 39:89-97.
28. Goldfarb, M., Shimizu, K., Perucho, M., and Wigler, M. (1982): Nature, 296:404-409.
29. Gonda, M.A., Young, H.A., Elser, J.E., Rasheed, S., Talmadge, C.B., Nagashima, K., Li, C.C., and Gilden, R.V. (1982): J. Virol., 44:520-529.
30. Goubin, G., Goldman, D.S., Luce, J., Neiman, E., and Cooper, G.M. (1983): Nature, 302:114-119.
31. Green, M.R., Treisman, R., and Maniatis, T. (1983): Cell, 35:137-148.
32. Greenberg, M., and Ziff, E. (1984): Nature, 311:433-437.
33. Greenberger, J.S. (1979): J. Nat. Cancer Inst., 62:337-344.
34. Groffen, J., Stephenson, J.R., Heisterkamp, N., DeKlein, A., Bartram, C.R., and Grosfeld, G. (1984): Cell, 36:93-99.
35. Gross, L. (1970): Oncogenic Viruses, second edition. Pergamon Press, Oxford.
36. Hall, A., Marshall, C.J., Spurr, N.K., and Weiss, R.A. (1983): Nature, 303:396-400.
37. Hankins, W.D., and Scolnick, E.M. (1981): Cell, 26:91-97.
38. Heisterkamp, N., Groffen, J., Stephenson, J.R., Spurr, N.K., Goodfellow, P.N., Solomon, B., Garrit, B., and Bodmer, W.F. (1982): Nature, 299:747-749.
39. Jainchill, J.L., Aaronson, S.A., and Todaro, G.J. (1969): J. Virol., 4:549-553.
40. Kelly, K., Cochran, B.H., Stiles, C.D., and Leder, P. (1983): Cell, 35:603-610.
41. Konopka, J.B., Watanabe, S.M., Witte, O.N. (1984): Cell, 37:1035-1042.
42. Kraus, M., Yuasa, Y., and Aaronson, S.A. (1984): Proc. Nat. Acad. Sci., USA, 81:5384-5388.
43. Lane, M.A., Sainten, A., and Cooper, G.M. (1981): Proc. Nat. Acad. Sci., USA, 78:5185-5189.
44. Lane, M.A., Sainten, A., and Cooper, G.M. (1982): Cell, 28:873-880.
45. Land, H., Parada, L.F., and Weinberg, R.A. (1983): Nature, 304:596-602.
46. Manolov, G., and Manolova, Y. (1972): Nature, 237:33.
47. Nowell, P.C., and Hungerford, D.A. (1960): Science, 132:1497.
48. Pierce, J.H., and Aaronson, S.A. (1982): J. Exp. Med., 156:873-887.
49. Powers, S., Kataoka, T., Fasano, O., Goldfarb. M., Strathern, J., Broach, J., and Wigler, M. (1984): Cell, 36:607-612.

50. Pulciani, S., Santos, E., Lauver, A.V., Long, L.K., Aaronson, S.A., and Barbacid, M. (1982): Nature, 300:539-542.
51. Pulciani, S., Santos, E., Lauver, A.V., Long, L.K., Robbins, K.C., and Barbacid, M. (1982): Proc. Nat. Acad. Sci., USA, 79:2845-2849.
52. Reddy, E.P., Reynolds, R.K., Santos, E., and Barbacid, M. (1982): Nature, 300:149-152.
53. Rowley, J.D. (1973): Nature, 243:290-291.
54. Rowley, J.D. (1983): Cancer Invest., 1:267-280.
55. Ruley, H.E. (1983): Nature, 304:602-606.
56. Santos, E., Tronick, S.R., Aaronson, S.A., Pulciani, S., and Barbacid, M. (1982): Nature, 298:343-347.
57. Schechter, A.L., Stern, D.F., Vaidyanathan, L., Decker, S.J., Drebin, J.A., Green, M.I., and Weinberg, R.A. (1984): Nature, 312:513-516.
58. Shih, C., Weinberg, R.A. (1982): Cell, 29:161-169.
59. Shih, T.Y., Williams, D.R., Weeks, M.O., Maryak, J.M., Vass, W.C., and Scolnick, E.M. (1978): J. Virol., 27:45-55.
60. Shilo, B.Z., and Weinberg, R.A. (1981): Nature, 289:607-609.
61. Shimizu, K., Birnbaum, D., Ruley, M.A., Fasano, O., Suard, Y., Edlund, L., Taparowsky, E., Goldfarb, M., Wigler, M., (1983): Nature, 304:497-500.
62. Shimizu, K., Goldfarb, M., Perucho, M., Wigler, M. (1983): Proc. Nat. Acad. Sci., USA, 80:383-387.
63. Sukumar, S., Notario, V., Martin-Zanca, D., and Barbacid, M. (1983): Nature, 306:658-661.
64. Tabin, C., Bradley, S.M., Bargmann, C.I., Weinberg, R.A., Papageorge, A.G., Scolnick, E.M., Dhar, R., Lowy, D.R., and Chang, E.H. (1982): Nature, 300:143-149.
65. Taparowsky, E., Shimizu, K., Goldfarb, M., Wigler, M. (1983): Cell, 34:581-586.
66. Taparowsky, E., Suard, Y., Fasano, O., Shimizu, K., Goldfarb, M., Wigler, M. (1983): Nature, 300:762-765.
67. Taub, R., Kirsch, I., Morton, C., Lenoir, G., Swan, D., Tronick, S., Aaronson, S., and Leder, P. (1982): Proc. Nat. Acad. Sci., USA, 79:7837-7841.
68. Waterfield, M.D., Scrace, G.T., Whittle, N., Stroobant, P., Johnsson, A., Wasteson, A., Westermark, B., Heldin, C.-H., Huany, J.D., and Deuel, T.F. (1983): Nature, 304:35-39.
69. Weinberg, R.A. (1982): Cell, 30:3-4.
70. Weiss, R.A., Teich, N., Varmus, H.E., and Coffin, J., editors (1982): RNA Tumor Viruses. The Molecular Biology of Tumor Viruses, second edition. Cold Spring Harbor Laboratory, Cold Spring Harbor, New York.
71. Weissman, B.E., and Aaronson, S.A. (1983): Cell, 32:599-606.
72. Yuasa, Y., Srivastava, S.K., Dunn, C.Y., Rhim, J.S., Reddy, E.P., and Aaronson, S.A. (1983): Nature, 303:775-779.

73. Yuasa, Y., Gol, R.A., Chang, A., Chiu, I.-M., Reddy, E.P., Tronick, S.R., and Aaronson, S.A. (1984): Proc. Nat. Acad. Sci., USA, 81:3670-3674.
74. Zech, L., Haglund, V., Nilsson, N., and Klein, G. (1979): Int. J. Cancer, 17:47-52.

Carcinogenesis, Vol. 10, edited by E. Huberman and S. H. Barr. Raven Press, New York © 1985.

Evolution and Cancer

M. Calvin

Department of Chemistry, University of California, Berkeley, California 94720

This essay is to be taken as a review of the state of our knowledge before the death of Dr. Charles Heidelberger. Much of what has happened since then deepens and details the situation already visible at that time.

In our laboratory we entered the cancer problem by way of the energy problem. The burning of coal produces, among other things, unburned hydrocarbons, which we now know to be very potent chemical carcinogens. We then asked how a chemical acts as a carcinogen. These questions arose, on the one hand, by virtue of our energy activity and, on the other, by our interest in evolution, which stemmed from the recognition that all life as we know it today is dependent upon the activities of green plants. I long ago concluded that plants could not have been the first organisms on the earth--there had to have been some others. My interest is basically in what I call "chemical evolution," the evolutionary period before the organization of complicated reactions began (1). Those two lines of interest, which started from different parts of science, have just come together in the last few years. This essay, which focuses on the nature of the cancer problem, will describe how chemicals play a role and what that knowledge led to in our thinking with respect to the way in which chemicals work (2,4), which, in turn, leads to the question of how the ability of the chemicals to induce malignancy has survived the whole evolutionary period.

CHEMICAL CARCINOGENESIS

It seems that we long ago recognized that the the most universal chemical characteristic of all the carcinogens is electropholic reactivity (3). The electrophilic properties of a group of chemical carcinogens (Fig. 1) indicate that the methyl group becomes a cationic methyl carbonium ion that methylates the DNA. The methylene of the cyclic lactone becomes a carbonium ion and alkylates the DNA and RNA. The nitrosamines and

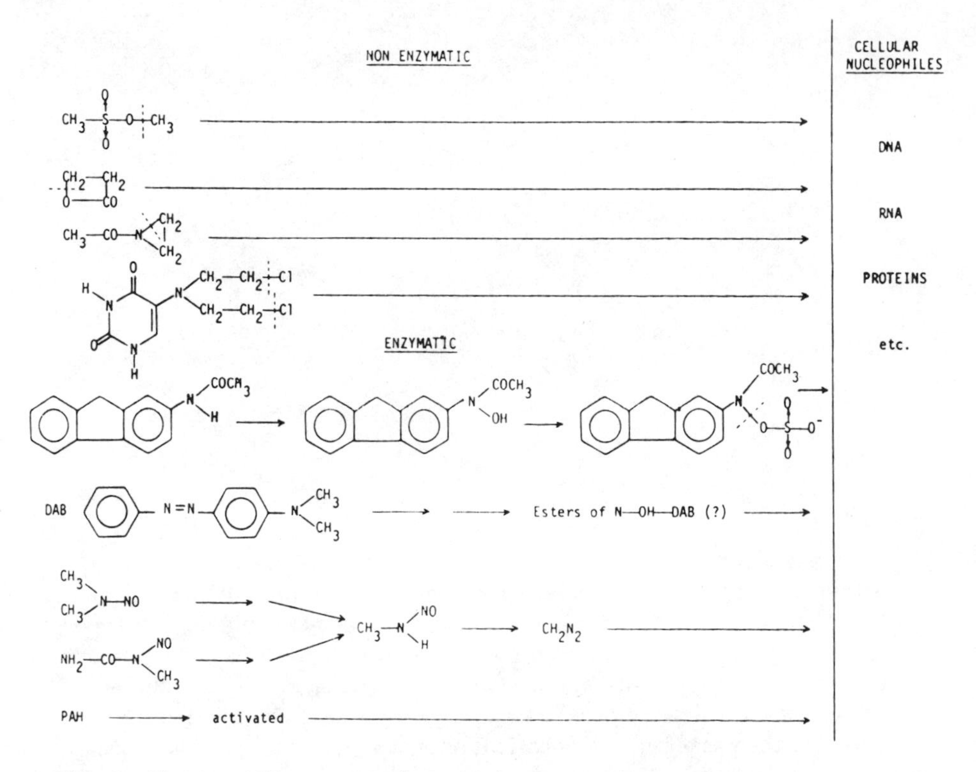

FIG. 1. Electrophilic nature of the ultimate chemical carcinogens.

polycyclic aromatic hydrocarbons (PAHs), such as benzo[a]pyrene (BAP), are activated to produce alkylating agents, which usually become more toxic before they become less toxic on the way to detoxification in preparation for elimination. This activation appears to be one of the mechanisms by which PAHs are eliminated by oxidation, during which time they become electrophilic reagents with a specific "address" attached to them (Fig. 2).

We made a thorough study of BAP because this material was easy to investigate. When whole cells were exposed to BAP and the products extracted, the BAP was converted in the whole cells and attached to cellular macromolecules, primarily nucleic acids. Through a series of chemical transformations, hydrolyses, and other analytical procedures we were able to demonstrate that the BAP was actually in the form of 7,8,9,10-tetrahydro-benzo[a]pyrene. The mechanism of the reaction is as follows: First a 7,8-epoxidation occurs, followed by hydration of the

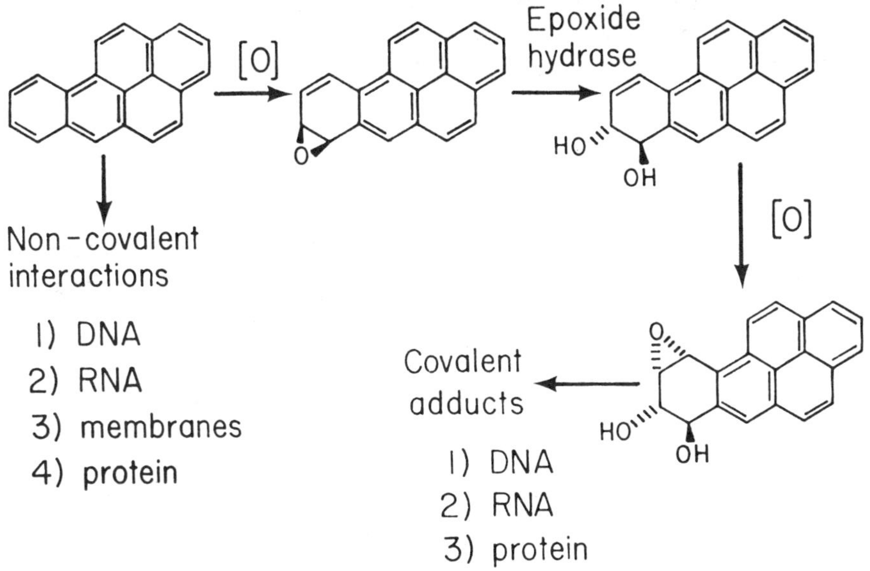

[O]

Epoxide hydrase

HO

OH

[O]

Non-covalent interactions

1) DNA
2) RNA
3) membranes
4) protein

Covalent adducts

1) DNA
2) RNA
3) protein

FIG. 2. Activation of benzo[a]pyrene.

epoxide to give a 7,8-diol, which is then epoxidized again at the 9,10-position to create a very reactive 10-position that can react with all of the macromolecules of living cells in a covalent way, generally by an electrophilic reaction (3).

The analytical procedure used to confirm the the reaction sequence was high-pressure liquid chromatography (HPLC). We found each of the components: BAP, tetrahydrodiol epoxide, and compounds with guanine, cytosine, adenine, and thymine. All of these materials were identified, particularly the guanine, which was the most important. We also showed that the two different guanine compounds formed in this reaction were optical isomers of each other, simply a pair of two different optical isomers that occur when the epoxy ring is opened at the 10-position. Since there are other optical centers present, which are formed enzymatically, the final pair are separable diasteromers.

The consequence of such a chemical reaction in the cell, which I postulated in 1973 (3) (Fig. 3) has now been experimentally confirmed in many different ways, as follows: The polycyclic aromatic hydrocarbon BAP creates damage in the chromosome of the cell. That damaged cell, or damaged chromosome, can then be repaired by the repair mechanism of the cell, an endonuclease, which finds the "bump" that the BAP has made in the double helix, cuts the bump out, and repairs it. You can see in Fig. 3 the endonuclease cutting out a particular piece, followed by resynthesis of the missing section. The nucleic acid is thus repaired and gives back the correct information. In the presence of some foreign information, which can come from virus, plasmid, or some other parts of the genome, it is possible that some piece of that foreign information could be inserted into the spot that was left when the endonuclease removed the error and did not immediately reinsert its correction. This mechanism can be tested at this point by adding exogenously a piece of virus of some particular consequence, that is, an oncogenic virus. It can be demonstrated that a piece of that viral gene has been integrated into the new chromosome that is generated. When normal replication occurs, a normal cell results, but if foreign information is present, either as a virus, plasmid, or transposable genome, there is a certain probability that it will be integrated into the vacancy left by the endonulease that takes out the error created by the BAP. When that happens, excision repair occurs, resulting in a chromosome that contains a piece of the foreign information, with the creation of transformed cells--which may be malignant, depending upon the nature of the integrated DNA (3).

Tissue culture can be an assay for the transformed cell (3). The foci of transformed cells can be counted very easily per million cells on a plate. The effect of action, either by virus, chemical, or some combination, can be easily quantitated by simply counting the number of foci. If you take one of these foci, which is now a transformed cell (a cell no longer subject to contact inhibition of growth), and inject it into a healthy animal, a tumor usually results. So, the calibration of the transformation assay as a tumor-producing assay has been accomplished (3,5).

Let us discuss a kind of environmental experiment that was done without knowing how or why it works. This experiment was done backwards, so to speak, and was first published by Stich (12). He used a tumor-producing virus, adenovirus-7, in combination with a carcinogenic chemical, 4-nitroquinoline-N-oxide (4-NQNO). If you put 4-NQNO into the system and immediately (within 1-1/2 h) add the adenovirus, you get 435 foci. On the other hand, if you begin the experiment with adenovirus alone, with no 4-NQNO, you get only 17 foci per 2 million cells. If you add the adenovirus to the system 48 h after the 4-NQNO, the cell system has returned to a normal level of transformation. In 48 h the excision repair mechanism for the DNA initiated by the chemical has finished its activity. Therefore, if you use adenovirus at the beginning of the experiment, when

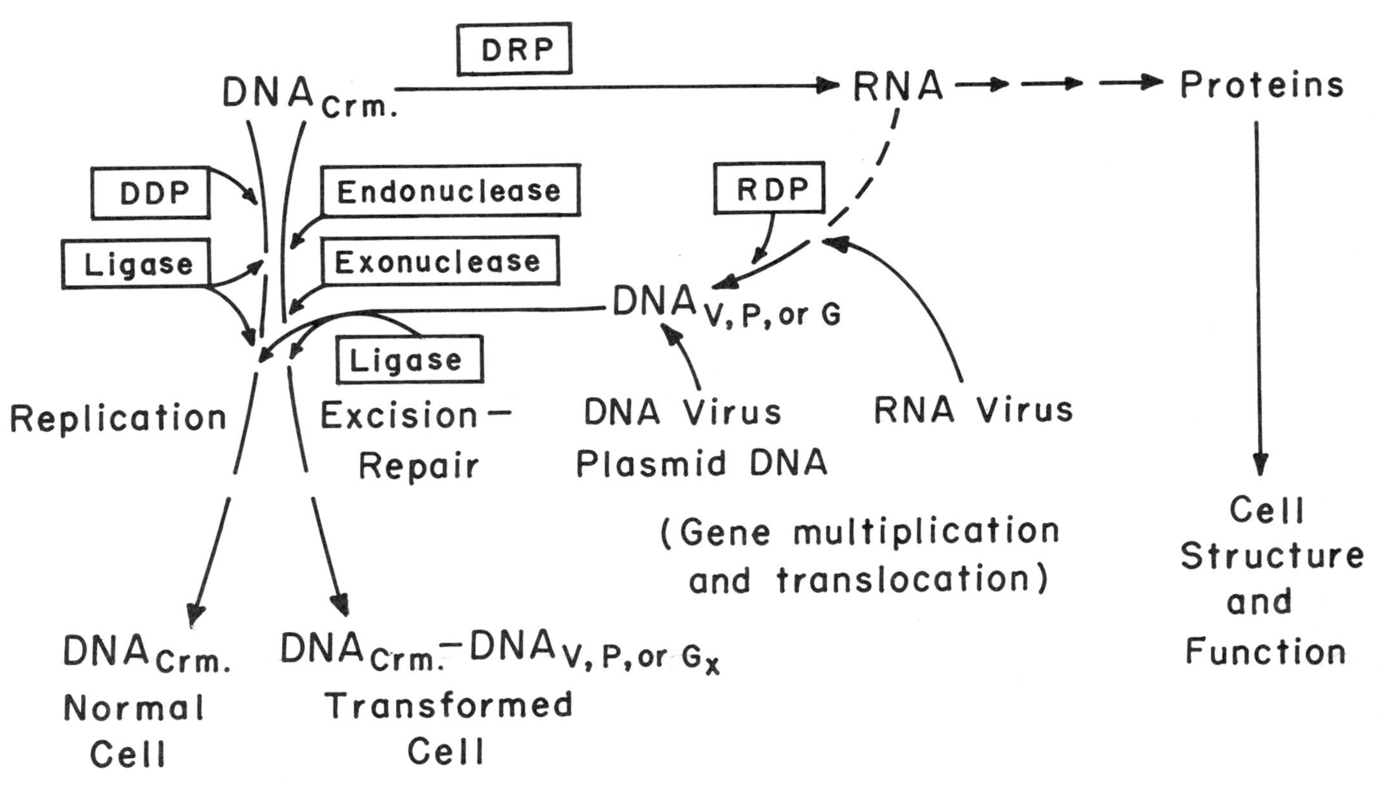

FIG. 3. Scheme for cell transformation, including chemical function.

the 4-NQNO is added, there are 435 foci (transformations). The synergism between the adenovirus and the 4-NQNO is the difference between 17 and 435 foci. If you wait for the repair mechanism to finish its work with all the DNA back in the normal condition, you get a nearly normal effect of the virus.

Stich continued to explore this type of experiment with results that confirm the synergism effect. He measured the transformation frequency, which seemed to be the easiest thing to do because it is possible to see the foci with a naked eye. The adenovirus alone produces 13 transformations/2 million cells, the 4-NQNO produces 0 transformations, and the 4-NQNO and adenovirus together produce 300 transformations. This result also exhibits the synergism between virus and chemical. You put the chemical in first, the virus immediately thereafter; if the experiment is done the other way, then the chemical also damages the virus. The assay can be performed with different carcinogenic viruses to provide the source of the oncogenic information. It is possible to test the theory by whole animal experiments to show that the focus is indeed a malignancy, followed by a hybridization experiment to show that the malignancy has indeed integrated some of the viral information from the cell, which is indeed the case. The experiments just described also indicated that the structure of the chromosomes is affected. There is an enormous synergism because of the repair system that is induced by the chemical damage. In the presence of oncogenic information, there is a real possibility that the foreign information is being incorporated.

The mechanisms are shown in Figs. 4 and 5, describing how BAP produces deformation of the double helix, with the endonuclease cutting out the pieces. When there is some foreign information present, there is a certain probability that the foreign information can be integrated into the DNA. This foreign DNA is indicated in Fig. 5 by darker lines. It may be formed from a larger piece by the same (or related) enzyme that removed the error in the first place and thus produces the piece with the same "sticky ends" such that it may be easily inserted (5). Eventually the transformed DNA is formed with a permanent change in the genes of the cell. The cell becomes a malignant cell because this oncogenic information from an external source has been introduced. The chemical itself has no information content. The chemical is merely a trigger for the introduction of oncogenic information from some external source: a virus, a plasmid, or a transposable gene. When this piece is present during repair, it can slide in between the sticky ends and be hooked together by a ligase. There now exists a piece of DNA with an abnormal piece of information in it, inserted into what was originally a normal chromosome. That abnormal piece of information is called an oncogene, a cancer-producing gene (6).

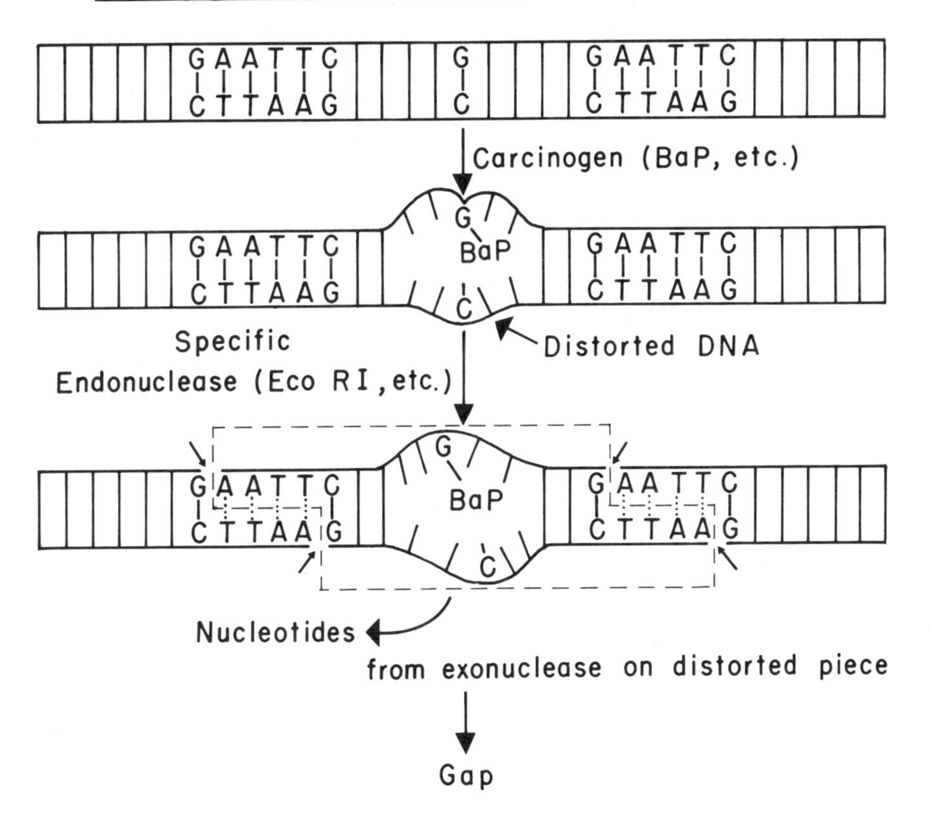

FIG. 4. Schematic proposal for collaboration of chemical carcinogen and oncogenic information (part I).

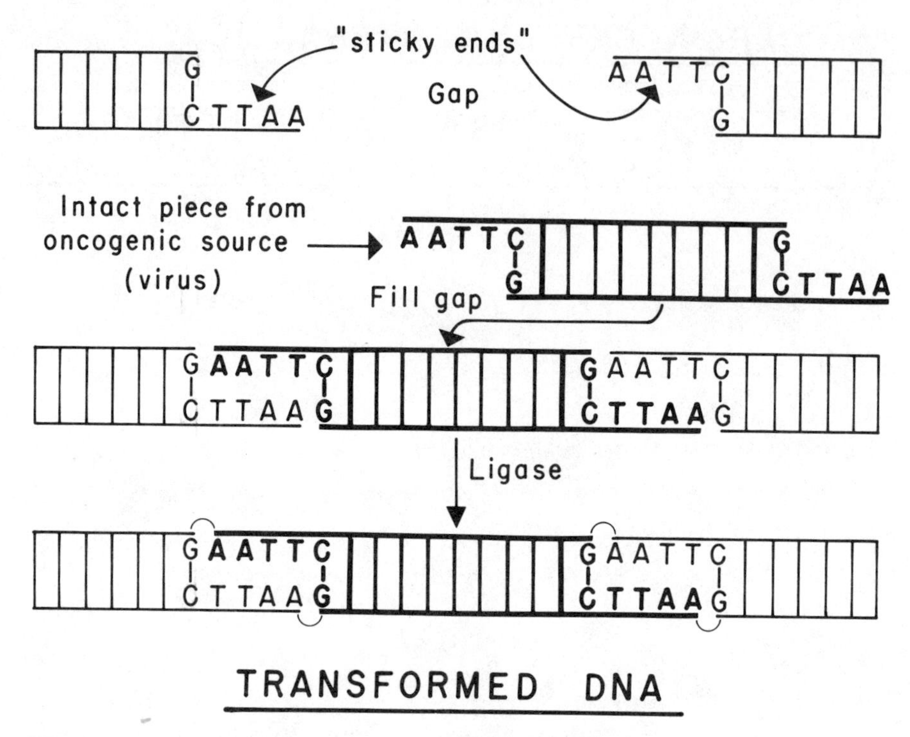

FIG. 5. Transformed DNA. Schematic proposal for collaboration of chemical carcinogen and oncogenic information (part II).

EVOLUTION OF ONCOGENES

Where do oncogenes come from and how do they survive? It is believed now that the viruses pick up these pieces, which are present in normal cells--but not all at the same time or in the same place. When a virus infects a normal cell, the oncogenes are eventually liberated from that cell and have pieces of chromosomal DNA attached. If you put two or more of these into the same cell in the right places, the cell can be transformed. In general, the experimental method uses tissue cultures that have been carried for a long period of time and have been changed in some unknown ways, partly already toward this malignant condition, and when this last piece of information is inserted, the transformation occurs (11).

Why should that oncogenic information have survived evolution for so many billions of years? You would have thought it would have been lost in the course of evolutionary history, either from the virus or from the animals that carry it. The answer turns out to be, I think, that the identical pieces, which we call oncogenic information, by themselves have a real positive function, as shown in Tables 1 and 2 (8).

The viral oncogenes produce proteins, just as any other gene does. When you learn what the viral oncogenes do, you realize they fulfill an important function. For example, the epidermal growth factor (EDF) receptor is one case. Apparently all cells require a growth factor, a protein, which, when bound to the cell surface at the right site, triggers the cell to divide. When there is only one growth factor for a given cell, it behaves normally. If several different factors are put into the same cell, uncontrolled growth can occur. This uncontrolled growth is what results when several growth factors are inserted into one cell. We know that these growth factor genes make proteins, but we have not yet sorted all of them out nor do we yet know what all their functions are (7).

The idea exists that malignant transformation is the result of the incorporation of several (two or more) of the oncogenes, several of which are known to be growth factors of various kinds corresponding to existing known growth factors (epidermal growth factor, platlet-derived growth factor, etc.) (11). Figure 6 shows how the growth factor receptor and the growth factor collaborate (9). When the growth factor binds to the receptor site, a signal is sent to the center of the cell nucleus and it divides. This normal procedure is required for normal cell division. Too many growth factors in the same cell can produce malignancy. Instead of having one copy of a growth factor gene, you might have hundreds or thousands of copies, thus making manyfold more growth factors, and the cell becomes malignant.

The idea of what the oncogenes, which were defined many years ago, do--they trigger the transformation of a cell--has been confirmed (10). We did not know then why they have persisted over evolutionary time. Now you can see they have a normal function--when they are expressed at the right level and there are not too many in one cell. The oncogenes actually have a natural function of keeping the cell going. The reason that the oncogenes have actually survived evolution and prospered is that their natural, normal function is to promote cell division. Without the oncogenes there would be no cell division. The presences of too many oncogenes in one cell or too many copies of an oncogene in one cell can be induced either by a combination of vertical transmission from the parent and horizontal infection or by vertical transmission to the parent and chemical induction of expression. By adding carcinogen, it is possible to move the control factors around the chromosome, so instead of being expressed normally, genes will be expressed in a thousand times normal fashion, which also leads to malignancy.

Malignancy can be produced in several ways. The chemical induction in this process triggers the damage, which leads to reorganization of the chromosomes. The reorganized chromosome can involve the existing genes, new inserted genes, or the control factors.

TABLE 1. Viral oncogenes. I.

Oncogene	Virus and species of origin	Function of products
abl	Abelson murine leukemia virus (mouse)	Tyrosine kinases
fes[a]	ST feline sarcoma virus (cat)	
fps[a]	Fujinami sarcoma virus (chicken)	
fgr	Gardner-Rasheed feline sarcome virus (cat)	
ros	UR II avian sarcoma virus (chicken)	
src	Rous sarcome virus (chicken)	
yes	Y73 sarcoma virus (chicken)	
erbB	Avian erythroblastosis virus (chicken) (EGF receptor)	Structure of a tyrosine kinase but no kinase activity detected
fms	McDonough feline sarcoma virus (cat)	
raf[a]	3611 Murine sarcome virus (mouse)	
mil(mht)[a]	MH2 virus (chicken)	
mos	Avian myeloblastosis virus (chicken)	

[a]From Marx (8).

TABLE 2. Viral oncogenes. II.

Oncogene	Virus and species of origin	Function of products
sis	Simian sarcoma virus (woolly monkey) (PD growth factor)	Growth factor
Ha-ras	Harvey murine sarcome virus (rat)	Bind guanosine triphosphate
Ki-ras	Kirsten murine sarcoma virus (rat)	
fos	FBJ osteosarcoma virus (mouse)	Nuclear location-bind DNA?
myb	Avian myeloblastosis virus (chicken)	
myc	MC29 myelocytomatosis virus (chicken)	
erbA	Avian erythroblastosis virus (chicken)	Cytoplasmic location-function unknown
ets	E26 virus (chicken)	
rel	Reticuloendotheliosis virus (turkey)	
ski	Avian SKV 770 virus (chicken)	

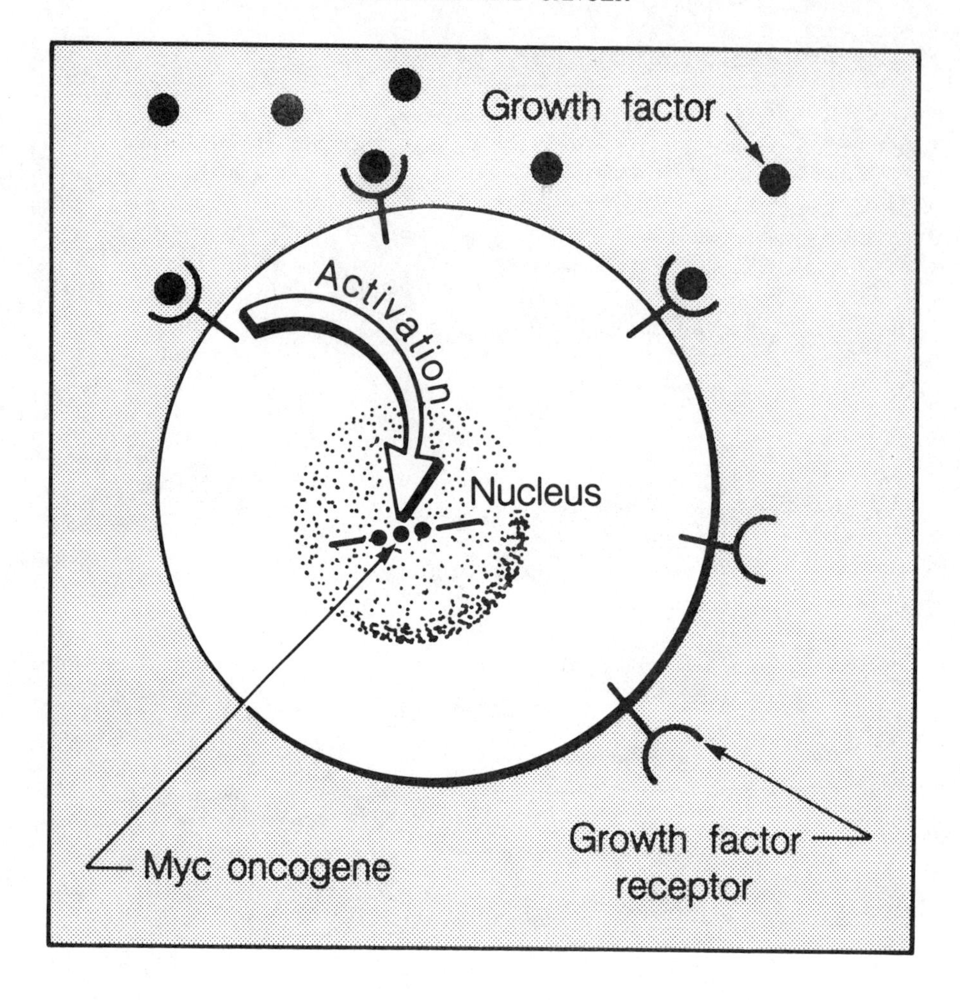

FIG. 6. Initiation of the cell division program. (New Scientist, February 16, 1984.)

REFERENCES

1. Calvin, M. (1969): Chemical Evolution: Molecular Evolution Towards the Origins of Living Systems on Earth and Elsewhere. Oxford University Press, New York.
2. Calvin, M. (1972): Radiat. Res., 50:105-119.
3. Calvin, M. (1973): In: Viral Replication and Cancer, edited by J.L. Melnick, S. Ochoa, and J. Oro, pp. 195-218. Editorial Labor, S.A., Barcelona.
4. Calvin, M. (1975): Naturwissenschaften, 62:404-413.
5. Calvin, M. (1978): Prog. Biochem. Pharmacol., 14:6-27.
6. Calvin, M. (1984): In: Concepts in Toxicology, Vol. 1, edited by F. Homburger, pp. 20-33. S. Karger, Basel, Switzerland.
7. Hunter, T. (1984): Sci. Amer., 251:70-79.
8. Marx, J.L. (1984): Science, 223:673-676.
9. New Scientist, Feb. 16, 1984, p. 21.
10. Sattaur, O. (1984): New Scientist, November 22, 1984, pp. 12-16.
11. Slamon, D.J., deKernion, J.B., Verma, I.M., and Cline, M.J. (1984): Science, 224:256-262.
12. Stich, H. (1972): In: Topics in Chemical Carcinogenesis, edited by W. Nakahara, S. Takayama, T. Sugimura, and S. Odashima, pp. 17-23. University Park Press, Tokyo.

Carcinogenesis, Vol. 10, edited by E. Huberman
and S. H. Barr. Raven Press, New York © 1985.

Genetics and Epigenetics of Neoplasia: Facts and Theories

H. C. Pitot, L. E. Grosso, and T. Goldsworthy

McArdle Laboratory for Cancer Research, University of Wisconsin Medical School, Madison, Wisconsin 53706

Neoplasia is a disease that has disturbed the general population and intrigued the scientific community for more than a century. The individual whom we honor in this symposium, although initially trained as an accomplished organic chemist, early in his scientific career oriented his research efforts toward studies in chemical carcinogenesis and later in chemotherapy. In no small way, these decisions were influenced by Harold P. Rusch, then Director of the McArdle Laboratory.

A scientific collaboration with Dr. Heidelberger and one of us (H.P.) began in the early 1960s and concerned itself with the subject, hotly debated in the McArdle Laboratory, of the nature of the basic and critical lesion in the neoplastic process. Specifically, is neoplastic transformation the result of one or more structural alterations in the genomic DNA of the cell, or can neoplasia result from changes in the cell that do not involve any structural genomic change but rather evolve from heritable changes in the expression of one or more genes? Today our knowledge of the molecular biology of carcinogenesis, especially viral oncogenesis, has strongly supported the former mechanism, but other data have provided support for the nongenomic nature of the neoplastic transformation. This paper briefly reviews some aspects of these dichotomous views; they represent a dilemma that occupied a significant amount of the scientific interest of Charlie Heidelberger.

THE GENETICS AND EPIGENETICS OF NEOPLASIA

A neoplasm (and thus neoplasia) can reasonably be defined as "a heritably altered, relatively autonomous growth of tissue." This definition to some extent points out the dichotomous nature of the consideration of the basic mechanisms of neoplastic transformation. No one would argue that the neoplastic phenotype is heritably transmitted from cell to cell.

Usually such heredity bespeaks a specific genomic constitution, but the examples of "somatic heredity" seen in differentiation of specific tissues and related processes in lower animal forms do not necessarily support such a blanket interpretation. On the other hand, the regulation of the expression of the genome whose alteration in neoplasia is implied by the term "relatively autonomous" is usually considered to be the result of environmental influences, although the genetic basis for many regulatory mechanisms is well known (26).

The principal evidence for the structural alterations in genomic DNA as the basis for neoplastic transformation comes from our knowledge of the nature of carcinogenic agents and their probable method of action and a knowledge of genetic structure in well-established neoplasms. The work of the Millers (22) has clearly demonstrated the necessity for "active" forms of chemicals, almost all of which are highly mutagenic, for the initiation of neoplastic transformation. The mutagenicity of carcinogenic radiation, both ultraviolet and ionizing, has now been well documented (13,20). All oncogenic viruses appear to exert their transformation capabilities by additions of viral information to the host genome either as a stable intragenomic insertion or as a stable episomal particle (3,36). As first argued by Boveri (5), almost all malignant neoplasms exhibit some form of structural abnormality within their karyotype (45). In addition, a number of specific karyotypic alterations are characteristic of specific histogenetic types of neoplasms (45). Furthermore, with an increased understanding of the structure of onc genes in transforming viruses and a similarity of many such genes to endogenous cellular genes, specific lesions of the genome resulting in the neoplastic transformation have been postulated (3).

THE EPIGENETICS OF NEOPLASIA--FACTS AND THEORIES

In view of the almost overwhelming evidence, some of which is cited above, in support of structural DNA alterations as the basic mechanism for neoplastic transformation, it is difficult to rationalize any mechanism not involving such alterations. However, a significant amount of information on carcinogenesis and tumor biology is difficult to reconcile with the purely genetic basis of neoplasia. The major lines of evidence for this are listed in Table 1. The earliest biological evidence was that of the now-classical nuclear transplantation studies of King and his associates (18) in which nuclei from renal carcinoma cells were transplanted into enucleate eggs in the frog. Although very few adult frogs resulted from such micromanipulations, many such hybrid eggs developed into normal tadpoles. The important studies of Mintz and Illmensee (23), following those of Papaioannou et al. (28), demonstrated that teratocarcinoma cells in the

TABLE 1. Some lines of evidence suggesting an epigenetic basis for neoplasia.

1. Nuclear transplantation experiments in amphibians.
2. Blastula and early embryonic regulation of teratoma and leukemic cells.
3. "Epigenetic" chemical carcinogens.
4. Chemical and physical induction of differentiation in neoplastic cells.

environment of the early embryo apparently lose their neoplastic potential and differentiate as normal cells. Dr. Sachs has reported similar experiments in this volume in which neoplastic cells of specific tissues are placed into the fetal environment at appropriate times during embryogenesis. Since it is presumed that such "forced" differentiation of neoplastic cells is the result of cell-to-cell contact or diffusible mediators, it is not surprising that chemicals and/or hormones are capable of inducing differentiation in neoplastic cells such that the neoplastic phenotype is lost because of terminal or near terminal differentiation (17,27).

In chemical carcinogenesis, it is now apparent that there are a significant number of chemicals capable of inducing neoplasms, both benign and malignant, when they are administered chronically to animals, but yet such chemicals exhibit no evidence of mutagenesis either by themselves or through any of their metabolites (44). All of these instances suggest that epigenetic factors may well play a major or a significant role in the development of neoplasms or in the alteration of neoplastic cells themselves.

Just as one can envisage a variety of theoretical mechanisms involving structural alterations in one or more genes resulting in neoplasia, one may suggest theoretical mechanisms for a heritable, epigenetic mode of carcinogenesis. Such theoretical considerations resulted from the collaboration with Dr. Heidelberger (34), referred to above. One theoretical regulatory "circuit" is reproduced in Fig. 1. The basis for such circuitry is the regulatory mechanisms commonly seen in prokaryotes, as originally elucidated by Jacob and Monod (16). In this theoretical context, a brief reaction of the carcinogenic agent with the protein product of a specific gene inactivates the regulatory product, thereby opening a circuit that remains open even following removal of the carcinogenic stimulus. There

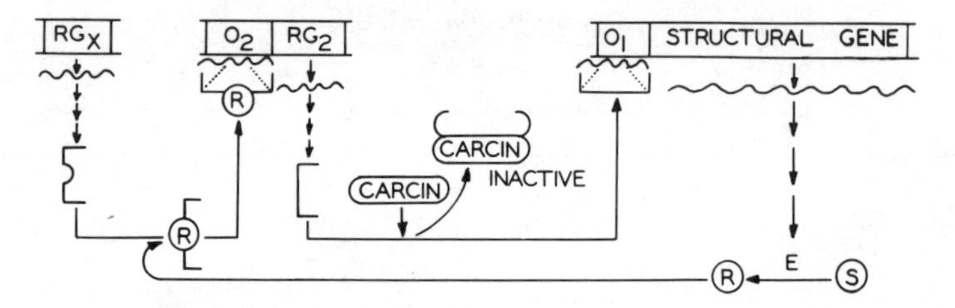

FIG. 1. Modified product perpetuation circuit. Individual genes are shown as boxes, their messenger RNAs as wavy lines, and the protein products of the regulatory genes, RG_x and RG_2, as brackets. Small molecular weight effectors are designated by R, while the enzyme product of the structural gene is E, which acts on the substrate S, converting it to R. Carcin = carcinogen, which combines with and inactivates the product of RG_2, which in turn allows synthesis of E, thus maintaining RG_2 in the inactive state. Note that the carcinogen need act for only a short period of time until sufficient E is synthesized to completely repress the synthesis of the RG_2 product.

are obviously a variety of other such theoretical circuits that could be envisaged, but evidence that such concepts have any reality in mammalian cells has not been forthcoming during the past two decades. Many other theoretical concepts supporting a heritable, nongenomic mechanism for the neoplastic transformation have been proposed but will not be discussed here. Rather, in the remainder of this discussion we will attempt to relate the types of evidence presented in Table 1 to the overwhelming evidence that the basic lesion of the neoplastic transformation involves a structural alteration in genomic DNA.

THE NATURAL HISTORY OF NEOPLASTIC DEVELOPMENT--GENETIC AND EPIGENETIC FACTORS

In attempting to reconcile the strong evidence indicating that heritable structural alteration(s) in genomic DNA are the basis for the neoplastic transformation with the more indirect but still significant evidence of an epigenetic basis of the neoplastic transformation (Table 1), it is important to consider the natural history of neoplastic development, the stages involved, and particularly the characteristics of such stages. It is now well recognized that neoplastic development may be divided into at least three separate and distinct stages beginning with initiation. The initiation process may be defined as follows:

> Initiation is that process occurring within a cell resulting from the action of a chemical, physical, or biological agent that alters in an irreversible manner the heritable structure(s) of the cell. Such an initiated cell has the potential of developing into a clone of neoplastic cells.

The critical items of this definition are the irreversibility of the process, its heritable nature, and its potential. Without any one of these characteristics, initiation of the neoplastic process probably does not occur. The characteristics of initiation (Table 2) can almost all be related to an effect on genomic DNA. On the other hand, it is clear that initiation is dependent on such epigenetic factors as the phase of the cell cycle at the time of the initiation event, the ability of the target cell to "activate" a chemical carcinogen to a form that can alter the structure of genomic DNA, and the apparent need for cell replication following the interaction of the carcinogen with the cell (31).

The second stage of neoplastic development, following initiation, is promotion.

> Promotion is that stage in the development of neoplasia that, if existent, is characterized by a chronic alteration in the expression of the genetic information of an initiated cell resulting from the action of one or more agents (promoters).

Characteristics of this stage are presented in Table 3 (31), which illustrates that virtually all of the characteristics of tumor promotion do not directly involve genomic DNA or any structural change therein. On the other hand, recent studies (2,10) have demonstrated that at least one effective promoting agent, tetradecanoyl phorbol acetate (TPA), and several of its congeners induce sister chromatid exchanges, DNA strand breaks, and chromosome aberrations in cells *in vitro*, probably indirectly through the stimulation of the formation of active oxygen radicals. Further evidence for such mechanisms hinges on the finding that antioxidants and other inhibitors of active oxygen formation inhibit tumor promotion and complete carcinogenesis *in vivo* (43). However, it is also clear that a number of promoting agents such as certain hormones, detergents, and iodoacetate do not exhibit such metabolic reactivity although they are effective promoting agents (33). Therefore, it is reasonable to argue that the stage of tumor promotion is largely if not entirely associated with epigenetic factors that directly or indirectly alter the expression of genomic DNA within the cell.

TABLE 2. Genetic and epigenetic characteristics of initiation.[a]

1.	All known agents, chemical, physical, or biological, capable of initiating the process of neoplasia can directly alter the structure of DNA either in their native form or in an "activated" form.
2.	Initiation is irreversible, inherited by progeny of the affected cell, and additive in its effects.
3.	Initiation appears to involve single-hit kinetics in the cases of irradiation *in vivo*, infection by oncogenic viruses *in vitro*, and chemical carcinogenesis *in vivo*, when the early progeny of initiated cells are monitored.
4.	Initiation does not exhibit a readily measurable threshold or maximal response within the limits of toxicity to the exposed cell populations.
5.	Initiation *in vitro* appears to be dependent on the phase of the cell cycle at the time of the initiation event; cell replication is probably required for the event to occur at all.

[a]Modified from Pitot et al. (33).

The final stage of neoplastic development is progression:

> Progression is characterized either by demonstrable changes in the number and/or arrangement of genes, as evidenced from nucleic acid sequence and/or hybridization studies, or by visible karyotypic alterations, as evidenced by light microscopic techniques, within a majority of the neoplastic cells that make up the tumor.

These alterations in turn are associated with increased growth rate, increased invasiveness, metastases, and alterations in biochemical and morphologic characteristics of the neoplasm. The critical characteristics of this stage revolve around demonstrable alterations in the structure of the genome. These alterations include a marked increase in the effectiveness of transfection with extracellular genetic material, an increase in gene amplification, especially in the presence of agents favoring the induction of such amplification (drugs, etc.), and, most important, an extreme degree of karyotypic instability as demonstrated by a wide variety of karyotypes within a given population of malignant cells (31).

TABLE 3. Genetic and epigenetic characteristics of promotion.[a]

1. Effects of individual exposures to a promoting agent are reversible and not additive.

2. Promoting agents are incapable of initiation *per se* but may promote fortuitously initiated cells (background).

3. Promotion is effective only following initiation.

4. Promotion is modulated by environmental factors such as age, sex, and diet.

5. No evidence for significant metabolism or "activation" is required for the effect of a promoting agent. Most promoting agents are not mutagenic.

6. All promoting agents alter genetic expression either in a positive or negative manner, many acting through a receptor mechanism, although some act nonspecifically.

7. The yield of neoplasms following promotion exhibits a measurable threshold and a maximal response usually short of toxicity.

8. At least one effective promoting agent, TPA, and its congeners induce sister chromatid exchanges, DNA strand breaks, and chromosome aberrations in cells *in vitro* by indirectly causing the formation of active oxygen radicals.

9. Antioxidants and other inhibitors of active oxygen formation inhibit tumor promotion and complete carcinogenesis *in vivo*.

[a]Modified from Pitot et al. (33).

In reconciling the types of evidence listed in Table 1 with the overwhelming evidence for structural changes in genomic DNA as the basis for neoplastic transformation, it is now possible to relate such epigenetic factors to the natural history of neoplastic development. The remainder of this paper will concern itself with the reconciliation of the genetic and epigenetic bases of the neoplastic transformation in light of the natural history of neoplastic development.

"EPIGENETIC" CHEMICAL CARCINOGENS

Several investigators (41,44) have pointed out that there are a significant number of chemical carcinogens for which there is no evidence of any mutational inductive capacity but which, when administered to animals for extended periods of time, cause a significant increase in one or more types of neoplasms. Although a variety of potential mechanisms could be suggested in the light of our knowledge of the characteristics of the stages of neoplastic development, it is quite likely that such "epigenetic" carcinogens exert their effects by the promotion of fortuitously initiated cells, termed the background, which are likely to be found universally in living organisms (31,40). Experimental evidence for such fortuitously initiated cells has come from a variety of sources, one of the more interesting of which is the rat liver system (40).

In our laboratory we have demonstrated the first two stages of hepatocarcinogenesis in rat liver (initiation, promotion) and have identified the cells involved (32). The immediate progeny of single initiated cells are small foci or islands of phenotypically unique cell populations that have the critical characteristic of irreversibility and under appropriate circumstances may also be transplanted from one host to another (30). Others have also demonstrated such foci of altered cells following carcinogen administration (6,14), but in some cases, with the model systems used, such foci are not stable (11) and thus probably do not represent *in toto* the immediate progeny of initiated cells. On the other hand, several investigators (25,35) have demonstrated in older rats (usually over 12 months of age) the spontaneous occurrence of such foci. In particular, Schulte-Herrmann and his associates (40) have shown that, as the animal ages, the area and thus presumably the volume of these few spontaneous foci increases rather dramatically after the animal is one year old.

We have developed a technique for quantitating in three dimensions such foci within the rat liver (7). Using a protocol seen in Fig. 2, we attempted to determine the level of promotable endogenous or background foci in the liver of rats from 3 weeks to 12 months of age. At each of the times indicated, animals were placed on diets containing 0.05% phenobarbital, an efficient promoter of hepatocarcinogenesis, as first demonstrated by Peraino and his associates (29). Animals were then sacrificed after 6 months of the promotion regimen, and the total number of foci in the liver was determined through the use of stereologic calculations developed in our laboratory (7). The results of this experiment (Table 4) show that the number of promotable endogenous foci was relatively low at 3 and 6 weeks of age, but by 3 months of age a plateau of the number of such foci was reached and essentially did not change for the remainder of the experiment, possibly for the further lifetime of the animal.

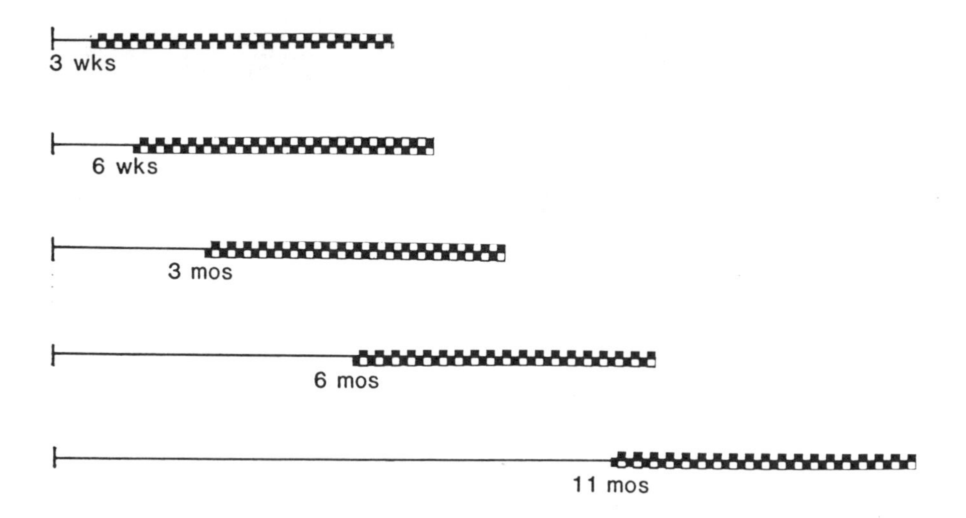

FIG. 2. Format for promotion of endogenous (background) enzyme-altered foci in livers of rats of the ages designated in the figure. When the animals reached the ages designated, groups were placed on diets containing 0.05% phenobarbital for a further 6 months; then the rats were sacrificed, and the number of foci per liver and the volume occupied by such foci were calculated. From Campbell et al. (7).

We have interpreted these studies as indicating that during the early weeks of the life of the rat, cells within the liver become initiated. Such initiation is "fixed" by cell division, which occurs at a relatively high level during the early life of the animal. After three months, however, when cell division within liver parenchymal cells decreases to extremely low levels for the remainder of the animal's life, no further initiation is seen, and a maximal number of foci, to be expected in the case when a promoting agent is administered, is seen for the remainder of the life-span. If phenobarbital is continuously given for the rest of the animal's life, a few hepatocellular carcinomas occur, as would be expected from the promotion of an occasional initiated clone (38). Similarly, other well-known promoting agents both for liver and skin have been shown, with prolonged administration, to induce the appearance of neoplasms (1,19). Thus, we

TABLE 4. The relationship of age to the number and volume of enzyme-altered foci in rat liver following phenobarbital (Pb) administration.[a]

Age at the onset of Pb feeding	No. of foci/liver	% Liver volume of foci
3 weeks	51 ± 27	0.012 ± .005
6 weeks	82 ± 25	0.045 ± .018
3 months	395 ± 181	0.048 ± .015
6 months	270 ± 72	0.039 ± .010
11 months	295 ± 60	0.068 ± .020

[a]The values are the mean ± the standard error from 8 to 22 rats per group.

feel that the so-called "epigenetic" carcinogens are actually promoting agents that exert their effects on fortuitously initiated cells and not by some more theoretical mechanism of chemical carcinogenesis. Table 3 summarizes the genetic and epigenetic characteristics of promotion.

DIFFERENTIATION AS AN EPIGENETIC CHARACTERISTIC OF NEOPLASTIC DEVELOPMENT

Perhaps the more striking evidences for an epigenetic basis for neoplasia, or at least certain neoplasms, are obtained from the nuclear transplantation experiments, the teratoma experiments, and the forced terminal differentiation of neoplastic cells by chemical mediation, as listed in Table 1. Such experiments were concerned with neoplasms that are largely in the stage of progression, although the experiments of Mintz and Illmensee (23) and those of Papaioannou et al. (28) were carried out with euploid teratoma cells. Even more striking from the potentially therapeutic viewpoint is the ability of certain chemicals to alter the program of genetic expression of neoplastic cells in vitro such that these cells tend toward normal development and, in most cases, a terminal differentiation resulting in a nondividing cell population. Such an effect was first noted in cultured Friend erythroleukemia cells of the mouse following the addition of dimethylsulfoxide (DMSO) to the culture (12). This treatment resulted in the terminal differentiation of the malignant erythroblasts and the production of normoblasts containing large amounts of hemoglobin. Similar effects have now been described in a number of other cell systems derived from leukemia cells from the mouse (39) and the human (15) and cells from solid tumors such as the neuroblastoma (17).

Our studies in this area have been largely concerned with the human promyelocytic cell line, HL-60, which can be induced to differentiate towards myelocytes by DMSO and toward monocytes by phorbol esters (8,24).

In this area we have been concerned primarily with the forced differentiation of the HL-60 line, which we have in our laboratory. As a model for this system we have attempted to determine whether the expression of an "activated" oncogene, the c-myc gene, may be regulated in a manner similar to that of other regulatory mechanisms seen during the process of differentiation in normal cells. In our line of HL-60 cells, the c-myc oncogene is activated by gene amplification to a level threefold that of normal (six copies of the gene rather than the normal two). We have been unable to see any distinction among any of the six gene copies present in the cell. Furthermore, on differentiation of the HL-60 cells induced by either TPA or DMSO, the rate of transcription of the c-myc gene decreases by 85-100% as measured by the technique of in vitro transcription of HL-60 nuclei. Earlier studies by Reitsma et al. (37) demonstrated a decrease in the cellular mRNA levels of the c-myc gene in whole cells treated with a metabolite of vitamin D and implied a transcriptional control of this gene.

Other evidence for the regulation of differentiated functions in cells comes from studies on the methylation of active and inactive genes (9), as well as the presence and absence of nuclease-sensitive sites within genes being actively transcribed (21). We find in our investigations of the c-myc gene in the HL-60 cell line that, both under circumstances of growth without differentiation and of forced differentiation induced by either of the two effectors mentioned above, the 5' end of the gene is hypomethylated whereas the 3' end is hypermethylated. This situation has been found for a number of mammalian genes (9). No change in methylation was evidenced by treatment with at least two isoschizomeric restriction enzymes, MspI and HpaII. These enzymes recognize and cleave the DNA sequence CCGG in the absence of base methylation of the internal cytidine residue. If this cytidine is methylated at the 5 position, then only MspI will cleave the DNA. No distinction could be seen in restriction enzyme cleavage in the differentiated and nondifferentiated cells. On the other hand, S_1 nuclease-sensitive sites in the c-myc gene of the HL-60 do exist (42). Figure 3 presents a map of the gene, including the three clones that we have been using to study its structural changes. The authors wish to express their appreciation to Drs. Wong-Staal and Gallo of the National Cancer Institute and to Dr. J. Michael Bishop of the University of California at San Francisco for the gift of the clones from the cellular gene seen in Fig. 3. The four S_1 nuclease-sensitive sites found in the gene of undifferentiated HL-60 cells are essentially the same as those reported by Tuan and London (42) as DNase-I-hypersensitive sites in HL-60 cells. A single S_1 sensitive site shown by the large open arrow is present in the undifferentiated HL-60 cell genome but is absent following induction of differentiation by either DMSO or TPA. Thus, it appears that this activated cellular oncogene reacts to forced differentiation by chemical mediators by mechanisms quite similar to those seen in the process of normal differentiation. This finding argues that neoplastic cells, even in the full state of tumor progression, may, by mechanisms as yet not understood, be forced to differentiate by a process that may be mechanistically quite similar to the normal process of differentiation found in the developmental biology of mammalian cells.

RECONCILIATION OF GENETIC AND EPIGENETIC MECHANISMS IN THE LIGHT OF THE NATURAL HISTORY OF NEOPLASTIC DEVELOPMENT

Although we still do not understand the mechanism of the forced differentiation of neoplastic cells, even during the stage of progression, it is quite reasonable both from experimental investigations and from our knowledge of human neoplasia (especially of the neuroblastoma and several other human tumors in the young) (4), that the natural history of neoplastic development may possess alternate pathways: induced differentiation or spontaneous differentiation. Although the spontaneous differentiation of neoplasms is an extremely unusual event, it is now becoming apparent, both through the use of elegant systems of developmental biology and by the chemically induced differentiation of neoplastic cells, that this pathway should be included in any discussion of the natural history of neoplastic development.

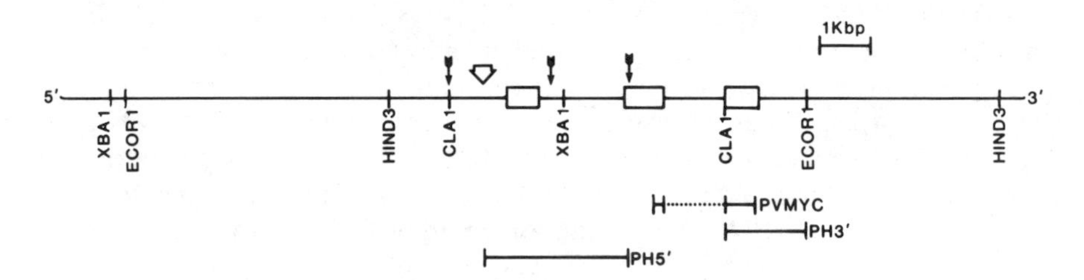

FIG. 3. Partial restriction enzyme map of the human c-myc gene in the HL-60 cell line. The three exons of the gene and the regions homologous to the myc probes used are indicated. The solid arrows indicate S_1-sensitive regions in control, TPA-, and DMSO-treated HL-60 cells. The open arrow shows the S_1-sensitive site that is present only in undifferentiated cultures and disappears on treatment with TPA and DMSO.

The pathway of differentiation of neoplasms does, in fact, represent a fourth stage in the natural history of neoplastic development. This path may arise from and occur concomitant with any of the three stages of initiation, promotion, and progression and in some instances lead to a termination of further neoplastic development. Therefore, "epigenetic carcinogenesis" from the evidence indicated in Table 1 may actually be considered to be a component of the normal natural history of neoplastic development. It, in all likelihood, initially arises from a structural alteration in genomic DNA, which is then irreversibly fixed, possibly through cell replication, to result in a cell that has the potential of becoming a malignant neoplasm, a benign terminally differentiated tumor, or a cell that never expresses these potentials during the lifetime of the host.

REFERENCES

1. Astrup, E.G., and Iversen, H. (1983): Acta Path. Microbiol. Immunol. Scand., 91:103-113.
2. Birnboim, H.C. (1982): Science, 215:1247-1249.
3. Bishop, J.M. (1983): Annu. Rev. Biochem., 52:301-354.
4. Bolande, R.P. (1971): Amer. J. Dis. Child., 122:12-14.
5. Boveri, T. (1914): Zur Frage der Entstehung maligner Tumorgen, Gustav Fischer, Jena.
6. Cameron, R., Kellen, J., Kolin, A., Malkin, A., and Farber, E. (1978): Cancer Res., 38:823-829.
7. Campbell, H.A., Pitot, H.C., Potter, V.R., and Laishes, B.A. (1982): Cancer Res., 42:465-472.
8. Collins, S.J., Ruscetti, F.W., Gallagher, R.E., and Gallo, R.C. (1979): J. Exper. Med., 149:969-974.
9. Doerfler, W. (1983): Annu. Rev. Biochem., 52:92-124.
10. Emerit, I., and Cerutti, P.A. (1981): Nature, 293:144-146.
11. Enomoto, K., and Farber, E. (1982): Cancer Res., 42:2330-2335.
12. Friend, C., and Freedman, H.A. (1978): Biochem. Pharmacol., 27:1309-1313.
13. Fry, R.J.M., Ley, R.D., Grube, D., and Staffeldt, E. (1982): Carcinogenesis, 7:155-165.
14. Goldfarb, S., and Pugh, T.D. (1981): Cancer Res., 41:2092-2095.
15. Honma, Y., Fujita, Y., Kasukabe, T., Hozumi, M., Sampi, K., Sakurai, M., Tsushima, S., and Nomura, H. (1983): Eur. J. Cancer Clin. Oncol., 19:251-261.
16. Jacob, F., and Monod, J. (1961): Cold Spring Harbor Symposium, Quant. Biol., 26:193-209.
17. Kimhi, Y., Palfrey, C., Spector, I., Barak, Y., and Littauer, U.Z. (1976): Proc. Nat. Acad. Sci., USA, 73:462-466.
18. King, T.J., and McKinnell, R.G. (1960): In: Cell Physiology of Neoplasia, pp. 591-617. University of Texas Press, Austin.

19. Kociba, R.J., Keyes, D.G., Beyer, J.E., Carreon, R.M., Wade, C.E., Dittenber, D.A., Kalnins, R.P., Frauson, L.E., Park, C.N., Barnard, S.D., Hummel, R.A., and Humiston, C.G. (1978): Toxicol. Appl. Pharmacol., 46:279-303.
20. Kohn, H.I., and Fry, R.J.M. (1984): New Engl. J. Med., 310:504-511.
21. Larsen, A., and Weintraub, H. (1982): In: Gene Structure and Regulation in Development, pp. 241-268. Alan R. Liss, Inc., New York.
22. Miller, E.C. (1978): Cancer Res., 38:1479-1496.
23. Mintz, B., and Illmensee, K. (1975): Proc. Nat. Acad. Sci., USA, 72:3585-3589.
24. Murao, S., Gemmell, M.A., Callaham, M.F., Anderson, N.L., and Huberman, E. (1983): Cancer Res., 43:4989-4996.
25. Ogawa, K., Onoe, T., and Takeuchi, M. (1981): J. Nat. Cancer Inst., 67:407-412.
26. Ohno, S. (1972): J. Med. Genet., 9:254-263.
27. Olsson, I. (1983): Acta Med. Scand., 214:261-272.
28. Papaioannou, V.E., McBurney, M.W., and Gardner, R.L. (1975): Nature, 258:70-73.
29. Peraino, C., Fry, R.J.M., Staffeldt, E., and Kisieleski, W.E. (1973): Cancer Res., 33:2701-2705.
30. Pitot, H.C. (1980): In: Carcinogenesis: Fundamental Mechanisms and Environmental Effects, edited by B. Pullman, P.O.P. Ts'o, and H. Gelboin, pp. 219-233. D. Reidel Publishing Company, Holland.
31. Pitot, H.C. (1983): Cancer Surveys, 2:519-537.
32. Pitot, H.C., Barsness, L., and Goldsworthy, T. (1978): Nature, 271:456-458.
33. Pitot, H.C., Grosso, L., and Dunn, T. (1984): In: Genes and Cancer, edited by J.M. Bishop, J.D. Rowley, and M. Greaves, pp. 81-98. Alan R. Liss, Inc., New York.
34. Pitot, H.C., and Heidelberger, C. (1963): Cancer Res., 23:1694-1700.
35. Pollard, M., Luckert, P.H., and Adams, R.A. (1982): Lab. Animal Sci., 32:147-149.
36. Rapp, F., and Westmoreland, D. (1976): Biochim. Biophys. Acta, 458:167-211.
37. Reitsma, P.H., Rothberg, P.G., Astrin, S.M., Trial, J., Bar-Shavit, Z., Hall, A., Teitelbaum, S.L., and Kahn, A.J. (1983): Nature, 306:492-494.
38. Rossi, L., Ravera, M., Repetti, G., and Santi, L. (1977): Int. J. Cancer, 19:179-185.
39. Sachs, L. (1982): J. Cell. Physiol. Suppl., 1:151-164.
40. Schulte-Hermann, R., Timmermann-Trosiener, I., and Schuppler, J. (1983): Cancer Res., 43:839-844.
41. Stott, W.T., Reitz, R.H., Schumann, A.M., and Watanabe, P.G. (1981): Fd. Cosmet. Toxicol., 19:567-576.

42. Tuan, D., and London, I.M. (1984): Proc. Nat. Acad. Sci., USA, 81:2718-2722.
43. Wattenberg, L.W. (1978): J. Nat. Cancer Inst., 60:11-18.
44. Weisburger, J.H., and Williams, G.M. (1981): Science, 214:401-407.
45. Yunis, J.J. (1983): Science, 221:227-236.

Carcinogenesis, Vol. 10, edited by E. Huberman
and S. H. Barr. Raven Press, New York © 1985.

Single-Hit Versus Multi-Hit Concept of Hepatocarcinogenesis

C. Peraino and *S. D. Vesselinovitch

*Division of Biological and Medical Research, Argonne National Laboratory, Argonne, Illinois 60439; *Departments of Radiology and Pathology, The Pritzker School of Medicine, The University of Chicago, Chicago, Illinois 60637*

The objective of this presentation is to define a new, nonevolutionary, "single-hit" concept of hepatocarcinogenesis, which postulates the following: (i) Each of the various types of carcinogen-induced lesions (from the most to the least differentiated) is initiated by a single interaction ("hit") at a specific genetic locus. (ii) Differences in the frequencies, and times to phenotypic expression, of these lesions result from qualitative differences in the characteristics of the carcinogen-sensitive loci controlling the development of the lesions. (iii) All lesions, irrespective of their levels of deviation from normalcy, develop in a mutually independent manner. In the following discussion we compare this concept with the current evolutionary, or "multi-hit," concept of carcinogenesis, which specifies the following: (i) Different numbers of initiation events are responsible for the production of different types of carcinogen-induced lesions. (ii) The relative frequency and latency of a given lesion reflect the relative frequency of the appropriate combination of initiation events required for its production, as well as the time required for the accumulation of these events. (iii) Less-differentiated lesions evolve from those that are more differentiated.

EVOLUTIONARY, MULTI-HIT CONCEPT OF CARCINOGENESIS

The multi-hit concept, which has been described in numerous publications (e.g., 4-10,17), views carcinogenesis as a stepwise process involving the time-dependent accumulation of several mutations. Thus, following the first carcinogen-induced mutation ("initiation"), cell division would lead to the production of clones of initiated cells that may (6,7) or may not (17) have altered phenotypes. In this context, the "promotion" phase of carcinogenesis is viewed as the expansion of this initiated cell population via proliferative stimulation (7,12). The "progression" of initiated cells to malignancy via the consecutive development of new and irreversible phenotypic properties (8) would involve a series of additional mutations (4,17) that may be self-generating (4,7).This concept is based on the following three major lines of evidence:

1. The consecutive emergence of increasingly deviated lesions at successively lower frequencies during carcinogenesis, implying a lineal relationship among these lesions.

Foulds (8,9) however, ascribed only limited significance to this type of evidence, citing observations of the de novo emergence of malignant neoplasms. Moreover, observations that relative yields of benign and malignant skin tumors were strongly influenced by the type of initiation protocol used (1,16,18) suggest that the characteristics of each type of tumor are determined at the time of initiation rather than by time-dependent evolutionary changes.

2. "Material continuity" between suspected sequentially generated lesions (6,10), implying that a more advanced lesion develops directly from cells within a lesion that is less advanced.

Such continuity has been observed most readily in lesions generated during virally induced skin and mammary neoplasia (8-10) in which the initiating agent is continuously present in some form (11), and in chemical carcinogenesis studies involving high and/or repeated doses of carcinogen (6,8). However, the latter treatment regimens also increase the de novo production of carcinomas (8-10), arguing against an obligatory precursor-product relationship between benign and malignant neoplasms.

3. A greater than first-order kinetic relationship between carcinogen dose rate and the formation of malignant tumors, suggesting that a given increment in tumor formation requires more than an equivalent increment in exposure to carcinogen (3).

This relationship implies, in turn, that the production of each tumor requires more than one direct carcinogen-mediated initiation event or "hit"; a requirement for as many as seven hits has been estimated for malignant tumor formation (3). This interpretation, however, is based on a concatenation of carcinogen-dosage-dependent and time-dependent events occurring during tumor induction. More recent experiments, in which the contributions of carcinogen dosage and time to tumor yield were considered separately, suggest a single carcinogen hit followed by multiple, time-dependent expression events as the model for malignant tumor formation (19, discussed below).

NONEVOLUTIONARY, SINGLE-HIT CONCEPT OF HEPATOCARCINOGENESIS

Experimental Evidence

The development of this concept was engendered by the growing realization that the evolutionary concept described above did not adequately account for the characteristics of lesions obtained in chemical hepatocarcinogenesis studies involving limited exposure to carcinogen. In one such study (14), weanling rats were fed a 0.02% acetylaminofluorene (AAF) diet for 2 weeks and then changed to food containing various concentrations of the tumor promoter phenobarbital. When fed at dietary concentrations that did not affect body weight gain, phenobarbital elicited dose-dependent increases in the final incidence (plateau level) of hepatic tumors (Fig. 1). However, these phenobarbital treatments did not affect the time to onset or to cessation (attainment of the plateau phase) of tumorigenesis. Moreover, none of the phenobarbital treatments affected the relative growth rates (as denoted by the similar tumor size distributions irrespective of phenobarbital dosage, Fig. 2) or the

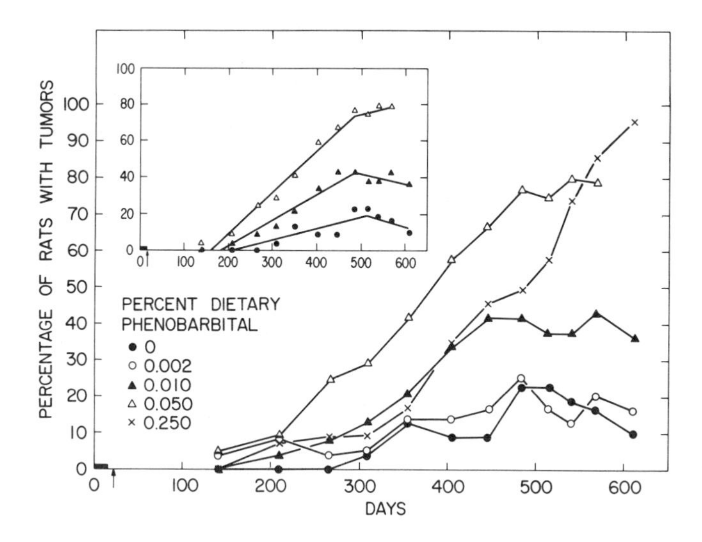

FIG. 1. Tumor incidences in rats chronically fed different dietary percentages of phenobarbital beginning (arrow on abscissa) 7 days after feeding AAF for 14 days (black bar on abscissa) at a dietary concentration of 0.02%. Inset, computer-generated plot of the data for the 0.00, 0.01, and 0.05% dietary phenobarbital groups fitted with lines by the method of least squares. From Peraino et al. (14).

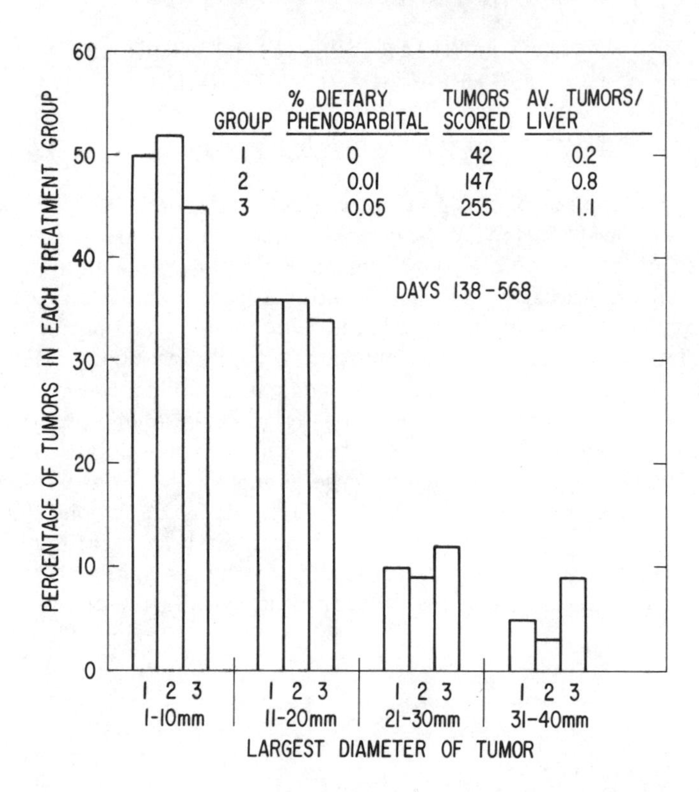

FIG. 2. Tumor size distribution in those rats in Fig. 1 that received AAF alone or AAF followed by 0.01 or 0.05% dietary phenobarbital. The tabulated data indicating the overall tumor frequency were calculated for the entire experimental interval monitored (138 to 568 days, as designated). From Peraino et al. (14).

morphological properties (as denoted by the similar ranges of tumor morphologies in all treatment groups, Fig. 3) of the more than 900 tumors produced over the 600-day duration of this study.

It is evident, therefore, that appropriate conditions (limited initiating stimulus followed by promotion) highly conducive to tumor formation did not induce tumor progression. This apparent phenotypic stability of tumors is at variance with the evolutionary concept of carcinogenesis described above, with its presumption of intrinsic genetic instability and consequent continual phenotypic progression of carcinogen-induced lesions.

Recently we extended our examination of the phenotypic behavior of carcinogen-induced hepatic lesions to include an analysis of the relationship between hepatic tumor formation and the characteristics of foci of altered hepatocytes that invariably appear in carcinogen-treated livers

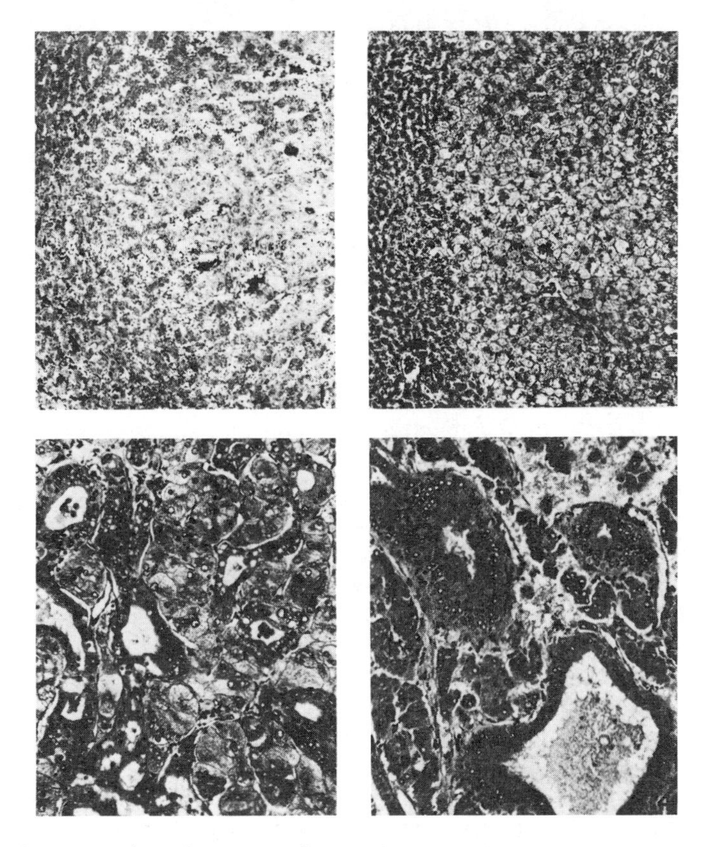

FIG. 3. Comparative histology of hepatic tumors in rats receiving AAF alone (left side of figure) or followed by 0.25% dietary phenobarbital (right side of figure). Upper two panels, neoplastic nodule; lower two panels, hepatocellular carcinoma, trabecular type with glands. From Peraino et al. (14).

prior to the emergence of grossly observable tumors (13). The experimental protocol involved the intraperitoneal injection of a single dose of carcinogen [diethylnitrosamine (DEN) or benzo[a]pyrene (BAP) at dosages of 0.15 and 0.59 micromoles per gram of body weight, respectively] into rats within one day after birth, followed by dietary exposure to promoter (0.05% dietary phenobarbital) beginning at weaning and continuing throughout the experiment. These rats were then killed at intervals, and their livers were examined for tumors and for histochemically detectable altered hepatocyte foci. Foci containing between one and six histochemical markers were identified, and the average focus diameters were calculated through the use of serial frozen-sectioning techniques and computer-assisted image analysis. The same complement of histochemical tests was applied to the primary hepatic tumors observed in this study. The histochemical markers assayed were increased gamma glutamyltranspeptidase, decreased glucose-6-phosphatase, increased diaphorase, increased alkaline phosphatase, iron exclusion, and altered glycogen deposition. As discussed in a recent review (12), all of these histochemical markers have been used to detect altered hepatocyte foci in the livers of carcinogen-treated rats.

The principal findings in this study were as follows:

1. Both the DEN and BAP treatments were tumorigenic. They produced foci with similar phenotypic properties (numbers and identities of histochemical markers), although more foci and tumors were produced by the DEN treatment than by the BAP treatment. This phenotypic consistency suggests that the foci represent a coherent response to carcinogenic stimuli, irrespective of the structure, metabolism, and tumorigenic effectiveness of these agents in liver.

2. The relative growth rates of the foci and their growth capacities (ranges of possible growth rates) were directly related to focus phenotypic complexity level (numbers of markers per focus). This positive correlation suggests a mechanistic linkage between processes controlling the number of markers expressed and those controlling the proliferative behavior of the cells that constitute the foci.

3. Individual foci were phenotypically stable, i.e., they neither gained nor lost markers. Some of the data (14) supporting this conclusion are illustrated in Fig. 4, which depicts the time-dependent changes in the relative frequencies of the foci phenotypic complexity levels.

The data in Fig. 4 show a decline in the relative frequency of single marker foci. The decline resulted from essentially uniform increases in the relative frequencies of foci that were phenotypically more complex. If individual foci were progressively losing or acquiring markers, this kinetic picture would have been quite different; marker loss would have produced a progressive relative enrichment of lower complexity foci, whereas marker acquisition would have had the opposite effect. Evidently, therefore, the phenotypic relationships shown in Fig. 4 denote the early appearance of the phenotypically simplest foci followed by the later _de novo_ emergence of foci with higher complexity levels. This interpretation is supported by additional data showing the maintenance of a constant number of the simplest foci over time and parallel changes in the average phenotypic complexity and frequency of the foci (13). Moreover, given the consistently low relative frequency of foci with six markers, in comparison with the relative frequencies of lower complexity foci (Fig. 4) and the extremely low frequency ratio of tumors to foci (15), the relatively constant frequencies of foci across phenotypic complexity levels cannot be attributed to a continuous flow of foci phenotypes through successive complexity levels into tumors. Instead, as indicated above, the data suggest that the phenotypes of individual foci remain constant.

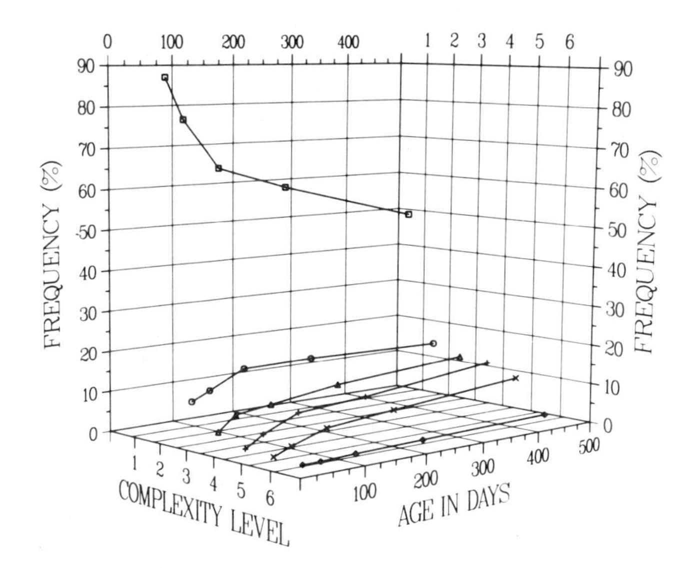

FIG. 4. Time-dependent changes in the relative frequencies (foci at each complexity level as a percentage of the total number of foci) of the six complexity levels in the total focus population. Symbols for the increasing complexity levels are: (□), Level 1; (○), Level 2; (△), Level 3; (+), Level 4; (×), Level 5; (◇), Level 6.

4. A substantial fraction of the tumors observed in this study had fewer markers than the most complex foci. Figure 5 compares the patterns of phenotypic complexity for the total pooled population of foci and tumors. It is apparent that most foci were relatively simple (approximately 80% showing no more than two markers), whereas most tumors were relatively complex (approximately 70% showing five or more markers). However, the tumors did not invariably have more complex phenotypes than did the foci, as would be predicted by the evolutionary, multi-hit concept of tumorigenesis. Instead, approximately 10% of the tumors showed as few as three markers, whereas approximately 6% of the foci showed five markers or more.

On the basis of these observations, we suggest that foci emerge as the result of a specific set of genetic changes inducible solely by carcinogenic stimuli, but the foci do not evolve through progressively more deviated forms into tumors. Instead, we postulate that the induction of tumors involves a specific initiation event that is distinct from the initiation event leading to focus formation.

Evidence that single, rather than multiple, initiation events are responsible for the production of tumors, as well as foci, was also

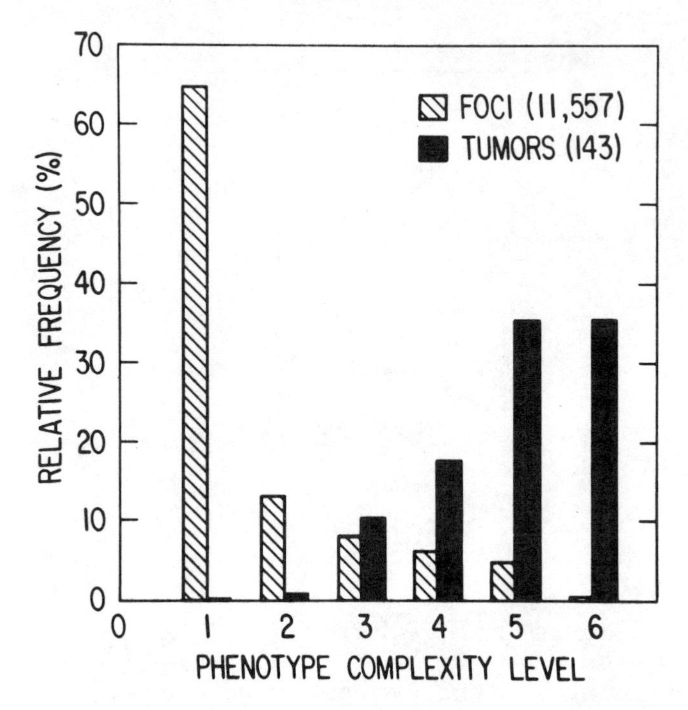

FIG. 5. A comparison of the phenotypic complexity distribution patterns in the total pooled populations of foci and of tumors. From Peraino et al. (13).

obtained in kinetic and dose-response studies of DEN-induced hepatocarcinogenesis in mice (19,20). Data from these studies show first-order relationships between carcinogen dosage and frequencies of basophilic foci and of hepatocellular carcinomas when these dose-response measurements were made at a given time after a single treatment with carcinogen (Fig. 6). However, the time-response kinetics showed that the development of basophilic foci and carcinomas was related to the time factor by a power of two for foci and a power of four for carcinomas (Fig. 6). Collectively, these data suggest that (i) single qualitatively different events are required for the initiation of foci and carcinomas and (ii) twice as many postinitiation ("expression") events are required for the development of hepatocellular carcinomas as for the development of basophilic foci.

Formulation of Single Hit Concept

To summarize our overall findings, we have observed that a minimal treatment with carcinogen followed by prolonged exposure to promoter

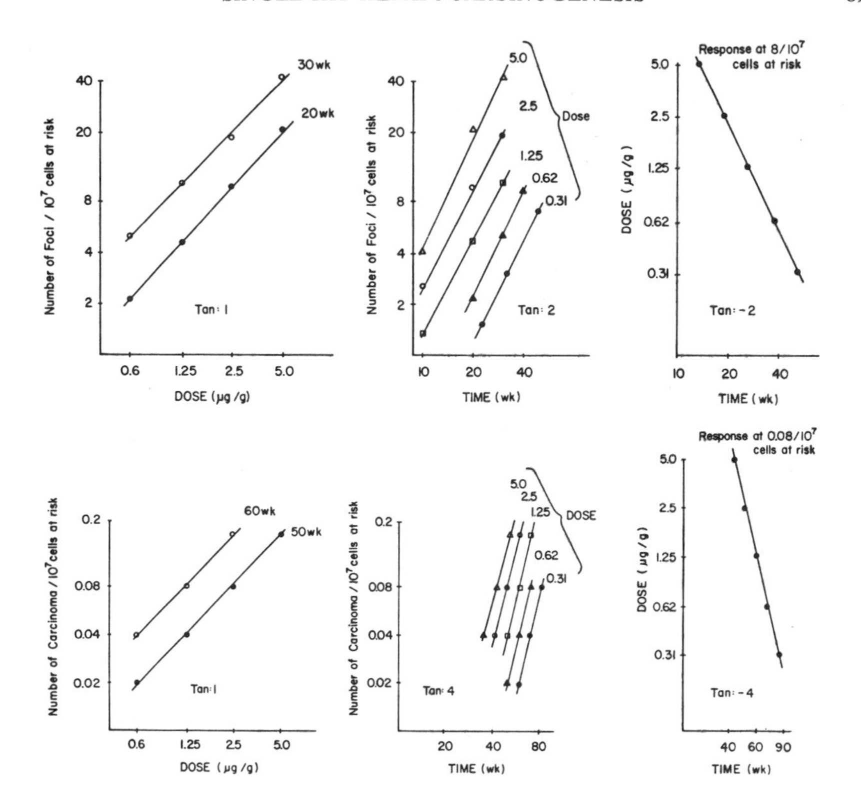

FIG. 6. Comparative dose-response and kinetic relationships in the development of basophilic foci and hepatocellular carcinomas. Left panels, DEN dose (on body weight basis) versus yield of basophilic foci (upper panel) or carcinomas (lower panel) at two times after DEN administration. Middle panels, time response for basophilic focus induction (upper panel) and carcinoma induction at different doses of DEN. Right panels, DEN dose versus time to the appearance of eight basophilic foci (upper panel) or 0.08 hepatocellular carcinomas (lower panel) per given number of cells at risk. From Vesselinovitch and Milhailovich (19).

produces large numbers of early-appearing nonneoplastic altered hepatocyte foci and much lower frequencies of later-appearing liver cell tumors (13). The major attribute of both types of lesions appears to be their phenotypic stability (13,14).

In connection with the mechanism of promoter action, the maintenance of the phenotypic stability of foci and tumors in the presence of continued promoter administration strengthens the argument that promoters facilitate the phenotypic expression of initiator-induced genetic changes without exerting any influence on the character of these changes. With reference to relationships among carcinogen-induced lesions, the major implication of this phenotypic stability is that the members of the different classes of lesions induced by limited carcinogen treatment develop independently. The existence of this independence, coupled with the observation that different classes of lesions can be produced by a single exposure to carcinogen (13,19,20), suggests that the potentiality for all the phenotypic attributes of each class of lesions may be dictated by specific carcinogen-mediated genetic changes at the time of initiation. Separation of dosage- and time-dependent effects of carcinogen treatment on the induction of foci and carcinomas showed that malignant, as well as nonmalignant, lesions are initiated by single carcinogen-mediated events, but the expression of the malignant phenotype requires more postinitiation events than does the expression of the nonmalignant phenotype (19,20).

Figure 7 is a conceptual presentation of the foregoing relationships and their proposed relevance to hepatocarcinogenesis. According to this concept, specific hierarchically related gene classes are involved in the development of the different types of carcinogen-induced lesions. The different gene classes are considered to be similar with respect to the magnitude of the carcinogenic insult required to initiate their neoplastic function but dissimilar (i) in their accessibility for interaction with the carcinogen and (ii) in the manifestation of the effects of this interaction, as expressed in the induction of specific types of lesions. Thus, the decreasing frequency of successively more deviated lesions may reflect the decreasing probability that the genes relevant to the development of these lesions will be affected by a given dose of carcinogen.

The sequential emergence of foci, nonmalignant tumors, and malignant tumors could reflect the longer intervals (i.e., larger numbers of cell divisions) needed for the completion of the increasingly complex repertoires of molecular events mediated by genes controlling the formation of the less-differentiated lesions. In the latter context, the occurrence of cooperation among genes in the expression of malignancy, as suggested by recent oncogene studies (2,17,21), may involve hierarchies of regulation in which the direct modification of higher-order differentiation-controlling genes by the initiator engenders the subsequent secondary activation of related lower-order genes, with the cooperative interaction of all genes in the hierarchy being necessary for the expression of the fully malignant phenotype. The monodirectional nature of this relationship would preclude the development of more-advanced lesions following an initiation event that is restricted to a lower member of this gene hierarchy.

With regard to the emergence of more-advanced lesions within those that are less deviated (material continuity), the single-hit concept suggests that this phenomenon results from subsequent additional independent initiation events--resulting from "background" hits (19,20) or from continued exposure to known initiators (6, 8-11)--that affect remain-

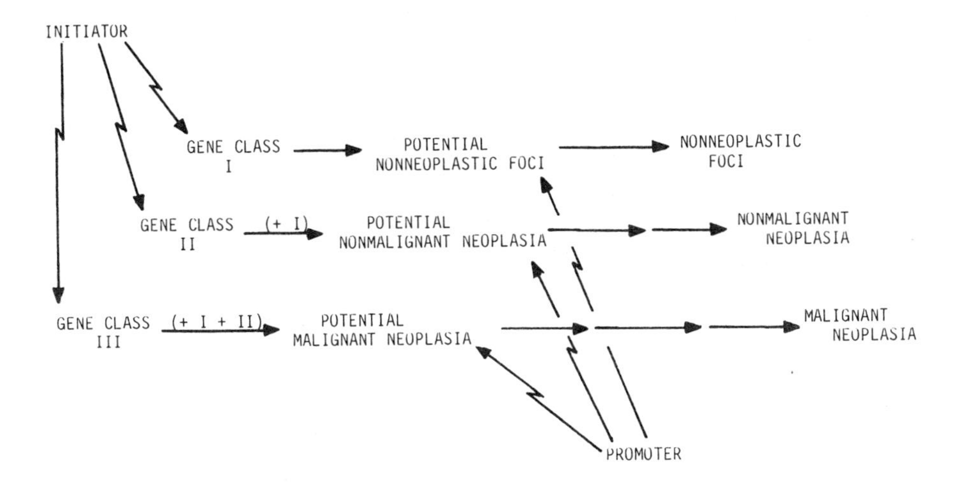

FIG. 7. Diagram of single-hit, multi-event concept of hepatocarcinogenesis.

ing available genes further along the gene hierarchy than those whose prior activation is responsible for the existing lesions. The probability for such events would be greater in the proliferating cell populations within such lesions than in the relatively quiescent surrounding cells.

Finally, the model we have described appears promising from an experimental viewpoint because it suggests that with the appropriate animal treatment and cellular isolation protocols it should be possible to separate the different classes of carcinogen-induced genetic alterations that are essential for the expression of malignancy into stable and distinguishable cell populations.

ACKNOWLEDGMENTS

This work was supported by the U. S. Department of Energy under contract No. W-31-109-ENG-38 and Public Health Service Grant No. CA-25552, from the National Cancer Institute, U.S. Department of Health and Human Services.

REFERENCES

1. Burns, F.J., Albert, R.E., and Altshuler, B. (1984): In: Mechanisms of Tumor Promotion, Vol. 2, edited by T. J. Slaga, pp. 17-39. CRC Press, Boca Raton, Florida.
2. Cooper, G.M., and Lane, M. (1984): Biochim. Biophys. Acta, 738:9-20.

3. Emmelot, P., and Scherer, E. (1977): Cancer Res., 37:1702-1708.
4. Emmelot, P., and Scherer, E. (1980): Biochim. Biophys. Acta, 605:247-304.
5. Farber, E. (1973): In: Methods in Cancer Research, Vol. 7, edited by H. Busch, pp. 345-375. Academic Press, New York.
6. Farber, E. (1980): Biochim. Biophys. Acta, 605:149-166.
7. Farber, E. (1984): J. Cell. Physiol. Suppl., 3:123-125.
8. Foulds, L. (1954): Cancer Res., 14:327-339.
9. Foulds, L. (1969): Neoplastic Development, Vol. 1. Academic Press, New York.
10. Foulds, L. (1975): Neoplastic Development, Vol. 2. Academic Press, New York.
11. Gross, L. (1983): Oncogenic Viruses, Vol. 1. Pergamon Press, New York.
12. Peraino, C., Richards, W.L., and Stevens, F.J. (1983): In: Mechanisms of Tumor Promotion, Vol. 1, edited by T.J. Slaga, pp. 1-53. CRC Press, Boca Raton, Florida.
13. Peraino, C., Staffeldt, E.F., Carnes, B.A., Ludeman, V.A., Blomquist, J.A., and Vesselinovitch, S.D. (1984): Cancer Res., 44:3340-3347.
14. Peraino, C., Staffeldt, E.F., Haugen, D.A., Lombard, L.S., Stevens, F.J., and Fry, R.J.M. (1980): Cancer Res., 40:3268-3273.
15. Pitot, H.C., Barsness, L., Goldsworthy, T., and Kitagawa, T. (1978): Nature, 271:456-458.
16. Saffiotti, U., and Shubik, P. (1956): J. Nat. Cancer Inst., 16:961-969.
17. Temin, H.M. (1984): J. Cell. Physiol. Suppl., 3:1-11.
18. Turosov, V., Dey, N., Andrianov, L., and Jain, D. (1971): J. Nat. Cancer Inst., 47:105-111.
19. Vesselinovitch, S.D., and Mihailovich, N. (1983): Cancer Res., 43:4253-4259.
20. Vesselinovitch, S.D., and Mihailovich, N. (1984): In: Mouse Liver Neoplasia. Current Perspectives, edited by T. J. Slaga, pp. 61-83. Hemisphere Publishing Corp., New York.
21. Weinstein, I.B., Gattoni-Celli, S., Kirchmeier, P., Lambert, M., Hsiao, W., Backer, J., and Jeffrey, A. (1984): J. Cell. Physiol. Suppl., 3:127-137.

Carcinogenesis, Vol. 10, edited by E. Huberman and S. H. Barr. Raven Press, New York © 1985.

Sulfuric Acid Esters as Ultimate Electrophilic and Carcinogenic Metabolites of Some Alkenylbenzenes and Aromatic Amines in Mouse Liver

E. C. Miller, J. A. Miller, E. W. Boberg, K. B. Delclos, C.-C. Lai, T. R. Fennell, R. W. Wiseman, and A. Liem

McArdle Laboratory for Cancer Research, University of Wisconsin Medical School, Madison, Wisconsin 53706

Two of us (E.C.M. and J.A.M.) had the privilege of working with Charles Heidelberger and of knowing him well from the time that he joined the staff of the McArdle Laboratory for Cancer Research in 1948. During the nearly 30 years that he was a member of the McArdle staff, Charlie was a major stimulant. He read widely, he thought incisively, and he always sought aggressively for new ideas and new frontiers of research. Charlie was critical of both his ideas and those of others, and he discussed new insights clearly and with enthusiasm. On the basis of these characteristics and his intensive approach to research, Charles Heidelberger became a major leader in cancer research early in his career and remained a leader until his untimely death in 1983. We are very pleased to contribute to this symposium honoring his life and research.

Research in chemical carcinogenesis in the past few decades has resulted in the description of the induction of cancer as a multistage process (Fig. 1). In the early studies of Mottram, Rous and his colleagues, and Berenblum and Shubik, carcinogenesis induced by polycyclic aromatic hydrocarbons or tars on the skin of mice and rabbits was divided into two stages termed initiation and promotion (6). More recent studies have provided experimental data to extend this concept to a number of tissues, including the mammary tissue, liver, thyroid, colon, and urinary bladder of rodents (6,37). Multistage processes, which include initiation and promotion, are now accepted as the typical situation in chemical carcinogenesis.

The initiating activity or complete carcinogenic activity of a wide variety of carcinogenic chemicals has been correlated with the metabolism of these chemicals to strong electrophilic reactants (26). Much evidence now strongly indicates that the initiation of carcinogenesis results from reaction of these electrophilic carcinogens with bases in the

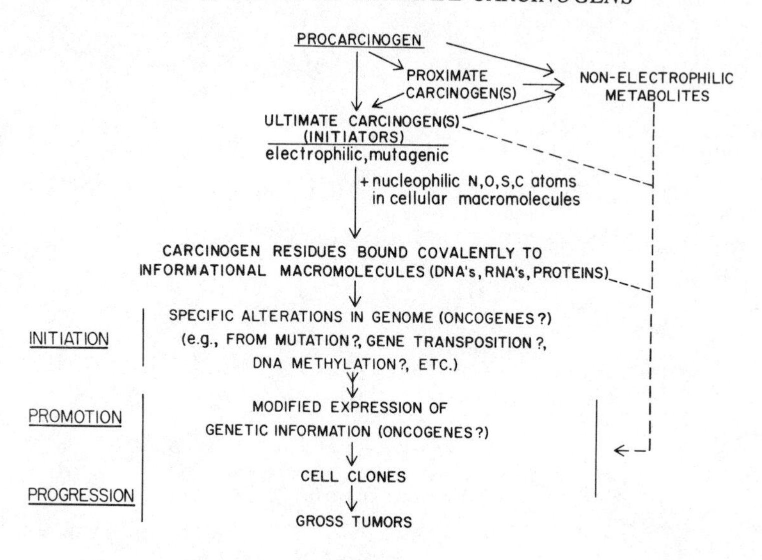

FIG. 1. General pathways for the activation of chemical carcinogens and for tumor induction by chemical carcinogens.

cellular DNA and that products of these reactions give rise to hereditary alterations in DNA structure and, accordingly, in genetic potential. Recent studies implicate the ras proto-oncogene family as at least one such genetic target. Thus, tumor cells induced by chemical carcinogens contain an "activated" ras gene, which has acquired the ability to transform indicator cells in culture (4,5,16,41). In several cases, a single point mutation within a ras coding sequence appears to be sufficient to induce this transforming potential. This "activation" event can also be demonstrated by transfection of NIH-3T3 cells with c-H-ras DNA modified in vitro by reaction with benzo[a]pyrene diolepoxide (23) or sodium bisulfite (13). The role of chemical carcinogens in tumor promotion and progression is less well understood. The classical view is that tumor promotion results from altered gene expression in initiated cells (8), but the molecular nature of this process is still not defined.

Our laboratory has directed much of its attention to the identification of ultimate electrophilic metabolites of chemical carcinogens and of their reactions with DNA in vivo (26). One of the goals of these studies has been to identify the specific electrophilic metabolites and the DNA adducts involved in the initiation of carcinogenesis. Our recent studies have indicated that sulfuric acid esterification of members of each of three classes of proximate carcinogens converts them to ultimate electrophilic and carcinogenic metabolites in mouse liver. These studies form the subject of this report.

1'-HYDROXYSAFROLE

Alkenylbenzene derivatives, which differ in the number and locations of methylenedioxy and methoxy substituents on the benzene ring, occur naturally in a variety of plant species (30). Safrole (1-allyl-3,4-methylenedioxybenzene), which is a major component of oil of sassafras and a minor component of some other essential oils, herbs, and spices, was found to be hepatocarcinogenic for rats and mice in 1960, and its carcinogenicity and metabolic activation have been studied in considerable detail (29). Estragole (1-allyl-4-methoxybenzene) and methyleugenol (1-allyl-3,4-dimethoxybenzene) have carcinogenic activity in mouse liver similar to that of safrole, but more extensively substituted analogs have so far shown little or no activity (27).

Safrole and estragole are oxidized to their 1'-hydroxy derivatives by the cytochrome P-450 monooxygenase system of liver, and the 1'-hydroxy derivatives are more potent carcinogens than the parent compounds in mouse liver (29,30). In turn, the 1'-hydroxy metabolites are metabolized to three classes of electrophilic metabolites in liver, i.e., 1'-esters, 2',3'-epoxides, and α,β-unsaturated ketones (14,29,30). Studies in our laboratory by Phillips et al. (32,33) showed that the DNA adducts formed in the livers of mice treated with either 1'-hydroxysafrole or 1'-hydroxyestragole cochromatographed during high-performance liquid chromatography with adducts formed by nonenzymatic reaction of the synthetic 1'-acetic acid esters with deoxyguanosine or deoxyadenosine (Fig. 2). The principal deoxynucleoside adducts formed both by nonenzymatic reaction of the acetic acid esters or in mouse liver from the 1'-hydroxy derivatives were characterized as 3'-(deoxyguanosin-N^2-yl)-*trans*-isosafrole and 3'-(deoxyguanosin-N^2-yl)-*trans*-isoestragole; the analogous deoxyadenosin-N^6-yl adducts were formed as minor products. 1'-(Deoxyguanosin-N^2-yl) - safrole, formed *in vivo* and nonenzymatically *in vitro*, has now been resolved into two diastereomers; 1'-(deoxyguanosin-N^2-yl)-estragole formed from 1'-acetoxyestragole has similarly been resolved, but the adduct formed *in vivo* has not been further examined (44). More detailed studies have tentatively identified two other adducts, both from 1'-hydroxysafrole in mouse liver and from nonenzymatic reactions of deoxyguanosine with 1'-acetoxysafrole, as 3'-(deoxyguanosin-7-yl)- and 3'-(deoxyguanosin-8-yl)-*trans*-isosafrole (44). On the other hand, the nucleoside adducts formed nonenzymatically from 1'-hydroxysafrole-2',3'-oxide or 1'-oxosafrole or from the corresponding estragole derivatives did not cochromatograph with the adducts obtained from the hepatic DNA of 1'-hydroxysafrole- or 1'-hydroxyestragole-treated mice (32,33). Thus, the major electrophilic precursors for the DNA adducts formed in mouse liver appeared to be esters of 1'-hydroxysafrole or 1'-hydroxyestragole. Sulfuric acid esters were candidates for the reactive forms, since Wislocki et

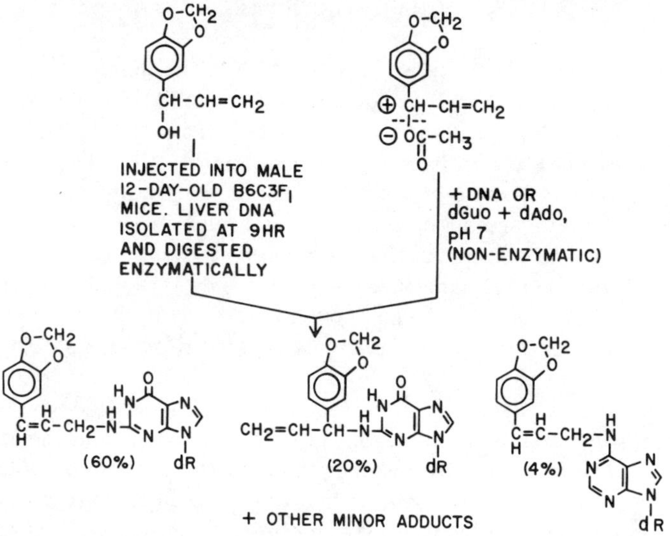

FIG. 2. Structures of adducts formed in mouse liver DNA from 1'-hydroxysafrole or nonenzymatically by reaction at neutrality of 1'-acetoxysafrole with deoxynucleosides.

al. (45) demonstrated sulfotransferase activity for 1'-hydroxysafrole in rat and mouse liver cytosols; acetyl coenzyme A-dependent acetyltransferase activity was not detected.

Male (C57BL/6 x C3H/He)F_1 ($B6C3F_1$) mice, which develop multiple gross hepatomas by 12 months of age after being given a single injection of 1'-hydroxysafrole at 12 days of age, were used to probe the role of 1'-sulfoöxysafrole (the sulfuric acid ester of 1'-hydroxysafrole) in hepatic carcinogenesis by 1'-hydroxysafrole (7). The single dose of carcinogen apparently initiates the hepatocarcinogenic response, while some feature(s) of normal maturation and growth in the male, probably including hormonal factors, apparently provide the tumor-promoting stimulus. The sulfotransferase activity for 1'-hydroxysafrole in livers of $B6C3F_1$ mice reached a maximum at 2-3 weeks of age and then decreased markedly. This transferase activity was dependent on 3'-phosphoadenosine-5'-phosphosulfate (PAPS) and was strongly inhibited by pentachlorophenol, a relatively specific inhibitor of sulfotransferases (31). Inhibitions of 70 and 90% were observed at 1 and 10 μM concentrations, respectively.

The levels of DNA adducts in the livers of male 12-day-old $B6C3F_1$ mice given a dose of [2',3'-^{3}H]1'-hydroxysafrole 45 min after a single dose of 0.04 μmol of pentachlorophenol per gram of body weight were reduced to 15% of the level observed for mice not pretreated with pentachlorophenol (Table 1). The same pretreatment with pentachlorophenol reduced the average number of hepatomas per mouse at 10 months to less than 10% of the number observed in the absence of the inhibitor (Table 1). In contrast, pentachlorophenol pretreatment had no inhibitory effect on hepatoma induction by diethylnitrosamine, a carcinogen that is not activated through an esterification reaction (22). Thus, the very marked inhibition of both hepatic DNA adduct formation and hepatoma formation from 1'-hydroxysafrole in mice pretreated with pentachlorophenol strongly suggested that the formation of 1'-sulfoöxysafrole was required for both events.

The use of brachymorphic mice provided a second approach to assessing the role of sulfuric acid esterification in hepatic DNA adduct formation and tumor formation. As elucidated by Sugahara and Schwartz (39, 40), the recessive trait of brachymorphism is a consequence of the limited capacity of these mice to synthesize PAPS and the resultant undersulfation of the glycosaminoglycans in their cartilage. Lyman and Poland (21) showed that the deficiency of PAPS in brachymorphic mice also resulted in reduced hepatic sulfation of p-nitrophenol and estrone. When [2',3'-^{3}H]1'-hydroxysafrole was administered to male B6C3F$_2$ brachymorphic mice at 12 days of age, the level of DNA-safrole adducts in the livers was only 15% as high as in the livers of the phenotypically normal B6C3F$_2$ mice given the same treatment (Table 2). Likewise, the multiplicity of hepatomas that developed in the brachymorphic mice at about 1 year was no more than 10% that observed in the normal littermates. On the other hand, the average number of hepatomas in the brachymorphic mice given a single dose of diethylnitrosamine was about 40% as great as in their phenotypically normal littermates. The factors associated with this lower tumor multiplicity in the male brachymorphic mice, as compared to the phenotypically normal mice, after treatment with diethylnitrosamine are not known. Throughout their lives, the brachymorphic mice, both carcinogen-treated and controls, had average body weights that were about 20% less than those of the phenotypically normal mice, but the weights of individual mice at the termination of the experiment were not correlated with the number of hepatomas (7). In any case, the difference between the essential lack of any tumorigenic response to 1'-hydroxysafrole in the brachymorphic mice, as compared to the reduction to 40% on treatment with diethylnitrosamine, was striking.

Thus, all of the data point to 1'-sulfoöxysafrole as a critical metabolite for the formation of both the hepatic DNA adducts and of hepatomas on administration of 1'-hydroxysafrole to male preweanling B6C3 hybrid mice (Fig. 3). Similar strong inhibition of hepatic tumor formation by pentachlorophenol was observed when a single dose of estragole, 1'-hydroxyestragole, or 1'-hydroxy-2',3'-dehydroestragole, a more potent hepatocarcinogen than 1'-hydroxyestragole, was administered to male 12-day-old B6C3F$_1$ mice (R. W. Wiseman, E. C. Miller,

TABLE 1. Effect of pretreatment with pentachlorophenol on hepatic DNA adduct formation and hepatoma initiation by 1'-hydroxysafrole.[a]

Carcinogen (μmol/g body (weight)	Penta-chloro-phenol	DNA adducts (pmol/mg)	Hepatomas Incidence (%)	Hepatomas Av. no./liver
1'-Hydroxysafrole				
(0.2)	-	190	97	4.4
(0.2)	+	24	10	0.1
(0.1)	-	68	86	2.2
(0.1)	+	9	12	0.2
Diethylnitrosamine				
(0.01)	-		100	12
	+		100	14
None	- or +		12	0.1

[a]12-day-old male B6C3F$_1$ mice were injected intraperitoneally with 0.04 μmol of pentachlorophenol per gram body weight 45 min before the intraperitoneal injection of carcinogen. The DNA adducts were assayed at 9 h, and the hepatomas were enumerated at 10 months. See (7) for further details.

J. A. Miller, and A. Liem, unpublished data). Furthermore, pretreatment with 0.04 μmol per gram body weight of pentachlorophenol markedly inhibited the formation of DNA-bound 1'-(deoxyguanosin-N^2-yl)-2',3'-dehydroestragole in the livers of mice administered [^{3}H]1'-hydroxy-2',3'-dehydroestragole at 12 days of age (T. R. Fennell, J. A. Miller, and E. C. Miller, unpublished data). Our data on the carcinogenicity of estragole and the finding of Randerath et al. (34) that pentachlorophenol similarly inhibited hepatic DNA adduct formation from safrole in adult female CD-1 mice provide further evidence that estragole and safrole are metabolized for adduct formation via 1'-hydroxylation and subsequent sulfuric acid esterification.

N-HYDROXY-2-ACETYLAMINOFLUORENE

The metabolic activation of 2-acetylaminofluorene has been studied extensively over the past twenty years. Its conversion to the proximate carcinogenic metabolite N-hydroxy-2-acetylaminofluorene by the cytochrome P-450 monooxygenases has been assigned a central role in its activation, but the electrophilic metabolites that are critical for the carci-

TABLE 2. The formation of hepatic DNA adducts and of hepatomas on administration of 1'-hydroxysafrole to brachymorphic and phenotypically normal B6C3F$_2$ mice.[a]

Carcinogen (μmol/g body weight)	Phenotype	DNA adducts (pmol/mg)	Hepatomas Incidence (%)	Hepatomas Av. no./ liver
1'-Hydroxysafrole				
(0.2)	Normal	110	43	1.2
(0.2)	Brachymorphic	16	6	0.1
(0.15)	Normal		56	1.9
(0.15)	Brachymorphic		0	0.0
Diethylnitrosamine				
(0.02)	Normal		80	11.6
(0.02)	Brachymorphic		52	5.0
None	Normal		11	0.1
	Brachymorphic		2	0.02

[a]12-day-old male B6C3F$_2$ mice were injected intraperitoneally with trioctanoin solutions of 1'-hydroxysafrole or diethylnitrosamine. The DNA adducts were assayed at 9 h, and the hepatomas were enumerated at 15 mo (1'-hydroxysafrole) or 9 mo (diethylnitrosamine). See (7) for further details.

nogenicity of the N-hydroxy derivative have not been clearly identified (26). The two acetylated adducts N-(deoxyguanosin-8-yl)- and 3-(deoxyguanosin-N^2-yl)-2-acetylaminofluorene comprise about 40% of the adducts in the hepatic DNA of rats given a single dose of N-hydroxy-2-acetylaminofluorene; the N-(deoxyguanosin-8-yl)-2-aminofluorene adducts account for the remaining 60%. The acetylated adducts may arise from the reaction of N-sulfoöxy-2-acetylaminofluorene formed by PAPS-dependent sulfotransferase activity in the cytosol (10,18,24) and/or from the N-acetoxy-2-acetylaminofluorene obtained on dismutation of free nitroxide radicals formed through the action of one-electron oxidants (e.g., peroxidase-H_2O_2) systems (2,3). The formation of the aminofluorene adducts in the DNA has been ascribed primarily to reaction of N-acetoxy-2-aminofluorene formed from N-hydroxy-2-acetylaminofluorene by a cytosol-catalyzed acyltranferase reaction (1,17). Although reaction of N-hydroxy-2-aminofluorene is very limited at neutral pH (19), this metabolite has also received consideration as a possible precursor of the aminofluorene adduct.

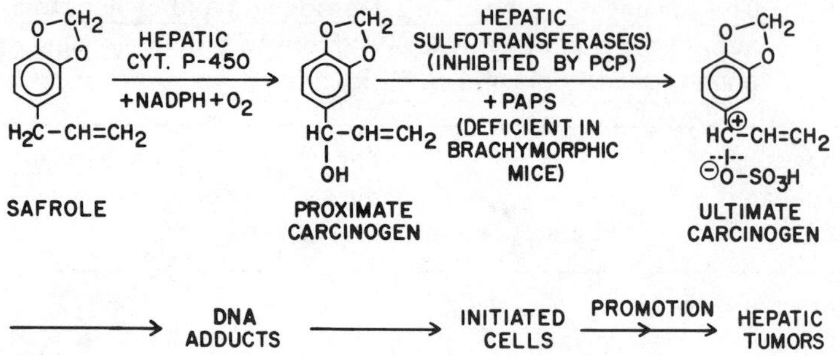

SAFROLE — HEPATIC CYT. P-450, +NADPH+O_2 → PROXIMATE CARCINOGEN — HEPATIC SULFOTRANSFERASE(S) (INHIBITED BY PCP), +PAPS (DEFICIENT IN BRACHYMORPHIC MICE) → ULTIMATE CARCINOGEN

→ DNA ADDUCTS → INITIATED CELLS —PROMOTION→ HEPATIC TUMORS

FIG. 3. The metabolic activation of safrole for the formation of hepatic DNA-bound adducts and for the initiation of hepatomas in male 12-day-old B6C3F$_1$ mice.

In contrast to the situation in rat liver DNA, N-(deoxyguanosin-8-yl)-2-aminofluorene accounted for at least 90% of the adducts formed from N-hydroxy-2-acetylaminofluorene in the liver DNA of male 12-day-old B6C3F$_1$ mice (20). Assays showed sulfotransferase activity for N-hydroxy-2-aminofluorene in hepatic cytosols of male 12-day-old B6C3F$_1$ mice and further showed that this activity was inhibited more than 90% by 10 μM pentachlorophenol. In studies analogous to those described above for 1'-hydroxysafrole, both the level of the N-(deoxyguanosin-8-yl)-2-aminofluorene adducts in the hepatic DNA 9 h after administration of N-hydroxy-2-acetylaminofluorene and the average number of hepatomas 10 months later were inhibited about 90% in male B6C3F$_1$ mice pretreated with pentachlorophenol. Thus, for mice given 0.06 μmol of N-hydroxy-2-acetylaminofluorene per gram body weight the level of N-(deoxyguanosin-8-yl)-2-aminofluorene adducts decreased from 2.9 to 0.3 pmol per mg of DNA, and the average number of hepatomas per liver decreased from 10 to 1 on pretreatment with 0.04 μmol of pentachlorophenol per gram body weight. Similarly, on administration of N-hydroxy-2-acetylaminofluorene to brachymorphic mice the level of the adducts in the liver DNA was only 25% of that observed in the liver DNA of the phenotypically normal mice. The average number of hepatomas per liver was only 10% as high in the brachymorphic mice treated with N-hydroxy-AAF as in the phenotypically normal mice. This reduction in hepatoma formation in the brachymorphic mice was comparable to that observed on administration of 1'-hydroxysafrole and far greater than that seen in diethylnitrosamine-treated brachymorphic mice.

The levels of cytosolic acyltransferase activity for N-hydroxy-2-acetylaminofluorene and of the acetyl coenzyme A-dependent acetylation of N-hydroxy-2-aminofluorene were similar to that of the sulfotransferase activity for N-hydroxy-2-aminofluorene in the livers of these young mice

(C.-C. Lai, J. A. Miller, and E. C. Miller, unpublished data). However, the acyltransferase reaction and the acetyl coenzyme A-dependent reaction were very much less sensitive to inhibition by pentachlorophenol than was the sulfotransferase reaction. Furthermore, the microsomal activity for the deacetylation of N-hydroxy-2-acetylaminofluorene was 10-fold higher in the livers of the male 12-day-old $B6C3F_1$ mice than in the livers of adult male rats. Accordingly, for male 12-day-old $B6C3F_1$ mice the major metabolic pathway for the formation of electrophilic metabolites that react with the liver DNA and that result in the induction of hepatomas appears to be the deacetylation of N-hydroxy-2-acetylaminofluorene to N-hydroxy-2-aminofluorene and the subsequent sulfuric acid esterification of this hydroxylamine (Fig. 4).

4-AMINOAZOBENZENE

The N-methyl and N,N-dimethyl derivatives of 4-aminoazobenzene have been recognized for many years as strong hepatocarcinogens in the rat when fed in the diet for several weeks (28). In contrast, 4-aminoazobenzene has shown little or no carcinogenic activity in the rat (28). However, on injection intraperitoneally into 12-day-old male $B6C3F_1$ mice, 4-aminoazobenzene and its N-methyl derivatives were equally strong hepatocarcinogens (11). Furthermore, administration of 4-aminoazobenzene to these mice resulted in the formation in the liver DNA of essentially only one adduct, N-(deoxyguanosin-8-yl)-4-aminoazobenzene (11). The same adduct was formed nonenzymatically on incubation of N-hydroxy-4-aminoazobenzene and an excess of acetic anhydride with deoxyguanosine or DNA at pH 7; N-acetoxy-4-aminoazobenzene was the presumed reactive intermediate (Fig. 5). N-(deoxyguanosin-8-yl)-4-aminoazobenzene was also formed in a PAPS-dependent reaction when deoxyguanosine and N-hydroxy-4-aminoazobenzene were incubated with liver cytosols from male 12-day-old $B6C3F_1$ mice (K. B. Delclos, J. A. Miller, E. C. Miller, and A. Liem, unpublished data). Unlike the sulfotransferase activities discussed above, but like the human platelet sulfotransferase activity for dopamine (35), the sulfotransferase activity for N-hydroxy-4-aminoazobenzene was only poorly inhibited by pentachlorophenol. At levels of 10 and 100 μM, pentachlorophenol inhibited the sulfotransferase activity for N-hydroxy-4-aminoazobenzene by only 20 and 70%, respectively; under the same conditions 10 μM pentachlorophenol inhibited the sulfotransferase activity for N-hydroxy-2-aminofluorene by greater than 95%. Similarly, when 0.04 μmol of pentachlorophenol per gram body weight was administered to male $B6C3F_1$ mice 45 min before the injection of 0.05 or 0.1 μmol of 4-aminoazobenzene per gram body weight, the formation of N-(deoxyguanosin-8-yl)-4-aminoazobenzene adducts in the hepatic DNA was inhibited by only 30-40%. Hepatoma formation by 9 months was inhibited only 50% under these conditions.

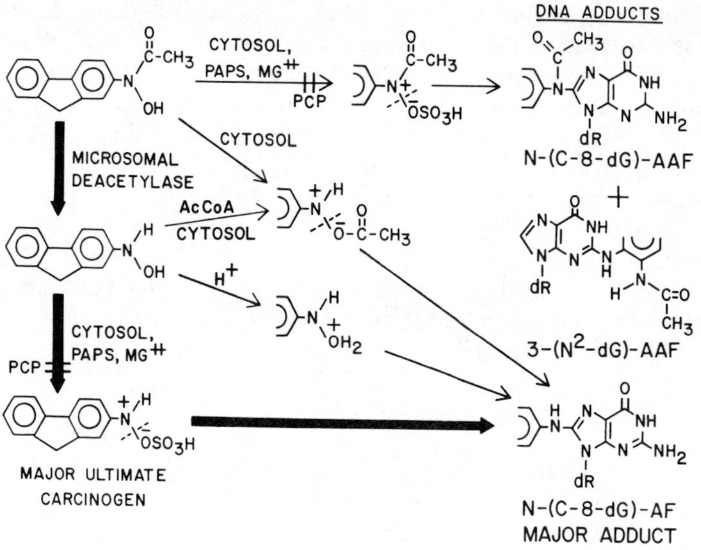

FIG. 4. The metabolic activation of N-hydroxy-2-acetylaminofluorene for the formation of hepatic DNA-bound adducts and for the initiation of hepatomas in male 12-day-old B6C3F$_1$ mice.

On administration of 0.15 μmol per gram body weight of 4-aminoazobenzene to brachymorphic mice, the level of adducts in the liver DNA at 8 h was reduced to 12% of that found in the liver DNA from their phenotypically normal littermates. Similarly, as observed on administration of 1'-hydroxysafrole or N-hydroxy-2-acetylaminofluorene, hepatoma formation in brachymorphic mice administered 4-aminoazobenzene was no more than 10% of that in phenotypically normal littermates that received the same dose. This marked inhibition of hepatoma formation and of the formation of N-(deoxyguanosin-8-yl)-4-aminoazobenzene adducts from 4-aminoazobenzene in the PAPS-deficient mice provided strong evidence that N-sulföoxy-4-aminoazobenzene is the critical electrophilic metabolite for both DNA adduct and hepatoma formation. Even though pentachlorophenol was a relatively weak inhibitor of the sulfotransferase activity for N-hydroxy-4-aminoazobenzene, the parallelism between this limited inhibition and the limited reduction in both hepatic DNA adduct formation and of hepatoma formation from 4-aminoazobenzene on pretreatment of the mice with pentachlorophenol is consistent with a critical role for sulfotransferase activity in both of these events.

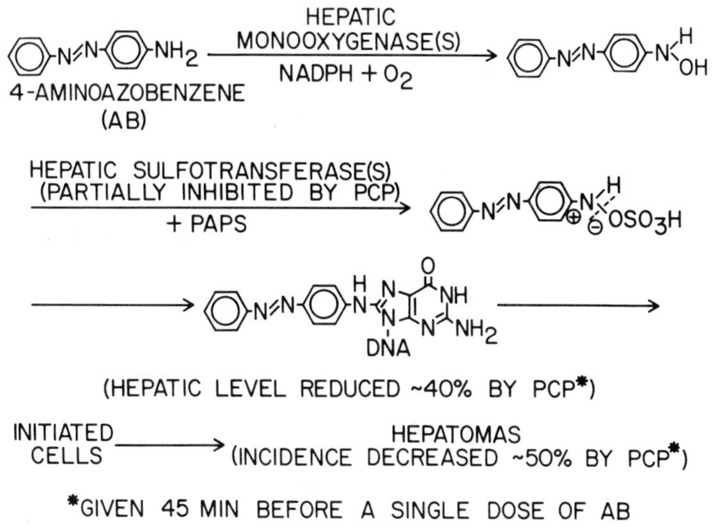

HEPATIC MONOOXYGENASE(S)
NADPH + O_2
4-AMINOAZOBENZENE (AB)

HEPATIC SULFOTRANSFERASE(S) (PARTIALLY INHIBITED BY PCP)
+ PAPS

DNA

(HEPATIC LEVEL REDUCED ~40% BY PCP*)

INITIATED CELLS → HEPATOMAS (INCIDENCE DECREASED ~50% BY PCP*)

*GIVEN 45 MIN BEFORE A SINGLE DOSE OF AB

FIG. 5. The metabolic activation of 4-aminoazobenzene in livers of male 12-day-old B6C3F$_1$ mice for the formation of DNA-bound N-(deoxyguanosin-8-yl)-4-aminoazobenzene and for the initiation of hepatomas.

PERSPECTIVES

The above data provide the first demonstrations that sulfuric acid esters of certain carcinogens or their proximate carcinogenic metabolites are responsible both for the formation of DNA adducts of these carcinogens in the target tissue and for the initiation of tumor formation. These data also provide some of the most direct evidence available to correlate the formation of a specific electrophilic metabolite with the formation of adducts in the target tissue DNA and with the initiation of tumors under identical conditions. This demonstration depended on the use of a system in which a single dose of carcinogen initiates tumor formation and on specific inhibition of the formation of the ultimate electrophilic reactant. The two independent in vivo approaches to inhibition of the formation of the electrophilic reactant, i.e. by inhibition of the sulfotransferase system by pentachlorophenol and by the limitation of the essential cofactor PAPS in brachymorphic mice, provided a clearer conclusion than would have been permitted by the use of either approach alone.

The extent to which the formation of sulfuric acid esters of these proximate carcinogenic metabolites, as compared to other possible activation reactions, is involved in hepatic carcinogenesis in other animals (e.g., of different ages, strains, or species) or under other conditions (e.g., repetitive administration) is not known. However, comparison of the data from Meerman and his colleagues (24,25) with our data suggest that there are major differences in the metabolic activation of N-hydroxy-2-acetylaminofluorene between infant $B6C3F_1$ mice and adult rats. Thus, Meerman et al. found that pretreatment of adult male rats with pentachlorophenol before a single dose of N-hydroxy-2-acetylaminofluorene reduced the levels of the acetylated adducts formed in the liver DNA but did not alter the level of the N-(deoxyguanosin-8-yl)-2-aminofluorene adducts. These data and the inhibition of the rat liver sulfotransferase activity for N-hydroxy-2-acetylaminofluorene by pentachlorophenol (25) suggest that N-sulfoöxy-2-acetylaminofluorene is a major precursor of the N-acetylated adducts in adult rat liver. However, in contrast to our results on infant $B6C3F_1$ mice, N-sulfoöxy-2-aminofluorene does not appear to be an important precursor of the aminofluorene adducts in the livers of adult rats (25).

Sulfuric acid esterification may be of importance in the metabolic activation of other carcinogens. Oxidation of 7-methylbenz[a]anthracenes or of 6-methylbenzo[a]pyrene to hydroxymethyl derivatives and subsequent sulfuric acid esterification has been suggested as an alternative pathway to epoxidation for the formation of ultimate carcinogenic metabolites of these hydrocarbons (9,42,43). Although representatives of these esters have shown strong electrophilic, mutagenic, and carcinogenic activities and are formed in PAPS-dependent reactions catalyzed by liver cytosols, there are as yet no data that relate these findings to the carcinogenic activities of any of the methylated hydrocarbons. It is evident that the natures of the ultimate carcinogenic metabolites and of the active forms of other xenobiotic chemicals reflect both the intrinsic activities of the compounds and their metabolites and a variety of other biochemical and pharmacological parameters that determine the effective levels of the active forms at critical sites in the cells of target tissues.

The predominantly cytosolic location of the known sulfotransferases for xenobiotic chemicals, steroids, and other low molecular weight compounds (36) raises the question of how labile sulfates reach the nuclear DNA, the presumed reaction site critical to the initiation of hepatomas. A corollary question is the possible occurrence of sulfotransferases with similar properties in the nucleus. Sulfotransferase activities for cerebrosides are localized in the membranes of the microsomes and Golgi apparatus (12,15), and Stöhrer and his colleagues (38) reported preliminary observations several years ago on the apparent occurrence of sulfotransferase activity for N-hydroxy-2-acetylaminofluorene and purine-N-oxides in rat liver nuclei. Preliminary studies on the PAPS-dependent esterification of 1'-hydroxysafrole by mouse liver nuclei indicate that, if it occurs, the level is very low. However, further searches for liver nuclear sulfotransferase activities for the proximate carcinogens considered here are in progress.

The identification of the ultimate carcinogenic metabolites of these and other chemical carcinogens makes feasible more detailed studies, such as those of Marshall et al. (23), on the effects of reaction of critical electrophilic metabolites with the cellular DNA, including its proto-oncogenes. In cases, such as those studied here, where the ultimate carcinogenic metabolites are unstable and have not yet been synthesized chemically, model compounds of similar reactivity can be utilized. In addition, information on the specific metabolic pathways involved in the activation and detoxification of chemical carcinogens provides a firmer basis for studies on the possible reduction of the levels of ultimate chemical carcinogens to which humans may be exposed.

ACKNOWLEDGMENT

This research was supported by Public Health Service Grants CA-07175, CA-09135, CA-09020, and CA-22484 from the National Cancer Institute, U.S. Department of Health and Human Services.

REFERENCES

1. Bartsch, H., Dworkin, M., Miller, J.A., and Miller, E.C. (1972): Biochim. Biophys. Acta, 286:272-298.
2. Bartsch, H., and Hecker, E. (1971): Biochim. Biophys. Acta, 237:567-578.
3. Bartsch, H., Miller, J.A., and Miller, E.C. (1972): Biochim. Biophys. Acta, 273:40-51.
4. Balmain, A., and Pragnell, I.B. (1983): Nature, 303:72-74.
5. Balmain, A., Ramsden, M., Bowden, G.T., and Smith, J. (1984): Nature, 307:658-660.
6. Berenblum, I. (1974): Carcinogenesis as a Biological Problem. Elsevier/North-Holland, Amsterdam.
7. Boberg, E.W., Miller, E.C., Miller, J.A., Poland, A., and Liem, A. (1983): Cancer Res., 43:5163-5173.
8. Boutwell, R.K., Verma, A.K., Ashendel, C.L., and Astrup, E. (1978): Carcinogenesis, Cocarcinogenesis and Biological Effects of Tumor Promoters, Vol. 7, edited by E. Hecker, N.E. Fusenig, W. Kunz, F. Marks, and H.W. Thielmann, pp. 1-12. Raven Press, New York.
9. Cavalieri, E., Roth, R., and Rogan, E. (1979): In: Polynuclear Aromatic Hydrocarbons, edited by P.W. Jones and P. Leber, pp. 517-529. Ann Arbor Science, Ann Arbor, Michigan.
10. DeBaun, J.R., Miller, E.C., and Miller, J.A. (1970): Cancer Res., 30:577-595.
11. Delclos, K.B., Tarpley, W.G., Miller, E.C., and Miller, J.A. (1984): Cancer Res., 44:2540-2550.

12. Farrell, D.F., and McKhann, G.M. (1971): J. Biol. Chem., 246:4694-4702.
13. Fasano, O., Aldrich, T., Tamanoi, F., Taparowsky, E., Furth, M., and Wigler, M. (1984): Proc. Nat. Acad. Sci., USA, 81:4008-4012.
14. Fennell, T.R., Miller, J.A., and Miller, E.C. (1984): Cancer Res., 44:3231-3240.
15. Fleischer, B., and Smigel, M. (1978): J. Biol. Chem., 253:1632-1638.
16. Guerrero, I., Calzada, P., Mayer, A., and Pellicer, A. (1984): Proc. Nat. Acad. Sci., USA, 81:202-205.
17. King, C.M., and Allaben, W.T. (1980): In: Enzymatic Basis of Detoxication, Vol. 2, edited by W.B. Jakoby, pp. 187-197. Academic Press, Inc., New York.
18. King, C.M., and Phillips, B. (1968): Science, 159:1351-1353.
19. Kriek, E. (1965): Biochem. Biophys. Res. Commun., 20:793-799.
20. Lai, C.-C., Miller, E.C., Miller, J.A., and Liem, A. (1984): Proc. Am. Assoc. Cancer Res., 25:85.
21. Lyman, S.D., and Poland, A. (1983): Biochem. Pharmacol., 22:3345-3350.
22. Magee, P.N., Montesano, R., and Preussmann, R. (1976): In: Chemical Carcinogens, edited by C.E. Searle, pp. 491-635. American Chemical Society, Washington, DC.
23. Marshall, C.J., Vousden, K.H., and Phillips, D.H. (1984): Nature, 310:586-589.
24. Meerman, J.H., Beland, F.A., and Mulder, G.J. (1981): Carcinogenesis, 2:413-416.
25. Meerman, J.H., van Doorn, A.B.D., and Mulder, G.J. (1980): Cancer Res., 40:3772-3779.
26. Miller, E.C., and Miller, J.A. (1981): Cancer, 47:2327-2345.
27. Miller, E.C., Swanson, A.B., Phillips, D.H., Fletcher, T.L., Liem, A., and Miller, J.A. (1983): Cancer Res., 43:1124-1134.
28. Miller, J.A., and Miller, E.C. (1953): Adv. Cancer Res., 1:339-396.
29. Miller, J.A., and Miller, E.C. (1983): Br. J. Cancer, 48:1-15.
30. Miller, J.A., Swanson, A.B., and Miller, E.C. (1979): In: Naturally Occurring Carcinogens-Mutagens and Modulators of Carcinogenesis, edited by E.C. Miller, J.A. Miller, I. Hirono, T. Sugimura, and S. Takayama, pp. 111-123. Japan Scientific Societies Press, Tokyo.
31. Mulder, G.J., and Scholtens, E. (1977): Biochem. J., 165:553-559.
32. Phillips, D.H., Miller, J.A., Miller, E.C., and Adams, B. (1981): Cancer Res., 41:176-186.
33. Phillips, D.H., Miller, J.A., Miller, E.C., and Adams, B. (1981): Cancer Res., 41:2664-2671.
34. Randerath, K., Haglund, R.E., Phillips, D.H., and Reddy, M.V. (1984): Cancer Res., in press.
35. Rein, G., Glover, V., and Sandler, M. (1982): Biochem. Pharmacol., 31:1893-1897.

36. Roy, A.B. (1981): In: Sulfation of Drugs and Related Compounds, edited by G.J. Mulder, pp. 83-130. CRC Press, Boca Raton, Florida.
37. Slaga, T.J., Sivak, A., and Boutwell, R.K., editors (1978): Carcinogenesis, Mechanisms of Tumor Promotion and Cocarcinogenesis, Vol. 2. Raven Press, New York.
38. Stöhrer, G., Harmonay, L.A., and Brown, G.B. (1979): Proc. Am. Assoc. Cancer Res., 20:285.
39. Sugahara, K., and Schwartz, N.B. (1979): Proc. Nat. Acad. Sci., USA, 76:6615-6618.
40. Sugahara, K., and Schwartz, N.B. (1982): Arch. Biochem. Biophys., 214:602-609.
41. Sukumar, S., Notario, V., Martin-Zanca, D., and Barbacid, M. (1983): Nature, 306:658-661.
42. Watabe, T. (1983): J. Toxicological Sciences, 8:119-131.
43. Watabe, T., Ishizuka, T., Isobe, M., and Ozawa, N. (1982): Science, 215:403-405.
44. Wiseman, R.W., Miller, J.A., Miller, E.C., Drinkwater, N.R., and Blomquist, J.C. (1984): Proc. Am. Assoc. Cancer Res., 25:85.
45. Wislocki, P.G., Borchert, P., Miller, J.A., and Miller, E.C. (1976): Cancer Res., 36:1686-1695.

Carcinogenesis, Vol. 10, edited by E. Huberman
and S. H. Barr. Raven Press, New York © 1985.

Role of Pharmacokinetics and DNA Dosimetry in Relating *In Vitro* to *In Vivo* Actions of N-Nitroso Compounds

P. N. Magee

Fels Research Institute, and Department of Pathology, Temple University School of Medicine, Philadelphia, Pennsylvania 19140

The demonstration of in vitro transformation with chemical carcinogens by Leo Sachs and his colleagues first with polycyclic hydrocarbons (7) and subsequently with a nitrosamine (35) was a landmark in the development of knowledge of mechanisms of chemical carcinogenesis. Charles Heidelberger recognized the great importance of this work and applied his formidable intellectual gifts to the advancement of this new field, both as a major contributor, with many colleagues, to new insights and as a writer of indispensable review articles, the most recent appearing in 1983 (32,33). The availability of reproducible methods for in vitro transformation of cells has been a major factor in the recent dramatic advances in the understanding of the molecular biology of carcinogenesis, including the discovery of the oncogenes, recently reviewed by Bishop (8). It has also led to the development of various tests for chemical carcinogens with transformation as the end point, discussed in relation to other short-term bioassays in a recent book (18). The advantages of the in vitro cellular and organ systems for the elucidation of transformation mechanisms are obvious, but these systems are not subject to the many hormonal, immunological and other influences that are present in the intact animal. It may be useful, therefore, to attempt to assess the validity of conclusions drawn from some in vitro studies by comparing them with corresponding findings in the intact animal. The following presentation reviews some relevant experiments with N-nitroso compounds carried out in our laboratory, many of which were not deliberately planned with this end in view.

The N-nitroso compounds, including the nitrosamines (Fig. 1), which require metabolic activation, and a group of directly acting compounds, exemplified by N-methylnitrosourea, are powerfully mutagenic and carcinogenic in many organs of a wide range of animal species (47). There is considerable evidence that the nitroso compounds act by alkylating DNA in target organs and that the alkylation of oxygens may be crucial for biological activity (54,60). The methylating N-nitroso compounds, both *in vitro* and *in vivo*, produce a characteristic ratio of about 0.1 of methylation on the O^6-position of guanine to that on the 7-position, which is relevant to some of what follows.

The following presentation will be divided into two parts, the first giving two examples of apparent discrepancies between conclusions to be drawn from *in vitro* studies and *in vivo* tests for carcinogenesis and the second suggesting that some conclusions from *in vitro* transformation experiments regarding the somatic mutation hypothesis of carcinogenesis may be inconsistent with the results of some experiments in the intact animal.

APPARENT DISCREPANCIES BETWEEN *IN VITRO* STUDIES AND *IN VIVO* STUDIES IN THE INTACT ANIMAL WITH N-NITROSO PROLINE AND N-NITROSOCIMETIDINE

N-nitrosoproline

N-nitrosoproline (Fig. 2) is the nitrosated derivative of the naturally occurring amino acid proline, which is a component of many proteins. Its decarboxylation product, N-nitrosopyrollidine (Fig. 2) is a potent carcinogen, inducing tumors at various sites in the rat, mouse, and Syrian golden

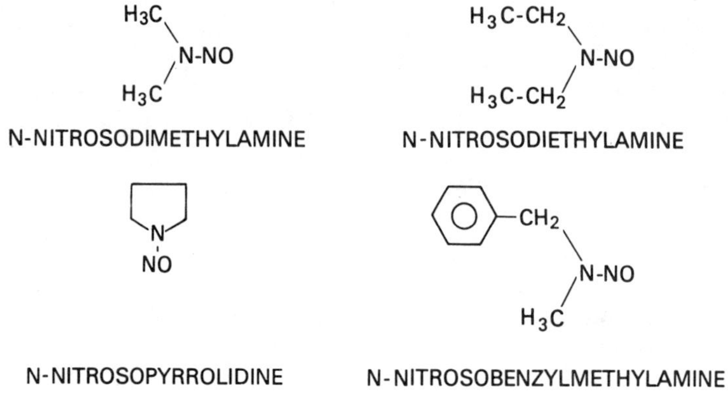

H_3C, H_3C N-NO

N-NITROSODIMETHYLAMINE

H_3C-CH_2, H_3C-CH_2 N-NO

N-NITROSODIETHYLAMINE

N NO

N-NITROSOPYRROLIDINE

CH_2, H_3C N-NO

N-NITROSOBENZYLMETHYLAMINE

FIG. 1. Some carcinogenic nitrosamines.

hamster (19,47). In contrast, nitrosoproline has been reported to be non-carcinogenic in rats or mice by several groups of investigators (25,50,52). Proline, which is readily nitrosated, (51), is an ubiquitous product of protein hydrolysis and a component of a mammalian gastric juice. It is perhaps not surprising, therefore, that N-nitrosoproline is not carcinogenic, since endogenous proline is available for nitrosation in the stomach. This potential for intragastric nitrosation has been exploited by Ohshima and Bartsch (53), who have demonstrated that nitrosation of proline does, in fact, occur in the human stomach; they have extended their work to develop a test, which is being used throughout the world, for endogenous nitrosation in human beings. In this test the subject ingests a standard quantity of proline, and the extent of endogenous intragastric nitrosation is monitored by measurement of the excretion of nitrosoproline in the urine. The latter compound is known to be, for the most part, excreted unchanged (14,16). The question of the potential carcinogenicity of N-nitrosoproline is therefore a matter of some practical concern since the use of the nitrosation test in human subjects could lead to an increased exposure to the compound. This possibility of hazard, together with the intrinsic interest of the contrast between the behavior of nitrosoproline and the effectively carcinogenic nitrosopyrollidine, led us in collaboration with Cecilia Chu to investigate the metabolism of nitrosoproline and to investigate its potential for formation of DNA adducts in vivo (14).

Rats were given [^{14}C]nitrosoproline labeled either generally or in the carboxyl group and their production of $^{14}CO_2$ and excretion of the labeled compound in the urine was measured. In agreement with the earlier studies of Dailey and his colleagues (16), very little radioactive carbon dioxide was detected, and virtually all of the generally labeled compound was found in the urine. Extension of the study to the compound labeled only in the carboxyl group gave similar results, indicating that only a very small proportion, if any, had been decarboxylated to yield the carcinogen N-nitrosopyrollidine. The rate of disappearance of nitrosoproline from the circulating blood of rats in vivo after an intraperitoneal dose of 10 mg/kg body weight was shown to be very rapid, the concentration falling to about 10% of the initial level after 60 min and becoming almost zero at 120 min. The possibility of formation of liver DNA adducts was explored by giving rats [U-^{14}C]nitrosoproline orally (75 μCi) and extracting the DNA from their livers 4 or 8 h later. Assay of the DNA by scintillation counting showed that the level of radioactivity was identical with that of the same amount of calf thymus DNA, indicating no detectable formation of DNA adducts (14). These findings were consistent with the other metabolic studies and with the reported lack of carcinogenicity of nitrosoproline. They were at variance, however, with the findings of Archer and Eng (3), who studied a range of nitrosamines and other N-nitroso compounds for their capacity to react with 4-nitrobenzylpyridine, a trapping agent for alkyl groups, in the presence of an inorganic

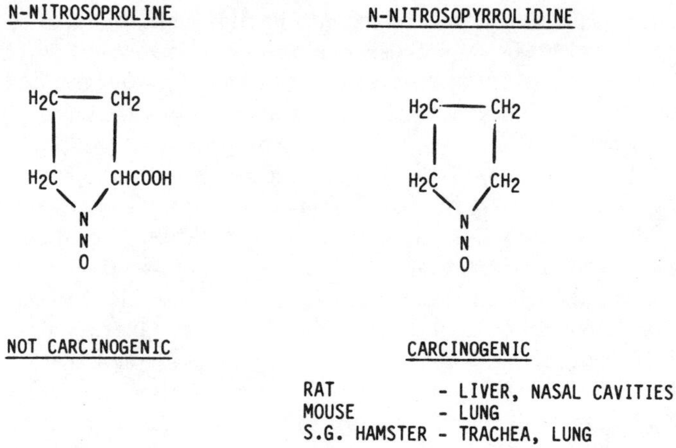

N-NITROSOPROLINE

```
H2C—— CH2
 |      |
H2C    CHCOOH
  \   /
    N
    N
    O
```

NOT CARCINOGENIC

N-NITROSOPYRROLIDINE

```
H2C—— CH2
 |      |
H2C    CH2
  \   /
    N
    N
    O
```

CARCINOGENIC

RAT - LIVER, NASAL CAVITIES
MOUSE - LUNG
S.G. HAMSTER - TRACHEA, LUNG

FIG. 2. N-nitrosoproline and N-nitrosopyrrolidine, carcinogenic properties.

activating system consisting of ascorbic acid, ferrous ions, EDTA, and molecular O_2. Rather surprisingly, nitrosoproline showed an activity comparable to that of diethylnitrosamine, suggesting that, under these conditions, the nitrosoproline can be activated to form an alkylating species. These in vitro findings are thus in sharp contrast with the pharmacokinetic and metabolic observations in the intact rat. A possible explanation could be that the compound is eliminated from the body so rapidly that there is insufficient time for detectable metabolic activation to occur in the liver of the intact animal. Studies on possible activation of N-nitrosoproline by isolated hepatocytes or subcellular preparations of liver in vitro would clearly be of interest.

N-nitrosocimetidine

N-nitrosocimetidine is the nitrosation product of the drug cimetidine (Fig. 3). Cimetidine (Tagamet) is a histamine H-2 receptor antagonist (12) that is very effective in the treatment of disorders of the esophagus, stomach, and duodenum, notably peptic ulcer. Many millions of human beings have been treated with the drug throughout the world and the suggestion made in 1979 by Elder and his colleagues (21) that exposure to it might be causally related to the occurrence of stomach cancer excited widespread interest. The same authors also pointed out the similarity of part of the structure of cimetidine with that of N-methyl-N'-nitroguanidine, and suggested that the drug might undergo nitrosation to yield a nitrosated product of structure similar to that of N-methyl-N'-nitro-N-nitrosoguanidine, a very powerful mutagen and effective chemical carcinogen capable of inducing stomach tumors in experimental animals. Although the clinical evidence in support of a role for cimetidine in the etiology of human gastric cancer was relatively weak and severely criticized, the suggestion that it could be nitrosated proved to be correct (6,23), and the biological properties of nitrosocimetidine have been studied by several groups of investigators.

N-nitrosocimetidine is mutagenic in bacteria and in mammalian cells (5,36,55), and it has been shown to induce bacterial mutations in nitrite-enriched human gastric juice (17). It causes DNA strand breaks (34,59), DNA repair synthesis (34), and sister chromatid exchanges and chromosome aberrations in cultured mammalian cells (4), as well as the adaptive response to alkylating agents in *E. coli* (1). Finally, the compound induces *in vitro* transformation of mammalian cells (5). There is thus very strong evidence from a battery of *in vitro* short-term tests that N-nitrosocimetidine should be carcinogenic in experimental animals, and thus by extrapolation, in human beings. Surprisingly, however, this does not appear to be the case. The compound has been tested independently in long-term bioassays in rats in two laboratories and found to be without carcinogenic activity under conditions where methylnitronitrosoguanidine showed its expected strong carcinogenic action (26,43). A possible explanation for this apparent discrepancy has emerged from our studies at the Fels Research Institute with David Jensen and our colleagues.

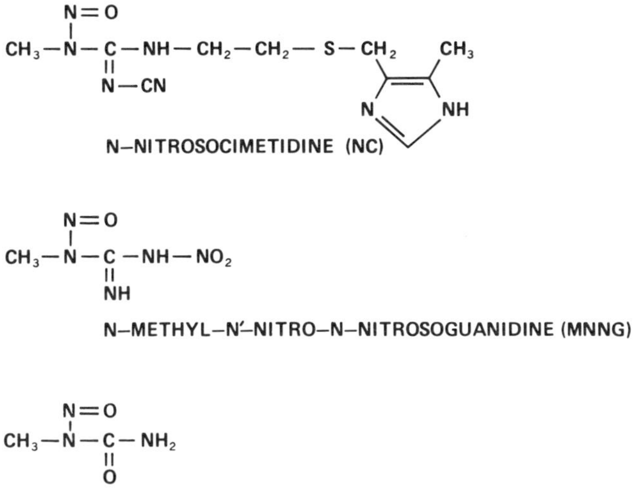

N=O
CH_3—N—C—NH—CH_2—CH_2—S—CH_2 CH_3
N—CN
N NH

N–NITROSOCIMETIDINE (NC)

N=O
CH_3—N—C—NH—NO_2
NH

N–METHYL–N'–NITRO–N–NITROSOGUANIDINE (MNNG)

N=O
CH_3—N—C—NH_2
O

FIG. 3. Some carcinogenic N-nitroso compounds not requiring metabolic activation.

N-nitrosocimetidine was shown to react with calf thymus DNA in vitro to yield a methylated product, which was compared with the products of reaction with N-methyl-N'-nitro-N-nitrosoguanidine or with N-nitrosomethylurea. Under the conditions used (pH 7.4, 23°C for 34 h) all of the compounds yielded 7-methylguanine, 3-methyladenine, and O^6-methylguanine after mild acid hydrolysis of the DNA. The ratios of 3-methyladenine to 7-methylguanine and of O^6-methylguanine to 7-methylguanine were virtually identical with all three compounds, suggesting that nitrosocimetidine reacted by the same mechanism as the other two to produce the pattern of DNA methylation associated with strong carcinogenic activity. Cimetidine itself, the unnitrosated drug, did not react with DNA at all under the same conditions of incubation (39). A similar pattern of DNA base methylation was found when nitrosocimetidine was incubated with human (Raji) cells in vitro (37). These results, like those of the short-term biological tests described above, also suggest that nitrosocimetidine should be carcinogenic. The capacity of N-nitrosocimetidine to methylate the DNA of organs of intact rats in vivo was then examined by Gombar and Magee (24) and compared with that of N-methyl-N'-nitro-N-nitrosoguanidine and with cimetidine itself, each compound being labeled with radioactive carbon in the methyl group. As in the in vitro studies, no trace of DNA methylation was detected after treatment with cimetidine. Both nitroso derivatives showed DNA methylation in some organs, in all cases, the extent of reaction with nitrosocimetidine being lower than that with N-methyl-N'-nitro-N-nitrosoguanidine. As anticipated, the stomach showed the highest level of DNA methylation by the nitrosated guanidine with liver having the next highest level. Readily detectable methylation was found in small intestine, colon, lung, and kidney, with only the brain DNA showing none. In contrast, nitrosocimetidine, given at an equimolar dose, produced lower levels of DNA methylation in all organs, sometimes considerably lower. The highest level of methylation by nitrosocimetidine occurred in the small intestine, where it was lower than that found with methylnitronitrosoguanidine by a factor of only two. In the kidney and liver the ratios were seven and nine and in the the stomach and lung much higher, 27 and 36 respectively, in favor of methylnitronitrosoguanidine. These differences in DNA methylation suggest that the failure of nitrosocimetidine to induce tumors in vivo may be because inadequate levels of DNA interaction were reached under the test conditions for carcinogenesis. Further investigations by David Jensen (38) have suggested a plausible explanation for the relatively low DNA binding by nitrosocimetidine and its lack of carcinogenicity in the intact rat. Nitrosocimetidine decomposes very rapidly when incubated with thiol compounds in neutral pH buffer, much of the decomposition being denitrosation. The compound also rapidly decomposes in whole blood isolated from rats, about 70% of this decomposition producing cimetidine. In solution with purified rat hemoglobin, approximately 90% of the degradation of nitrosocimetidine results from denitrosation, cysteine residues of the hemoglobin being implicated. Treatment of rats with [^{14}C]nitrosocimetidine by intravenous injection was followed by excretion of approximately 85% of the radioactivity in the urine during the succeeding 48 h, of which 70% was identified as cimetidine. Essentially identical results were obtained when radioactive cimetidine was injected intravenously. Taken together, these findings suggest that nitrosocimetidine is rapidly denitrosated in the circulating blood of rats, the reaction being mediated by hemoglobin (38). Jensen further showed that the partition of nitrosocimetidine between water and octanol was dependent on the pH of the aqueous phase, remaining almost entirely therein at pH values of less than 5. Thus, nitrosocimetidine given orally may be poorly absorbed from the stomach, where the luminal contents are acidic, which may explain the relatively higher levels of DNA methylation in the small intestine than in the stomach as found by Gombar and Magee (24). It seems possible, therefore, that the reported lack of carcinogenicity of nitrosocimetidine in the rat may be related to its very rapid decomposition in the circulating blood with consequent failure to achieve the necessary concentrations for the induction of cancer in the organs.

DNA DOSIMETRY IN CARCINOGENESIS AND MUTAGENESIS BY N-NITROSO COMPOUNDS

In the second part of this presentation we will discuss some implications of quantitation of DNA adducts, i.e., methyl or ethyl groups in carcinogenesis and mutagenesis by nitroso compounds in relation to the somatic mutation theory of carcinogenesis. The pros and cons of the somatic mutation idea have been discussed by a number of authors since it was first proposed by Boveri many years ago (11). A major objection raised by Burdette (13), which is that some powerful chemical carcinogens are not in fact mutagenic, has now been satisfactorily explained by the findings of James and Elizabeth Miller that many of the major types of chemical carcinogens require metabolic activation (49). Subsequent extensive studies by Ames and his colleagues have firmly established that, in many cases, "carcinogens are mutagens" (2). There is substantial evidence of various kinds, including the probable clonal origin of many tumors (22), that favors this hypothesis (10,44,61), and a mutational origin of cancer is also consistent with the accumulating evidence that DNA is the crucial target for the activated forms of chemical carcinogens (41,45). The possibility that many cancers arise by mutation does not, of course, invalidate the possibility that some do not and may arise by various epigenetic or nongenetic mechanisms (69). Various arguments have been raised, however, against the proposition that cancer could result from cellular mutation, notably by Markert (48), Rubin (56,57), and Weinhouse (68). A major objection has been the lack of quantitative correlation between the concentrations of the same chemical required to induce mutations and malignant transformation in vitro, that required for transformation being considerably less than that required for mutation. In what follows, the levels of DNA alkylation associated with the induction of tumors in vivo by single doses of some N-nitroso compounds will be compared with the levels associated with the induction of mutations in mammalian cells in vitro, with the data derived from both our earlier and more recent studies. The models of in vivo carcinogenesis to be considered are the kidney tumors induced in rats by a single dose of dimethylnitrosamine (46) and the tumors at various other sites inducible by single doses of methyl or ethylnitrosourea (20,42). The model of mutagenesis will be the induction of ouabain resistance in BHK cells by N-methyl-N'-nitro-N-nitrosoguanidine (5).

The renal tumors induced in the rat by a single dose of dimethylnitrosamine may become very large (Fig. 4), are sometimes bilateral, and are usually associated with the death of the animals about 10 to 14 months after the treatment. They are of two distinct histological types: epithelial (including adenomas and adenocarcinomas) and secondary mesenchymal (i.e., connective tissue), the cell of origin being either the cortical interstitial fibrocyte or possibly a vascular cell (29,30). The predominant type of tumor depends on the age at which the carcinogen is given. When a carcinogen is injected in the newborn period, the tumors are almost entirely mesenchymal, whereas treatment around the age of three months results mainly in epithelial tumors (28). Exposure at intermediate ages produces tumors of both types, sometimes each occurring in the same kidney. The pathogenesis of these renal tumors has been studied in great detail by Gordon Hard and his colleagues who have demonstrated the presence of cells in the kidneys of the rats within three weeks after nitrosamine treatment that have the morphological characteristics of the established mesenchymal tumors (30). Furthermore, cells put into culture from the kidneys of rats during the first week after carcinogen treatment take on the characteristics of morphological transformation by five subcultures (9,31), and, after subsequent passage, grow as tumors with similar morphology to the primary mesenchymal tumors after intrarenal implantation in immunologically suppressed rats (27).

It should, perhaps, be emphasized that there is no requirement for any further (promoting) stimulus with these tumors, the single application of the carcinogen being sufficient for the entire carcinogenic process. The same applies to the induction of tumors at other sites by single doses of methylnitrosourea (20), which decomposes in the animal body within 10 min after administration (63).

A dose-response study on the induction of the rat kidney tumors (66) showed that, under the conditions of the experiment, a dose of 50-60 mg/kg body weight of dimethylnitrosamine resulted in a tumor incidence approaching 100%. This dose is lethal for some of the treated rats, which

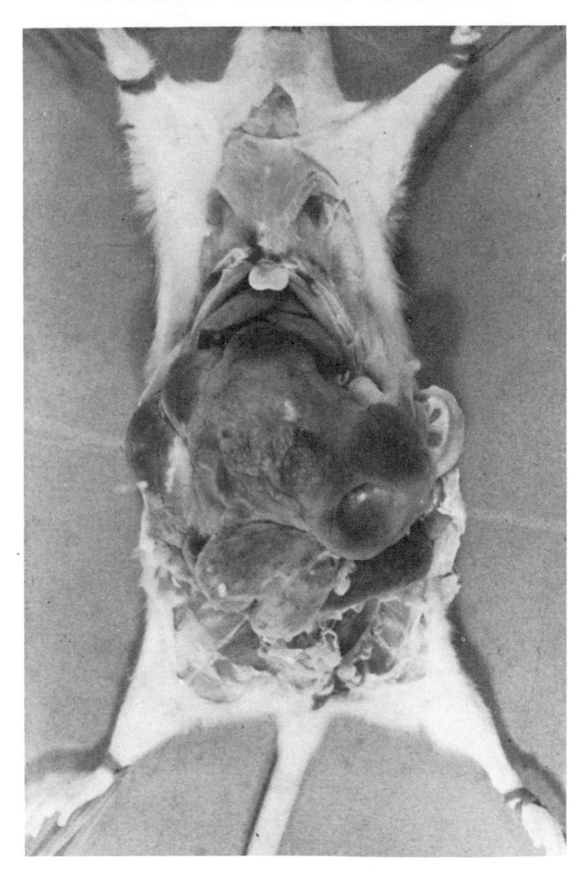

FIG. 4. Rat with large kidney tumor induced by one dose of dimethylnitrosamine (30 mg/kg body weight).

die from massive acute liver damage. It is thus not possible to exceed this dose level, and the concentrations of dimethylnitrosamine in the animal body and the levels of DNA-adduct formation (DNA methylation) in the kidney are the highest compatible with tumor induction in vivo. Extrapolation of the dose response curve to 16 mg/kg body weight of dimethylnitrosamine indicated a kidney tumor incidence of 5%, which agreed well with the observed incidence of 6% (66). Since the extents of DNA methylation in the kidneys of rats treated with various single doses of dimethylnitrosamine have been established in several laboratories, we decided to compare these levels with those found in experiments where mutations were induced in mammalian cells by N-methyl-N'-nitro-N'-nitrosoguanidine. The extent of DNA methylation in rat kidney by dimethylnitrosamine in vivo, as reported by Swann and Magee (64) was compared with that by methylnitronitrosoguanidine in BHK cells in vitro as found by Barrows and his colleagues (5) both at toxic levels of the compounds. Dimethylnitrosamine given at 50 mg/kg body weight to rats resulted in kidney DNA methylation of 1963 micromoles 7-methylguanine per mole guanine. Methylnitronitrosoguanidine, at a concentration of 50 μM, which killed approximately 99% of the plated BHK cells, gave rise to 1078 micromoles 7-methylguanine per mole guanine. These values are of the same order of magnitude and indicate an upper limit of DNA methylation by those compounds above which survival of the animals or the cells is not possible. These results show that the dose and concentration levels for carcinogenesis and mutagenesis by these agents are not arbitrarily chosen. Levels of DNA methylation corresponding to mutation to ouabain resistance by methylnitronitrosoguanidine in the concentration range 2-20 μM were calculated from the measurements of Barrows and his colleagues (5) at higher concentrations of the mutagen, assuming linearity of DNA methylation with concentration. The values obtained were 43-430 micromoles 7-methylguanine per mole guanine and the level of DNA methylation in rat kidney by a dose of 16 mg/kg body weight dimethylnitrosamine, which induced a kidney tumor incidence of 6%, was 1000 micromoles 7-methylguanine/mole of guanine. Although there are obviously many uncertainties in making such a comparison, the results of the calculations suggest that similar extents of DNA modification by methylating N-nitroso compounds are required for carcinogenesis in the intact animal and for mutagenesis in vitro, which is not inconsistent with a mutagenic mechanism for the initiation of cancer. The same conclusion is, of course, strongly supported by the work of Russell and his colleagues (58) who have shown that ethylnitrosourea is a very effective specific locus mutagen in mice at dose levels at which this compound is carcinogenic in this and other species. Swann and Magee (65) had earlier reported that this directly acting carcinogen ethylates the DNA of a range of organs in the rat after intravenous injection to about the same extent.

N-methylnitrosourea, also a directly acting carcinogen, is almost uniformly distributed throughout the body of the rat after intravenous injection, as shown by Kleihues and Patzschke (40) through whole body autoradiography, and the extent of DNA methylation in all organs studied was closely similar (6). Thus, under these conditions, virtually all of the cells in the body of the rat are exposed to this powerful mutagen at concentrations and levels of DNA methylation that are known to result in mutagenesis. Nevertheless, the animals rarely develop more than two or three macroscopically detectable tumors (42) despite the very large numbers of cells that are potentially at risk of mutation. These observations suggest that cancer induction in vivo may be a less frequent event than malignant transformation in vitro by the same carcinogen/mutagen.

The recent work of Sukumar and her colleagues (62) is also consistent with a mutational mechanism for the initiation of cancer. These workers reported that each of nine mammary carcinomas induced by a single injection of methylnitrosourea in rats contained a transforming H-ras-1 gene as demonstrated by transfection into NIH/3T3 cells. They demonstrated the presence of a point mutation at the twelfth codon of the gene, which is the type of mutation expected to result from exposure to N-methylnitrosourea. These findings certainly do not prove that the tumors necessarily resulted from the mutation. However, oncogenes are currently thought to be relatively few in number, and the chance of such a mutation occurring is therefore very low, which is consistent with the very small numbers of tumors actually observed after so many cells in the animal have been exposed to mutagenic concentrations of the carcinogen. It is emphasized that these thoughts apply only to the initiating event in carcinogenesis. For example, in the rat mammary tumor system, removal of the ovaries prevents cancer induction at the same dose level of methylnitrosourea, liver tumors are only induced in the rat by this compound with the additional stimulus of partial hepatectomy (15), and thyroid tumors are only induced by this compound when it is given in combination with methylthiouracil (67) (a thyrotropic agent).

CONCLUSIONS

Absorption, organ distribution, and excretion of some nitroso compounds may profoundly influence their carcinogenic activities. Pharmacokinetic and metabolic studies may be crucial in explaining apparent discrepancies between in vitro and in vivo (intact animal) studies.

Pharmacokinetic studies and DNA dosimetry in the intact animal are not inconsistent with the hypothesis that initiation of carcinogenesis by nitroso compounds may involve a somatic mutation.

ACKNOWLEDGMENTS

This work has been supported by Public Health Service Grants CA-12227 and CA-23451 from the National Institutes of Health, U. S. Department of Health and Human Services, by Grant SIG-6 from the American Cancer Society, and by grants from the National Foundation for Cancer Research and from the Samuel S. Fels Fund of Philadelphia.

REFERENCES

1. Alldrick, A.J., Rowland, I.R., and Gangolli, S.D. (1984): Mutat. Res., 139:111-114.
2. Ames, B.N., Durston, W.E., Yamasaki, E., and Lee, F.D. (1973): Proc. Nat. Acad. Sci., USA, 70:2281-2285.

3. Archer, M.C., and Eng, V.W.S. (1981): Chem. Biol. Interact., 33:207-214.
4. Athanasiou, K., and Kyrtopoulos, S.A. (1981): Cancer Lett., 14:71-75.
5. Barrows, L.R., Gombar, C.T., and Magee, P.N. (1982): Mutat. Res., 102:145-158.
6. Bavin, P.M.G., Durant, G.J., Miles, P.D., Mitchel, R.C., and Pepper, F.S. (1980): J. Chem. Res., (S), 212-213.
7. Berwald, Y., and Sachs, L. (1963): Nature, 200:1182-1184.
8. Bishop, J.M. (1983): Annu. Rev. Biochem., 52:301-354.
9. Borland, R., and Hard, G.C. (1974): Eur. J. Cancer, 10:177-184.
10. Bouck, N., and diMayorca, G. (1976): Nature, 264:722-727.
11. Boveri, T. (1914): Zur Frage der Entstehung maligner Tumoren, pp. 64. Gustave Fischer, Jena.
12. Brimblecombe, R.W., Duncan, W.A.M., Durant, G.J., Emmett, J.C., Ganellin, C.R., Leslie, G.B., and Parsons, M.E. (1978): Gastroenterology,74:339-347.
13. Burdette, W.J. (1955): Cancer Res.,15:201-226.
14. Chu, C., and Magee, P.N. (1981): Cancer Res., 41:3653-3657.
15. Craddock, V.M. (1976): In: Liver Cell Cancer, edited by H.M. Cameron, C.A. Linsell, and G.P. Warwick, pp. 153-201. Elsevier/North Holland, Amsterdam.
16. Dailey, R.E., Braunberg, R.C., and Blaschka, A.M. (1975): Toxicology, 3:23-28.
17. DeFlora, S., and Picciotto, A. (1980): Carcinogenesis, 1:925-930.
18. Douglas, J.F. (1984): In: Carcinogenesis and Mutagenesis Testing, pp. 335. Human Press, Clifton, New Jersey.
19. Druckrey, H., Preussmann, R., Ivankovic, S., and Schmahl, D. (1967): Z. Krebsforsch, 69:103-201.
20. Druckrey, H., Steinhoff, D., Preussmann, R., and Ivankovic, S. (1963): Naturwissenschaften, 50:735.
21. Elder, J.B., Ganguli, P.C., and Gillespie, I.E. (1979): Lancet, 1:1005-1006.
22. Fialkow, J. (1976): Biochim. Biophys. Acta, 458:283-321.
23. Foster, A.B., Jarman, M., Manson, D., and Schulten, H.R. (1980): Cancer Lett., 9:47-52.
24. Gombar, C.T., and Magee, P.N. (1982): Chem. Biol. Interact., 40:149-157.
25. Greenblatt, M., and Lijinsky, W. (1972): J. Nat. Cancer Inst., 48:1389-1392.
26. Habs, M., Schmahl, D., Eisenbrand, G., and Preussmann, R. (1982): Banbury Rep., 12:403-405.
27. Hard, G.C. (1978): Cancer Res., 38:1974-1978.
28. Hard, G.C. (1979): Cancer Res., 39:4965-4970.
29. Hard, G.C., and Butler, W.H. (1970): Cancer Res., 30:2796-2805.
30. Hard, G.C., and Butler, W.H. (1971): Cancer Res., 31:337-347.

31. Hard, G.C., King, H., Borland, R., Stewart, B.W., and Dobrostanski, B. (1977): Oncology, 34:16-19.
32. Heidelberger, C. (1973): Adv. Cancer Res., 18:317-366.
33. Heidelberger, C., Freeman, A.E., Pienta, R.J., Sivak, A., Bertram, J.S., Casto, B.C., Dunkel, V.C., Francis, M.W., Kakunaga, T., Little, J.B., and Schechtman, L.M. (1983): Mutat. Res., 114:283-385.
34. Henderson, E.E., Basilio, M., and Davis, R.M. (1981): Chem. Biol. Interact., 38:87-98.
35. Huberman, E., Salzberg, S., and Sachs, L. (1968): Proc. Nat. Acad. Sci., USA, 59:77-82.
36. Ichinotsubo, D., MacKinnon, E.A., Liu, C., Rice, S., and Mower, H.F. (1981): Carcinogenesis, 2:261-264.
37. Jensen, D.E. (1981): Biochem. Pharmacol., 30:2864-2867.
38. Jensen, D.E. (1983): Cancer Res., 43:5258-5267.
39. Jensen, D.E., and Magee, P.N. (1981): Cancer Res., 41:230-236.
40. Kleihues, P., and Patzschke, K. (1971): Z. Krebsforsch., 75:193-200.
41. Lawley, P.D. (1980): Br. Med. Bull., 36:19-24.
42. Leaver, D.D., Swann, P.F., and Magee, P.N. (1969): Br. J. Cancer, 23:177-187.
43. Lijinsky, W., and Reuber, M.D. (1984): Cancer Res., 44:447-449.
44. Magee, P.N. (1977): In: Progress in Genetic Toxicology, edited by D. Scott, B.A. Bridges, and F.H. Sobels, pp. 15-27. Elsevier/North Holland, Amsterdam.
45. Magee, P.N. (1982): In: Accomplishments in Cancer Research 1981, General Motors Cancer Research Foundation, edited by J.G. Fortner and J.E. Rhoads, pp. 202-215. Lippincott, Philadelphia.
46. Magee, P.N., and Barnes, J.M, (1962): J. Path. Bact., 84:19-31.
47. Magee, P.N., Montesano, R., and Preussmann, R. (1976): In: Chemical Carcinogens, A.C.S. Monograph 173, edited by C.E. Searle, pp. 491-625. American Chemical Society, Washington, DC.
48. Markert, C.L. (1968): Cancer Res., 28:1908-1914.
49. Miller, E.C., and Miller, J.A. (1966): Pharmacol. Rev., 18:805-838.
50. Mirvish, S.S., Bulay, O., Runge, R.G., and Patil, K. (1980): J. Nat. Cancer Inst., 64:1435-1440.
51. Mirvish, S.S., Sams, J., Fan, T.Y., and Tannenbaum, S.R. (1973): J. Nat. Cancer Inst., 51:1833-1839.
52. Nixon, J.E., Wales, J.H., Scanlan, R.A., Bills, D.D., and Sinnhuber, R.O. (1976): Fd. Cosmet. Toxicol., 14:133-135.
53. Ohshima, H., and Bartsch, H. (1981): Cancer Res., 41:3658-3662.
54. Pegg, A.E. (1977): Adv. Cancer Res., 25:195-269.
55. Pool, B.L., Eisenbrand, G., and Schmahl, D. (1979): Toxicology, 15:69-72.
56. Rubin, H. (1980): J. Nat. Cancer Inst., 64:995-1000.
57. Rubin, H. (1983): Science, 219:1170.

58. Russell, W.L., Kelly, E.M., Hunsicker, P.R., Maddux, S.C., and Phipps, E.L. (1979): Proc. Nat. Acad. Sci., USA, 76:5818-5819.
59. Schwarz, M., Hummel, J., and Eisenbrand, G. (1980): Cancer Lett., 10:223-228.
60. Singer, B. (1979): J. Nat. Cancer Inst., 62:1329-1339.
61. Straus, D.S. (1981): J. Nat. Cancer Inst., 67:233-241.
62. Sukumar, S., Notario, V., Martin-Zanca, D., and Barbacid, M. (1983): Nature, 306:658-661.
63. Swann, P.F. (1968): Biochem. J., 110:49-52.
64. Swann, P.F., and Magee, P.N. (1968): Biochem. J., 110:39-47.
65. Swann, P.F., and Magee, P.N. (1971): Biochem. J., 125:841-847.
66. Swann, P.F., Magee, P.N., Mohr, U., Reznik, G., Green, U., and Kaufman, D.G. (1976): Nature, 263:134-136.
67. Thomas, C., and Bollmann, R. (1974): Z. Krebsforsch., 81:243-249.
68. Weinhouse, S. (1983): In: Advances in Enzyme Regulation, edited by G. Weber, pp. 369-386. Pergamon Press, New York.
69. Williams, G.M. (1983): Ann. N.Y. Acad. Sci., 407:328-333.

Carcinogenesis, Vol. 10, edited by E. Huberman and S. H. Barr. Raven Press, New York © 1985.

Analogous Patterns of Benzo[a]Pyrene Metabolism in Human and Rodent Cells

J. K. Selkirk

Biology Division, Oak Ridge National Laboratory, Oak Ridge, Tennessee 37831

The study of polyaromatic hydrocarbons and their role in carcinogenesis has grown in logarithmic proportions over the last decade. This growth has been largely due to more sensitive methods of analysis, better synthetic schemes, and a whole series of bacterial and mammalian biological assays for toxicity, mutagenicity, and carcinogenicity (10). It is generally assumed that the parent molecules of environmentally prevalent chemical carcinogens are structurally stable and relatively inactive metabolically (12). This assumption is not unreasonable from a teleological point of view, since one would expect labile chemical substances to be rapidly degraded or oxidized due to sunlight and weather, if released into the open environment. Most reactive carcinogens and mutagenic intermediates, are synthesized in the laboratory and have relatively short half-lives in aqueous media or when presented with a nucleophilic binding site.

The burgeoning data base from biochemical and carcinogenesis studies from all classes of chemical carcinogens has resulted in the generalization that all carcinogenic chemicals appear to be electrophilic reagents that readily alkylate nucleophilic sites within the cell (7). In addition, it would appear for polyaromatic and heterocyclic carcinogens that the same types of oxygenated intermediates and final detoxification products are formed in cells, both those susceptible and those resistant to malignant transformation (15). A universal scheme for the metabolic processing of any chemical carcinogen is seen in Fig. 1. After a toxic chemical enters the cell, it will be conjugated by cytoplasmic transferases if it contains reactive functional groups and be rapidly excreted into the surrounding fluid as an inert material. However, if the xenobiotic is relatively stable and hydrophobic, such as a polyaromatic molecule, the cellular detoxification machinery will attempt to metabolically deactivate it by linking it to a cytoplasmic moiety to render it more water soluble and thereby decrease its toxic potential. Unfortunately, this detoxification process forms the carcinogenically active intermediate. This activation is accomplished by

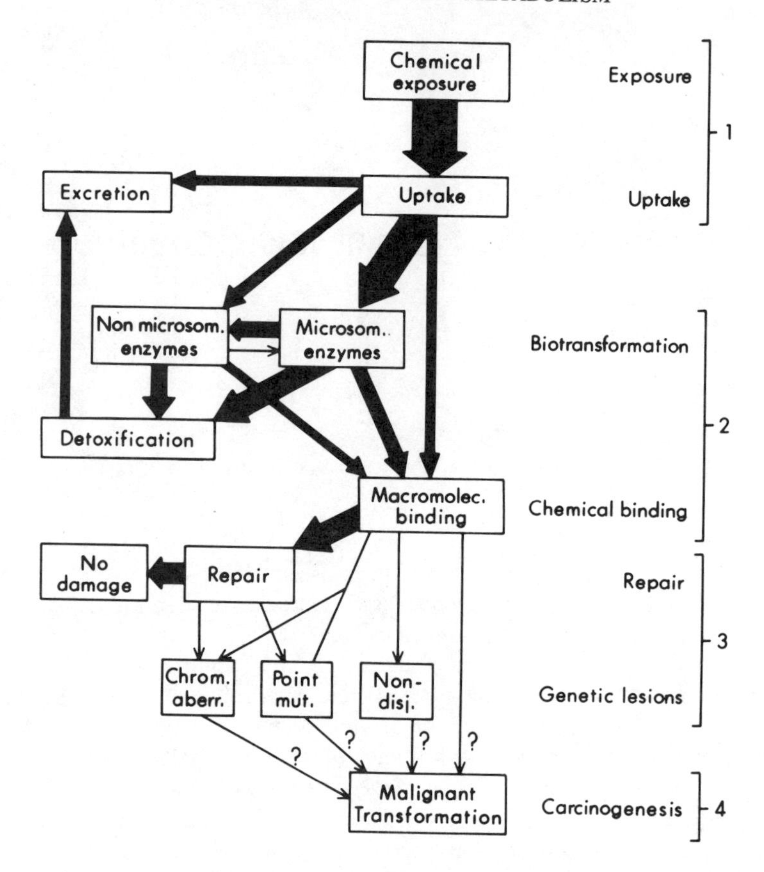

FIG. 1. Metabolic processing of foreign chemicals. Heavy arrows indicate the major pathways taken by cells to deactivate and remove toxic chemicals. When macromolecular binding occurs, it can be repaired, leaving the cell unharmed. However, when repair does not occur or is faulty, a series of genetic lesions can lead to mutation and cancer.

the microsomal monooxygenases that transform the parent molecule into a highly reactive electrophile (1) that can interact with cellular nucleophiles such as DNA, RNA, and protein. Benzo[a]pyrene (BAP) has been the most widely studied polyaromatic carcinogen; it is an ubiquitous substance in the environment that is created by the incomplete pyrolysis of carbonacious material (13). BAP is metabolized to a series of oxygenated derivatives, whose structures are seen in Figs. 2 and 3. These derivatives represent the major known metabolites found in all eukaryotic systems. They are isolated either in their free oxygenated form or as conjugated derivatives to glutathione, glucuronide, or sulfate.

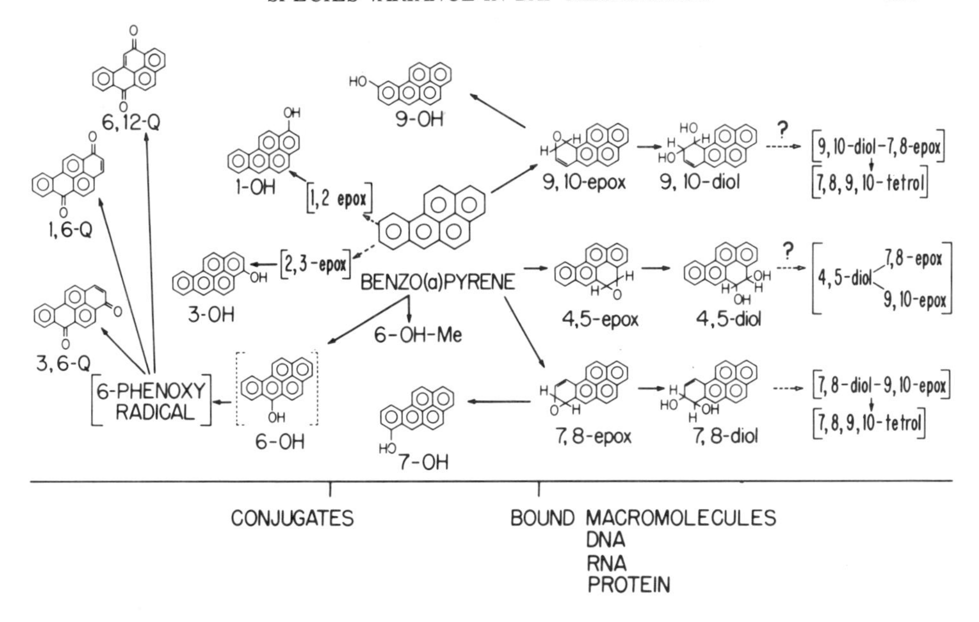

FIG. 2. Metabolism of BAP--composite of oxygenated derivatives formed by tissue monooxygenases. Bracketed structures have not been isolated with the exception of the 7,8,9,10-tetrols.

It is currently felt that the pathway critical for carcinogenesis and mutagenesis is that shown in the lower right-hand corner of Fig. 2 where BAP forms the intermediate 7,8-epoxide, which becomes hydrated by microsomal epoxide hydrase and then is reactivated by the monooxygenases to form the 7,8-diol-9,10-epoxide (17). This labile diol epoxide intermediate opens to form an electrophilic trihydroxycarbonium, which has been identified as the major DNA binding species for BAP. Additional studies have shown that approximately 80% of this binding is found at the exocyclic amino group of guanine (18). Intact cell metabolism is significantly different from cell-free metabolism, where the latter produces phenols as the major metabolites due to the absence of cytoplasmic transferase activity. Intact cells use cytoplasmic transferases to remove toxic metabolites (14). In all species, dihydrodiols appear to be less active substrates for conjugation, resulting in their accumulation during metabolism. Dihydrodiol accumulation allows a greater probability for remetabolism by the monooxygenases to the aforementioned reactive diol-epoxides. Figure 3 shows that mouse, hamster, rat, and human fibroblasts all form, to various degrees, the three dihydrodiols as the major metabolic products, with the lowest activity exhibited by the human foreskin fibroblasts. Since BAP forms the identical profile of metabolites for all

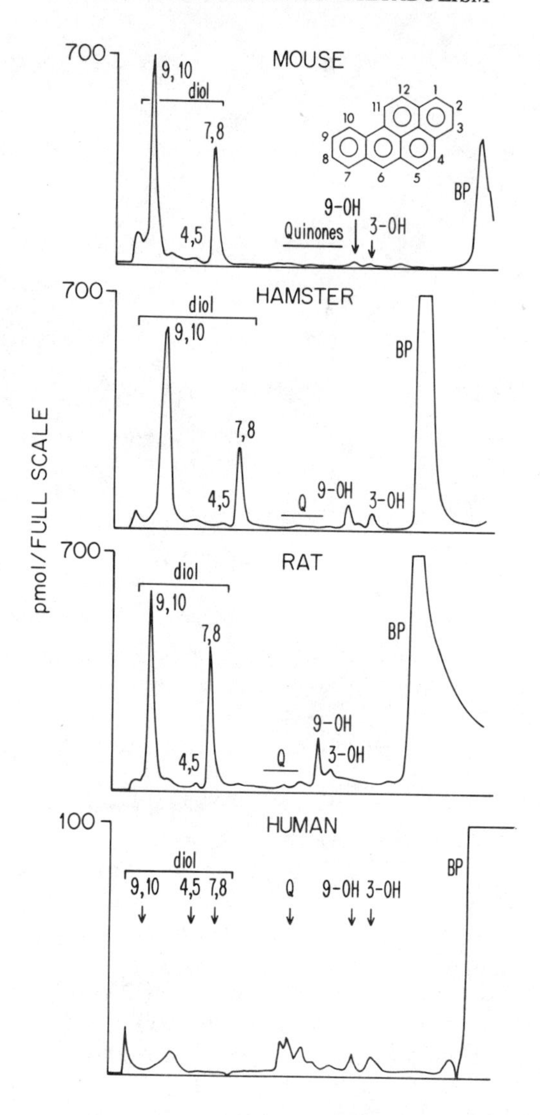

FIG. 3. Metabolism of BAP. Species comparison of fibroblast metabolism by HPLC analysis of the organic-solvent-soluble metabolites. Note absence of phenols and quinones, which are removed as conjugates by cytoplasmic transferases.

species, as well as binding to the same sites on DNA, it becomes increasingly important to understand the relationship between the microsomal processing of the carcinogen in the cytoplasm, and the relative extent of binding to DNA between susceptible and resistant tissues. Metabolism studies in the laboratory, for the most part, make use of experimental animals where the major parameters can be controlled, while human studies almost always are based on biopsy and/or autopsy specimens from patients of various ages, both sexes, and often uncertain medical histories. However, while human metabolism studies may show a large quantitative variation in drug metabolism between tissue samples, as a function of the constitutive level or relative inducibility of monooxygenase activity in each tissue sample, qualitative metabolite profiles are identical and ratios between the respective metabolites are fairly constant (16). Figure 4 represents the range of values and arithmetic means for a composite of each organic-solvent-soluble BAP metabolite produced from four different tissues from each of eight donors. Metabolite ratios were comparable in all four tissues with the tetrols, the 9,10-dihydrodiol, and the quinones as major metabolites. In general, human tissue profiles were quite similar to the profiles of a majority of nonhuman intact cell systems and included the formation of the 9,10-dihydrodiol as the major cellular metabolite. However, there were significant variations in binding levels of BAP to DNA among the four tissues as seen in Fig. 5. The results clearly show wide variability in specific activity among donors, although there appears to be some degree of consistency of binding in the four tissues from an individual patient. Patient #41 exhibited high specific activity for skin, bronchus, and bladder. Conversely, patients #79 and #80 were consistently low in specific activity in all four tissues. While a relatively high variance among patients might be expected due to their relative medical histories and physical condition immediately prior to death, the relative level of metabolic activity found in all four tissues from a single person may consistently be either high or low. The relatively similar metabolic activity of the tissues from several individuals in this study may reflect the genetic variance at the aryl hydrocarbon hydroxylase (AH) locus from monooxygenase induction.

It is not yet known whether tumor origin involves subpopulations and is truly clonal in origin or whether all cells in a tissue are equally susceptible. The concept of progression through qualitatively different stages of irreversible change reported by Foulds suggests subclones of cells within the tissue with an "independent progression of characters" (5). In a number of studies, Fidler showed that it was possible to obtain tumor lines with increasing metastatic potential through successive transplantation and intermittent in vivo culturing (2,3,4). Also, increased colonizing ability has been shown to be a function of increased metastatic potential of the original primary tumor (6). While these studies deal with phenotypic expression of tumor cells, they all suggest the presence of variant cell populations with altered genetic and biochemical characteristics. We have studied the mouse tumor line, Hepa-1c1c7, and three variants MuL-12,

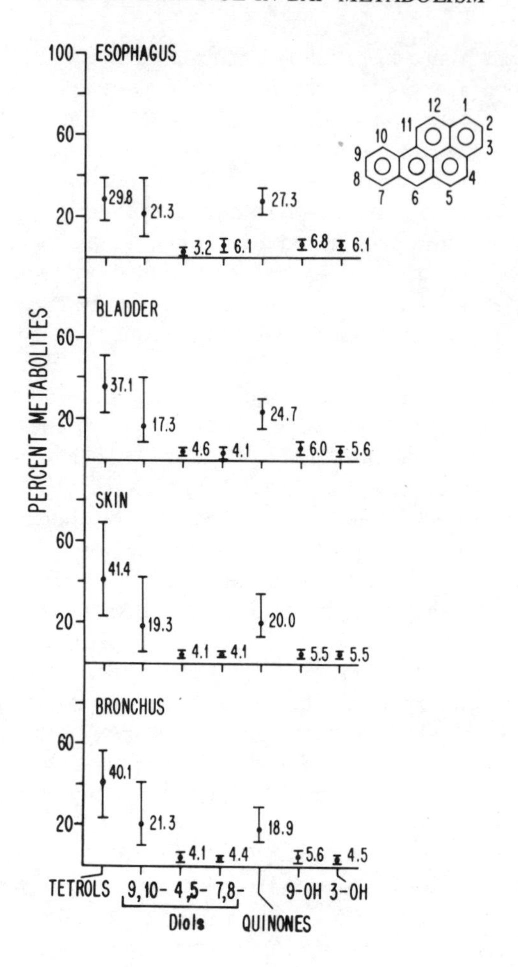

FIG. 4. Human tissue explant metabolism. Composite of BAP metabolites averaged from eight donors. The numbers represent the arithmetic mean of the respective tissue incubations from all eight patients.

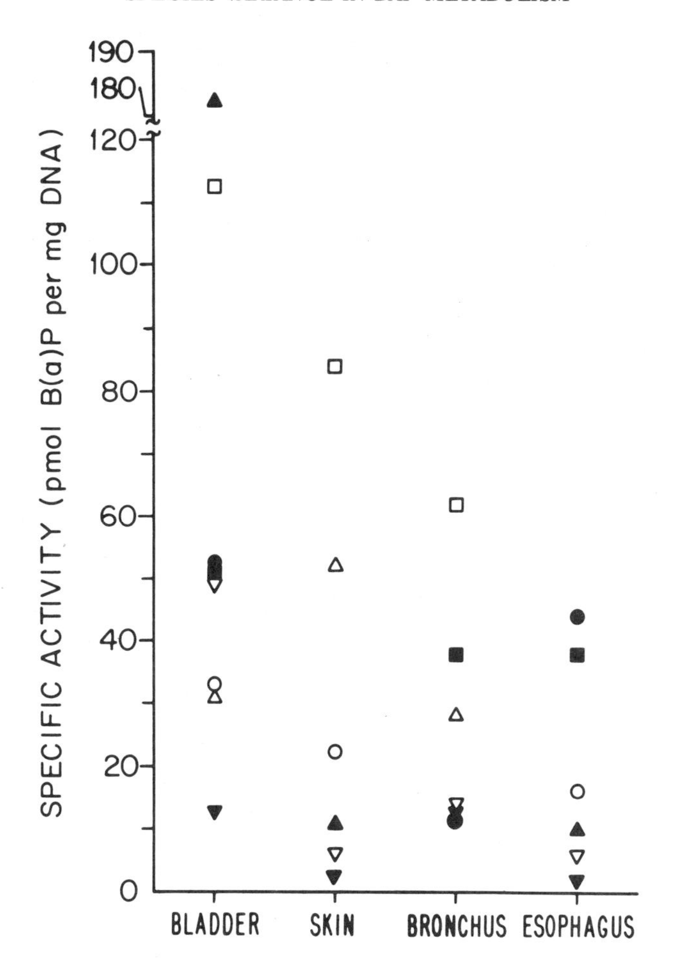

FIG. 5. Comparative binding of BAP to DNA in human tissue explants. Interindividual variation of BAP binding to four tissues. Symbols correspond to patient number: □,41; ■,45; △,52; ▲,56; ○,64; ●,69; ▽,79; ▼,80.

BP^rcl and TAOcl/BP^rcl, separated by a fluorescence-activated cell sorter with an argon laser, which selected the cells based on the amount of BAP fluorescence present in the cytoplasm after short incubations (18). These cells are defined as variants, not mutants, since they are all taken from the parent clone, and by definition, remixing would essentially reconstitute the intact parent cell line. The parent line and variants were incubated with tritiated BAP for extended periods of time and the entire profile of organic-soluble and water-soluble metabolites was measured by high-performance liquid chromatography. Figure 6 shows a typical distribution of radioactivity for Hepa-1c1c7 cells and the three variants. Overall BAP metabolism (4 μM) was the lowest for the BP^rc1 variant, with most of the organic-solvent material representing unused BAP. In contrast, MuL-12 had metabolized most of the BAP during the first 12 h. Studies carried to 24 h showed the Hepa-1c1c7, TAOc1/BP^rc1 and the MuL-12 completing metabolism while the BP^rc1 variant continued to show no inducible activity at 24 h. When these results were normalized to cell number (pmole/10^6 cells) as shown in Table 1, we observed a 16-fold metabolism differential between the high and the low variants. These experiments demonstrate an altered biochemical potential in subpopulations of cells that is most likely masked in studies that use cells in mass cultures in a composite parent clone. The capability of isolating these variants and studying them in terms of total metabolic processing from the moment the carcinogen molecule enters the cytoplasm through its interaction with critical cellular macromolecules or its detoxification and excretion opens a new avenue of investigation. We can now attempt to understand susceptibility and resistance in variant cells that may possess unique metabolic perturbations that directly predispose the cell to carcinogen risks.

TABLE 1. Total accumulation of ethyl acetate-soluble metabolites in extracellular medium.

Variant	Hours of Incubation 12	24
Hepa-1c1c7	3.4[a]	10.1
MuL12	6.7	11.0
TAOc1BP^rc1	4.8	5.1
BP^rc1	0.6	0.7

[a]pmol/10^6 cells.

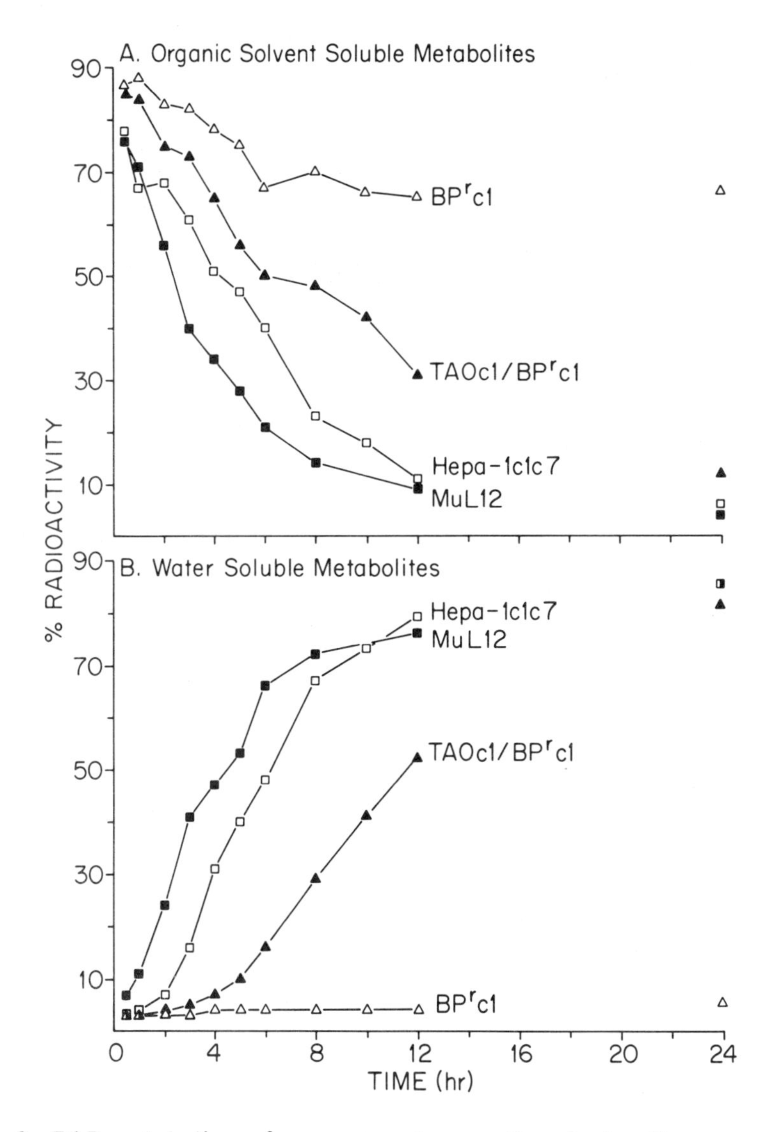

FIG. 6. BAP metabolism of mouse hepatoma cell variants. Upper panel represents the ethyl acetate-soluble compounds shown in Fig. 2. The lower panel represents a composite of water-soluble products including glucuronides, glutathiones, and other uncharacterized water-soluble moieties.

ACKNOWLEDGMENTS

The research described here was sponsored jointly by Public Health Service Grant No. R01-CA30355 from the National Cancer Institute, U.S. Department of Health and Human Services, and by the Office of Health and Environmental Research, U.S. Department of Energy, under contract DE-AC05-840R21400 with Martin Marietta Energy Systems, Inc. I thank Dr. James P. Whitlock, Stanford University School of Medicine, for giving us the Hepa-1c1c7 cells and the three variants. I also thank Ms. Betty K. Mansfield for review of the manuscript.

REFERENCES

1. I. A. R. C. Monograph 32 (1983): Evaluation of the Carcinogenic Risk of Chemicals to Humans, Polynuclear Aromatic Compounds, Part 1, pp. 33-68. World Health Organization, Lyon, France.
2. Fidler, I.J. (1973): Nature New Biol., 242:148-149.
3. Fidler, I.J. (1975): Cancer Res., 35:219-224.
4. Fidler, I.J., and Nicolson, G.L. (1976): J. Nat. Cancer Inst., 57:1199-1202.
5. Foulds, L. (1954): Cancer Res., 14:327-339.
6. Liotta, L.A., Tryggvason, K., Garbisa, S., Hart, I., Foltz, C.M., and Shafie, S., (1980): Nature, 284:67-68.
7. Miller, J.A., and Miller, E.C. (1977): In: Origins of Human Cancer, Volume B, edited by H.H. Hiatt, J.D. Watson, and J.A. Winsten, pp. 605-627. Cold Spring Harbor Laboratory, Cold Spring Harbor, New York.
8. Nebert, D.W. (1978): Biochimie, 60:1019-1028.
9. Public Health Service Document No. 83-2607 (1983): U.S. Government Printing Office, Washington, DC.
10. Preston, R.J., Au, W., Bender, M.A., Brewen, J.G., Carrano, A.V., Heddle, J.A., McFee, A.F., Wolff, S., and Wassom, J.S. (1981): Mutat. Res., 87:143-188.
11. Schaefer, G., and Selkirk, J.K., submitted.
12. Selkirk, J.K., MacLeod, M.C., Moore, C.J., Mansfield, B.K., Nikbakht, A., and Dearstone, K. (1982): In: Mechanisms of Chemical Carcinogenesis, edited by C.C. Harris and P.A. Cerutti, pp. 331-349. Alan R. Liss, Inc., New York.
13. Selkirk, J.K., and MacLeod, M.C. (1983): BioScience 32, 7:601-605.
14. Selkirk, J.K. (1977): Nature, 270:604-607.
15. Selkirk, J.K., MacLeod, M.C., Mansfield, B.K., Nikbakht, P.A., and Dearstone, K.C. (1983): In: Organ and Species Specificity in Chemical Carcinogenesis, edited by R. Langenbach, S. Nesnow, and J.M. Rice, pp. 283-294. Plenum Press, New York.

16. Selkirk, J.K., Nikbakht, A., and Stoner, G.D. (1983): Cancer Lett., 18:11-19.
17. Sims, P., Grover, P.L., Swaisland, A., Pal, K., and Hewer, A. (1974): Nature, 252:326-328.
18. Weinstein, I.B., Jeffrey, A., Jennette, K., Blobstein, S.H., Harvey, R.G., Harris, C., Autrup, H., Kasai, H., and Nakanishi, K. (1976): Science, 193:592-595.

Carcinogenesis, Vol. 10, edited by E. Huberman
and S. H. Barr. Raven Press, New York © 1985.

In Vitro Systems to Study Organ and Species Differences in the Metabolic Activation of Chemical Carcinogens

R. Langenbach

Cellular and Genetic Toxicology Branch, National Institute of Environmental Health Sciences, Research Triangle Park, North Carolina 27709

The organ and species differences that occur in response to chemical carcinogens pose intriguing problems, and understanding the basis for these differences may aid in understanding the carcinogenesis process as a whole, as well as in extrapolating rodent carcinogenesis data to humans. Many chemicals exist in a chemically nonreactive form and must be metabolically activated to manifest their biological activity (3,22). Therefore, metabolism is one of the major factors contributing to the organ and species specificity of chemical carcinogens. I have been interested in developing in vitro systems to study the organ and species differences in the bioactivation of chemical carcinogens to genotoxic intermediates (4,11-20,23,24,34). Although the approach described does not, at present, take into account factors such as nature of DNA adducts formed, DNA repair, pharmacodynamics, etc., which also contribute to species or organ specificity, certain of these parameters could be measured with only slight modifications of the system described. Thus, a data base that includes several biochemical parameters and various biological end points could be generated for individual chemicals. Such a data base obtained in vitro with cell types from various species (including humans) could facilitate the estimation of a chemical's hazard to humans. In the studies summarized here, primary cells were obtained from various organs (i.e., liver, lung, kidney, and/or bladder) of different species (i.e., rat, hamster, dog, or cow) and used to metabolically activate carcinogens from various chemical classes. Mutation of Chinese hamster V79 cells, induction of sister chromatid exchanges (SCEs) in V79 cells, or reversions of Salmonella are the biological end points that were measured. Additionally, we measured metabolism of some chemicals by high-performance liquid chromatography (HPLC) and attempted to correlate the metabolic profiles with the genetic end points.

MATERIALS AND METHODS

Because the methodologies have previously been reported in detail elsewhere, only references will be cited here. Methods for obtaining primary rat liver, lung, bladder, and kidney cells were as described (11-17,34). Bovine liver and bladder cells were obtained as described by Hix et al. (4). All primary cell types were more than 85% viable at the time of seeding and displayed an epithelial morphology in culture. The medium in all studies was Williams Medium E plus 10% heat-inactivated fetal bovine serum supplemented with L-glutamine, penicillin, and streptomycin. Cells were maintained at 37°C in humidified incubators with an atmosphere of 5% CO_2 and air. The protocols for V79 cell mutagenesis and SCE induction were as described (11,29). The _Salmonella_ assay was performed basically according to Ames (1) except for the use of intact cells in place of S9 and the modifications cited by Hix et al. (4). Benzo[a]pyrene (BAP) metabolism measured by HPLC was as reported (13,14), and 2-acetylaminofluorene (AAF) metabolism was carried out by the HPLC method of Smith and Thorgeirsson (31).

RESULTS AND DISCUSSION

Cell-Mediated Mutagenesis of V79 Cells

Huberman and Sachs first described a cell-mediated mutagenesis system in which rodent embryonic fibroblasts were used for metabolic activation and Chinese hamster V79 cells were used to measure mutagenesis (5). We have since expanded the approach to use different cell types for metabolic activation and also different target cells. A review of cell-mediated systems has recently been published (18). Figure 1 illustrates the protocol for our cell-mediated mutagenesis assay. Due to slower attachment, lung and kidney cells are seeded first and allowed to attach, then the V79 cells are added. In the liver and bladder cell-mediated system, the V79 cells were seeded first. In all systems, the ratio of activating cells to V79 cells is about 3 to 1. In the hepatocyte system for example (see Fig. 1), the hepatocytes attach on top of the V79 cells, forming a near monolayer. The medium is then changed to fresh medium containing the chemical, and the co-cultivation is continued for 48 h. The V79 cells are then reseeded to determine toxicity, mutation, and/or SCE induction. The activating cells (hepatocytes, kidney cells, and bladder cells) do not attach and grow after reseeding, and the lung cells are irradiated prior to cocultivation to prevent their growth or reseeding (34). Primary cells are used for metabolic activation because it is believed they have the metabolic capability most nearly resembling the metabolism as it occurs in the respective cell type _in vivo_. We have

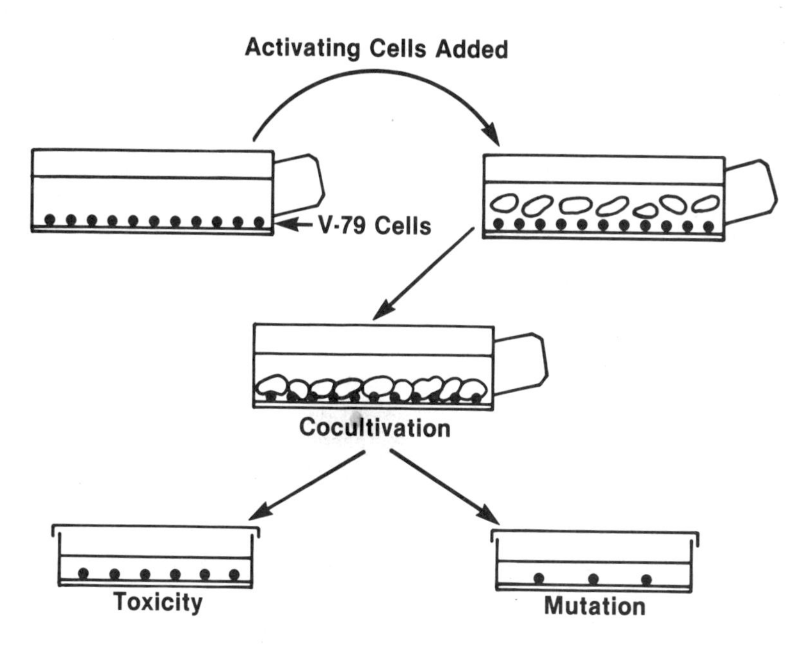

FIG. 1. Protocol for conducting the cell-mediated mutagenesis system.

shown that subculture or prolonged culture of cells alters the metabolic capability of cells, with activation of some chemicals increasing (i.e., BAP in the lung cells) and activation of other chemicals decreasing [i.e., dimethylnitrosamine (DMN) in hepatocytes].

We have previously suggested that the hepatocyte to V79 cell arrangement shown in Fig. 1 may be an in vitro model of cell arrangement in tissues as it occurs in vivo, and therefore, is useful for studying the transport of reactive intermediates between cells (17). That is, the terminally differentiated type of cell (represented by the hepatocyte) carries out the metabolic functions of the organ but is not capable (or has limited capability) of further cell division to express the genetic damage. Less-well-differentiated cells, such as stem cells (represented by the V79 cells), may not have the metabolic capability to activate the chemical but are capable of cell division. Therefore, progeny cells can express the genetic damage caused in the stem cell by a reactive intermediate transported from the activating cell. This model is not intended to exclude the possibility that some cells may be capable of both metabolic activation and further cell replications to express the altered genotype.

The most often used in vitro metabolic activation system to study organ specificity of carcinogen activation has been tissue homogenates (S9). However, although S9 preparations are convenient to prepare and use, they do not necessarily mimic in vitro metabolism (for review, see reference 18). The profiles of carcinogen metabolites and the DNA adducts formed when studied with intact cells and tissue homogenates have been found to differ--with intact cells better mimicking the in vivo situation. In intact cells, the combination of activation and detoxification enzymes presumably retain their normal relative activities, whereas conjugation processes in S9 fractions probably do not occur because the appropriate cofactor levels are decreased. For screening purposes, lower levels of conjugation could be advantageous as reactive intermediate production would be maximized; however, these aberrant conditions may not be appropriate for mechanistic studies into in vivo events. Also, with broken-cell preparations, there may be a differential loss of the enzymes that contribute to the activation of different classes of chemical carcinogens. Additionally, certain enzymes that participate in carcinogen activation in extrahepatic tissues may be present at reduced levels in liver S9 preparations, or the substrates necessary for their enzyme activity are not supplied in the standard cofactor mix. An example of such an enzyme is prostaglandin synthetase (35). This enzyme should be present and functional in certain types of intact cells. Therefore, we believe mechanistic phenomena such as organ and species differences in carcinogen activation may be studied with greater relevance to in vivo events when intact cells are used for metabolic activation.

In early studies (12), we showed a cell-type specificity in the activation of chemical carcinogens in the cell-mediated mutagenesis system. BAP, a potent skin and lung carcinogen, and aflatoxin B_1 (AFB_1), a potent liver carcinogen, were analyzed for mutagenic activity with rat embryonic fibroblasts or adult rat hepatocytes used for metabolic activation and V79 cells used as the mutable target cell. BAP was mutagenic in the fibroblast-mediated system, but in the hepatocyte-mediated system, no significant increase in mutagenic activity was observed with the BAP doses studied. In contrast, AFB_1 was not mutagenic to V79 cells with fibroblasts as the activating cells but was mutagenic in the hepatocyte-mediated system. These studies demonstrated that we could observe in vitro a cell-type specificity that agreed with the in vivo tumorogenic response of fibroblasts and hepatocytes to the carcinogens BAP and AFB_1.

We have extended the cell-mediated approach to investigate the organ specificity of carcinogen activation with intact adult male rat epithelial cells from liver, lung, bladder, and kidney tissues. The data in Table 1 show the relative mutagenic activity of chemicals from four different classes when primary cultures of intact cells from these organs were used as the source of metabolic activation. Anthracene, the noncarcinogenic hydrocarbon, was not mutagenic to V79 cells in any of the cell-mediated

TABLE 1. Mutagenic activities to V79 cells of chemicals in various rat cell-mediated mutagenesis systems.

Chemical (μg/mL)	Activating cell type			
	Liver	Lung	Kidney	Bladder
Anthracene (3)[a]	1	1	1	1
BAP (1)	2	56	5	6
DMN (100)	95	2	19	11
AFB_1[b] (3)	42	4	ND[c]	ND
AAF (50)	6	ND	ND	1

[a]Numbers in parentheses are references.
[b]Data are expressed as number of times the mutation frequency has increased over the background frequency with ouabain used as the selective agent for the liver and lung cell-mediated systems and 6-thioguanine as the selective agent in the kidney and bladder cell-mediated systems. The background mtuation frequency for ouabain was 1 per 10^6 survivors; for 6-thioguanine the background frequency was 3 per 10^6 survivors. The values have also been normalized per 1.5 x 10^6 activating cells.
[c]ND = not determined.

systems. In the liver cell-mediated system, BAP appeared inactive at the concentration studied while DMN and AFB were mutagenic (Table 1). Somewhat surprisingly, the liver carcinogen AAF appeared only weakly mutagenic in the rat hepatocyte-mediated V79 cell system. In contrast to hepatocytes, rat lung cells extensively activated BAP but did not activate DMN. Kidney cells activated BAP and DMN to mutagenic intermediates. Rat bladder cells also activated BAP and DMN but did not appear to activate AAF to intermediates mutagenic to V79 cells. As described below, the low level of AAF mutagenicity in V79 cells when activated by liver and bladder cells (both organs are target organs for this carcinogen) is not due to the absence of genotoxic intermediates produced by these activating cells. Furthermore, treatment of V79 cells with the chemicals in the absence of activating cells did not produce mutations of the V79 cells. In general, the mutagenicity of the chemicals shown in Table 1 with activating cells from the various organs is in agreement with the relative carcinogenicity of the chemical to that organ.

In addition to determining the agreement between mutagenicity (Table 1) and organ-specific carcinogenicity of the chemicals studied, the embryonic fibroblast cell-mediated approach has been used to demonstrate correlations between the mutagenicity of several polycyclic aromatic hydrocarbons and their carcinogenicity in vivo (8,27,28,30). For nitrosamines, it has also been reported that mutagenicity correlates with tumorigenicity when hamster (16) or rat hepatocytes (9) are used for metabolic activation with V79 cells as the mutable target. However, although correlations between cell-mediated mutagenicity and carcinogenicity exist, it should be emphasized that cellular activation systems are only providing the metabolic activation component of the specificity, and other organ-specific functions such as rates of DNA repair, cell turnover, etc. are not accounted for by the cell-mediated approach.

After measuring the mutagenic activities of the carcinogens with activating cell types from different organs (Table 1), we investigated carcinogen metabolism by these cells to determine if correlations between metabolism and mutagenesis could be discerned. The HPLC profiles of BAP metabolites from primary cells from the four organs were previously reported (13-15). All the cell lines produced metabolites that were soluble in organic solvents or conjugated to glucuronic acid or sulfate. Table 2 shows the amounts of two metabolites produced by the primary cells in culture. These particular metabolites were chosen for discussion because of their potential involvement in the transformation or mutation process. The BAP-4,5-diol probably results from the BAP-4,5-oxide, a species capable of interacting with cellular macromolecules, although the 4,5-diol itself is inactive or weakly active as a mutagen to V79 cells (6,7). The BAP-7,8-diol is a precursor to the BAP-7,8-diol-9,10-epoxide, the presumed active intermediate of BAP (6). As can be seen in Table 2, the liver produces the greatest amount of the BAP-4,5-diol, while the BAP-7,8-diol is formed to the greatest extend in the lung cells where BAP is mutagenic (Table 1). Bladder and kidney cells, which also activate BAP to mutagenic intermediates, also produced significant levels of the 7,8-diol. However, liver cells, which do not activate BAP to mutagenic intermediates (Table 1), also produced measurable BAP-7,8-diol. Therefore, the absolute amount of BAP-7,8-diol produced does not directly cor-

TABLE 2. BAP-diol metabolites formed by liver, lung, bladder, and kidney cells.

Cell type	pmoles formed/48 h/10^6 cells	
	7,8-diol	4,5-diol
Lung	750	90
Liver	212	217
Bladder	415	108
Kidney	115	50

relate with the mutagenic response in the cell-mediated system. However, it may be the next step of BAP-7,8-diol metabolism that shows tissue specific differences, i.e., the conversion to BAP-7,8-diol-9,10-epoxide, which occurs to a greater extent in lung cells than in the liver cells. This correlation has not been determined and would be difficult to accomplish because of the reactivity of the diol-epoxide with cellular constituents. Thus, a weakness in quantitatively relating metabolic profiles and genetic toxicity is the inability to measure the ultimate activation step(s) from metabolism profiles alone.

DMN metabolism studies were also conducted to determine if a difference between rat hepatocytes and rat lung cells could be observed (14). The DMN metabolism studies were carried out with cells from the two tissues immediately after their isolation, so that loss of enzyme capability should have been minimized. DMN was metabolized to CO_2 by the liver cells. However, lung cells, which did not metabolize DMN to mutagenic intermediates (Table 1), also did not metabolize the DMN to detectable levels of CO_2. The lack of DMN metabolism by lung cells may account for its lack of mutagenicity in the lung cell-mediated system.

It was of surprise and interest that AAF, a known liver and bladder carcinogen, was only weakly active in the liver cell-mediated system and inactive in the bladder cell system (Table 1). Subsequent studies indicated that although AAF was not mutagenic in a hepatocyte-mediated V79 cell system, AAF did induce dose-dependent increases in SCEs in the cocultivated V79 cells (19). Furthermore, when hepatocytes were combined with Salmonella, mutagenesis of the Salmonella by AAF was readily detectable (4,24). Therefore, although V79 cell mutagenesis is not a suitable indicator for hepatocyte activation of AAF, SCE induction in V79 cells and bacterial mutagenesis did provide evidence for the genotoxic activity of AAF with hepatocyte activation. The SCE data do indicate that genotoxic intermediates are interacting with V79 cell DNA, a concern that we did have on the basis of the data in Table 1. Several interesting observations can be derived from these results. SCEs and mutations in V79 cells may arise from different DNA adducts or differential repair of adducts. Mutation of V79 cells and reversion of Salmonella could also result from different adducts or differential repair. Further activation of certain metabolites in Salmonella or inactivation of certain metabolites by V79 cells are also possibilities. Although we have not investigated the possible causes, the data do indicate the sensitivity and utility of using a cell-mediated Salmonella system for studying the activation of aromatic amines. Thorgeirsson and colleagues had previously reported the development of a hepatocyte-Salmonella system (22).

Cell-Mediated Mutagenesis of Salmonella

The cell-mediated Salmonella approach has been used to gain insight into the relative roles of the liver and bladder in the activation of aromatic amines, a class of chemicals known to cause bladder cancer (4,23, 24). In general, there have been two theories on the activation of chemicals that cause bladder cancer. One theory has postulated that carcinogen metabolism, including conjugation, occurred in the liver with subsequent enzymatic or pH-dependent hydrolysis in the urine leading to the formation of the active intermediate in the bladder (10,25,26). Alternatively, activation of the carcinogen or metabolite could occur in the bladder epithelium itself (15,20). These theories are not necessarily mutually exclusive, and, in fact, we believed that the relative contributions of the two mechanisms is possibly species dependent. Furthermore, this species difference may be dependent on the relative levels of an enzyme such as prostaglandin synthetase in the bladders of different species.

Because of the limitations of V79 cells to detect aromatic amines stated above, bovine bladder and bovine liver cell-mediated Salmonella systems were developed to study these aromatic amines (4,23,24). Bovine tissues, rather than tissues from rat or other rodent species for which there is carcinogenesis data, were studied initially to assure that ample numbers of bladder cells could be obtained to develop the methodology. As described below, once the methodology was developed, it was used with rat and dog bladder cells.

The data shown in Table 3 demonstrate that bovine bladder cells can activate aromatic amines to mutagenic intermediates, and, in fact, bovine bladder cells are more active than bovine liver cells. Several doses of each chemical were tested in both S. typhimurium tester strains TA100 and TA98 (4), but only results with two doses of each chemical and only the effects on strain TA98 are shown. On a per cell basis, bladder cells were about 4 to 10 times, respectively, more active than the liver cells in activating AAF and aminofluorene (AF) to mutagens. Additionally, although bovine bladder cells could activate 4-aminobiphenyl (ABP), benzidiene (BZ), and 2-naphthylamine (2NA) to mutagenic intermediates, bovine liver cells did not appreciably activate these chemicals to mutagens under the conditions we used. The noncarcinogenic aromatic amine, 1-naphthylamine (1NA), was not mutagenic with either bladder or liver cell activation and illustrates the ability of the bovine bladder cell system to discriminate between carcinogenic and noncarcinogenic aromatic amines.

To gain further insight into the differences of aromatic amine activation by these types of cells, we studied AAF metabolism by these cell types using HPLC (unpublished results). It was observed that AAF was readily metabolized by both bovine bladder and liver cells. However, the bladder cells were about 10 times more effective in converting AAF to N-hydroxy AAF than were the liver cells. As N-hydroxylation is considered an obligatory step in the conversion of AAF to mutagenic/ carcinogenic intermediates (21,32), this correlation is in good agreement

TABLE 3. Mutagenicity to S. typhimurium of aromatic amines activated by intact bovine bladder or liver cells.

Chemical	(μg/plate)	TA98 Revertants/plate[a]	
		Bladder	Liver
AF	10	3098	305
	20	3546	390
AAF	20	2375	534
	40	2739	709
4ABP	10	195	8
	20	409	0
BZ	20	100	21
	40	222	4
2NA	20	40	0
	80	100	0
1NA	20	0	0
	80	4	0

[a]Data are expressed per 6 x 10^6 bladder or liver cells/plate with backgrounds subtracted. The background for TA98 was 23.

with the biological effects of AAF with the two activating cell types. However, further metabolism steps, including sulfation or acetylation (21,32) may also play a role in the differences between these types of cells to activate the aromatic amines to genotoxic intermediates.

During the course of our studies with bovine bladder cell-mediated mutagenesis, we observed that bladder cells obtained from different animals produced different levels of Salmonella mutagenesis (4). This individual variation observed with the animals is illustrated in Table 4.

The range for ABP activation for the data presented shows about a fourfold difference between high and low values, and this range holds true for most animals we have studied. However, about 1 out of 20 animals show a level of activation that is about 10% of the mean activity shown in Table 4. Whether this finding is an artifact of the cell isolation procedure or a real phenomenon remains to be determined. For bovine hepatocytes, about a fourfold variation among animals has also been observed (unpublished). At present, it is not known which factors account for this individual variation, but factors such as diet, age of the animal, sex of the animal, breed of the animal, and possible past drug treatment of the animal may all contribute to the variation observed.

We have also investigated the relative activation potential of liver and bladder from different species to gain insight into possible species differences for target organ metabolic activation. The preliminary data on the

TABLE 4. Variation in the ability of freshly isolated bovine bladder cells to activate 4-aminobiphenyl.

Animal Number	Revertants/plate
1	674
2	1168
3	548
4	302
5	795

Experiments were performed with 2 x 10^6 bladder cells per plate and 20 μg ABP per plate. The background value for S. typhimurium TA100 was 118, and this number has been subtracted for all values presented.

relative activation of AAF to mutagens by liver and bladder cells from bovine, dog, and rat indicate that species differences do exist (Table 5). While bovine bladder cells have a high capability of activating AAF to intermediates mutagenic to the S. typhimurium, rat bladder cells and dog bladder cells have substantially less activity on a per cell basis. By contrast, hepatocytes from these species all have the ability to activate AAF, with dog hepatocytes being the most active followed closely by bovine and rat hepatocytes with approximately 70% and 50% of that activity, respectively. Metabolism studies with AAF have indicated that bovine bladder has the greatest ability to N-hydroxylate AAF, with dog and rat bladder cells having about 10% of bovine N-hydroxylase activity. These data indicate that the relative ability of bladder cells to activate bladder carcinogens is species dependent. The enzyme systems responsible for liver and bladder activation in the different species are currently under investigation.

TABLE 5. Species differences in the activation of AAF to mutagens by bovine, dog, and rat hepatocytes and bladder cells.

	Species		
Organ	Bovine	Dog	Rat
	(revertants per plate)		
Bladder	580 ± 190	35 ± 10	56 ± 17
Hepatocytes	170 ± 90	240 ± 40	115 ± 30

TA 98 was the tester strain used, and the data are expressed per 4 x 10^6 cells per plate.

Species Differences in Data from Liver Cell-Mediated Assays

By combining data accumulated over the past five to six years, it is possible to compare the relative ability of hepatocytes from hamster, rat, mouse, dog, and cow to activate DMN to mutagens for V79 cells. In summary, results from these studies indicate that the hamster hepatocytes have the highest level of DMN activation (approximately 400 ouabain resistant V79 cells per 10^6 cells), with mouse having about 50% that activity and rat hepatocytes about 25% of the activity of the hamster. Dog and bovine hepatocytes are approximately equal to rat hepatocytes in the activation of DMN. Interestingly, by reference to the literature we can, in a limited sense, relate these results to the situation for human hepatocytes. Strom, et al. (33) has recently reported that human and rat hepatocytes have about the same activation potential for diethylnitrosamine. Therefore, this in vitro approachmay be useful in aiding in the extrapolation of rodent carcinogenesis data to potential effects of a chemical in humans. However, as variables other than metabolism contribute to species differences, the development of additional systems and measurement of other parameters will increase the accuracy of the extrapolation.

REFERENCES

1. Ames, B.N., McCann, J., and Yamasaki, E. (1975): Mutat. Res., 31:347-363.
2. Dybing, E., Soderlund, E., Haug, L.T., and Thorgeirsson, S.S. (1979): Cancer Res., 39:3268-3275.
3. Heidelberger, C. (1975): Annu. Rev. Biochem., 44:79-121.
4. Hix, C., Oglesby, L., MacNair, P., Sieg, M., and Langenbach, R. (1983): Carcinogenesis, 4:1401-1407.
5. Huberman, E., and Sachs, L. (1974): Int. J. Cancer, 13:326-333.
6. Huberman, E., and Sachs, L. (1976): Proc. Nat. Acad. Sci., USA, 73:188-192.
7. Huberman, E., Sachs, L., Yang, S.K., and Gelboin, H.V. (1976): Proc. Nat. Acad. Sci., USA, 73:607-611.
8. Huberman, E., and Slaga, T.J. (1979): Cancer Res., 39:411-414.
9. Jones, C.A., Marlino, P.J., Lijinsky, W., and Huberman, E. (1981): Carcinogenesis, 2:1075-1077.
10. Kadlubar, F.F., Miller, J.A., and Miller, E.C. (1977): Cancer Res., 37:805-814.
11. Langenbach, R., Freed, H.J., and Huberman, E. (1978): Proc. Nat. Acad. Sci., USA, 75:2864-2867.
12. Langenbach, R., Freed, H.J., Raveh, D., and Huberman, E. (1978): Nature, 276:277-280.

13. Langenbach, R., Nesnow, S., Malick, L., Gingell, R., Tompa, A., Kuszynski, C., Leavitt, S., Sasseville, K., Hyatt, B., Cudak, C., and Montgomery, L. (1981): In: Polynuclear Aromatic Hydrocarbons, edited by M. Cooke and A.J. Dennis, pp. 75-84. Batelle Press, Columbus, Ohio.
14. Langenbach, R., Nesnow, S., Tompa, A., Gingell, R., and Kuszynski, C. (1981): Carcinogenesis, 2:851-858.
15. Langenbach, R., Malick, L., and Nesnow, S. (1981): J. Nat. Cancer Inst., 66:913-917.
16. Langenbach, R., Kuszynski, C., Gingell, R., Lawson, T., Nagel, D., Pour,P., and Nesnow, S. (1983): In: Structure Activity as a Predictive Tool in Toxicology, edited by L. Golberg, pp. 241-256. Hemisphere Publishing Co., New York.
17. Langenbach, R., and Nesnow, S. (1983): In: Organ and Species Specificity in Chemical Carcinogenesis, edited by R. Langenbach, S. Nesnow, and J.M. Rice, pp. 377-389. Plenum Press, New York.
18. Langenbach, R., and Oglesby, L. (1983): In: Chemical Mutagens, edited by F. de Serres, pp. 55-93. Plenum Press, New York.
19. Langenbach, R., Leavitt, S., Hix, C., Sharief, Y., and Allen, J. (1984): submitted.
20. Malick, L.E., and Langenbach, R. (1970): J. Cells Biol., 83:112a.
21. Miller, E.C. (1978): Cancer Res., 38:1479-1496.
22. Miller, J.A. (1970): Cancer Res., 30:559-576.
23. Oglesby, L., Hix, C., MacNair, P., Sieg, M., Snow, L., and Langenbach, R. (1983): Environ. Health Perspect., 49:147-154.
24. Oglesby, L., Hix, C., MacNair, P., Sieg, M., Snow, L., and Langenbach, R. (1983): Cancer Res., 43:5194-5199.
25. Poupko, J.M., Hearn, W.L., and Radomski, J.L. (1979): Toxicol. Appl. Pharmacol., 50:479-484.
26. Radomski, J.L., and Brill, E. (1970): Science, 167:992-993.
27. Raveh, D., Slaga, T.J., and Huberman, E. (1982): Carcinogenesis, 3:763-766.
28. Reiners, J.J., Yotti, L.P., McKeown, C.K., Nesnow, S., and Slaga, T.J. (1983): Carcinogenesis, 4:321-326.
29. Sharief, Y., Campbell, J., Leavitt, S., Langenbach, R., and Allen, J.W. (1984): Mutat. Res., 126:159-167.
30. Slaga, T.J., Huberman, E., DiGiovanni, J., Gleason, G., and Harvey, R.G. (1979): Carcinogenesis, 2:81-88.
31. Smith, C.L., and Thorgeirsson, S. (1981): Anal. Biochem., 113:62-67.
32. Staiano, N., Erickson, L.C., Smith, C.L., Marsden, E., and Thorgeirsson, S.S. (1983): Carcinogenesis, 4:161-167.
33. Storm, S., Novicki, D., Novotny, A., Jiretle, R., and Michalopoulos, G. (1983): Carcinogenesis, 4:683-686.
34. Tompa, A., and Langenbach, R. (1979): In Vitro, 15:569-578.
35. Zenser, T.V., Armbrecht, M.B., and Davis, B.B. (1980): Cancer Res., 40:2839-2845.

Carcinogenesis, Vol. 10, edited by E. Huberman
and S. H. Barr. Raven Press, New York © 1985.

Interindividual Differences in the Metabolism of Xenobiotics

A. H. Conney and *A. Kappas

*Department of Experimental Carcinogenesis and Metabolism, Hoffmann-La Roche Inc., Nutley, New Jersey 07110; *The Rockefeller University Hospital, New York, New York 10021*

Person-to-person differences in the response of human beings to drugs and environmental pollutants are important problems in medicine and environmental toxicology. Indeed, variability in the responsiveness of all living organisms to environmental chemicals has general importance in biology. Individuality in the responsiveness of humans to xenobiotics is caused, in part, by differences in the rates and pathways of chemical biotransformations in different individuals. Figure 1 shows a 35-fold difference in steady-state plasma concentrations of the antidepressant drug, desmethylimipramine, in 11 subjects given the same daily dose of this drug (32). The two individuals with the highest plasma concentrations of the antidepressant experienced some undesirable side effects. Other studies demonstrated 10- to 30-fold differences in the steady-state plasma concentrations of chlorpromazine and nortriptyline in different individuals, and 20-, 10-, 10-, 7-, and 6-fold interindividual differences were observed for the plasma half-lives of ethylbiscoumacetate, bishydroxycoumarin, diphenylhydantoin, antipyrine, and phenylbutazone, respectively (16). Variability in the metabolism of medicinal agents in different people is one of the reasons why it is necessary for physicians to individualize the dose of drugs administered to their patients. The wide range in dosage needed for the proper therapy of patients with a few representative drugs is shown in Table 1.

This manuscript is dedicated to the memory of Dr. Charles Heidelberger. He was a remarkable man who made many important contributions to science, humanity, and his colleagues.

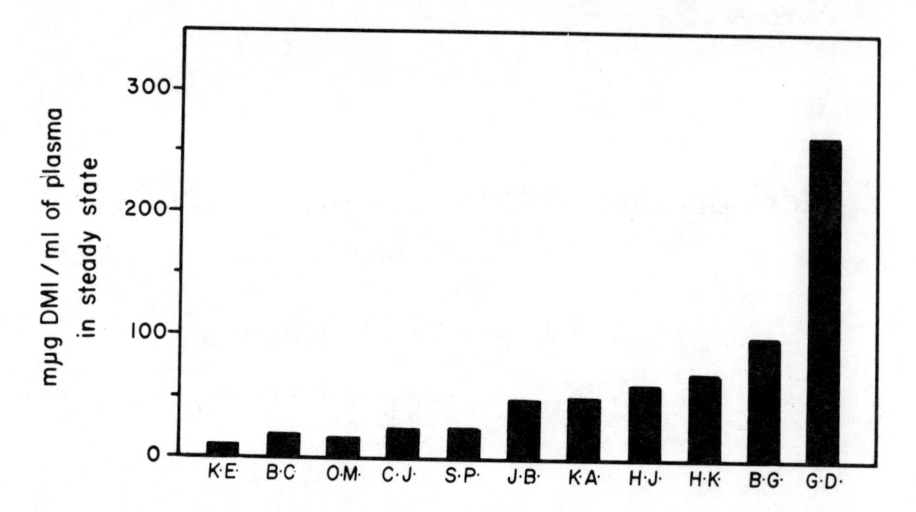

FIG. 1. Steady-state plasma levels of desmethylimipramine (DMI) in various human subjects who received 25 mg of the drug three times daily for 16 days. From Hammer et al. (32).

An example of interindividual differences in the in vivo metabolism of a carcinogen was described 20 years ago by Weisburger and his associates who found that although four cancer patients were equally able to metabolize 2-acetylaminofluorene by ring hydroxylation, the extent of metabolism to the highly carcinogenic N-hydroxy-2-acetylaminofluorene dif-

TABLE 1. Variable dose requirements encountered for some common therapeutic agents.

Drug	Daily dose requirements (mg)
Warfarin	1 - 20
Imipramine	10 - 225
Bethanidine	20 - 200
Debrisoquine	20 - 400
Dicoumarol	25 - 150
Propranolol	40 - 2000
Levodopa	125 - 8000
Ethyl biscoumacetate	150 - 900

From Idle et al. (42).

fered by threefold among the different subjects studied (78). Shortly after the above studies with 2-acetylaminofluorene, the metabolism of some drugs and carcinogens by samples of human liver obtained during abdominal surgery was reported, and variability in the catalytic activity of samples from different individuals was found (58). More recent studies indicated a ninefold difference in the metabolism of aflatoxin B_1 to mutagens and a sevenfold difference in the metabolism of benzo[a]pyrene 7,8-dihydrodiol to mutagens among ten liver biopsy samples that were studied (17,21). Similar differences were observed in the metabolism of benzo[a]pyrene to phenolic metabolites by the different liver samples (17,21).

Large interindividual differences were found for the metabolism of benzo[a]pyrene by human placenta (79), cultured human skin (59), macrophages (13), lymphocytes (12,53), monocytes (10), bronchial cells (35,36), and pancreatic duct cells (36). Large interindividual differences were also observed for the metabolism of dimethylhydrazine by human colon explants (8) and for the metabolism of dimethylnitrosamine by human liver (22). A 75-fold difference was observed for the metabolism of benzo[a]pyrene to DNA-bound adducts by cultured bronchial explants from different individuals (35,37), and a 100- to 150-fold difference was observed for the metabolism of 1,2-dimethylhydrazine, aflatoxin B_1, and dimethylnitrosamine to DNA-bound adducts by cultured colon explants from different individuals (8,37). The studies described above indicate that individuality in the biotransformation of environmental carcinogens to reactive metabolites and to nontoxic metabolites may help explain individual differences in the sensitivity of people to the carcinogenic action of these chemicals.

DRUGS AS PROBES FOR PREDICTING CARCINOGEN METABOLISM IN DIFFERENT INDIVIDUALS

Since large interindividual differences exist for the metabolism of xenobiotics in human beings, it would be desirable to be able to test people for their capacity to metabolize environmental chemicals to toxic and nontoxic products. The results of these studies might enable us to identify individuals who are at unusually high risk when exposed to potentially toxic chemicals. We suggested earlier that the determination of plasma levels or half-lives of a nontoxic drug that was metabolized by the same enzyme system(s) that metabolizes an environmental carcinogen may help predict rates and pathways of metabolism of the carcinogen in different individuals; these studies may help determine the potential hazards of exposure of different individuals to the carcinogen (17,49). Since steroid hormones and other normal body constituents are also metabolized by the same monooxygenases that metabolize foreign compounds (15), the determination of profiles of metabolites obtained from endogenous substrates may represent another approach for predicting rates and pathways of metabolism of xenobiotics in different individuals.

During the past several years, studies have been initiated on the comparative oxidative metabolism of drugs and carcinogens by multiple human liver samples as an approach for identifying drugs that are metabolized by the same enzymes that metabolize carcinogens (17,49). The metabolism of benzo[a]pyrene to fluorescent phenols, the hydroxylation of antipyrine in the 4-position and the hydroxylation of coumarin in the 7-position were studied with whole homogenates prepared from 32 livers obtained at autopsy and from 16 surgical biopsy samples. The data in Fig. 2 indicate a good correlation between the hydroxylation of benzo[a]-pyrene and that of antipyrine ($r = 0.91$) among the different livers, whereas the hydroxylation of benzo[a]pyrene was rather poorly correlated with the hydroxylation of coumarin ($r = 0.65$). The lack of perfect correlations between the metabolism of substrate pairs in the different livers suggests that multiple monooxygenases in human liver are responsible for the metabolism of the chemicals studied and that the relative amounts of the multiple monooxygenases are different in the various liver samples studied.

The metabolism of aflatoxin B_1 and benzo[a]pyrene 7,8-dihydrodiol to mutagens was studied with microsomes from 10 human liver samples obtained by surgical biopsy. The metabolism of aflatoxin B_1 to mutagens by the 10 liver samples was highly correlated with the metabolism of benzo[a]pyrene 7,8-dihydrodiol to mutagens ($r = 0.96$) and with the metabolism of hexobarbital to 3'-hydroxyhexobarbital ($r = 0.92$), but the metabolic activation of aflatoxin B_1 was poorly correlated with the hydroxylation of zoxazolamine (Fig. 3). The metabolism of benzo[a]pyrene 7,8-dihydrodiol to mutagens by the 10 liver samples was highly correlated with the hydroxylation of benzo[a]pyrene to fluorescent phenols and with the 3'-hydroxylation of hexobarbital, but the metabolic activation of benzo[a]-pyrene 7,8-dihydrodiol was poorly correlated with the hydroxylation of zoxazolamine (Fig. 3).

Similar studies on the comparative metabolic hydration of 11 arene and alkene oxides by microsomes from human liver samples were also described (48). Plotting epoxide hydrolase activity obtained for one substrate against epoxide hydrolase activity for another substrate for nine human liver samples revealed excellent correlations for all combinations of the 11 substrates studied. These data suggest the presence in human liver of a single microsomal epoxide hydrolase with broad substrate specificity or similar regulatory control for several epoxide hydrolases.

Recent studies have shown that the *in vivo* hydroxylation of debrisoquine in the 4-position is defective in about 10% of the British population and that these individuals also have impaired oxidative metabolism of several other drugs (43). It is possible that the extent of debrisoquine hydroxylation may be associated with the extent of oxidative

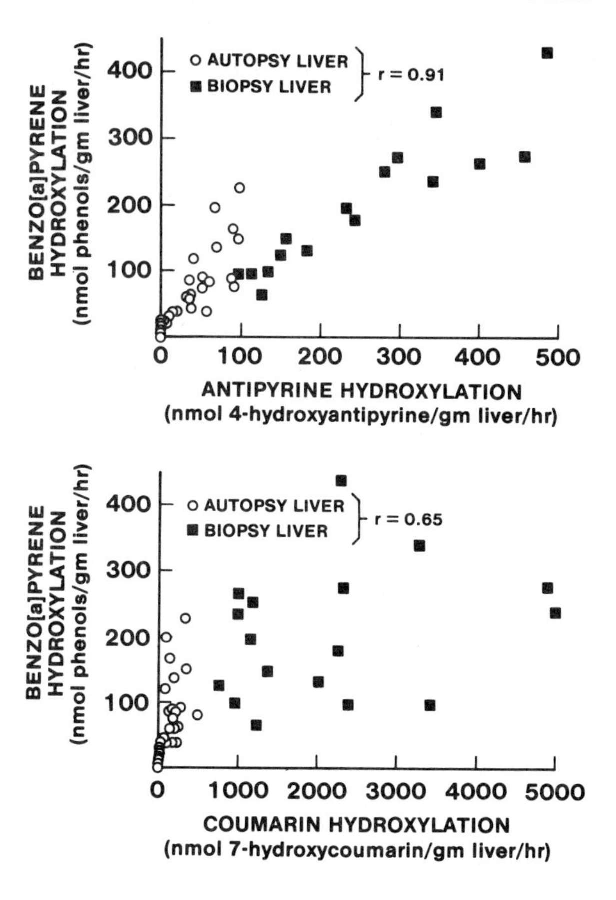

FIG. 2. Comparative rates of metabolism of benzo[a]pyrene, antipyrine and coumarin by samples of 16 human livers obtained during abdominal surgery and by samples of 32 human livers obtained at autopsy (17,49).

metabolism of certain carcinogens in tobacco smoke and with lung cancer in cigarette smokers. Ayesh and his associates reported that 234 cigarette smokers with recently diagnosed lung cancer had a higher proportion of extensive metabolizers of debrisoquine than did a carefully matched control population of 245 cigarette smokers without lung cancer (9). The mean ratio of debrisoquine to 4-hydroxydebrisoquine in the urine of the control subjects given debrisoquine was 1.4 whereas this ratio was 0.5 in the lung cancer patients ($p < 0.00001$). The possible use of drugs as probes for determining the capacity of different individuals to metabolize carcinogens is receiving increasingly more attention.

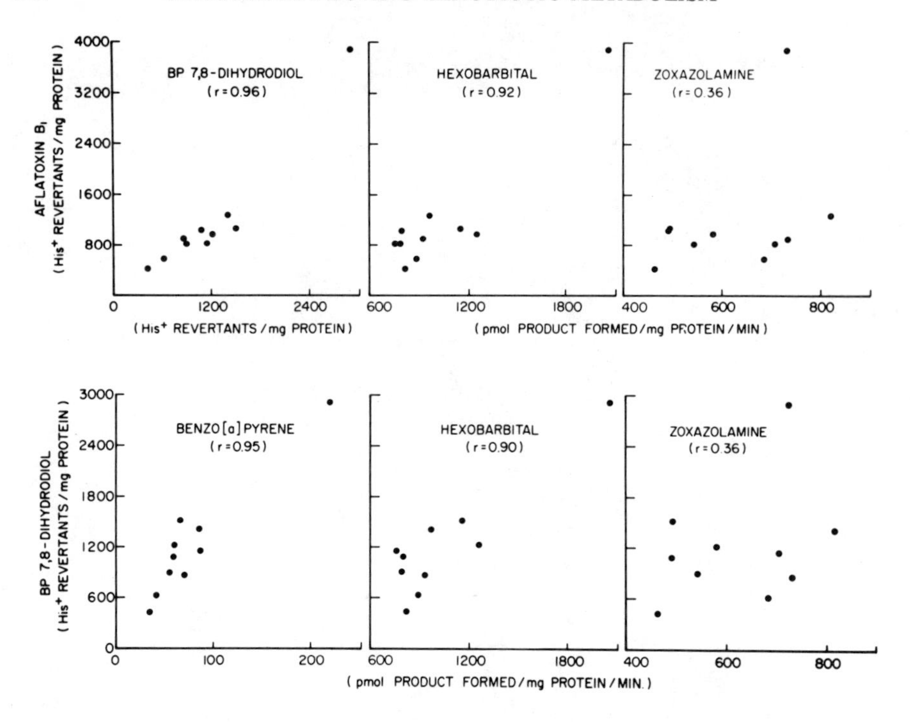

FIG. 3. Comparative rates of metabolism of aflatoxin B_1 and benzo[a]-pyrene 7,8-dihydrodiol to mutagens, the metabolism of benzo[a]pyrene to fluorescent phenols, the metabolism of hexobarbital in the 3'-position and the metabolism of zoxazolamine in the 6-position by microsomes from 10 surgical biopsy samples of human liver (17).

FACTORS INFLUENCING XENOBIOTIC METABOLISM

Both genetic and environmental factors control the rates and pathways of xenobiotic biotransformation in human beings, but it is difficult to quantify the relative contributions of genes and environment because it is not possible to study one of these variables in the absence of the other. Interindividual variations in the *in vivo* metabolism of antipyrine, phenylbutazone, and nortriptyline are greater in fraternal than in identical twins (1,75,76), which suggests that genetic factors are important in the control of oxidative drug metabolism. The importance of genetic factors in the regulation of human drug metabolism has also been emphasized by recent studies on the oxidative metabolism of debrisoquine (43,81) and sparteine (28).

Variability in the *in vivo* metabolism of a prototype drug that occurs when it is given to an individual on several occasions is an approach that we have used for assessing the influence of environment and life style on the metabolism of xenobiotics in human beings. In these studies, we found that the amount of day-to-day variation in the in vivo metabolism of three prototype drugs by several normal volunteers who were allowed to pursue their normal life styles and to eat unrestricted diets depended on both the drug and the subject studied (4,21). This approach, however, tends to underestimate the role of environment in regulating human drug metabolism since the presence of a potent environmental modifier of drug metabolism would remain undetected unless the degree of exposure to the modifier changed during the study. Factors that influence the metabolism of xenobiotics in experimental animals and man include age, disease states, dietary and nutritional factors, hormonal changes in the body, ingestion of medicinal agents, and exposure to environmental chemicals.

Many examples of medicinal agents that inhibit or stimulate human drug metabolism are now known, and treatment with these drugs can contribute to interindividual differences in the metabolism and action of xenobiotics. Bishydroxycoumarin, chloramphenicol, phenyramidol, and cimetidine are examples of drugs that inhibit human drug metabolism, and phenobarbital, diphenylhydantoin, anticonvulsant mixtures, and rifampicin are examples of therapeutic agents that stimulate human drug metabolism. Chemicals in the workplace can also influence the oxidative metabolism of drugs in humans. Exposure of people to carbon disulfide inhibits the oxidative metabolism of aminopyrine (62). In contrast to these results, people exposed to DDT (72), polychlorinated biphenyls (3,55), Kepone (31), combinations of DDT and lindane (54), or combinations of phenoxy acids and chlorophenols (25) in chemical factories have enhanced rates of oxidative drug metabolism. Similarly, anesthesiologists who are exposed to large amounts of volatile anesthetic chemicals in the operating room (27,34), gasoline station attendants who are exposed to high levels of petroleum (33), and greenhouse gardeners exposed to a mixture of pesticides and organic solvents (25) have enhanced rates of drug metabolism. These results indicate that people who are occupationally exposed to enzyme inducers or inhibitors in their environment or who are treated with drugs that influence the metabolism of xenobiotics will have altered rates of metabolism of some drugs; it is likely that these individuals will also have altered rates of metabolism of environmental carcinogens and endogenous substrates that are metabolized by the same enzymes that metabolize drugs.

STIMULATORY EFFECT OF CIGARETTE SMOKING ON THE BIOTRANSFORMATION OF XENOBIOTICS AND STEROID HORMONES

Polycyclic aromatic hydrocarbons (PAH) are potent inducers of microsomal monooxygenases in the rat (18,19), and these hydrocarbons are ubiquitous environmental contaminants (14) that very likely contribute to interindividual differences in the metabolism of xenobiotics in man. It is estimated that humans ingest an average of 2-16 micrograms PAH per day in food, air, and water with most of the exposure coming from the food (14), and some people are exposed to much larger amounts of these hydrocarbons than are other people. Since cigarette smoke and charcoal-broiled beef contain substantial amounts of PAH (Table 2), we studied the effects of cigarette smoking and the ingestion of charcoal-broiled beef on xenobiotic metabolism in experimental animals and human beings (18,20, 69,79). Treatment of rats with PAH in cigarette smoke has a marked stimulatory effect on benzo[a]pyrene hydroxylase activity in the placenta (Table 3). Exposure of pregnant rats to cigarette smoke stimulated the hydroxylation of benzo[a]pyrene in maternal liver, lung, and intestine, and in the placenta. Increased monooxygenase activity was also found in fetal liver, indicating that enzyme inducers in cigarette smoke pass through the placenta and reach the fetus. Investigations on the effects of cigarette smoking on the oxidative metabolism of several drugs and carcinogens in human placenta at full term revealed that cigarette smoking markedly stimulated benzo[a]pyrene hydroxylase, aminoazo dye N-demethylase and zoxazolamine hydroxylase activities (18,79,80). Cigarette smoking caused a smaller stimulatory effect on the 0-dealkylation of 7-ethoxycoumarin and did not change the oxidative aromatization of 4-androstene-3,17-dione to estradiol and estrone. These results suggested the presence in human placenta of several monooxygenases that are under different regulatory control. Benzo[a]pyrene hydroxylase activity varied about sixfold among the placentas from nonsmokers and about 80-fold among the placentas from smokers. The highest benzo[a]pyrene hydroxylase

TABLE 2. Concentration of polycyclic aromatic hydrocarbons in cigarette smoke and charcoal-broiled beef.

Polycyclic hydrocarbon	μg/kg of steak	μg/pack of cigarettes
Benzo[a]pyrene	8-24	0.2-1.0
Chrysene	1.4	1.0
Benz[a]anthracene	4.5	1.1
Dibenz[a,h]anthracene	0.2	0.8
Anthracene	4.5	---
Pyrene	18	2.5
Fluoranthene	20	3.6

From Lijinsky and Shubik (61), Hecht et al. (39) and Hoffmann et al. (40).

activity in the placentas from the smoking group was at least 400-fold higher than the lowest benzo[a]pyrene hydroxylase activity in the placentas from the nonsmoking group. Among the subjects who smoked 15 to 20 cigarettes per day, placental benzo[a]pyrene hydroxylase activity varied over a 70-fold range. It would be of considerable interest to know what causes the marked individuality in the induction of benzo[a]pyrene hydroxylase activity in cigarette smokers and to know whether individuality in the induction of carcinogen-metabolizing enzymes can help explain why some people develop cancer when exposed to cigarette smoke and others do not.

Since the metabolism of phenacetin to its major metabolite, N-acetyl-p-aminophenol, occurs by a polycyclic hydrocarbon-inducible enzyme system in rat liver and intestine (66), we investigated the effect of cigarette smoking on the *in vivo* metabolism of phenacetin in man. We found that cigarette smoking lowered the plasma levels of orally administered phenacetin without changing its plasma half-life or the plasma levels of total N-acetyl-p-aminophenol (68,69). The ratio of the concentration of total N-acetyl-p-aminophenol in plasma to that of phenacetin in plasma was markedly increased in cigarette smokers, suggesting that cigarette smoking stimulated the metabolism of phenacetin in the gastrointestinal

TABLE 3. Stimulatory effect of polycyclic aromatic hydrocarbons in cigarette smoke and charcoal-broiled beef on benzo[a]pyrene metabolism in rat placenta.

Hydrocarbon administered	Benzo[a]pyrene hydroxylase activity (nmol 3-HO-BP/g placenta/h)
Control	0.9
Benz[a]anthracene	16.0
Dibenz[a,h]anthracene	14.2
Benzo[a]pyrene	14.1
Chrysene	13.0
Anthracene	5.5
Pyrene	4.9
Fluoranthene	4.5

Pregnant rats were given an oral dose of hydrocarbon (40 mg/kg). Placental benzo[a]pyrene hydroxylase activity was measured 21 hours later. Fluorescent phenolic metabolites were measured, and the data are expressed as amount of 3-hydroxybenzo[a]pyrene (3-HO-BP) formed. From Welch et al. (80).

tract and/or during its first pass through the liver. Although cigarette smoking stimulated the metabolism of phenacetin in most subjects studied, some cigarette smokers did not have enhanced phenacetin metabolism. Additional studies revealed that cigarette smokers have shorter plasma half-lives of antipyrine (38), theophylline (44), and caffeine (71) than do nonsmokers, but cigarette smoking does not stimulate the metabolism of phenytoin, meperidine, or nortriptyline. Since enzyme induction by PAH in experimental animals enhances the in vivo metabolism of several xenobiotics and can markedly influence the toxicity of these compounds (15,18), it is likely that cigarette smokers have an altered toxic response to some environmental xenobiotics.

Recent studies indicate that cigarette smoking influences the metabolism of testosterone in dogs. Long-term cigarette smoking by beagles caused a threefold stimulation in hepatic testosterone 6β-hydroxylase activity, a 46% decrease in serum testosterone levels, and a 56% decrease in prostate size (63). These observations may help explain reports indicating that human cigarette smokers have a lower serum level of testosterone (74) and a lower incidence of benign hypertrophy of the prostate (65) than do nonsmokers. The possible effects of cigarette smoking on the metabolism and action of androgens and other hormones in man requires further investigation.

EFFECTS OF DIET ON THE BIOTRANSFORMATION OF XENOBIOTICS AND STEROID HORMONES

Dietary factors are known to influence the metabolism of xenobiotics by cytochrome P-450 dependent enzymes in experimental animals, and recent studies have shown that dietary factors also influence the metabolism of xenobiotics in man. Charcoal-broiled beef, a food that contains high concentrations of PAH (Table 2), is eaten by large numbers of people. Because of the possible effect of this food on drug metabolism, we studied the effect of feeding charcoal-broiled beef on the metabolism of phenacetin in rats and in man. The in vitro metabolism of phenacetin by intestine was stimulated severalfold when rats were fed a diet containing charcoal-broiled beef (Table 4). In vivo studies revealed that feeding a charcoal-broiled beef diet for several days enhanced the oxidative metabolism of phenacetin, theophylline, and antipyrine in man, but the conjugation of acetaminophen was not altered (7,20, 50). The effect of feeding a diet containing charcoal-broiled beef on phenacetin plasma levels in human beings is shown in Fig. 4. In these studies, feeding charcoal-broiled beef for 4 days markedly lowered the plasma levels of orally administered phenacetin and increased the ratio of N-acetyl-p-aminophenol to phenacetin in the plasma. Marked interindividual differences occurred in the plasma concentrations of phenacetin among the nine subjects who had been fed the control diet; there were also large individual differences in the

TABLE 4. Effect of a diet containing charcoal-broiled beef on the metabolism of phenacetin in rat intestine.

Diet	N-acetyl-p-aminophenol formed (μg/5 cm intestine/90 min)
Control	1.35
Cooked ground beef	1.97
Charcoal-broiled ground beef	15.88

Adult male rats were fed a nutritionally complete semisynthetic diet or a 3:1 ratio of beef in semisynthetic diet for 7 days (67).

responsiveness of the subjects to the charcoal-broiled beef diet (Fig. 5). Switching from the control diet to a charcoal-broiled beef diet resulted in a decreased area under the plasma concentration of phenacetin vs. time curve for seven of the nine subjects. The two subjects who did not respond to charcoal-broiled beef feeding had very low plasma concentrations of phenacetin while on the control diet. The reason(s) for these low concentrations throughout the study is unknown, but a low concentration of phenacetin in plasma may have resulted from genetic and/or environmental factors. One of the two subjects worked as a carpenter and may have been exposed to volatile oil inducers of drug metabolism, which are present in certain soft woods.

Increasing the ratio of protein to carbohydrate or fat in the diet or feeding a diet containing cabbage and brussels sprouts stimulated the oxidative metabolism of xenobiotics in tissues of experimental animals and stimulated human drug metabolism in vivo (2,5, 51,70). Although the average half-lives of antipyrine and theophylline increased 63% and 46%, respectively, when six subjects were shifted from a high-protein, low-carbohydrate diet to an isocaloric high-carbohydrate, low-protein diet, there was considerable individuality in response to this alteration in the diet. The increase in antipyrine half-lives among the six subjects ranged from no change in one subject to 111% in another subject. The increase in theophylline half-lives ranged from 14% to 71% in the different subjects.

The administration of certain methylxanthines or the ingestion of methylxanthine-containing beverages inhibited human drug metabolism (26,64). The chewing of cola nuts, which contain high concentrations of caffeine and smaller amounts of theobromine, was reported to inhibit the metabolism of antipyrine in west African villagers (29), but this effect was not confirmed in Caucasian subjects in south central Pennsylvania (77).

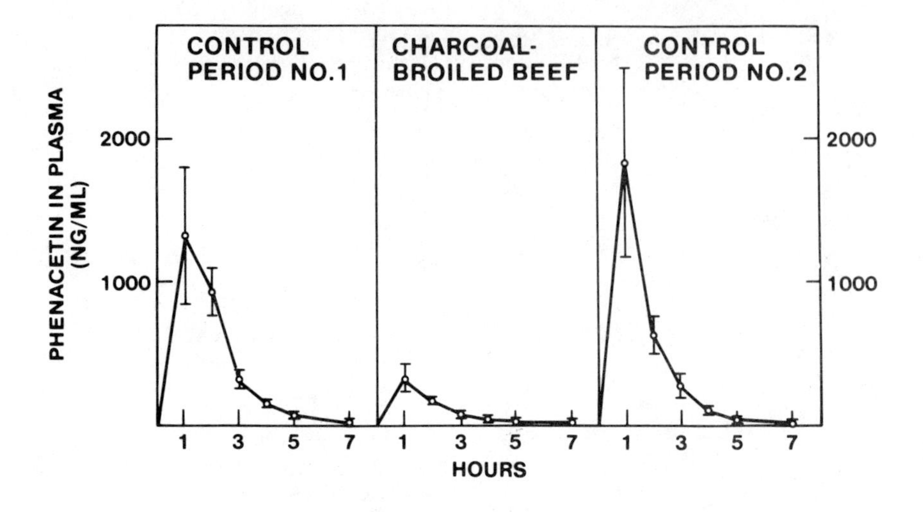

FIG. 4. Effect of a charcoal-broiled beef diet on plasma levels of phenacetin in man. Nine subjects were given an oral dose of 900 mg of phenacetin after 7 days on control diet, 4 days on a charcoal-broiled beef diet, and 7 days on the control diet a second time (20).

Alcoholics, when sober, usually show a tolerance for drugs to which they are highly sensitive when inebriated. Indeed, deaths have occurred after inebriated individuals have taken sedatives or hypnotic drugs. The tolerance to drugs that is observed in alcoholics may be explained by an increase in the activity of certain monooxygenases in human liver after chronic ingestion of ethanol and by the enhanced rates of drug metabolism measured in vivo in alcoholics (60). In contrast to the stimulatory effect of chronic ingestion of alcohol on drug metabolism, the administration of large amounts of ethanol immediately before administration of meprobamate or pentobarbital increased the plasma half-lives of these two drugs by two- to fourfold (60). This inhibitory effect of acute ethanol administration on human drug metabolism in vivo helps explain the dangerous and synergistic central depression that has been observed when ethanol and a sedative or hypnotic drug are ingested together.

The fasting of obese subjects for 7-10 days does not influence the metabolism of antipyrine or tolbutamide (73). Although early studies suggested that malnourished individuals do not have a markedly abnormal half life of antipyrine in the plasma (56), more recent studies indicate that protein-calorie malnutrition is associated with decreased rates of oxidative metabolism of some xenobiotics (41,57).

Since steroid hormones are metabolized by the same cytochrome P-450 enzymes that metabolize xenobiotics, we investigated the effects of alterations in dietary protein and carbohydrate on the metabolism of the

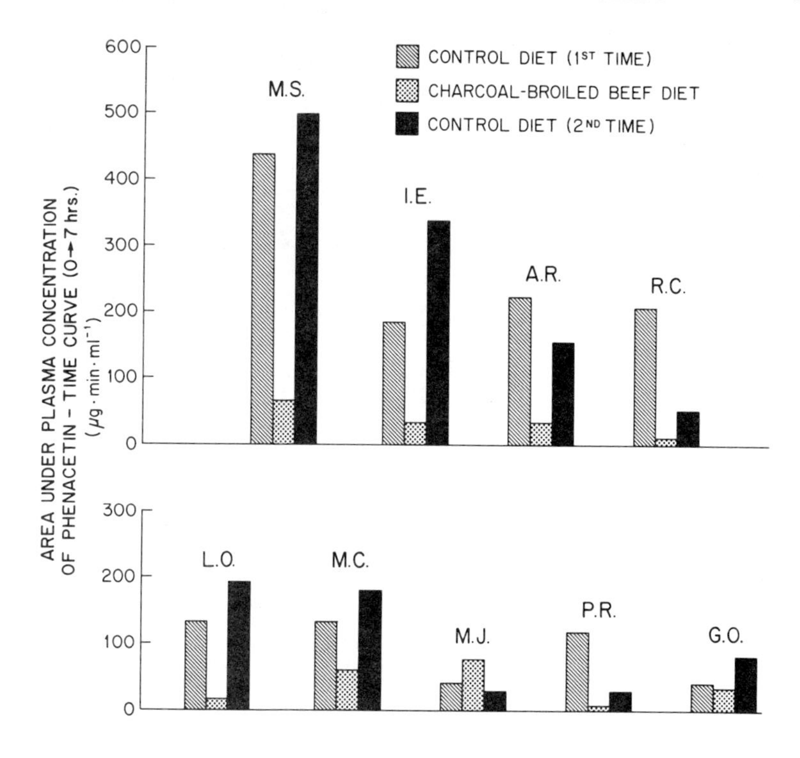

FIG. 5. Effect of a charcoal-broiled beef diet on the area under the curve for plasma concentration of phenacetin vs. time. Nine subjects were treated as described under Fig. 4 (20).

prototype steroids, estradiol and testosterone (6,52). Increasing the ratio of protein to carbohydrate in the diet stimulated the 2-hydroxylation of estradiol, had little or no effect on the 16α-hydroxylation of estradiol, and inhibited the 5α-reduction of testosterone relative to the 5β-reduction of this steroid (Table 5). The ratio of androsterone to etiocholanolone in the urine was increased about 50% when the subjects were switched from a high-protein low-carbohydrate diet to a low-protein high-carbohydrate diet. During these studies, we observed interindividual differences in the metabolism of testosterone and estradiol, as well as substantial differences in the response of different individuals to changes in the ratio of protein to carbohydrate in the diet (Fig. 6). Since steroid hormones are believed to play a role in the carcinogenic process, factors influencing the metabolism of these endogenous substrates may influence tumorigenesis.

XENOBIOTIC METABOLISM IN MIGRANT POPULATIONS

There are several examples of differences in the metabolism of xenobiotics in different ethnic populations (45-47), and there are a few examples with a limited number of subjects where xenobiotic metabolism in an ethnic population in its home country was compared with xenobiotic metabolism in a similar population living in another country. In one study, antipyrine clearance rates were measured in British subjects, Sudanese living in Britain, and Sudanese living in their home villages (11). All of the subjects were healthy nonsmokers. The Sudanese living in their home villages had a slower antipyrine clearance rate than the Sudanese living in Britain or the British subjects; the latter two groups had similar clearance rates. These observations suggested that the Sudanese who migrated to Britain acquired the xenobiotic-metabolizing capacity of their new British neighbors. In a second study, Asians living in London had a slower antipyrine clearance rate than a comparable group of Asians living in their home villages (23,24, 30). Although the above studies provide evidence that migrant populations change their xenobiotic-metabolizing capacity when they move from one country to another with a different life style and with different dietary customs, the number of subjects studied was small; further research is needed to determine xenobiotic metabolism in the same individuals before and after a move to another country.

SUMMARY AND CONCLUDING REMARKS

Substantial interindividual differences occur for the metabolism of drugs, carcinogens, and steroid hormones, and these person-to-person differences are caused by genetic and environmental factors. It is likely

TABLE 5. Effect of changes in dietary protein and carbohydrate on the metabolism of antipyrine, estradiol, and testosterone.

Diet	Antipyrine half-life (h)	Estradiol hydroxylations (percent of dose)		Testosterone reduction (ratio of androsterone to etiocholanolone in urine)
		C-2	C-16	
High protein[a]	9.1 ± 0.8	44.0 ± 2.6	13.2 ± 0.6	1.1 ± 0.2
High carbohydrate[b]	12.6 ± 0.8	33.1 ± 0.8	11.3 ± 1.2	1.6 ± 0.3
P values (paired t test)	< 0.001	< 0.005	N.S.	< 0.025

[a]44% protein, 35% carbohydrate, 21% fat
[b]10% protein, 70% carbohydrate, 20% fat
Eight normal men consumed a high protein-low carbohydrate diet or an isocaloric high carbohydrate-low protein diet for 2 weeks (6,52).

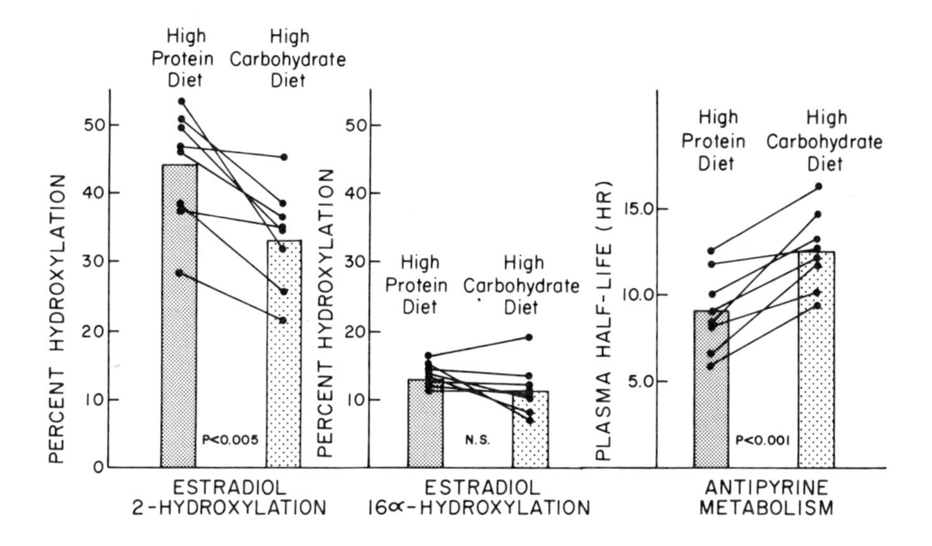

FIG. 6. Eight normal men consumed a high-protein/low-carbohydrate diet or an isocaloric high-carbohydrate/ low-protein diet for two weeks as described in Table 5 (6,52).

that interindividual differences in the metabolism of xenobiotics and steroid hormones play a role in explaining interindividual differences in the initiation and progression of some human cancers. Factors that influence the metabolism and action of xenobiotics in human beings include age, disease states, hormonal changes in the body, ingestion of medicinal agents, exposure to environmental chemicals, and changes in life style, including factors such as cigarette smoking, alcohol consumption, and diet. Some individuals have a much larger response to environmental perturbations than do other individuals, and further research is needed to elucidate the reasons for differences in the responsiveness of people to modulators of chemical biotransformations. Immigration to a new country can represent a substantial change in a person's diet, life style, and environment, and these changes may alter the metabolism of xenobiotics and endogenous hormones that play a role in the carcinogenic response.

ACKNOWLEDGMENT

We thank Ms. Maria Devenney for her help in the preparation of this manuscript.

REFERENCES

1. Alexanderson, B., Evans, D.A.P., and Sjöqvist, F. (1969): Br. Med. J., 4:764-768.
2. Alvares, A.P., Anderson, K.E., Conney, A.H., and Kappas,A. (1976): Proc. Nat. Acad. Sci., USA, 73:2501-2504.
3. Alvares, A.P., Fischbein, A., Anderson, K.E., and Kappas, A. (1977): Clin. Pharmacol. Ther., 22:140-146.
4. Alvares, A.P., Kappas, A., Eiseman, J.L., Anderson, K.E., Pantuck, C.B., Pantuck, E.J., Hsiao, K.-C., Garland, W.A., and Conney, A.H. (1979): Clin. Pharmacol. Ther., 26:407-419.
5. Anderson, K.E., Conney, A.H., and Kappas, A. (1979): Clin. Pharmacol. Ther., 26:493-501.
6. Anderson, K.E., Kappas, A., Conney, A.H., Bradlow, H.L., and Fishman, J. (1984): J. Clin. Endocrinol. Metab., 59:103-107.
7. Anderson, K.E., Schneider, J., Pantuck, E.J., Pantuck, C.B., Mudge, G.H., Welch, R.M., Conney, A.H., and Kappas, A. (1983): Clin. Pharmacol. Ther., 34:369-374.
8. Autrup, H., Harris, C.C., Schwartz, R.D., Trump, B.F., and Smith, L. (1980): Carcinogenesis, 1:375-380.
9. Ayesh, R., Idle, J.R., Ritchie, J.C., Crothers, M., and Hetzel, M. (1984): Nature, in press.
10. Bast, R.C., Jr., Whitlock, J.P., Jr., Miller, H., Rapp, H.J., and Gelboin, H.V. (1974): Nature, 250:664-665.
11. Branch, R.A., Salih, S.Y., and Homeida, M. (1978): Clin. Pharmacol. Ther., 24:283-286.
12. Busbee, D.L., Shaw, C.R., and Cantrell, E.T. (1972): Science, 178:315-316.
13. Cantrell, E.T., Warr, G.A., Busbee, D.L., and Martin, R.R. (1973): J. Clin. Invest., 52:1881-1884.
14. Committee on Pyrene and Selected Analogues, Board on Toxicology and Environmental Health Hazards, Commission on Life Sciences, National Research Council (1983): Polycyclic Aromatic Hydrocarbons: Evaluation of Sources and Effects. National Academy Press, Washington, DC.
15. Conney, A.H. (1967): Pharmacol. Rev., 19:317-366.
16. Conney, A.H. (1973): Rev. Can. Biol., 32:163-170.
17. Conney, A.H. (1980): In: Microsomes, Drug Oxidations and Chemical Carcinogenesis, Vol. 2, edited by M.J. Coon, A.H. Conney, R.W. Estabrook, H.V. Gelboin, J.R. Gillette, and P.J. O'Brien, pp. 1103-1118. Academic Press, Inc., New York.
18. Conney, A.H. (1982): Cancer Res., 42:4875-4917.
19. Conney, A.H., Miller, E.C., and Miller, J.A. (1956): Cancer Res., 16:450-459.

20. Conney, A.H., Pantuck, E.J., Hsiao, K.-C., Garland, W.A., Anderson, K.E., Alvares, A.P., and Kappas, A. (1976): Clin. Pharmacol. Ther., 20:633-642.
21. Conney, A.H., Pantuck, E.J., Pantuck, C.B., Buening, M., Jerina, D.M., Fortner, J.G., Alvares, A.P., Anderson, K.E., and Kappas, A. (1979): In: The Induction of Drug Metabolism, edited by R.W. Estabrook and E. Lindenlaub, pp. 583-605. F.K. Schattauer Verlag, New York.
22. Czygan, P., Greim, H., Garro, A.J., Hutterer, F., Rudick, J., Schaffner, F.R., and Popper, H. (1973): J. Nat. Cancer Inst., 51:1761-1764.
23. Desai, N.K., Sheth, U.K., Mucklow, J.C., Fraser, H.S., Bulpitt, C.J., Jones, S.W., and Dollery, C.T. (1980): Br. J. Clin. Pharmacol, 9:387-394.
24. Dollery, C.T., Fraser, H.S., Mucklow, J.C., and Bulpitt, C.J. (1979): Drug. Metab. Rev., 9:207-220.
25. Dossing, M. (1982): Clin. Pharmacol. Ther., 32:340-346.
26. Drouillard, D.D., Vesell, E.S., and Dvorchik, B.H. (1978): Clin. Pharmacol. Ther., 23:296-302.
27. Duvaldestin, P., Mazze, R.I., Nivoche, Y., and Desmonts, J.-M. (1981): Anesthesiology. 54:57-60.
28. Eichelbaum, M., Spannbrucker, N., Steincke, B., Dengler, H.J. (1979): Eur. J. Clin. Pharmacol., 16:183-187.
29. Fraser, H.S., Bulpitt, C.J., Kahn, C., Mould, G., Mucklow, J.C., and Dollery, C.T. (1976): Clin. Pharmacol. Ther., 20:369-376.
30. Fraser, H.S., Mucklow, J.C., Bulpitt, C.J., Kahn, C., Mould, G., and Dollery, C.T. (1979): Br. J. Clin. Pharmacol., 7:237-243.
31. Guzelian, P.S., Vranian, G., Boylan, J.J., Cohn, W.J., and Blanke, R.V. (1980): Gastroenterology, 78:206-213.
32. Hammer, W., Ideström, C.M., and Sjöqvist, F. (1967): In: International Congress Series No. 122, Proceedings of the First International Symposium on Antidepressant Drugs, edited by S. Garattini and M.N.G. Dukes, pp. 301-310. Excerpta Medica, Amsterdam.
33. Harman, A.W., Frewin, D.B., and Priestly, B.G. (1981): Br. J. Ind. Med. 38:91-97.
34. Harman, A.W., Russell, W.J., Frewin, D.B., and Priestly, B.G. (1978): Anaesth. Intensive Care, 6:210-214.
35. Harris, C.C., Autrup, H., Connor, R., Barrett, L.A., McDowell, E.M., and Trump, B.F. (1976): Science, 194:1067-1069.
36. Harris, C.C., Autrup, H., Stoner, G., Yang, S.K., Leutz, J.C., Gelboin, H.V., Selkirk, J.K., Connor, R.J., Barrett, L.A., Jones, R.T., McDowell, E., and Trump, B.F. (1977): Cancer Res., 37:3349-3355.

37. Harris, C.C., Trump, B.F., Autrup, H., Hsu, I.-C., Haugen, A., and Lechner, J. (1982): In: Host Factors in Human Carcinogenesis, edited by B. Armstrong and H. Bartsch, pp. 497-514. International Agency for Research on Cancer, Lyon, France.
38. Hart, P., Farrell, G.C., Cooksley, W.G.E., and Powell, L.W. (1976): Br. Med. J., 2:147-149.
39. Hecht, S.S., Grabowski, W., and Groth, K. (1979): Fd. Cosmet. Toxicol., 17:223-227.
40. Hoffmann, D., Schmeltz, I., Hecht, S.S., and Wynder, E.L. (1978): In: Polycyclic Hydrocarbons and Cancer, Environment, Chemistry, and Metabolism, Vol. 1, edited by H.V. Gelboin and P.O.P. Ts'o, pp. 85-117. Academic Press, Inc., New York.
41. Homeida, M., Karrar, Z.A., and Roberts, C.J.C. (1979): Arch. Dis. Child., 54:299-302.
42. Idle, J.R., Oates, N.S., Ritchie, J.C., Shah, R., Sloan, T., and Smith, R.L. (1980): In: Advanced Medicine, Vol. 16, pp. 227-234. Royal College of Physicians, London.
43. Idle, J.R., and Smith, R.L. (1983): In: Proceedings of the Second World Conference on Clinical Pharmacology and Therapeutics, edited by L. Lemberger and M.M. Reidenberg, pp. 148-164. American Society for Pharmacology and Experimental Therapeutics, Bethesda, Maryland.
44. Jenne, J., Nagasawa, H., McHugh, R., MacDonald, F., and Wyse, E. (1975): Life Sci., 17:195-198.
45. Kalow, W. (1982): Clin. Pharmacokinet., 7:373-400.
46. Kalow, W. (1984): Fed. Proc., 43:2314-2318.
47. Kalow, W. (1984): Fed. Proc., 43:2326-2331.
48. Kapitulnik, J., Levin, W., Lu, A.Y.H., Morecki, R., Dansette, P.M., Jerina, D.M., and Conney, A.H. (1977): Clin. Pharmacol. Ther., 21:158-165.
49. Kapitulnik, J., Poppers, P.J., and Conney, A.H. (1977): Clin. Pharmacol. Ther., 21:166-176.
50. Kappas, A., Alvares, A.P., Anderson, K.E., Pantuck, E.J., Pantuck, C.B., Chang, R., and Conney, A.H. (1978): Clin. Pharmacol. Ther., 23:445-450.
51. Kappas, A., Anderson, K.E., Conney, A.H., and Alvares, A.P. (1976): Clin. Pharmacol. Ther., 20:643-653.
52. Kappas, A., Anderson, K.E., Conney, A.H., Pantuck, E.J., Fishman, J., and Bradlow, H.L. (1983): Proc. Nat. Acad. Sci., USA, 80:7646-7649.
53. Kellermann, G., Luyten-Kellermann, M., and Shaw, C.R. (1973): Am. J. Hum. Genet., 25:327-331.
54. Kolmodin, B., Azarnoff, D.L., and Sjöqvist, F. (1969): Clin. Pharmacol. Ther., 10:638-642.

55. Krampl, V., and Kontsekova, M. (1978): Bull. Environ. Contam. Toxicol., 20:191-198.
56. Krishnaswamy, K., and Naidu, A.N. (1977): Br. Med. J., 1:538-540.
57. Krishnaswamy, K., Kalamegham, R., and Naidu, N.A. (1984): Br. J. Clin. Pharmacol., 17:139-146.
58. Kuntzman, R., Mark, L.C., Brand, L., Jacobson, M., Levin, W., and Conney, A.H. (1966): J. Pharmacol. Exp. Ther., 152:151-156.
59. Levin, W., Conney, A.H., Alvares, A.P., Merkatz, I., and Kappas, A. (1972): Science, 176:419-420.
60. Lieber, C.S. (1982): Medical Disorders of Alcoholism: Pathogenesis and Treatment. W.B. Saunders Company, Philadelphia.
61. Lijinsky, W., and Shubik, P. (1964): Science, 145:53-55.
62. Mack, T., Freundt, K.J., and Henschler, D. (1974): Biochem. Pharmacol., 23:607-614.
63. Mittler, J.C., Pogach, L., and Ertel, N.H. (1983): J. Steroid Biochem., 18:759-763.
64. Monks, T.J, Caldwell, J., and Smith, R.L. (1979): Clin. Pharmacol. Ther., 26:513-524.
65. Morrison, A. S. (1978): J. Cron. Dis., 31:357-362.
66. Pantuck, E.J., Hsiao, K.-C., Kaplan, S.A., Kuntzman, R., and Conney, A.H. (1974): J. Pharmacol. Exp. Ther., 191: 45-52.
67. Pantuck, E.J., Hsiao, K.-C., Kuntzman, R., and Conney, A.H. (1975): Science, 187:744-746.
68. Pantuck, E.J., Hsiao, K.-C, Maggio, A., Nakamura, K., Kuntzman, R., and Conney, A.H. (1974): Clin. Pharmacol. Ther., 15:9-17.
69. Pantuck, E.J., Kuntzman, R., and Conney, A.H. (1972): Science, 175:1248-1250.
70. Pantuck, E.J., Pantuck, C.B., Garland, W.A., Min, B.H., Wattenberg, L.W., Anderson, K.E., Kappas, A., and Conney, A.H. (1979): Clin. Pharmacol. Ther., 25:88-95.
71. Parsons, W.D., and Neims, A.H. (1978): Clin. Pharmacol. Ther., 24:40-45.
72. Poland, A., Smith, D., Kuntzman, R., Jacobson, M., and Conney, A.H. (1970): Clin. Pharmacol. Ther., 11:724-732.
73. Reidenberg, M.M., and Vesell, E.S. (1975): Clin. Pharmacol. Ther., 17:650-656.
74. Shaarawy, M., and Mahmoud, K.Z. (1982): Fertil. and Steril., 38:255-257.
75. Vesell, E.S., and Page, J.G. (1968): Science, 159:1479-1480.
76. Vesell, E.S., and Page, J.G. (1968): Science, 161:72-73.
77. Vesell, E.S., Shively, C.A., Passananti, G.T. (1979): Clin. Pharmacol. Ther., 26:287-293.
78. Weisburger, J.H., Grantham, P.H., Vanhorn, E., Steigbigel, N.H., Rall, D.P., and Weisburger, E.K. (1964): Cancer Res., 24:475-479.

79. Welch, R.M., Harrison, Y.E., Conney, A.H., Poppers, P.J., and Finster, M. (1968): Science, 160:541-542.
80. Welch, R.M., Harrison, Y.E., Gommi, B.W., Poppers, P.J., Finster, M., and Conney, A.H. (1969): Clin. Pharmacol. Ther., 10:100-109.
81. Woolhouse, N.M., Andoh, B., Mahgoub, A., Sloan, T.P., Idle, J.R., and Smith, R.L. (1979): Clin. Pharmacol. Ther., 26:584-591.

Carcinogenesis, Vol. 10, edited by E. Huberman
and S. H. Barr. Raven Press, New York © 1985.

Chemical and Biochemical Dosimetry of Exposure to Genotoxic Chemicals

G. N. Wogan

Massachusetts Institute of Technology, Cambridge, Massachusetts 02139

The sensitivity of epidemiologic studies designed to evaluate the health significance of environmental chemicals is compromised by the lack of quantitative exposure data for individuals in exposed populations. Estimates of levels of compounds of interest in environmental media are often the only data available, and average exposure must therefore be calculated on a population basis. Biological monitoring (i.e., measurements on cells, tissues, or body fluids of exposed persons) has the objective of defining the so-called "internal dose" or "effective dose." Such measurements can be used to ensure that current or past exposure does not entail unacceptable health risks or to detect potentially excessive exposure before the appearance of adverse health effects. Results obtained through this approach can be interpreted on an individual basis and used to estimate for that individual the amount of chemical absorbed during a specific time interval or the amount bound to critical sites. These individual results, if they are grouped, may also be useful for characterization of community exposure. In this respect, biological monitoring data complement environmental measurements but have certain advantages in estimating health risks. For example, the data obtained are more directly related to adverse effects and thus provide a better estimate of risk than data obtained by environmental monitoring. Biological monitoring also takes into account absorption by all routes and integrates exposure from all sources. Therefore, it can be used as a basis for estimating total risk from multiple chemicals.

DIRECT CHEMICAL ANALYSIS

Quantification of carcinogens in body fluids (e.g., blood, breast milk, saliva, semen) or excreta has been carried out by direct chemical or immunological analysis or by bioassay for mutagenicity. Most of the existing chemical methods and available data relate to occupational exposure, since such measurements have been components of occupational hygiene programs for many years. Immunoassays and bioassays are more recent developments that have, as yet, found only limited application. With the recognition that most genotoxic chemicals require metabolic activation to electrophilic forms to express their effects, an additional strategy for chemical dosimetry has developed; this strategy is based upon the detection and quantification of covalently bound derivatives formed between activated chemicals and cellular macromolecules such as nucleic acids and proteins. Immunologic and chemical analytical methods, which are sufficiently sensitive to detect the consequences of ambient exposure in the workplace and general environment, have been developed for this purpose.

Prevention of excessive exposure to chemicals in industry has been approached traditionally by setting standards for the concentration of compounds in ambient air. Air monitoring has, therefore, constituted the principal means of assessing exposure. This method obviously takes into account only exposure via the pulmonary route and does not estimate total exposure. These shortcomings have stimulated much research aimed at development of biological monitoring methods for evaluating individual exposure. Over the past two decades, methods have been developed for many substances representing a variety of chemical classes to which people are exposed, principally in the workplace. Lauwerys (22) has summarized these methods from the perspective of their usefulness in biological monitoring programs; he described a total of 69 chemicals for which methodology was considered to be sufficiently well developed for application in worker surveillance programs. Lauwerys (22), as well as Baselt (2) and Linch (23), summarized the analytical methodology and provided additional pertinent information including representative values for each of the chemicals in body fluids.

In addition to their use in monitoring programs in the workplace, a few of these methods have also been applied to population studies involving large numbers of subjects. For example, chlorinated hydrocarbons have been extensively studied with respect to their storage and accumulation in adipose tissue and other body compartments. Hayes (12) summarized the very-extensive literature concerning levels of DDT and its derivatives in adipose tissue and other body compartments in large numbers of subjects studied over a period of two decades. In a related area, analysis of human milk has been used to determine exposure to chlorinated pesticides (39) and polychlorinated biphenyls (36,37).

With reference specifically to environmental carcinogens and mutagens, the field is in a more primitive state of development. The main focus of research has been on the development of analytical methods for detecting carcinogens that can occur as contaminants of food; consequently, most of the existing methods were intended for food analysis. Methods have been published for some N-nitroso compounds, polycyclic aromatic hydrocarbons, aromatic amines, vinyl chloride, and mycotoxins.

For the most part, the suitability of these methods for analysis of media other than foods has not been evaluated, and in many instances these methods cannot be used for analysis of metabolites without extensive modifications. These methods also require relatively sophisticated equipment, which limits their utility for surveillance studies requiring analysis of large numbers of samples.

A larger number of methods are available for detection of aflatoxins than for any other class of carcinogen. In this case, methods originally developed for analysis of oilseeds and grains have been adapted for analysis of edible tissues and milk of animals to minimize human exposure through residues of the parent compounds or metabolites. These methods have also been applied to analyze tissues of people suspected of aflatoxin exposure.

In evaluating the putative etiologic role of aflatoxins in cases of Reye's Syndrome, an acute highly fatal disease of children, aflatoxin B_1 was detected in samples of liver, stool, brain, and kidney at levels in the order of 100 ng/g body weight through application of a thin-layer chromatography (TLC) method with fluorescence detection (3,41). These results were extended and substantiated in more-recent studies in which high-performance liquid chromatography (HPLC) separation with fluorescence detection was used (28,38,42). Aflatoxins have also been detected in human tumors of the liver (29,43) and lung (7) with TLC methodology.

Studies in human populations consuming aflatoxin-contaminated peanut butter revealed the presence of aflatoxin M_1 in urine, as detected by TLC analysis (5), and aflatoxins were also found in the HPLC analysis of urine from Sudanese children suffering from kwashiorkor (15). Chemical screening methods for detection of aflatoxin and metabolites in human urine (24) and serum (21) have been reported.

Only a small number of chemical methods have been developed for other specific chemical carcinogens. Matsumoto et al. (26) have developed a method for determination of the carcinogen methylazoxymethanol-D-glucosiduronic acid in rat bile and urine. Although the method produces good recovery of carcinogen, it is relatively insensitive. Pylypiw and Harrington (33) reported a method for detecting N-nitroso-N-methylaniline in urine and serum at levels of 0.01 to 0.001 ppm.

The rapidly developing field of immunoassay has up to now principally been applied to the detection of covalent adducts of carcinogens with DNA. However, methods based on radioimmunoassays have been reported for two carcinogens.

Johnson et al. (19) developed a radioimmunoassay procedure to detect 4-acetamidobiphenyl, a metabolite of the carcinogen 4-aminobiphenyl, in urine. They produced rabbit polyclonal antibodies with high affinity (2.8×10^8 L/mol), which were capable of detecting the metabolite in human urine at levels of about 1 ng (4.8 pmol).

Sizaret et al. (43) similarly developed rabbit polyclonal antibodies, which detected not only aflatoxin B_1 but also cross-reacted with various aflatoxin metabolites including M_1, the principal urinary excretory product discussed above. The radioimmunoassay in which these antibodies were used was capable of detecting urinary metabolites of aflatoxin administered to rats at doses of 600 pmol or less. The authors thus propose that the assay would be suitable for use in human population studies.

DETECTION OF NUCLEIC ACID ADDUCTS

The rationale underlying the strategy of chemical dosimetry by determining levels of derivatives covalently bound to cellular macromolecules is based on current understanding of the mode of action of genotoxic carcinogens and mutagens. Chemicals that are active as carcinogens and mutagens have electrophilic properties or are metabolically converted into electrophiles. These reactive derivatives undergo attack by nucleophilic centers in nucleic acids and proteins, resulting in the formation of covalent adducts. Particular attention has been paid to DNA adducts because these are thought to represent initiating events leading to mutation and/or malignant transformation. Indeed, it has been empirically established that carcinogenic potency of a large number of chemicals bears a proportionality to their ability to bind to DNA--the so-called "covalent binding index"--when reacted in vivo with DNA (25). Covalent adducts formed in RNA and proteins have no putative mechanistic role in carcinogenesis but should relate quantitatively to total exposure and activation and could therefore contribute to dosimetry of both exposure and activating capability.

Covalent nucleic acid adducts have differing levels of stability, some being removed spontaneously through chemical depurination and others enzymatically, as in the process of DNA repair. A small proportion remain in DNA for long periods of time. In the few experimental models in which appropriate measurements have been made, adducts removed spontaneously or enzymatically from DNA are excreted in urine in amounts dependent upon total binding levels. In contrast, as pointed out in the preceding discussion, protein adducts are stable over the life-span of the protein and, therefore, accumulate with multiple exposures.

These properties collectively form the basis for several complementary approaches to development of chemical dosimeters, providing different kinds of information. Measurement of DNA adducts in situ in the DNA of cells should give the most direct evidence of genotoxic exposure. Measurement of DNA adducts (or products of them) in urine should give an indication of total, recent exposure. Protein adducts, by contrast, should provide an index of total exposure integrated over the life-span of the target proteins.

Two approaches are available to obtain experimental data on levels of DNA adducts formed in a given set of circumstances. On the one hand, the levels of DNA adducts derived from a chemical of interest in cells of an accessible tissue (e.g., white blood cells or biopsy or autopsy samples) can be measured. Providing that the chemical nature and stability of the DNA adducts for the compound of interest have been fully characterized, both qualitative and quantitative identification of adduct levels could provide for that individual an indication not only of exposure history but also of the capability to activate the carcinogen to DNA-binding forms.

A second approach to monitorinq DNA adducts takes advantage of the fact that adducts removed from cellular nucleic acids are excreted in urine. Their detection and measurement of their excretion rates would, in principle, provide information on the (recent) exposure history of the subject and also indicate that individual's capability for DNA repair. Thus, studies of urinary excretion of adducts provide data complementary to simultaneous measurement of adduct levels in cellular DNA.

ANALYSIS OF CELLULAR DNA

Several methods of analysis for carcinogen-DNA adducts are currently under development, including immunoassays, radiochemical labeling procedures, and physicochemical techniques. For immunological detection, antisera have been raised in rabbits against the RNA and DNA adducts of aromatic amines, polycyclic aromatic hydrocarbons, and methylating and ethylating carcinogens. The properties of these antisera have been reviewed by Poirier (32). High-affinity antisera were elicited with either nucleoside adducts covalently bound to a protein carrier or modified DNA electrostatically coupled to a protein carrier. These polyclonal antibodies were of generally high affinity (affinity constants in the range of 4×10^6 to 1.8×10^{10} L/mole) and were capable of detecting adducts at levels in the range of 10^{-4} to 10^{-10} molar (50% inhibition in competitive radioimmunoassays).

Monoclonal antibodies have also been produced that bind carcinogen-DNA adducts with high affinity. Carcinogens studied thus far inlude N-nitroso compounds (methylating, ethylating, and butylating agents), aromatic amines, benzo[a]pyrene, aflatoxin B_1, and ultraviolet radiation. In general, these antibodies are of high affinity (10^8-10^{10} L/mole) and high sensitivity (detection limits in the range of 5 picomoles to 10 femtomoles). [See Muller and Rajewsky (27) and IARC/IPCS Working Group Report (18) for details.]

Antibodies of both types have been used in the development of immunoassays (RIA, ELISA, USERIA) for the quantification of carcinogen adduct levels in DNA. Adducts studied to date include benzo[a]pyrene-DNA (17), aflatoxin B_1-DNA (9,16), and O^6-methylguanosine (46). The latter techniques may be useful in detecting DNA adducts in people exposed to environmental carcinogens under ambient conditions.

Determination of carcinogen-DNA adducts by immunologic procedures has certain advantages. The sensitivity is frequently better than that obtainable with radiolabeled carcinogens (which are useful, in any event, only for experimental purposes). Antibodies are specific for particular three-dimensional structures and can be used to probe the conformation of unknown adducts on DNA. Immunologic assays are rapid, highly reproducible, and relatively inexpensive. Their high sensitivity and capability for detecting nonradioactive adducts would therefore support their use in the monitoring of human tissues. In addition, immunologic techniques can be applied together with morphologic procedures (electron microscopy and immunofluorescence) to localize adducts in particular cells, subcellular compartments, or DNA molecules [e.g., intracellular localization of benzo[a]pyrene diol-epoxide (44) and acetylaminofluorene (20)].

Preliminary studies are in progress in which these assays are being used to determine the presence and levels of benzo[a]pyrene adducts in the white blood cells and tissues of smokers and individuals with suspected substantial exposure to this compound, such as roofers, shale oil workers, and lung cancer patients (31,40). Benzo[a]pyrene adducts have been detected in tissue and peripheral blood samples in these people, and the amounts of adduct measured were highly variable among individuals. These results indicate that the methodology has potential value for epidemiologic studies.

Postlabeling methods to detect and characterize carcinogen-DNA adducts have been described by Randerath et al. (34), Gupta et al. (10), and Haseltine et al. (11). The general experimental strategy and procedures involved in the method can be summarized as follows. Carcinogen-adducted DNA is subjected to enzymatic hydrolysis under conditions which, when carried to completion, produce a mixture of normal and adducted nucleotides, with the phosphate localized on the 3' position of deoxyribose. These nucleotides are then subjected to phosphorylation through the action of polynucleotide kinase, with γ ^{32}P-ATP used as the source of ^{32}P. This substrate can be obtained in extremely high specific activity, so that nucleotides are radiolabeled in the 5'-deoxyribose position also at a high specific activity. Unmodified nucleotides are removed by TLC or HPLC, and the mixture of adducted nucleotides is resolved on two-dimensional TLC and subjected to autoradiography. The presence of adducts and quantitative estimation of their levels is achieved by densitometry of the autoradiograms or by liquid scintillation counting of the chromatograms.

Gupta et al. (10) applied this method to studies of DNA modified *in vitro* by the bulky aromatic carcinogens 2-aminofluorene, acetylaminofluorene, and benzo[a]pyrene. The method had sufficient sensitivity to detect adducts at a frequency of one in 10^7-10^8 DNA bases. More recently, Reddy et al. (35) applied a modified version of the method to analysis of DNA adducts formed in vivo in livers and skin of mice and rats after exposure to 28 carcinogens, including seven arylamines and derivatives; three azocompounds; two nitroaromatics; 12 polycyclic aromatic hydrocarbons; and four methylating agents. The procedures described enabled the detection of one aromatic DNA adduct in about 10^8 nucleotides, while the limit of detection for methylated adducts was one adduct in about 6 x 10^5 nucleotides. This approach has great potential as a method for application to human population studies, but much further development will be required. For example, in its present form, the method is sufficiently sensitive for analysis only of DNA modified by bulky, aromatic carcinogens; new hydrolysis and separation techniques will be required to enable its application for detecting adducts in alkylated DNA. Furthermore, identification of individual adducts will be impossible until reference standards with known properties are available for each adduct. Since most carcinogens form complex mixtures of DNA adducts, this development will require much additional research effort.

URINARY EXCRETION OF ADDUCTS

Exploitation of the detection of DNA adducts in urine for dosimetry has been undertaken in two experimental models.

Bennett et al. (4) found that rats dosed with aflatoxin B_1 excreted in their urine a large fraction (about 35%) of the major N^7-guanine adduct of the carcinogen more than 48 h after a single injection. The adduct was isolated from urine by the combined use of preparative and analytical HPLC and was quantified by absorbance at 365 nm. The method allowed reproducible, quantitative measurement of adduct in urine from rats treated with doses as low as 0.125 mg/kg body weight. Application of the method in rats treated with different doses of carcinogen showed that the amount of adduct excreted bore a constant relationship to peak adduct levels in the livers of animals treated with the same doses. Thus, one condition of adequacy as a dosimeter was met, viz, quantitative reflection of adduct levels in target tissue.

However, the limit of sensitivity of the method was still inadequate to detect ambient exposure levels of aflatoxins known to occur in human populations, and further methodologic development was required. Donahue et al. (6) report improvements that approach that objective. These improvements consist of modifications in the chromatography, but more important, improved sensitivity by substituting for the absorbance measurement of the earlier method, a radiolabeling of the adduct with ^{3}H-dimethylsulfate and ultimately determination of the radioactive product, ^{3}H-9-methylguanine.

Experiments with similar objectives were carried out with the carcinogen dimethylnitrosamine (DMN) by Hemminki and Vainio (14) and

Hemminki (13). Rats were injected with ^{14}-C-dimethylnitrosamine and urine was collected over the succeeding five days. Radioactivity was extracted and separated by Sephadex G-10 chromatography, and the main DMN-derived adducts were tentatively identified. These included N-acetyl-S-methylcysteine, 1-methylhistamine, S-methylcysteine, and methionine, allantoin, and 7-methylguanine. Although dose-response experiments were not carried out, these results illustrate the potential applicability of the approach to human monitoring if suitable detection methods for the adducts can be devised. They also illustrate the complexity of the adduct mixtures formed by alkylating agents such as DMN.

Recently, Gombar et al. (8) studied the excretion of 7-methylguanine in urine of rats following the oral administration of DMN over a wide range of doses. They found that the excretion of 7-methylguanine was linearly related to the dose of DMN given and further established that the excretion rate of 7-methylguanine provided an accurate index of the endogenous formation of DMN when sodium nitrate and aminopyrine were administered simultaneously. These results suggest the possible utility of this assay for monitoring endogenous nitrosation in man.

In the only published study on excretion of carcinogen-DNA adducts in man, Autrup et al. (1) report the detection of the putative aflatoxin B_1-N^7-guanine adduct in the urine of people living in an area of Kenya in which contamination of foods with aflatoxin B_1 is known to occur frequently. Urine samples were analyzed by HPLC according to the method of Bennett et al. (4), and the putative adduct was detected by photon-counting fluorescence spectroscopy. Evidence for the presence of the adduct was found in six of 81 samples analyzed. These findings demonstrate the validity of the approach and point up the necessity for further development of methods suitable for routine application. It is clear that the complexities of this field necessitate extensive further research in method development and careful validation in animal models before interpretable data can be obtained from studies in human populations. However, the potential usefulness of the information to be gained justifies the additional research effort, as discussed in several recent reviews (18,30,45).

REFERENCES

1. Autrup, H., Bradley, K.A., Shamsuddin, A.K.M., Wakhisi, J., and Wasunna, A. (1983): Carcinogenesis, 4:1193-1195.
2. Baselt, R.C. (1980): Biological Monitoring Methods for Industrial Chemicals. Biomedical Publications, Davis, California.
3. Becroft, D.M.O., and Webster, D.R. (1972): Br. Med. J., 4:117.
4. Bennett, R.A., Essigmann, J.M., and Wogan, G.N. (1981): Cancer Res., 41:650-654.

5. Campbell, T.C., Caedo, J.P., Jr., Bulatao-Jayme, J., Salamat, L., and Engel, R.W. (1970): Nature, 227:403-404.
6. Donahue, P.R., Essigmann, J.M., and Wogan, G.N. (1982): In: Banbury Report 13: Indicators of Genotoxic Exposure, edited by B.A. Bridges, B.E. Butterworth, and I.B. Weinstein, pp. 221-229. Cold Spring Harbor Laboratories, Cold Spring Harbor, New York.
7. Dvorackova, I., Stora, C., and Ayuraud, N. (1981): J. Cancer Res. Clin. Oncol., 100:221-224.
8. Gombar, C.T., Zubroff, J., Strahan, G.D., and Magee, P.N. (1983): Cancer Res., 43:5077-5080.
9. Groopman, J.D., Haugen, A., Goodrich, G.R., Wogan, G.N., and Harris, C.C. (1982): Cancer Res., 42:3120-3124.
10. Gupta, R.C., Reddy, M.V., and Randerath, K. (1982): Carcinogenesis, 3:1081-1092.
11. Haseltine, W.A., Franklin, W., and Lippke, J.A. (1983): Environ. Health Perspect., 48:29-41.
12. Hayes, W.J., Jr., (1975): In: Toxicology of Pesticides. The Williams and Wilkins Company, Baltimore.
13. Hemminki, K. (1982): Chem.-Biol. Interact., 39:139-148.
14. Hemminki, K., and Vainio, H. (1982): In: Advances in Experimental Medicine and Biology, edited by R. Snyder, D.V. Parke, J.J. Kocsis, D.J. Jallow, C.G. Gibson, and C.M. Whitmer, pp. 1149-1156. Plenum Press, New York.
15. Hendrickse, R.G., Coulter, J.B.S., Lamplugh, S.M., MacFarlane, S.B.J., Williams, T.E., Omer, M.I.A., and Suliman, G.I. (1982): Br. Med. J., 285:843-846.
16. Hertzog, P.J., Smith, J.R., and Garner, R.C. (1982): Carcinogenesis, 3:825-828.
17. Hsu, I.-C., Poirier, M.C., Yuspa, S.H., Grunberger, D., Weinstein, I.B., Yolken, R.H., and Harris, C.C. (1981): Cancer Res., 41:1091-1095.
18. IARC/IPCS Working Group Report, (1982): Cancer Res., 42:5236-5239.
19. Johnson, H.J., Jr., Cernosek, S.F., Jr., Gutierrez-Cernosek, R.M., and Brown, L.L. (1980): J. Anal. Toxicol., 4:86-90.
20. Lang, M.C., de Murcia, D., Mazen, A., Fuchs, R.P.P., Leng, M., and Daune, M. (1982): Chem.-Biol. Interact., 41:83-93.
21. Lamplugh, S.M. (1983): J. Chromatogr., 273:442-448.
22. Lauwerys, R.R. (1983): In: Industrial Chemical Exposure: Guidelines for Biological Monitoring. Biomedical Publications, California.
23. Linch, A.L. (1974): Biological Monitoring for Industrial Chemical Exposure Control. CRC Press, Inc., Boca Raton, Florida.
24. Lovelace, C.E.A., Njapau, H., Salter, L.F., and Bayley, A.C. (1982): J. Chromatogr., 227:256-261.
25. Lutz, W.K. (1979): Mutat. Res., 65:289-356.

26. Matsumoto, H., Takata, R.H., and Ishizaki, H. (1981): J. Chromatogr., 211:403-408.
27. Muller, R., and Rajewsky, M.F. (1981): J. Cancer Res. Clin. Oncol., 102:99-113.
28. Nelson, D.B., Kimbrough, R., Landrigan, P.S., Hayes, A.W., Yang, G.C., and Benanides, J. (1980): Pediatrics, 66:865-869.
29. Onyemelukwe, C.G., Nirodi, C., and West, C.E. (1980): Trop. Geogr. Med., 32:237-240.
30. Perera, F.P., and Weinstein, I.B. (1980): J. Chronic Diseases, 35:581-600.
31. Perera, F.P., Poirier, M.C., Yuspa, S.H., Nakayama, J., Jaretzki, A., Curnen, M.M., Knowles, D.M., and Weinstein, I.B. (1982): Carcinogenesis, 33:1405-1410.
32. Poirier, M.C. (1981): J. Nat. Cancer Inst., 67:515-519.
33. Pylypiw, H.M., Jr., and Harrington, G.W. (1981): Anal. Chem., 53:2365-2367.
34. Randerath, K., Reddy, M.V., and Gupta, R.C. (1981): Proc. Nat. Acad. Sci., USA, 78:6126-6129.
35. Reddy, M.V., Gupta, R.C., Randerath, E., and Randerath, K. (1984): Carcinogenesis, 5:231-243.
36. Rogan, W., and Gladen, B.C. (1983): Environ. Health Perspec., 48:87-91.
37. Rogan, W.J., Gladen, B.C., McKinney, J.D., and Albro, P.W. (1983): J. Am. Med. Assoc., 249:1057-1058.
38. Ryan, N.J., Hogan, G.R., Hayes, A.W., Unger, P.D., and Siraj, M.Y. (1979): Pediatrics, 64:71-75.
39. Savage, E.P., Keefe, T.J., Tessari, J.D., Wheeler, H.W., Applehans, F.M., Goes, E.A., and Ford, S.A. (1981): Am. J. Epidemiol., 113:413-422.
40. Shamsuddin, A.K.M., Sinopoli, N.T., Vahakangas, K., Hemminki, K., Boesch, R.R., and Harris, C.C. (1983): Fed. Proc., 42:1042.
41. Shank, R.C., Bourgeois, C.H., Keschamras, N., and Chandavimol, P. (1971): Fd. Cosmet. Toxicol., 9:501-507.
42. Siraj, M.Y., Hayes, A.W., Unger, P.D., Hogan, G.R., Ryan, N.J., and Wray, B.B. (1981): Toxicol. Appl. Pharmacol., 58:422-430.
43. Sizaret, P., Malaveille, C., Montesano, R., and Frayssinet, C. (1982): J. Nat. Cancer Inst., 69:1375-1381.
44. Slor, H., Mizusawa, H., Neihart, N., Kakefuda, T., Day III, R.S., and Bustin, M.(1981): Cancer Res., 41:3111-3117.
45. Weinstein, I.B. (1983): Annual Reviews of Public Health, 4:409-413.
46. Wild, C.P., Smart, G., Saffhill, R., and Boyle, J.M. (1983): Carcinogenesis, 4:1605-1609.

Carcinogenesis, Vol. 10, edited by E. Huberman and S. H. Barr. Raven Press, New York © 1985.

Cell Culture Studies on the Mechanism of Action of Chemical Carcinogens and Tumor Promoters

I. B. Weinstein

Division of Environmental Sciences and Cancer Center, Institute of Cancer Research, Columbia University, New York 10032

Perhaps the greatest tribute to a scientist is to follow in his footsteps. Over the past twenty years, scientists in our laboratory have been fortunate to be able to follow in the footsteps of Charles Heidelberger. He was truly one of the great cancer researchers of the twentieth century. His brilliance, enthusiasm, and critical mind inspired several major advances in carcinogenesis research and cancer chemotherapy. In this paper I will review some of our recent findings on the mechanism of action of carcinogens and tumor promoters. It is sad that Charles Heidelberger is not present to provide his insightful and critical comments on these findings.

CELLULAR RESPONSES TO CARCINOGEN-INDUCED DNA DAMAGE

Charles Heidelberger was one of the first investigators to study the interaction of chemical carcinogens with cellular macromolecules (7,10). It is now well established that a variety of chemical carcinogens yield highly reactive intermediates that bind covalently to cellular DNA (31). This finding and others have led to the concept that these carcinogens act by producing mutations in somatic cells. We would caution, however, that carcinogenesis probably involves much more complex changes in DNA structure than simple point mutations at sites of carcinogen-induced DNA damage. Indeed, studies by C. Heidelberger and colleagues with chemical carcinogens in the C3H 10T1/2 cell line (4), and studies by A. Kennedy et al. (14) with radiation in the same cell line, indicate that the frequency of initiation is much higher than that of random point mutations. Furthermore, recent studies on activated oncogenes in tumors reveal not only point mutations but also more complex changes including oncogene amplification, sequence deletions and insertions, and chromosome translocations (for reviews, see 2,30).

We have found that the carcinogen benzo[a]pyrene (BAP) and its activated derivative benzo[a]pyrene 7,8 dihydrodiol 9,10-epoxide (BPDE) induce a marked increase in the asynchronous replication of polyoma DNA in transformed rat cells containing integrated polyoma virus DNA (17,18). Furthermore, this effect does not require direct carcinogen damage to the polyoma DNA since we can induce viral DNA replication by fusing normal cells previously exposed to BPDE to the polyoma-virus-transformed cells (17). In recent studies in which we used recombinant DNA constructs that contained either the bacterial drug resistance gene gpt or the mammalian dihydrofolate reductase gene dhfr linked to the polyoma DNA, we found that when cells carrying these constructs were exposed to BPDE, the gpt or dhfr genes also underwent asynchronous replication (18). These findings, and other evidence (17,18,20), suggest that carcinogen-induced damage to cellular DNA can induce the formation of a trans acting factor that can induce the asynchronous replication and amplification of specific genes (Fig. 1). This phenomenon may be relevant to the finding of amplified oncogenes, amplified genes related to drug resistance, and other amplified DNA sequences in tumors. It may also be relevant to the synergistic interactions seen between certain DNA viruses and environmental chemicals in cell transformation and cancer causation (31).

MECHANISMS OF ACTION OF TUMOR PROMOTERS

Membrane Effects and Activation of Protein Kinase C

Our studies on the mechanism of action of the phorbol ester tumor promoters were also stimulated by findings from Heidelberger's laboratory (21) indicating that these compounds enhanced the in vitro transformation of C3H 10T1/2 cells previously exposed to 3-methylcholanthrene. These findings suggested that one could use tissue culture systems to explore the cellular and molecular mechanisms of action of tumor promoters, thus greatly simplifying mechanistic studies. Eventually, studies from several laboratories led to the concept that, in contrast to the action of initiating carcinogens that function by damaging cellular DNA, the primary effects of the potent tumor promoter 12-0-tetradecanoyl-phorbol-13-acetate (TPA) and related compounds are due to changes in membrane structure and function (9,31,33). Very recent studies indicate that these effects may be mediated by the ability of TPA to bind to and enhance the activity of the phospholipid-dependent enzyme protein kinase C (PKC) (3,22).

We have recently studied the effects of various types of tumor promoters on the activity of PKC in vitro (1,24,32). The enzyme was partially purified from either rat or bovine brain and displayed a high dependence on added phospholipid and Ca^{++}. A striking finding is that

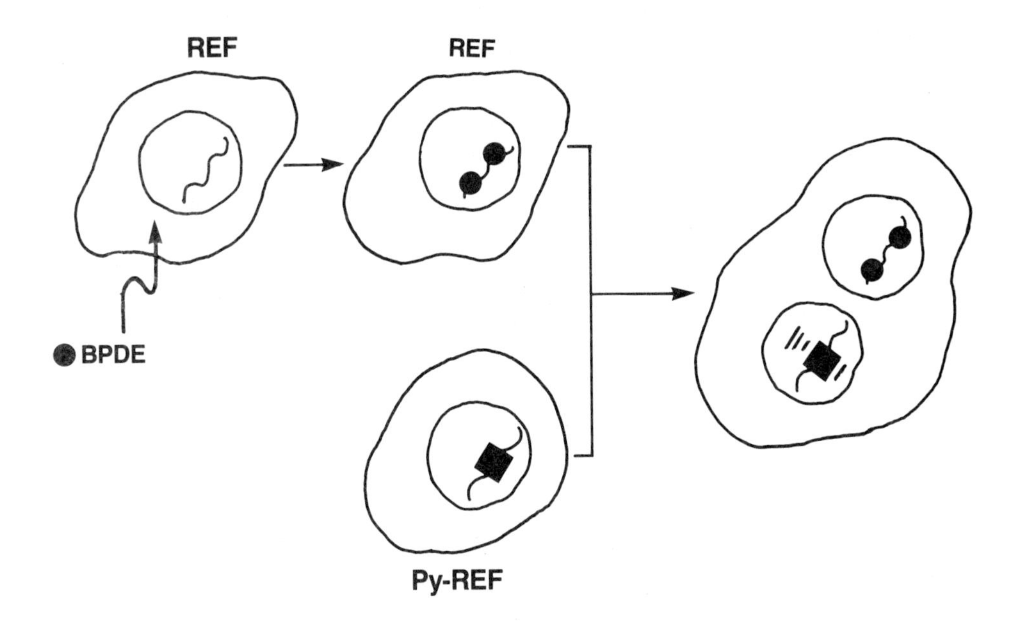

FIG. 1. Normal rat embryo fibroblasts (REF) were exposed to benzo[a]-pyrene trans-7,8-dihydrodiol-9, 10-epoxide (anti) (BPDE) for 3 h and then washed extensively with phosphate buffered saline. Rat embryo fibroblasts containing integrated polyoma DNA (Py-REF) that had not been exposed to BPDE were then cocultivated and fused with the BPDE-treated REF cells with polyethylene glycol (PEG). Extrachromosomal DNA extracted from the heterokaryons 48 h after cell fusion and analyzed by Southern blot hybridization showed a marked increase in extra chromosomal polyoma DNA, when compared with fused untreated controls (see 17). These results indicate that BPDE-induced DNA damage need not be on the same genome as the integrated polyoma DNA to enhance asynchronous replication of the viral DNA. Thus, they provide evidence for a "trans" acting factor.

maximum stimulation (>10-fold) of PKC activity by TPA occurs in the presence of phospholipid but in the absence of added Ca^{++}. In effect, nanomolar concentrations of TPA substitute for millimolar concentrations of added Ca^{++}, and the two agents are not synergistic. Biologically active analogs of TPA such as phorbol dibutyrate (PDBu), 12-0-hexadecanoyl-16-hydroxyphorbol-13-acetate (HHPA), and mezerein were also effective activators of PKC, as were the chemically unrelated tumor promoters teleocidin and aplysiatoxin, when tested at nanomolar concentrations in the absence of added Ca^{++}. On the other hand, the biologically inactive compounds phorbol and 4α-phorbol-12,13-didecanoate (4αPDD) did not affect PKC activity in the absence or presence of Ca^{++}. These and additional results are consistent with a stereochemical model (Fig. 2) in which the structurally similar hydrophilic domains of certain diterpenes, teleocidin, and aplysiatoxin interact specifically with a protein receptor (in this case PKC apoenzyme), while their less-specific hydrophobic domains interact with phospholipid, thus forming an enzymatically active ternary complex. In intact cells, these tumor promoters might bind first to lipid domains in cell membranes thus inducing changes in lipid structure that enhance binding to and activation of PKC. The ability of tumor promoters to substitute for added Ca^{++} in the activation of PKC may also be of significance in terms of their action in intact cells. There is previous evidence that TPA lowers the Ca^{++} requirement for the growth of certain cell types in cell culture (31).

Several findings suggest that there may be heterogeneity (or subclasses) of receptors for phorbol esters and related tumor promoters (29,32). Receptor heterogeneity could contribute to the tissue specificity and pleiotropic effects of these compounds. The basis for this heterogeneity is not known but it could involve the following mechanisms: (i) heterogeneity of PKCs; (ii) variations in lipid domains associated with PKC; and (iii) interactions of tumor promoters with other lipid-modulated enzymes, in addition to protein kinase C. Further studies are required to determine whether other chemical classes of tumor promoters also activate PKC or other protein kinases, perhaps via indirect mechanisms.

A fundamental area for future research is the identification of specific cellular proteins phosphorylated by PKC, particularly those that are critical to the process of tumor promotion. We have recently found that the synthetic nonapeptide Arg-Arg-Lys-Ala-Ser-Gly-Pro-Pro-Val is an *in vitro* substrate for PKC in the presence of phospholipid plus either Ca^{++} or TPA (24). We are currently examining other synthetic peptides to further define the amino acid sequence specificity of PKC substrates. This approach should provide clues to the types of cellular proteins that are substrates for PKC and, therefore, mediate tumor promotion. This approach might also lead to the design of polypeptide inhibitors of PKC that might be effective inhibitors of tumor promotion (24).

Synergy between an Activated Oncogene and Tumor Promoters

DNA transfection studies in which the NIH 3T3 cells are used as recipients have revealed activated oncogenes in a number of human tumors and tumor cell lines (2,30). This approach does not in itself, however, indicate the types of interactions that might occur between environmental and endogenous factors in the *de novo* transformation of normal

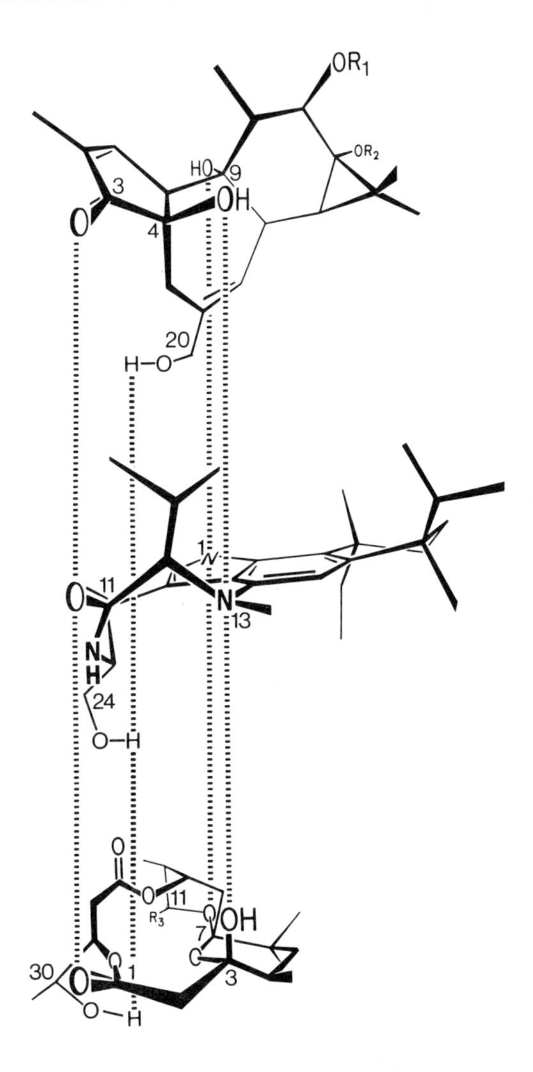

FIG. 2. Perspective drawings of TPA (top), dihydroteleocidin B (middle) and aplysiatoxin (bottom). The dotted lines connect heteroatoms whose spatial positions might correspond with one another and represent residues that could interact with protein kinase C (PKC) apoenzyme. The hydrophobic R_1 residue on TPA (myristate), the hydrophobic ring system on the right side of dihydroteleocidin B and the phenolic side chain of aplysiatoxin might interact with the phospholipid cofactor for PKC. The stereochemistries of dihydroteleocidin B and aplysiatoxin were chosen arbitrarily to maximize their similarity to TPA. Further studies are required to establish their actual stereochemistries. For additional details see ref. 32,33.

cells. We are intrigued with the possibility that during the multistage carcinogenic process tumor promoters might interact synergistically with cellular oncogenes, since promoters can induce mimicry of transformation, modulate differentiation, and enhance the transformation of cells previously exposed to chemical carcinogens, radiation, or certain DNA viruses (13,21,31). It was of interest, therefore, to see if tumor promoters would enhance the transformation of cell cultures transfected with an activated oncogene. For these studies, we used the mutated human c-*ras*^{H} oncogene, designated T24, cloned from a human bladder cancer cell line. The latter gene seemed particularly appropriate for such studies because its DNA sequence is well defined, and it is extremely effective in transforming NIH 3T3 cells (6,19,30). On the other hand, when used alone, it has weak or negligible activity in transforming primary cell cultures, suggesting to us that its activity might be enhanced by tumor promoters or growth factors. We used the C3H 10T1/2 cell line as recipients since cells from this line have a more uniform fibroblastic morphology and a lower saturation density than NIH-3T3 cells and thus appear to more closely resemble normal cells. Although they are aneupoid, the C3H 10T1/2 cells have a low incidence of spontaneous transformation and are not tumorigenic in nude mice (27). They are particularly well suited for studies on the action of phorbol ester tumor promoters since they contain an abundance of high-affinity receptors for these and related compounds and display striking changes in cell morphology and membrane-related properties in response to these agents (29).

We have indeed found that the potent tumor promoters TPA and teleocidin markedly enhance the transformation of C3H 10T1/2 mouse fibroblasts when these cells are transfected with T24 DNA (12). Transfection studies with the drug resistance marker *gpt*, time course studies, and Southern blot analyses of the extent of integration of the T24 DNA indicate that this enhancement is not simply an effect on the process of DNA transfection (12). These findings, together with parallel studies with NIH 3T3 and rat fibroblasts, also indicate that the "competence" of animal cells for DNA transfection is a function of the recipient cell line, the transfected marker, and the growth environment (12). Our results suggest that during two-stage carcinogenesis the initiating carcinogen might function by activating cellular proto-oncogenes, whereas tumor promoters might induce the expression of these genes, induce the outgrowth of the altered cells, and/or enhance the expression of other cellular genes that complement the function of the activated oncogenes.

ABERRATIONS IN GENE TRANSCRIPTION IN TUMOR CELLS

The complexity of multistage carcinogenesis predicts that the evolution and maintenance of malignant cancer cells involves changes in multiple types of cellular genes (31,33). Indeed, as emphasized earlier in this paper, there is accumulating evidence that tumors display a wide variety of alterations in the state of integration and/or expression of several cellular oncogenes.

We discovered that murine cells transformed by chemical carcinogens or radiation express constitutively a series of poly A^+ RNAs that contain murine leukemia virus LTR (long terminal repeat)-like sequences (15). Carcinogen-transformed murine cells also express RNAs homologous to endogenous VL30 (11) and intracisternal (IAP) (16) sequences, both of which are moderately repetitive retrovirus-like sequences present in the mouse genome. These findings have led us to suggest that carcinogenesis involves a disturbance in mechanisms controlling the function of multiple classes of cellular DNA promoter and enhancer sequences (15,33). In terms of targets for the action of chemical carcinogens it is worth noting that, although there are about 18 known cellular oncogenes, the normal mouse genome contains more than 1200 copies of various types of LTR-sequences per haploid genome. The structural resemblence of LTRs to known transposable elements (28), the ability of LTRs to enhance the transcription of a variety of genes (28), and the phenomena of viral LTR-promoter-insertion (8,25) and of insertion-mutation by endogenous intra-cisternal A-particle sequences (5,16,26), indicate the capacity of LTRs to act as mobile elements. Thus these elements, and other transcriptional enhancer sequences yet to be identified, constitute a major repertoire for potential aberrations in the control of gene expression during the course of chemical carcinogenesis. The switching on of transcription of retrovirus-like sequences may, in part, be related to alterations in the state of methylation of cytidine residues in DNA, which are often associated with carcinogen action and tumor formation (23). Furthermore, since endogenous retrovirus-like sequences can also code for reverse transcriptase, once their expression is switched on the RNA transcripts might undergo reverse transcription and the resulting DNA copies might then be reinserted into foreign sites in the genome, thus producing insertion mutations and further abnormalities in gene expression (Fig. 3). Thus, an epigenetic event (i.e., activation of transcription of endogenous retrovirus-like elements) could lead to a stable genetic event. We are currently examining whether this postulated mechanism plays a role in tumor promotion or progression.

SUMMARY

The evolution of a fully malignant tumor is a multistep process resulting from the action of multiple factors, both environmental and endogenous, and involves alterations in the function of multiple cellular genes. Chemical carcinogens that initiate this process appear to do so by damaging cellular DNA. In addition to producing simple point mutations, this damage appears to induce the synthesis of a trans acting factor that

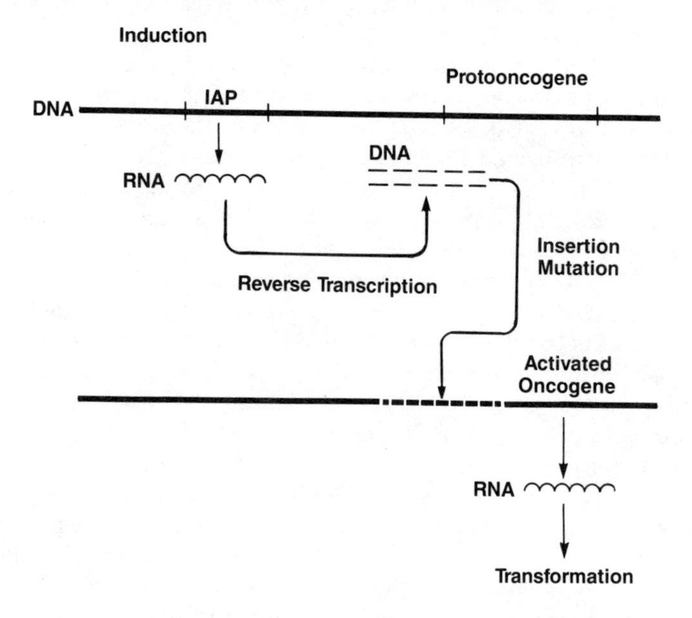

FIG. 3. Hypothetical mechanism by which transient activation of transcription of an endogenous intracisternal A particle sequence (IAP), perhaps in response to exposure to environmental chemicals, could lead to transposition of this sequence and thus a stable genetic change. Since the IAP sequence contains coding information for reverse transcriptase, the IAP transcript could be reverse transcribed yielding a free DNA copy that might insert into the genome of the same cell. If this insertion mutation occurs in the region of a cellular oncogene, it could activate its transcription, thus enhancing cell transformation. This type of mechanism could explain activation of the oncogene c-mos in chemically induced murine plasmacytomas (5).

can induce asynchronous DNA replication. This response may result in gene amplification and/or gene rearrangement. This phenomenon may also play a role in synergistic interactions between chemicals and viruses in the causation of certain cancers. The primary target of the tumor promoters TPA, teleocidin, and aplysiatoxin appears to be the cell membrane. All three of these agents act, at least in part, by enhancing the activity of the phospholipid-dependent enzyme protein kinase C. We have proposed a stereochemical model to explain the interaction of these amphiphilic compounds with the PKC system. We have found that TPA and teleocidin markedly enhance the transformation of C3H 10T1/2 mouse fibroblasts when these cells are transfected with the cloned H-ras human bladder cancer oncogene. Thus, tumor promoters can act synergistically with an activated oncogene to enhance cell transformation. Furthermore, carcinogen-transformed rodent cells display aberrations in the expression of various endogenous retrovirus-related sequences. Activation of some of these sequences may lead to insertion mutations and further aberrations in gene expression. Thus, multistage carcinogenesis may involve both changes in cellular oncogenes and aberrations in the function of DNA sequences that control gene transcription.

ACKNOWLEDGMENT

This research was supported by Public Health Service Grants CA 021111 and CA 26056, U.S. Department of Health and Human Services; funding from the National Foundation for Cancer Research; and funds from the Dupont Company and the Alma Toorock Memorial for Cancer Research. The author is indebted to the following colleagues for their valuable contributions to the studies described in this paper: Wen-Luan Hsiao, Michael Lambert, John Arcoleo, Catherine O'Brian, Sebastiano Gattoni-Celli, Alan M. Jeffrey, and Paul Kirschmeier.

REFERENCES

1. Arcoleo, J., and Weinstein, I.B. (1984): Proc. Amer. Assoc. Cancer Res., 25:142 (abstract).
2. Bishop, J.M. (1983): Cell, 32:1018-1020.
3. Castagna, M., Takai, U., Kaibuchi, K., Sano, K., Kikkawa, U., and Nishizuka, Y. (1982): J. Biol. Chem., 257:7847-7851.
4. Fernandez, A., Mondal, J., and Heidelberger, C. (1980): Proc. Nat. Acad. Sci., USA, 77:7272-7276.
5. Gattoni-Celli, S., Hsiao, W.-L., and Weinstein, I.B. (1983): Nature, 306:795.
6. Goldfarb, M., Shimizu, K., Perucho, M., and Wigler, M. (1982): Nature, 296:404-409.
7. Goshman, L.M., and Heidelberger, C. (1967): Cancer Res., 27:1678-1688.
8. Hayward, W.S., Neel, B.G., and Astrin, S.M. (1981): Nature, 290:475-480.
9. Hecker, E., Fusenig, N.E., Kunz, W., Marks, F., and Thielmann, H.W., editors (1982): Cocarcinogenesis and Biological Effects of Tumor Promoters. Raven Press, New York.
10. Heidelberger, C., and Moldenhauer, M.G. (1956): Cancer Res., 16:442-449.

11. Hodgson, C.P., Elder, P.K., Ono, T., Foster, D.N., and Gertz, M.J. (1983): Mol. and Cell. Biol., 3:2221-2234.
12. Hsiao, W.-L., Gattoni-Celli, S., and Weinstein, I.B. (1984): Science 226:552-555.
13. Kennedy, A.R., Mondal, S., Heidelberger, C., and Little, J.B. (1978): Cancer Res., 38:439-443.
14. Kennedy, A.R., Fox, M., Murphy, G., and Little, J.B. (1980): Proc. Nat. Acad. Sci., USA, 77:7262-7266.
15. Kirschmeier, P., Gattoni-Celli, S., Dina, D., and Weinstein, I.B. (1982): Proc. Nat. Acad. Sci., USA, 79:273-277.
16. Kuff, E.L., Feenstra, A., Lueders, K., Smith, L., Harvey, R., Hozumi, N., and Shulman, M. (1983): Proc. Nat. Acad. Sci., USA, 80:1992-1996.
17. Lambert, M.E., Gattoni-Celli, S., Kirschmeier, P., and Weinstein, I.B. (1983): Carcinogenesis, 4:587-594.
18. Lambert, M., Pelligrini, S., Gattoni-Celli, S., and Weinstein, I.B. (1984): Cold Spring Harbor Meeting on DNA Tumor Viruses, Cold Spring Harbor, New York (abstract).
19. Land, H., Parada, L.F., and Weinberg, R.A. (1983): Nature, 304:596-602.
20. Lavi, S., and Etkin, S. (1981): Carcinogenesis, 2:417-423.
21. Mondal, S., Brankow, D.W., and Heidelberger, C. (1976): Cancer Res., 2254-2260.
22. Nishizuka, Y. (1984): Nature, 304:648-651.
23. Nyce, J., Weinhouse, S., and Magee, P.N. (1983): Br. J. Cancer, 48:463-475.
24. O'Brian, C.A., Lawrence, D.S., Kaiser, E.T., and Weinstein, I.B. (1984): Biochem. Biophys. Res. Commun., 124:269-302.
25. Payne, G.S., Bishop, J.M., and Varmus, H.E. (1982): Nature, 295:209-214.
26. Rechavi, G., Givol, D., and Canaani, E. (1982): Nature, 300:607-611.
27. Reznikoff, C.A., Brankow, D.W., and Heidelberger, C. (1973): Cancer Res., 33:3231-3238.
28. Temin, H.M. (1982): Cell, 28:3-5.
29. Tran, P.L., Castagna, M., Sala, M., Vassent, G., Horowitz, A.D., Schachter, D., and Weinstein, I.B. (1983): Eur. J. Biochem., 130:155-160.
30. Weinberg, R.A. (1983): Sci. Am., 249:126-142.
31. Weinstein, I.B. (1981): J. Supramol. Struct. Cell. Biochem., 17:99-120.
32. Weinstein, I.B., Arcoleo, J., Backer, J., Jeffrey, A.M., Hsiao, W., Gattoni-Celli, S., and Kirschmeier, P. (1984): In: Cellular Interactions of Environmental Tumor Promoters, edited by H. Fujiki, E. Hecker, R.E. Moore, T. Sugimura, and I.B. Weinstein. Japan Scientific Society Press, Tokyo.

33. Weinstein, I.B., Gattoni-Celli, S., Kirschmeier, P., Lambert, M., Hsiao, W., Backer, J., and Jeffrey, A. (1984): In: Cancer Cells 1/The Transformed Phenotype, edited by A. Levine, G. Vande Woude, J.D. Watson, and W.C. Topp, pp. 229-237. Cold Spring Harbor Laboratory, Cold Spring Harbor, New York.

Carcinogenesis, Vol. 10, edited by E. Huberman
and S. H. Barr. Raven Press, New York © 1985.

Mechanisms Involved in Multistage Skin Tumorigenesis

T. J. Slaga

The University of Texas System Cancer Center, Science Park—Research Division, Smithville, Texas 78957

The mouse skin tumorigenesis model has contributed greatly to our understanding that chemical carcinogenesis is a multistage process. It is now well known that skin tumors in mice can be induced by the sequential application of a subthreshold dose of a carcinogen (initiation stage) followed by repetitive treatment with a weak or noncarcinogenic promoter (promotion stage). In the past several years, the multistage nature of chemical carcinogenesis has also been demonstrated in a number of other tissues such as the liver, bladder, respiratory system, colon, esophagus, stomach, mammary gland, and pancreas (20).

This report is not intended to be an exhaustive review on skin carcinogenesis; it will instead summarize some of the critical cellular and biochemical events of multistage skin tumorigenesis (initiation-promotion-progression).

CRITICAL TARGETS AND EVENTS IN SKIN TUMOR INITIATION AND PROMOTION

The tumor initiation stage in mouse skin appears to be an irreversible stage that probably involves a somatic mutation in some aspect of epidermal differentiation (20). Extensive data has revealed a good correlation between the carcinogenicity of many chemical carcinogens and their mutagenic activities (24). Most tumor-initiating agents either generate or are metabolically converted to electrophilic reactants, which bind covalently to cellular DNA and other macromolecules (12). Previous studies have demonstrated a good correlation between the skin tumor initiating activities of several polycyclic aromatic hydrocarbons and their abilities to bind covalently to DNA (18).

We have recently found that skin tumor initiation probably occurs in dark basal keratinocytes since a good correlation exists between the degree of tumor initiation and the number of dark basal keratinocytes present in the skin (T.J. Slaga and A.J.P. Klein-Szanto, unpublished results). The dark basal keratinocytes are present in the skin in large numbers during embryogenesis, in moderate numbers in newborns, in low numbers in young adults, and in very low numbers in old adults--an observation that suggests these cells may be epidermal stem cells (10,26). The initiating potential of mouse skin decreases with the age of the mouse to the point that it is very difficult to initiate skin from mice older than one year when few dark basal keratinocytes are present.

In a series of experiments, Goerttler and coworkers (6,7), transplacentally initiated fetal epidermis with low doses of 7,12-dimethylbenz[a]-anthracene (DMBA) at different days of gestation. The sensitivity to initiation, which was demonstrated postnatally by 12-O-tetradecanoyl-phorbol-13-acetate (TPA) topical treatments, varied markedly according to the prenatal day of initiation. Our recent results with SENCAR mice (Slaga et al., unpublished results) are similar to those reported by Goerttler. In addition, we found that the pattern of sensitivity to initiation was extremely similar to the pattern of dark cell distribution reported by Klein-Szanto and Slaga (10), i.e., gradual increases from day 12 of gestation, a maximum number at day 19 of gestation, and a sudden drop thereafter. Van Duuren et al. (28), using a two-stage carcinogenesis protocol (DMBA + TPA), showed that age at promotion plays a critical role in tumor production; i.e., they observed a general decrease in tumor incidence with increasing age at time of promotion. These coincidences point to the fact that dark cells, being less differentiated keratinocytes, are more susceptible to the action of carcinogens and promoters; therefore, the number of dark cells available at initiation and/or promotion could be a critical factor for efficient tumor induction. In this context, it is interesting to note that although carcinogenic doses of DMBA induce the appearance of a large number of dark cells, initiating doses (subcarcinogenic doses) do not alter the percentage of dark basal cells in the epidermis. These data suggest that the number of dark basal cells (stem cells) is correlated with the degree of tumor initiation.

The major effect of tumor promoters appears to be the specific expansion of the initiated cell population (dark basal keratinocytes) in a target tissue such as the skin. This effect appears to occur by both direct and indirect mechanisms. The skin tumor promoters appear to have a direct effect on the initiated cells possibly by mimicking some endogenous embryonic growth factor, which leads to a decreased differentiation capacity of the initiated cells (dark cells). An inhibition of cell-cell communication and stimulation of differentiation of noninitiated cells appear to be important indirect mechanisms of further expanding the initiated cell population (19).

A number of investigators using in vivo and cell culture systems have suggested that altered differentiation plays a critical role in tumor promotion and carcinogenesis in general. Tumor promoters have been found to transiently induce in epidermal and other cells a set of phenotypic changes that resemble those found in embryonic and malignant cells. Raick

(16) found that tumor promoting agents induced in basal cells certain morphologic changes resembling those found in embryonic, papilloma, and carcinoma cells. Klein-Szanto et al. (9) found that tumor promoters increased the number of dark basal keratinocytes, which were normally found in large quantities in embryonic skin, as well as in papillomas and carcinomas. Theoretically, modulation of the commitment to differentiation or differentiation potential of subpopulations of keratinocytes could result in the accumulation of subpopulations of initiated cells. Reiners and Slaga (17) found that tumor promoters induce a subpopulation of basal cells to commit to terminal differentiation and accelerate the rate of differentiation of committed cells. Yuspa et al. (30) also found that tumor promoters can induce subpopulations of basal cells in culture to differentiate. This mechanism could be important in the expansion of the initiated cell population. In this regard, Yuspa and Morgan (31) found that initiated epidermal cells in culture do not differentiate under a physiological stimulus to differentiate (high calcium). We have also recently found that a small subpopulation of very dense basal keratinocytes (which may be dark cells) isolated from initiated skin have a decreased capacity to differentiate in culture when high calcium is present (Morris, Fischer, and Slaga, unpublished results).

Skin tumor promoters bring about a number of other important epigenetic changes and possibly some genetic effects (21). Of the observed phorbol-ester-related effects on the skin, the induction of epidermal cell proliferation and the increases in the amounts of ornithine decarboxylase (ODC), polyamines, prostaglandins, and number of dark basal keratinocytes have the best correlation with promoting activity (11). In addition to the increase in the number of dark cells, which are normally present in large numbers in embryonic skin, there are many other embryonic conditions that appear in adult skin after treatment with tumor promoters (21).

It is difficult to determine which of the many effects associated with phorbol ester tumor promotion are in fact essential components of the promotion process. A good correlation appears to exist between promotion and epidermal hyperplasia induced by phorbol esters (27). However, other agents that induce epidermal cell proliferation do not necessarily promote carcinogenesis (18). Nevertheless, it should be emphasized that all known skin tumor promoters do induce epidermal hyperplasia (21). O'Brien, et al. (14) have reported an excellent correlation between the tumor promoting ability of various compounds (phorbol esters, as well as nonphorbol ester compounds) and their ability to induce ODC activity in mouse skin. However, mezerein, a diterpene similar to TPA but with weak promoting activity, was found to induce ODC activity to levels that were comparable to those induced by TPA (13). Raick found that phorbol ester tumor promoters induced the appearance of "dark basal cells" in the epidermis, whereas ethylphenylpropiolate (EPP), a nonpromoting epidermal hyperplasic agent, did not (16). Wounding caused a few dark cells to appear, which seemed to correlate with its ability to be a weak promoter (16). In addition, a large number of these dark cells are found in papillomas and carcinomas (9,16). Klein-Szanto, et al. (9) reported that TPA induced about three to five times the number of dark cells that were induced by mezerein--the first major difference found between these compounds.

Skin tumor promoters bring about a number of other important changes in the skin such as membrane and differentiation alterations, an increase in protease activity and phospholipid synthesis, and a decrease in the number of glucocorticoid receptors (21). This latter effect may be a critical event in tumor promotion because the glucocorticoids are potent inhibitors of skin tumor promotion, epidermal cell division, and promoter-induced epidermal hyperplasia and increases in the number of dark basal keratinocytes (3,4). A decrease in the number of glucocorticoid receptors in the epidermis could, therefore, lead to increased cell proliferation since an important negative feedback on epidermal cell division has been decreased. It should be emphasized that the glucocorticoid receptors appear to be totally absent from papillomas and carcinomas (4).

MULTISTAGE TUMOR PROMOTION

We have found that skin tumor promotion can be operationally and mechanistically further divided into at least two stages (23,24). Mezerein which is a weak or nonpromoting agent, nevertheless can induce many of the cellular events induced by TPA (13). We rationalized that TPA must be inducing additional cellular event(s) that mezerein could not effectively induce. If our supposition is true, then mezerein should show promoting activity in skin pretreated with TPA at a level too low to cause promotion by itself. This regimen was found to be an effective way of inducing tumors (23-25). Similar results were later found by Furstenberger and coworkers (5) using 12-retinoate-phorbol-13-acetate as a stage II promoter.

Table 1 summarizes the known characteristics of the first and second stages of promotion. Both stages of promotion show a good dose response relationship. It should be emphasized that only one application of a first-stage promoter is required and that this stage is irreversible for 4-6 weeks (24 and Slaga, unpublished results). Besides TPA and 12-deoxyphorbol-13-decanoate, which are known promoters, nonpromoters such as 4-O-methyl-TPA, calcium ionophore A23187, hydrogen peroxide, and wounding are effective stage I promoters (24).

Stage I of promotion can be inhibited by fluocinolone acetonide (FA) and tosylphenylalanine chloromethyketone (TPCK), which also counteracts the TPA-induced appearance of dark basal keratinocytes (25). Vitamin E also inhibits stage I of promotion, but its effect on dark cells is not

TABLE 1. Characteristics of the first and second stages of tumor promotion.

Stage I

1. A good dose-response exists for TPA and 12-deoxy-phorbol-13-decanoate as first stage promoters.
2. The nonpromoting agents, calcium ionophore A23187, 4-O-methyl-TPA, hydrogen peroxide, and wounding can act as stage I promoters
3. Only one application of a first-stage promoter is necessary.
4. Prostaglandins are important since PGE_2 can enhance stage I promotion by TPA.
5. Stage I is partially irreversible:
   a. Four to six weeks can separate first and second stages of promotion without a decrease in tumor response.
   b. There is an 80% decrease in tumor response if 10 weeks separate stage I and II of promotion.
6. An increase in the number of dark basal keratinocytes (stem cells) is important. This increase occurs directly because existing dark cells are stimulated to divide, thus converting some basal cells to dark cells and increasing the differentiation of some basal cells and differentiated cells.
7. Decrease in glucocorticoid receptors by TPA may help in proliferation of dark cells.
8. Stage I can be inhibited by FA and TPCK.

Stage II

1. A good dose-response exists for mezerein as a second stage promoter
2. The nonpromoting agent 12-deoxyphorbol-13-2,4,6-decatrienoate can act as a stage II promoter.
3. Multiple applications are required.
4. Stage II promotion is reversible for a relatively long period but later becomes irreversible.
5. Polyamines are important since putrescine can enhance stage II promotion by mezerein.
6. Most of the biochemical events shown to be important in promotion occur in this stage.
7. Mezerein can maintain dark cell proliferation and decrease in glucocorticoid receptors but cannot induce these events by itself.
8. Stage II promotion can be inhibited by FA, RA, DFMO, BHA, Vitamins E and C, and CPA.

BHA - butylated hydroxyanisole; CPA - cyproterone acetate; DFMO - difluoromethylornithine; FA - fluocinolone acetonide; RA - retinoic acid; TPA - 12-O-tetradecanoyl-phorbol-13-acetate; TPCK - tosylphenylalanine chloromethyketone.

not currently known (Slaga, unpublished results). Prostaglandin E_2 was also found to specifically enhance stage I of promotion. As discussed earlier, the dark cells appear to be the critical target cells of the tumor initiator and the first stage promoters stimulate these cells to divide rapidly by both a direct and indirect mechanism (Slaga and Klein-Szanto, unpublished results).

The second stage of tumor promotion is reversible at first but later becomes irreversible. A number of weak or nonpromoting agents, such as mezerein and 12-deoxyphorbol-13-2,4,6-decatrienoate, are effective second-stage promoters (24). Although mezerein alone cannot increase the number of dark cells or decrease the level of glucocorticoid receptors, it can effectively maintain these conditions after TPA treatment in a two-stage promotion protocol (4). Stage II of tumor promotion can be inhibited by FA, retinoic acid (RA), α-difluoromethylornithine (DFMO), butylated hydroxyanisole (BHA), cyproterone acetate (CPA), and vitamines E and C, which counteract either the mezerein-induced appearance of ODC, cell proliferation, and/or as yet unidentified events (24 and Slaga, unpublished results). Putrescine was found to specifically enhance Stage II of promotion (24). On the basis of these results with inhibitors, we conclude that putrescine and polyamines, epidermal cell proliferation, as well as some unidentified event(s), appear to be important events in Stage II of promotion.

Varshavsky (29) reported that phorbol ester tumor promoters were potent gene amplifying agents. Butler and Slaga (22 and unpublished results) have recently found that mezerein is much more effective than TPA in amplifying methotrexate resistance in 3T3 cells. Preliminary results also suggest that this amplification is a property of second-stage promoters and not of first-stage promoters. In addition, RA is a potent inhibitor of the mezerein effect on gene amplification.

CRITICAL EVENTS IN SKIN TUMOR PROGRESSION

A number of changes that occur very late in the carcinogenesis process are related to the conversion of benign to malignant tumors. We have found that all squamous cell carcinomas lack several differentiation product proteins such as high-molecular-weight (60,000-62,000) keratins and filaggrin but are positive for gamma glutamyltransferase (GGT). However, only about 20% of the papillomas generated by an initiation-promotion protocol exhibit similar conditions (11 and Mamrack and Slaga, unpublished results). Before visible tumors are observed in animals treated with an initiation-promotion protocol, these changes are not observed, suggesting that they are very late responses (11 and Slaga, et al., unpublished). Hennings and coworkers (8) recently reported that if mice with papillomas are treated repetitively with N-methyl-N-nitrosoguanidine (MNNG), a significant increase in the conversion of papillomas to carcinomas occurs. We have also found similar results with limited (either one or four applications) treatment of MNNG, as well as with ethylnitrosourea (ENU) and benzoyl peroxide (O'Connel, Fischer, and Slaga, unpublished results). This type of treatment (initiation-promotion-initiation) produces a carcinoma response similar to complete carcinogenesis, i.e., the repetitive application of a carcinogen such as DMBA or MNNG probably supplies both initiating and promoting influences continuously. The reason why a different type of promoter like benzoyl peroxide can increase the conversion of papillomas to carcinomas is currently not known.

Balmain and coworkers (1,2) have recently found that the levels of HA-ras transcripts in a percentage of papillomas and carcinomas in epidermis treated with DMBA-TPA were higher than those in nontreated skin. Furthermore, the tumor DNA was capable of malignantly transforming NIH 3T3 cells in DNA transfection studies (1). Studies in our laboratory (15 and unpublished results) indicate that initiation alone or repetitive TPA treatments do not turn on the expression of the Ha-ras oncogene in adult SENCAR mouse epidermis. Initiation followed by either one or six weeks of TPA treatment also failed to activate Ha-ras expression. Like Balmain, we observed elevated levels of Ha-ras RNA in a percentage of papillomas and carcinomas tested. At present, it still remains to be determined whether oncogene activation plays a critical role in multistage skin tumorigenesis.

Table 2 summarizes some of this information about papillomas and carcinomas. As an overall summary, Fig. 1 presents the various stages in skin carcinogenesis (initiation-promotion-progression), as well as some critical events.

TABLE 2. Characteristics of skin tumors.

Benign Papillomas

1. A large number of dark cells are present.
2. Glucocorticoid receptors are lost.
3. High levels of polyamines and prostaglandins are found.
4. Approximately 80% of the papillomas have high molecular weight keratins and filaggrin and are negative for GGT; 20% have reverse conditions.
5. Approximately 50% of papillomas express Ha-*ras* RNA.
6. Some papillomas are reversible, while others are irreversible.
7. Treatment of papillomas with MNNG, ENU, and benzoyl peroxide increases the conversion of papillomas to carcinomas.

Carcinomas

1. A large number of dark cells are present.
2. All lack glucocorticoid receptors.
3. High level of polyamines and prostaglandins are found.
4. All lack high molecular weight keratins and filaggrin.
5. All are positive for GGT.
6. Approximately 67% of carcinomas express Ha-*ras* RNA.
7. MNNG treatment of mice with carcinomas increases their metastatic potential.

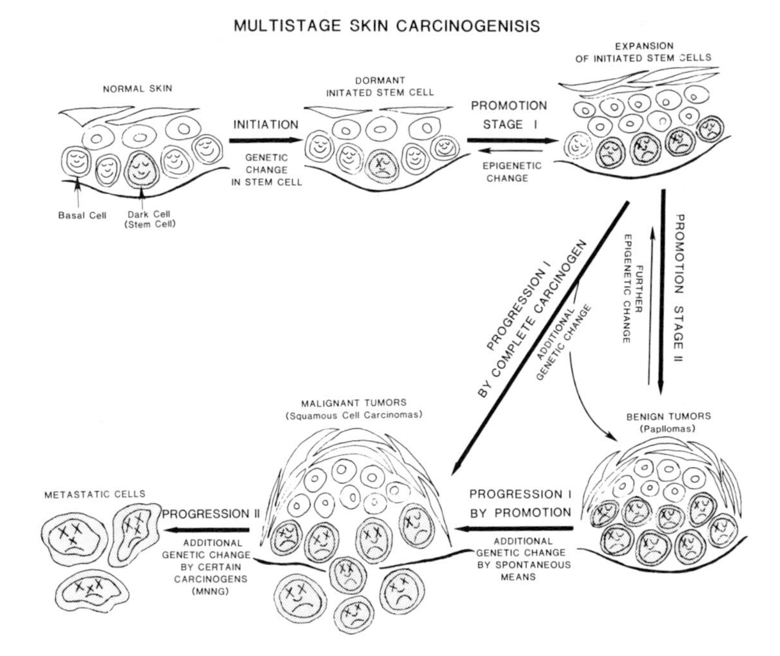

FIG. 1. A schematic of the various stages of skin carcinogenesis.

ACKNOWLEDGMENTS

The research was supported by Public Health Service Grants CA-34890, CA-34962, and CA-34521 from the National Cancer Institute, U.S. Department of Health and Human Services. I gratefully acknowledge all my past and present technicians, students, postdoctoral fellows, and collaborators who have contributed to these studies.

REFERENCES

1. Balmain, A., and Pregnell, I.D. (1983): Nature, 303:72-74.
2. Balmain, A., Ramsden, M., Bowden, G.T., and Smith, J. (1984): Nature, 307:658-660.
3. Davidson, K.A., and Slaga, T.J. (1983): J. Invest. Dermatol., 79:378-383.
4. Davidson, K.A., and Slaga, T.J. (1983): Cancer Res., 43:3847-3851.
5. Furstenburger, G., Berry, D.L., Sorg, B., and Marks, F. (1981): Proc. Nat. Acad. Sci., USA, 78:772.
6. Goerttler, K., Loehrke, H. (1976): Virchows Arch. A: Pathol. Anat. Histol., 372:29-38.

7. Goerttler, K., Loehrke, H., Schweizer, J., Hesse, B., (1980): J. Cancer Res. Clin. Oncol., 98:267-276.
8. Hennings, H., Shores, R., Wenk, M.L., Spangler, E.F., Tarone, R., and Yuspa, S.H. (1983): Nature, 304:67-69.
9. Klein-Szanto, A.J.P., Major, S.M., and Slaga, T.J. (1980): Carcinogenesis, 1:399-406.
10. Klein-Szanto, A.J.P., and Slaga, T.J. (1981): Cancer Res., 41:4437.
11. Klein-Szanto, A.J.P., Nelson, R.G., Shah, Y., and Slaga, T.J. (1983): J. Nat. Cancer Inst., 70:161-168.
12. Miller, E.C., and Miller, J.A. (1976): In: Chemical Carcinogens: The Metabolism of Chemical Carcinogens to Reactive Electrophiles and their Possible Mechanism of Action in Carcinogenesis, edited by C.E. Searle, pp. 737-762. American Chemical Society, Washington, DC.
13. Mufson, R.A., Fischer, S.M., Verma, A.K., Gleason, G.L., Slaga, T.J., and Boutwell, R.K. (1979): Cancer Res., 39:4791-4795.
14. O'Brien, T.G., Simsiman, R.C., and Boutwell, R.K. (1975): Cancer Res., 35:1662-1670.
15. Pelling, J.C., Hixson, D.C., Nairn, R.S., and Slaga, T.J. (1984): Proc. Am. Assoc. Cancer Res., Abstract 307, Toronto, Canada.
16. Raick, A.N. (1974): Cancer Res., 34:920-926.
17. Reiners, J.J., and Slaga, T.J. (1983): Cell, 32:247-255.
18. Slaga, T.J. (1980): In: Cancer, Etiology, Mechanisms and Prevention - A Summary, edited by T.J. Slaga, pp. 243-262. Raven Press, New York.
19. Slaga, T.J. (1983): Cancer Surveys, 2:595-612.
20. Slaga, T.J. (1983): Environ. Health Perspect., 50:3-14.
21. Slaga, T.J., editor (1984): Mechanisms of Tumor Promotion. Volumes 1-4, CRC Press, Inc., Boca Raton.
22. Slaga, T.J., and Butler, A.P. (1984): In: Cellular Interactions by Environmental Tumor Promoters, edited by H. Fujiki, E. Hecker, R.E. Moore, T. Sugimura, and I.B. Weinstein, pp. 291-301. Japan Scientific Society Press, Tokyo, Japan.
23. Slaga, T.J., Fischer, S.M., Nelson, K., and Gleason, G.L. (1980): Proc. Nat. Acad. Sci., USA, 77:3659-3663.
24. Slaga, T.J., Fischer, S.M., Weeks, C.E., Klein-Szanto, A.J.P., and Reiners, J. (1982): J. Cell. Biochem., 18:99-119.
25. Slaga, T.J., Klein-Szanto, A.J.P., Fischer, S.M., Weeks, C.E., Nelson K., and Major, S. (1980): Proc. Nat. Acad. Sci., USA, 77:2251-2254.
26. Slaga, T.J., and Klein-Szanto, A.J.P. (1983): Cancer Invest., 1:425-436.
27. Slaga, T.J., Scribner, J.D., Thompson, S., and Viaje, A. (1974): J. Nat. Cancer Inst., 52:1611-1618

28. Van Duuren, B.L., Sivak, A., Katz, C., Seidman, J., Melchionne, S., (1975): Cancer Res., 35:502-505.
29. Varshavsky, A. (1981): Cell, 25:561-572.
30. Yuspa, S.H., Hennings, H., Kulsease-Martin, M., and Lichti, U. (1982): In: Cocarcinogenesis and Biological Effects of Tumor Promoters, edited by E. Hecker, N.E. Fusenig, W. Kunz, F. Marks, and H.W. Thielmann, pp. 217-230. Raven Press, New York.
31. Yuspa, S.H., and Morgan, D.L. (1981): Nature, 293:72-74.

Carcinogenesis, Vol. 10, edited by E. Huberman
and S. H. Barr. Raven Press, New York © 1985.

Cellular and Molecular Changes During Chemical Carcinogenesis in Mouse Skin Cells

S. H. Yuspa

Laboratory of Cellular Carcinogenesis and Tumor Promotion, National Cancer Institute, Bethesda, Maryland 20205

The formation of tumors on mouse skin following topical or systemic administration of chemical agents has been the major experimental model from which the biological principles of chemical carcinogenesis have been derived (24,25). Studies with this model system first delineated operationally and mechanistically distinct stages in tumor development, termed initiation and promotion. These stages have more recently been confirmed with other models for epithelial cancers. Epidemiologic studies have implied that similar multistage events are important in neoplastic development in humans. A careful evaluation of the biology of tumorigenesis in the skin model is essential for an understanding of the mechanisms involved and for an improved understanding of the pathogenesis of cancer in humans.

The principal lesion that develops in an initiation-promotion protocol in skin is the papilloma, a premalignant tumor, which may regress. Carcinomas of skin are infrequent. They occur late in two-stage carcinogenesis and almost invariably arise in a preexisting papilloma. In contrast, repeated application of initiating agents to induce tumors on mouse skin leads to the appearance of carcinomas and papillomas in close temporal proximity and a greater frequency of malignant lesions.

Recent studies in our laboratory demonstrate the importance of recognizing the distinction between tumorigenesis associated with initiation followed by promotion and carcinogenesis associated with repeated carcinogen administration (6). Mice treated by a standard two-stage protocol, consisting of a single initiating dose of 7,12-dimethylbenzanthracene followed by 10 weeks of promotion with 12-O-tetradecanoyl phorbol-13-acetate (TPA) developed multiple papillomas as the only neoplastic lesions. Groups of papilloma-bearing mice were then continuously exposed to the TPA promoter or treated only with vehicle for an additional 40 weeks. Other groups (control and papilloma-bearing mice) received either topical N-methyl-N'-nitro-N-nitrosoguanidine, topical

4-nitroquinoline-n-oxide, or systemic urethane once per week for the same 40-week period. The conversion rate from papilloma to carcinoma was then monitored for the entire experimental period. In groups receiving either TPA or vehicle, the malignant conversion rate was low and the values were identical for both. Thus, the promoter did not enhance conversion from benign to malignant lesions. All three genotoxic carcinogens administered to papilloma-bearing mice markedly enhanced malignant conversion and also accelerated carcinoma development by as much as 12 weeks. Control groups without papillomas (no initiation or no promotion) developed very few carcinomas in response to the genotoxic agents. Identical results were achieved with Charles River CD-1 or SENCAR mice. These results indicate that malignant conversion is due to further genetic change in papilloma cells. The ineffectiveness of TPA as a converting agent points to a mechanism of action different from that of mutagenic carcinogens.

The foregoing data confirmed a number of reports that indicated that the papilloma is the relevant lesion for understanding initiation and promotion in skin. The morphology of an epidermal papilloma resembles that of normal epidermis in that there are multiple layers of cells in various stages of differentiation. There is an orderly progression of differentiation, but each stratum is generally represented in greater abundance than in normal skin. The profile of keratin proteins extracted from papillomas is very similar to the profile of proteins from normal skin, again reflecting the minimal nature of the deviation in these benign lesions (15,23). Some information on the biological alteration in papillomas can be obtained from autoradiographs of these lesions removed from animals injected with ^{3}H-thymidine (Fig. 1). The labeling index of papillomas is about 10-fold higher than that of normal skin. Most striking is the presence of proliferating cells in layers far removed from the basement membrane, a finding not observed in normal epidermis or epidermis in a nonneoplastic hyperplastic state. The results suggest that a major biological change in the benign lesion induced by initiators and promoters is a cellular alteration that uncouples the obligatory cessation of proliferation from the migration off of the basement membrane as epidermal cells commit to terminal differentiation. Thus, an alteration in the program of terminal differentiation is a fundamental change in tumor development and is a premalignant characteristic. It is this biology that has guided our cell culture studies designed to elucidate the cellular and molecular basis for initiation and promotion.

CELLULAR CHANGES ASSOCIATED WITH INITIATION

The discovery that extracellular ionic calcium is a key regulator of epidermal growth and differentiation (5) has facilitated experimental approaches to study initiation. In medium with reduced calcium concen-

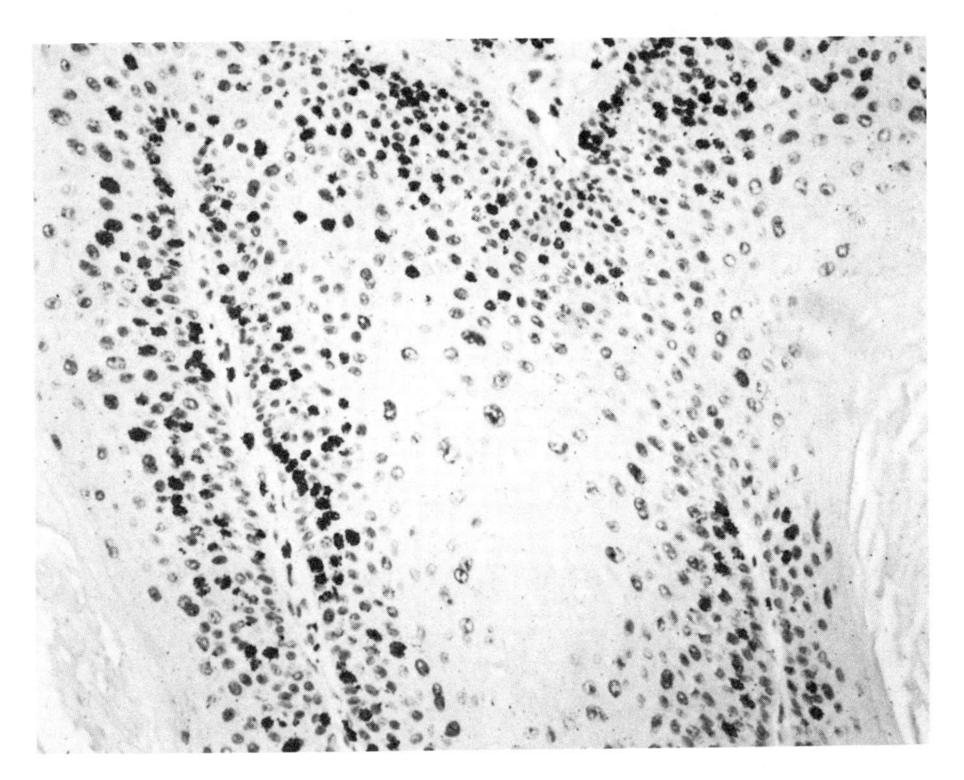

FIG. 1. ^{3}H-Thymidine autoradiography of mouse skin papilloma. Female HA/ICR mice bearing papillomas resulting from initiation-promotion protocols received an injection of 5 μCi of ^{3}H-thymidine and were sacrificed 30 minutes later. Papillomas were excised and processed for autoradiography as described by Burns et al. (3). Slide provided by Dr. Fred Burns, New York University.

trations (0.02-0.09 mM), epidermal basal cells are selectively cultivated. These cells have morphologic, cell kinetic, and marker protein characteristics of basal cells and grow as a monolayer with a high proliferation rate. When the calcium concentration of the culture medium is elevated to levels found in most commercial culture preparations (1.2-1.4 mM), proliferation ceases and terminal differentiation rapidly ensues, with squamous differentiation and sloughing of cells occurring by 72-96 h. The responses of cultured basal cells to increases in extracellular calcium resemble the changes observed in basal cells in vivo as they migrate away from the basement membrane and commit to terminal differentiation. In cell culture, therefore, in analogy to the in vivo situation, it might be expected that initiation would alter the basal cell response to calcium-induced terminal differentiation.

A number of experimental results from this laboratory have supported that conclusion. Basal cells exposed to carcinogens in vitro and subsequently induced to differentiate by calcium form foci that resist terminal differentiation (11). The association of this change with initiation is supported by additional findings: (i) There is a positive dose-response relationship for carcinogen exposure and focus number (9,11,14). (ii) Spontaneous foci develop in cell cultures from SENCAR mice, which are sensitive to papilloma development by promotion alone (8,31). (iii) Identical foci develop in vitro when epidermal cells obtained from initiated mice are cultivated in low calcium medium and challenged by high calcium medium (8,31). (iv) The focus-forming potential of initiated skin persists for at least 10 weeks after initiation (9). (v) Stronger initiators and higher doses of initiators yield more focus-producing cells after topical administration to skin in vivo or after exposure of epidermal cells in vitro (8,9). (vi) The number of foci directly correlates with the extent of DNA binding for several initiators after in vitro exposure (14). (vii) When a variety of initiators of different potency and from different chemical classes are tested in the focus assay, a positive correlation is observed between their

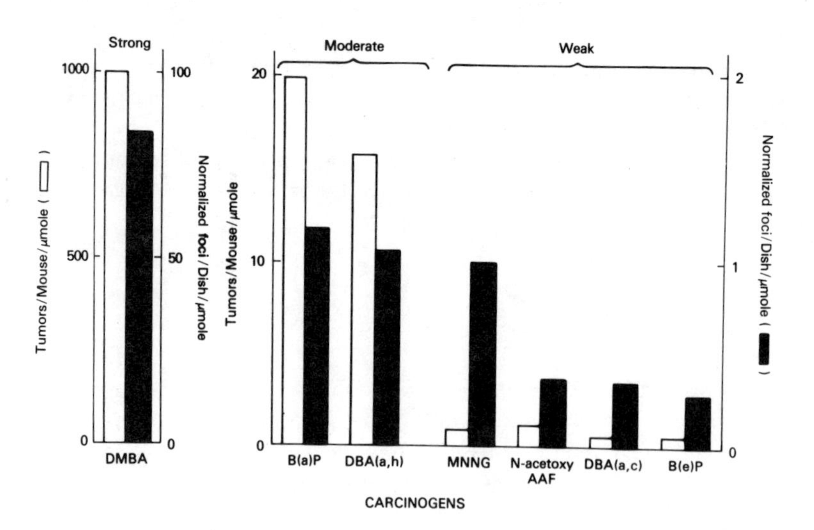

FIG. 2. Correlation of potency as an initiator in vivo with potency in induction of Ca^{++}-resistant foci in vitro. In vivo potency values were derived from (19). In vitro potency was derived from multiple assays with at least two concentrations of each agent. For experimental details see reference 11.

potency as initiators in vivo and their induction of foci in vitro (9, Fig. 2). This correlation is impressive since the mutagenic potency for each agent in prokaryotic test systems is not closely correlated with initiating ability, as discussed by Scribner and Suss (19). Thus, cellular foci resistant to terminal differentiation signals are analogous in many ways to initiated skin and papillomas.

We have tried to characterize cells isolated from these assays in several ways. Twenty-six isolated foci removed directly from primary culture and implanted into nude mice have failed to produce tumors after more than 6 months in vivo (9). Tumorigenic keratinocytes (28) removed from culture and implanted in an identical fashion produced tumors within one month. Further, tumorigenic keratinocytes show an identical resistance to Ca^{++}-induced differentiation in vitro (28). While further time will be required for final analysis, it appears that foci are not composed of highly malignant cells. We have also subcultured foci, and cells obtained in this way are capable of forming cell lines that grow and subculture repeatedly in low or high calcium medium (10). Some of these have progressed to malignancy, but no definitive cellular markers distinguish malignant from nonmalignant lines (10). These results suggest that resistance to Ca^{++}-induced terminal differentiation is an early consequence of carcinogen exposure and is a preneoplastic phenotype.

MOLECULAR CHANGES ASSOCIATED WITH INITIATION

The foregoing studies in mouse epidermal cell culture are consistent with the definition of initiation as a change that allows the altered cell and its progeny to proliferate under conditions where normal cells are obligated to differentiate terminally. It is apparent from Fig. 1 that this change in proliferation is an essential change in the papilloma, which allows for tumor development. Thus, the putative initiated cells defined by their resistance to Ca^{++}-induced terminal differentiation express the papilloma phenotype. In several reports, we have proposed a mechanism by which tumor promoters are required to change tissue kinetics to allow the initiated cell to form a tumor by selective clonal expansion (29,30).

The ability to isolate cells in culture that express the initiated (and papilloma) phenotype provides an ideal opportunity to analyze the molecular changes involved. Recently Balmain et al. (1,2) have reported that H-ras expression is increased in epidermal papillomas and carcinomas, and more than 60% of papillomas contain an activated ras gene when tested by 3T3 transfection studies. These studies are consistent with the notion that an activated ras gene causes papilloma formation. We have isolated RNA from chemically induced epidermal papillomas and carcinomas as well as from Ca^{++}-resistant cell lines in culture and have analyzed these RNAs by slot blotting for increased relative content of messages for the cellular homologues of a number of viral oncogenes (21). When we compared these RNA messages with those from normal epidermis or from basal cells in culture, we found no enhanced expression of H-ras, Ki-ras, fos, myc, abl, raf, myb, or fes. These results indicate that increased expression of these proto-oncogenes is not required for tumor formation or for the Ca^{++} resistance phenotype. However, we have not ruled out the possibility that an altered form of the proto-oncogene is expressed in the phenotypes studied and that it could serve as an active oncogene.

To test this possibility, we have studied the effect of an activated ras oncogene by infecting epidermal cells with Harvey or Kirsten sarcoma virus and monitoring parameters previously used to characterize cells exposed to chemical initiators (32). Infection of basal cells in 0.02 mM Ca^{++} causes a marked stimulation of basal cell proliferation. Unlike cultures of carcinogen-treated cells, basal cell cultures infected with sarcoma virus markedly reduce their proliferation rate in response to 0.5 mM Ca^{++} medium. However, infected cultures persist in the higher Ca^{++} medium for long periods without sloughing, although their morphology changes to that found at a later stage of differentiation. Infected cultures express much higher levels of p21 protein than controls, and studies with temperature-sensitive mutants indicate that persistent p21 function is required for the observed effects of the virus. Expression of p21 remains high in 0.5 mM Ca^{++} indicating that the proliferative block is not related to reduced expression of p21. Virus-infected cells placed in 0.5 mM Ca^{++} medium also demonstrate a rise in the differentiation-associated enzymatic activity of epidermal transglutaminase. In epidermis, therefore, an activated ras gene is closely linked to epidermal proliferation and alters the program of terminal differentiation at some step subsequent to the proliferative block but prior to loss of cell viability. The inability of virus-infected cells to continue to proliferate under conditions favoring differentiation, unlike chemically initiated cells, which do continue proliferation, would seem inconsistent with an activated ras gene being sufficient to initiate skin carcinogenesis (1,2). To clarify this discrepancy, we studied further the properties of keratinocytes expressing an activated ras gene transduced by virus infection (Table 1).

To identify the differentiation state achieved by virus-infected cells, we assayed two epidermal marker proteins. Previous studies have shown that the pemphigoid antigen, a 220 kd protein found in the basement membrane, is synthesized exclusively by basal cells in low Ca^{++} and the pemphigus antigen, a 210 kd glycoprotein found on the cell surface, is exclusively synthesized by differentiating cells in high Ca^{++} (20). However, virus-infected cells did not stop synthesizing pemphigoid antigen nor begin synthesizing pemphigus antigen when switched to high Ca^{++} medium, suggesting that these cells maintain basal cell properties even when signaled to differentiate. Further evidence supporting an early block in differentiation after virus infection was obtained when cells maintained in high Ca^{++} medium were returned to low Ca^{++}. Normal cells are incapable of resuming basal cell growth (4), but the virus-

TABLE 1. Phenotypic alterations in normal, carcinogen-treated, or sarcoma-virus-infected keratinocytes.

Phenotype	Normal keratinocytes	Carcinogen treated[a]	Virus infected
Synthesis of pemphigoid antigen			
Low Ca^{++}	+	N.D.[b]	+
High Ca^{++}	-	N.D.	+
Synthesis of pemphigus antigen			
Low Ca^{++}	-	N.D.	-
High Ca^{++}	+	N.D.	-
Ability to grow in low Ca^{++} after exposure to high Ca^{++}	-	+	+
Inducibility of ODC[c] by TPA in high Ca^{++}	-	+	+
Modulation of transglutaminase activity by TPA[d]			
Low Ca^{++}	↑	↔	↑
High Ca^{++}	↔	↔	↓

[a]Carcinogen-treated cells designate putative initiated cell lines rendered Ca^{++} resistant as a result of carcinogen treatment (10).
[b]N.D. = not determined for carcinogen treated cells but known for malignant keratinocyte cell lines that express pemphigoid antigen in both low and high Ca^{++} and pemphigus antigen in high Ca^{++} (28).
[c]ODC = Ornithine decarboxylase activity as assayed in (13).
[d]Transglutaminase activity was measured as described in (27).

infected cells resumed their basal cell morphology. Of considerable interest was the response of virus-infected cells to phorbol esters. Normal basal cells are highly sensitive to the induction of ornithine decarboxylase by TPA but lose sensitivity once they are exposed to high Ca^{++} (13). However, virus-infected cells remain responsive to this action of TPA even under high Ca^{++} conditions. This result further supports the basal cell nature of the virus-induced block in keratinocyte maturation. Phorbol ester exposure may even reverse the maturation state of virus-infected cells in high Ca^{++} as suggested by the finding that elevated transglutaminase activity, associated with differentiation of both normal and virus-infected keratinocytes (7,32), can be reduced in the virus-infected (but not normal) cells by exposure to TPA.

This new information on the biological state of keratinocytes infected with murine sarcoma viruses provides a biological framework to explain how an activated ras gene could be sufficient to initiate epidermal tumorigenesis. In keratinocytes, an activated ras gene provides the proliferative signal required to form a tumor, but the cells are unable to maintain vertical growth once they have migrated above the basement membrane. Since virus-altered cells fail to complete their differentiation program, they might be expected to accumulate within the epidermis and expand to a limited extent in the space available. Exposure to tumor promoters would selectively reverse the quiescent growth state of those cells with an activated ras gene while normal committed basal cells would be unaffected or stimulated to differentiate further (26). Repeated promoter exposure, such as that required for tumor promotion, would maintain the permissive environment allowing reactivated cells to grow in the suprabasal regions of the epidermis, thus accomplishing a requirement for tumor formation.

Why do these cells differ from putative initiated cells produced by chemical carcinogen exposure in vitro and in vivo? It is well known from skin tumorigenesis experiments that at least two classes of papillomas develop in protocols involving initiation and promotion (3). Depending on the mouse strain, the commonest type of tumor requires repeated promoter exposures for both expression and maintenance. These promoter-dependent tumors regress upon withdrawal of the promoting stimulus (3). A second class of tumors, once formed, do not require promoter exposure for maintenance. These autonomous papillomas are much more likely to progress to carcinomas (3,18). The biological changes we have defined for keratinocytes expressing an activated ras gene are most consistent with ras activation being involved in the formation of promoter-dependent tumors. In contrast, the characteristics of chemically altered cells, selected by virtue of their constitutive ability to grow in high Ca^{++}, are most consistent with changes in autonomous papillomas. Whether the latter cell type requires more than one genetic change (of which ras activation could be a component) or an alteration in a different genetic element remains to be determined.

Both chemically altered and virus-altered cells fail to produce epidermal carcinomas in nude mice (unpublished data). Thus, ras activation alone is not sufficient to complete the cancer process. Our in vivo studies have shown that at least two genetic events are required to produce carcinomas in mouse skin (6). Other studies have suggested that several viral oncogenes must work in concert to transform primary cultures of murine cells (12,16,17). Weissman and Aaronson (22) have shown that murine sarcoma viruses can elicit the transformed phenotype in aneuploid mouse epidermal cells. Thus, the combination of an activated ras gene and aneuploidy may satisfy additional requirements for transformation.

The general conclusion to be reached from these data is that carcinogenesis in a terminally differentiating lining epithelium requires multiple changes in genetic information, most likely in genes controlling both proliferation and differentiation processes. Further, the effects of some of these changes, for example the activation of a ras gene, can be modulated by epigenetic factors such as tumor promoters.

REFERENCES

1. Balmain, A., and Pragnell, I.B. (1983): Nature, 303:72-74.
2. Balmain, A., Ramsden, M., Bowden, G.T., and Smith, J. (1984): Nature, 307:658-660.
3. Burns, F.J., Vanderlaan, M., Sivak, A., and Albert, R.E. (1976): Cancer Res., 36:1422-1427.
4. Hennings, H., and Holbrook, K. (1982): In: Ions, Cell Proliferation and Cancer, edited by A.L. Boynton, W.L. McKeehan, and J.F. Whitfield, pp. 499-516. Academic Press, Inc., New York.
5. Hennings, H., Michael, D., Cheng, C., Steinert, P., Holbrook, K., and Yuspa, S.H. (1980): Cell, 19:245-254.
6. Hennings, H., Shores, R., Wenk, M., Spangler, E.F., Tarone, R., and Yuspa, S.H. (1983): Nature, 304:67-69.
7. Hennings, H., Steinert, P., and Buxman, M.M. (1981): Biochem. Biophys. Res. Commun., 102:739-745.
8. Kawamura, H., Strickland, J.E., and Yuspa, S.H. (1984): Cancer Res., in press.
9. Kilkenny, A.E., Morgan, D., Spangler, E.F., and Yuspa, S.H. (1984): submitted.
10. Kulesz-Martin, M., Kilkenny, A.E., Holbrook, K.A., Digernes, V., and Yuspa, S.H. (1983): Carcinogenesis, 4:1367-1377.
11. Kulesz-Martin, M., Koehler, B., Hennings, H., and Yuspa, S.H. (1980): Carcinogenesis, 1:995-1006.
12. Land, H., Parada, L.F., and Weinberg, R.A. (1983): Nature, 304:596-602.
13. Lichti, U., Patterson, E., Hennings, H., and Yuspa, S.H. (1981): J. Cell Physiol., 107:261-270.
14. Nakayama, J., Yuspa, S.H., and Poirier, M.C. (1984): Cancer Res., 44:4087-4095.
15. Nelson, K.G., and Slaga, T. (1982): Cancer Res., 42:4176-4181.
16. Newbold, R.F., and Overell, R.W. (1983): Nature, 304:648-651.

17. Ruley, E.H. (1983): Nature, 304:602-606.
18. Scribner, J.D., Scribner, N.K., McKnight, B., and Mottet, N.K. (1983): Cancer Res., 43:2034-2041.
19. Scribner, J.D., and Suss, R. (1978): Int. Rev. Exp. Pathol., 18:137-198.
20. Stanley, J.R., and Yuspa, S.H. (1983): J. Cell Biol., 96:1809-1814.
21. Toftgard, R., Roop, D.R., and Yuspa, S.H. (1984): submitted.
22. Weissman, B.E., and Aaronson, S.A. (1983): Cell, 32:599-606.
23. Winter, H., Schweizer, J., and Goerttler, K. (1980): Carcinogenesis, 1:391-398.
24. Yuspa, S.H. (1981): Prog. Dermatol., 15:1-10.
25. Yuspa, S.H. (1982): Prog. Dermatol., 16:1-10.
26. Yuspa, S.H., Ben, T., and Hennings, H. (1983): Carcinogenesis, 4:1413-1418.
27. Yuspa, S.H., Ben, T., and Lichti, U. (1983): Cancer Res., 43:5707-5712.
28. Yuspa, S.H., Hawley-Nelson, P., Koehler, B., and Stanley, J.R. (1980): Cancer Res., 40:4694-4703.
29. Yuspa, S.H., Hennings, H., Kulesz-Martin, M., and Lichti, U. (1981): In: Cocarcinogenesis and Biological Effects of Tumor Promoters, edited by E. Hecker, N.E. Fusenig, W. Kunz, F. Marks, and H.W. Thielman, pp. 217-230. Raven Press, New York.
30. Yuspa, S.H., Hennings, H., and Lichti, U. (1981): J. Supramol. Struct. Cell. Biochem., 17:245-247.
31. Yuspa, S.H., and Morgan, D.L. (1981): Nature, 293:72-74.
32. Yuspa, S.H., Vass, W., and Scolnick, E. (1983): Cancer Res., 43:6021-6030.

Carcinogenesis, Vol. 10, edited by E. Huberman
and S. H. Barr. Raven Press, New York © 1985.

Mechanisms of Chemically Induced Multistep Neoplastic Transformation in C3H 10T½ Cells

J. R. Landolph

Norris Cancer Hospital and Research Institute, Comprehensive Cancer Center, University of Southern California School of Medicine, Los Angeles, California 90033

Chemical carcinogenesis both in whole animals (reviewed in 30) and in cultured cells (2,25, reviewed in 9) is a multistep process, divided broadly into initiation and promotion steps. Initiation is widely believed to be a mutational event. Many chemical carcinogens are either already electrophilic or are activated by various cytochrome P450 enzymes to chemically reactive electrophiles (reviewed in 8) that bind covalently to DNA (12,13) and, hence, are mutagenic in bacterial (1) and mammalian cell (10,24, 27,36) systems. Other chemical carcinogens may exert their neoplastic effects in part by damaging chromosomes (3) or by provoking chromosomal aneuploidy (34). Chromosome aberrations and aneuploidy are also broadly defined as mutations, i.e., a change in cellular DNA content, although they are more drastic changes than the usual base substitution and frameshift mutations.

In this manuscript, I detail our efforts in studying mechanisms of the multistep process of chemical transformation in cultured C3H 10T1/2 cells.

DEVELOPMENT OF AN ASSAY FOR CHEMICALLY INDUCED MUTATION TO OUABAIN RESISTANCE IN C3H 10T1/2 CELLS

Early work from the laboratory of Dr. Charles Heidelberger resulted in the establishment of an aneuploid, contact-inhibited, permanent mouse cell line, C3H 10T1/2 (29), and the demonstration that carcinogenic polycyclic aromatic hydrocarbons could induce neoplastic transformation in these cells (28). Our early efforts involved development of an assay measuring mutation to ouabain resistance (Oua^r) in the same C3H 10T1/2 cells so we could compare and contrast the processes of mutagenesis and transformation in response to a series of chemical carcinogens to look for similarities and differences between the two processes (reviewed in 21). We hoped that this process would lend insight into the mechanisms of chemical transformation.

We, therefore, defined the survival curve for wild-type C3H 10T1/2 cells in ouabain and showed that treatment of cells with 3 mM ouabain for 16 days reduced the survival fraction to 10^{-6} (Fig. 1, ref. 15). The chemical carcinogen and mutagen N-methyl-N'-nitro-N-nitrosoguanidine (MNNG) caused a dose-dependent induction of Oua^r colonies (Fig. 2, ref. 15). We showed that there was no metabolic cooperation in the assay conditions we defined (16) and determined the maximal expression time to be 2 days after a 2-day treatment with the carcinogens MNNG and N-acetoxy-acetylaminofluorene (NAcO-AAF)(15).

These Oua^r colonies were 10^5-fold more resistant to 3 mM ouabain than wild-type C3H 10T1/2 cells (Fig. 3, ref. 15 and Fig. 4, ref. 17), and

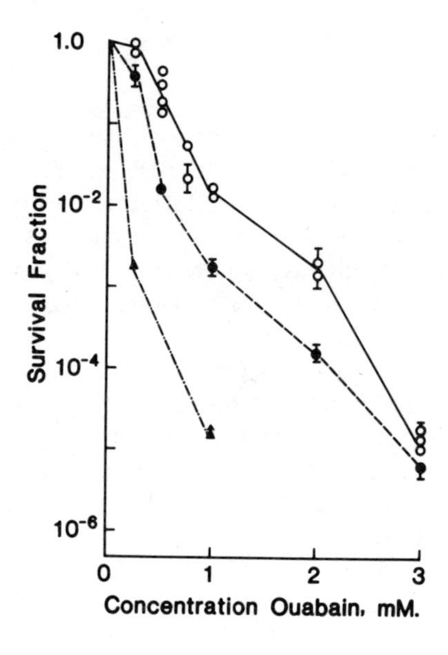

FIG. 1. Cytotoxicity of ouabain (Oua) to C3H 10T1/2 cells in the presence of various concentrations of potassium. Oua was added 1 day after plating, and the medium was changed on days 7 and 13. On day 16 the cells were fixed and stained, and the colonies were counted. O: medium containing 6.9 mM K^+;●: medium containing 3.7 mM K^+; and▲: medium containing 0.5 mM K^+. In this and all succeeding figures a symbol lacking error bars has a standard deviation less than the size of the symbol. A ^ over a symbol indicates that the survival fraction was less than or equal to the value indicated by the symbol.

Oua^r cell lines induced by MNNG, NAcO-AAF, and benzo[a]pyrene (BAP) retained their Oua^r stably when passaged in the absence of ouabain for prolonged periods of time (Table 1 and ref. 16). Oua^r cells were specifically Oua^r and not cross-resistant to nonrelated drugs like adriamycin and daunomycin (Fig. 4, ref. 17). Inhibition of K^+ transport is directly related to ouabain-induced cytotoxicity in C3H 10T1/2 cells (16). Chemically induced Oua^r C3H 10T1/2 cell lines have an ^{86}Rb uptake, which reflects (Na,K)ATPase-driven K^+ transport, which is Oua^r (16).

We also showed that MNNG, a known base substitution mutagen, induced Oua^r in C3H 10T1/2 cells (Fig. 2, ref. 15) and in Chinese hamster V79 cells (18). On the other hand, the frameshift mutagens, ICR 191 and ICR 170, induce 8-azaguanine resistance in V79 cells but do not induce Oua^r in either C3H 10T1/2 cells or in V79 cells (Fig. 5, ref. 18). The Oua^r mutation that results in viable Oua^r mutants is therefore probably a base substitution mutation. Frameshift mutations at this locus likely destroy vital functions of the protein involved in Oua^r (presumably a subunit of the Na, K adenosine triphosphatase) and result in nonviable Oua^r mutants.

In collaboration with my colleague R.E.K. Fournier at the University of Southern California, I showed that microcells containing discrete

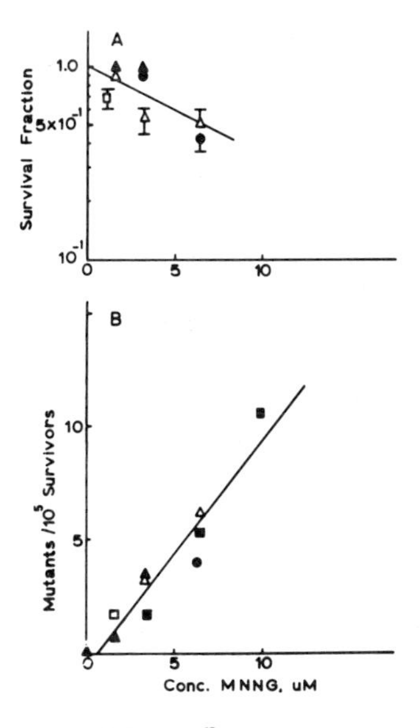

FIG. 2. (A) Cytotoxicity and (B) Oua^r mutation induction by MNNG in C3H 10T1/2 cells. Different symbols represent different experiments; an experiment is represented by the same symbol in both plots. Solid line in B is a linear least squares computer fit to the data; correlation coefficient, 0.87.

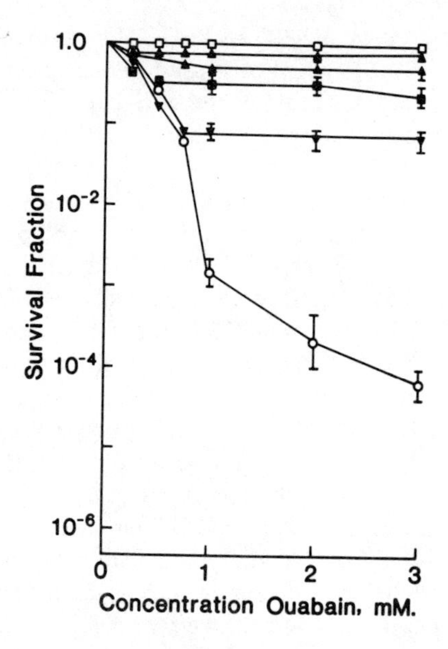

FIG. 3. Cytotoxicity to wild-type C3H 10T1/2 cells ( ) and various MNNG-induced Oua^r mutants (△,▲,■,▽,□) caused by a 9-day exposure to ouabain.

numbers of chromosomes could be prepared from chemically induced Oua^r cell lines and fused to human D98 or HT1080 cells and that Oua^r microcell hybrids could be selected (19). Most of the microcell hybrids in the HT1080 background contained only one intact mouse chromosome as shown with Hoescht fluorescent techniques. By sequential Giemsa-trypsin banding of murine chromosomes previously identified by Hoechst fluorescent staining techniques, we identified murine chromosome No. 3, and only murine chromosome No. 3, in a high percentage of metaphase spreads of Oua^r microcell hybrids examined (19). This identification proved that mouse chromosome No. 3 encodes the mutated locus responsible for conferring chemically induced Oua^r.

Hence, as a result of these studies, we have firm evidence that chemically induced Oua^r is due to a stable, specific, base substitution mutation at a locus affecting K^+ transport on murine chromosome No. 3.

We have another set of microcell hybrids in the D98 background that exhibit high-level Oua^r but do not contain cytologically detectable murine chromosomes. It is likely that in these microcell hybrids, the chromosome has broken, the centromere has been lost, and the Oua^r encoding locus has translocated to a human chromosome (19). To examine this question further, we have been conducting restriction enzyme-Southern gel

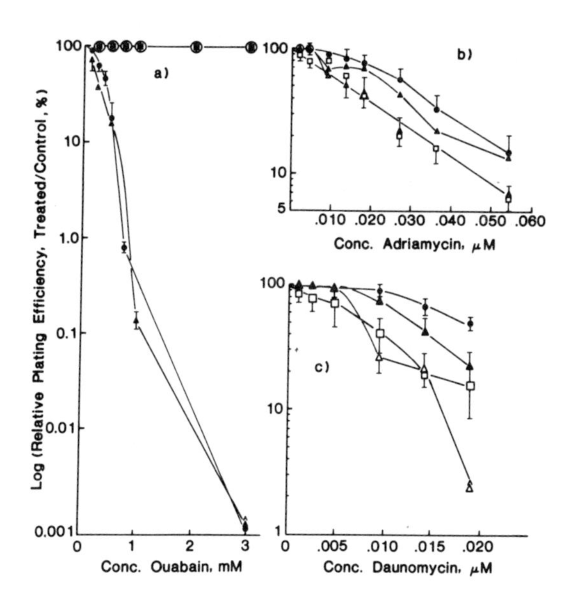

FIG. 4. a) relative cloning efficiency after a 9-day exposure to 3 mM ouabain in logarithmically growing C3H 10T1/2 Cl 8(▲), Cl 16 (●), and 3 Oua^r clones: MNNG Oua^r Cl 1(□); MNNG Oua^r Cl 2(△), and NAcO-AAF Oua^r Cl 1 (○). At 3 mM Oua, the^over the symbols indicates that the relative cloning efficiency was less than this value. b) relative cloning efficiencies after a 9-day exposure to adriamycin (ADM). Same symbols as in a. — is used for the last 2 clones because their survival in ADM was not significantly different. c) relative cloning efficiencies after a 9-day exposure to daunomycin (DM), with same clones and symbols as in b. In a, b, and c, each point is the mean of determinations from two experiments, and in each experiment, four dishes were used at each concentration of drug; bars: average deviation.

transfer analysis with mouse repetitive DNA sequences used as hybridization probes. We have shown that Oua^r microcell hybrids containing no cytologically detectable mouse chromosomes do contain mouse genetic material, and we have identified clones in which the amount of murine DNA is minimal compared to microcell hybrids containing the intact mouse chromosome No. 3. We are now cloning the affected gene(s), identifying it, and will ultimately define the DNA sequence changes in this gene that lead to chemically induced Oua^r. We are also isolating (Na,K) ATPase from wild-type and Oua^r C3H 10T1/2 cells and comparing the ability of ouabain to inhibit these enzymes and the mobility of wild-type and mutant enzymes in 2-dimensional gel electrophoretic analysis.

TABLE 1. Stability of the phenotype of Oua^r clones induced by MNNG.

Clone number[b]	Cloning efficiency in nonselective media	Ratio of cloning efficiency in 3 mM ouabain to cloning efficiency in nonselective media (%)	
		15 days[a] postselection	43 days postselection
1	22 ± 1	90 ± 7	90 ± 10
3	29 ± 1	68 ± 10	100 ± 20
4	26 ± 1	100 ± 6	104 ± 5
5	23 ± 2	40 ± 5	40 ± 5
6	23 ± 2	74 ± 3	110 ± 10
7	10 ± 1	50 ± 3	110 ± 10
8	26 ± 1	100 ± 10	120 ± 20
13	28 ± 1	100 ± 10	87 ± 6
14	27 ± 1	100 ± 10	100 ± 10
15	21 ± 2	90 ± 10	30 ± 10
16	26 ± 2	70 ± 5	90 ± 10
18	20 ± 2	25 ± 2	13 ± 2
20	17 ± 1	100 ± 10	100 ± 3
23	18 ± 2	90 ± 10	100 ± 2
Wild type (C3H 10T1/2, C18)	27 ± 2	$< 2 \times 10^{-4}$	$< 2 \times 10^{-4}$

[a]Cells were passaged approximately every 10 days.
[b]All these clones were obtained after C3H 10T1/2 Cl 8 cells were caused to mutate by 6.7 μM MNNG.

work should provide information as to whether mutations conferring Oua^r reside in the α or β subunits of the (Na,K) ATPase and will complement Oua^r cloning studies.

We have shown to date that this Oua^r assay in C3H 10T1/2 cells detects base substitution mutations induced by MNNG, BAP, 3-methylcholanthrene (MCA), NAcO-AAF, ±(trans)-7,8 dihydrodiol of BAP, and the ±(anti) 7,8 dihydrodiol-9,10 epoxide of BAP (Table 2, ref. 15). Hence, our

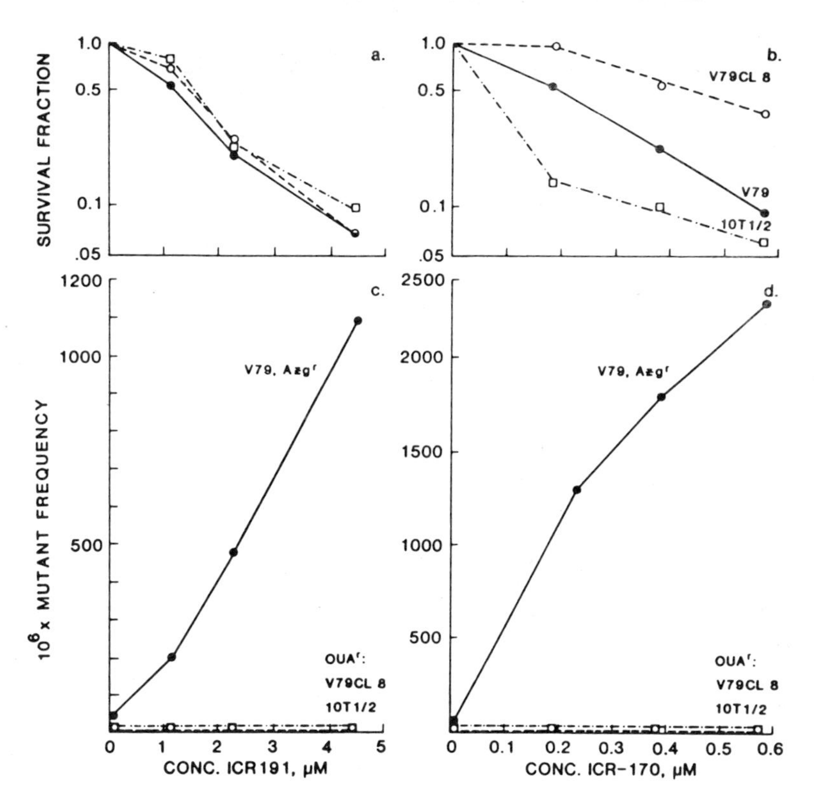

FIG. 5. a: cytotoxicity of ICR-191 to 10T1/2 cells (□), V79 cells (●), and V79 Cl 8 cells (○); b: cytotoxicity of ICR-170 to 10T1/2 cells (□), V79 cells (●), and V79 Cl 8 cells (○); c: mutagenicity of ICR-191 to 10T1/2 cells in an Oua^r assay (□), to V79 cells in an Azg^r assay (●), and to V79 Cl 8 cells in an Oua^r assay (○); d: mutagenicity of ICR-170 to 10T1/2 cells in an Oua^r assay (□), to V79 cells in an Azg^r assay (●), and to V79 Cl 8 cells in an Oua^r assay (○).

Oua^r assay is a well-defined assay that will detect the ability of chemical carcinogens to induce base substitution mutations (reviewed in 21).

Using similar procedures, we developed an assay detecting base substitution mutations to Oua^r in a cultured line of rat prostate epithelial cells (23). These mutants possess a 10^{-5}-fold greater resistance to 3 mM ouabain than wild-type cells and are stably Oua^r in the absence of ouabain. In addition, dome or hemicyst formation in wild-type rat prostate epithelial cells is reversibly inhibited by ouabain (23), whereas dome formation in these epithelial Oua^r mutants is Oua^r. We also defined the maximum expression times for this assay and conditions that minimize metabolic

TABLE 2. Mutagenic activities of various carcinogens.

Compound[a]	LD_{37} (μM)	Mutants/10^5 survivors per μM
MNNG	8.30	1.00
NAcO-AAF	15.00	0.29
BAP-anti-diol-epoxide	0.25	35.00
BAP	15.00	1.50
MCA	40.00	5.10

[a]Cells were exposed to carcinogen for 48 h and then reseeded. This procedure constitutes a 48-h exposure and expression time.
[b]Values are the slopes of the dose-response curves.

cooperation and maximize mutant recovery. We have further shown that the chemical carcinogens MNNG and aflatoxin B, cause dose-dependent induction of Oua^r mutants in these cells (23).

COMPARISON OF CHEMICAL TRANSFORMATION AND CHEMICALLY INDUCED MUTATION TO OUABAIN RESISTANCE IN C3H 10T1/2 CELLS

Having defined conditions that result in the induction and detection of chemically induced Oua^r mutations in C3H 10T1/2 cells, we proceeded to study both chemical transformation and chemically induced mutation to Oua^r in these cells. We showed that the chemical carcinogens benzo[a]-pyrene and NAcO-AAF caused dose-dependent increases in both the transformation frequency and in the Oua^r mutation frequency (Table 3, ref. 15). For both compounds, the transformation frequency and the Oua^r mutation frequency both increased in parallel over the same concentration ranges of carcinogen. Similarly, we found that BAP, the ± (trans) 7,8-dihydrodiol of BAP, and the ± (anti) 7,8-dihydrodiol-9,10 epoxide of BAP all induce strong mutation to Oua^r in these cells (Fig. 6, ref. 7) and also transform the cells over the same concentrations ranges (7). These findings suggest that mutagenicity is an intrinsic property of many chemical carcinogens directly related to their transforming abilities. In particular, I propose that the ability of mutagenic chemical carcinogens to mutate a discrete number of cellular loci is likely responsible in part for their ability to cause neoplastic transformation in cultured C3H 10T1/2 cells.

We found in these experiments that the ratio of the transformation to the Oua^r mutation frequency averaged between 20 and 50 (Table 3, ref. 15,

TABLE 3. Frequencies of transformation and mutation produced by NAcO-AAF and BAP.

Compound	Conc. (μM)	Oua^r mutants/ 10^5 survivors[a]	Type III foci/ 10^3 survivors	Ratio[b]
NAcO-AAF	0.00	< 0.47	< 0.099[c]	-
	7.50	1.10	0.290	26
	10.00	1.70	0.550	32
	15.00	4.03	5.280	129
	20.00	5.60	1.050	19
			Average	52
BAP	0.00	< 0.38	< 0.077	-
	3.00	8.85	2.800	31
	5.00	9.30	2.300	25
	15.00	28.00	6.400	23
			Average	26

[a]Each entry is the average of two experiments for NAcO-AAF and for BAP. All mutation frequencies were calculated by the Poisson formula.
[b]The ratio of transformation to mutation frequencies.
[c]Statistical calculation. There were no type III colonies in any of the controls.

reviewed in 21). Others have independently measured ratios similar to our values (2,4,11). It has been shown that DNA from chemically transformed C3H 10T1/2 cells can transfect foci into NIH3T3 cells (31). Genes identified by ability to transfect focus formation are grouped into a class known as oncogenes (35). Hence, I further propose that the transformation/ mutation frequency ratios of 20-50 suggest that the proto-oncogene loci, which number approximately 20 at present (35), and other genes regulating the expression of proto-oncogenes, are likely targets for chemical transformation by mutagenic chemical carcinogens and that mutation in these genes is causally associated with the initiation step of chemical transformation of C3H 10T1/2 cells. To address this question directly, we have begun to study oncogene expression in chemically transformed C3H 10T1/2 cell lines.

STUDIES OF CHEMICAL CARCINOGENS THAT ARE NOT SIGNIFICANTLY MUTAGENIC

In collaboration with Dr. Peter A. Jones of USC, I studied the mutagenicity of the nucleoside analog, 5-azacytidine (5-AzaCR), and related

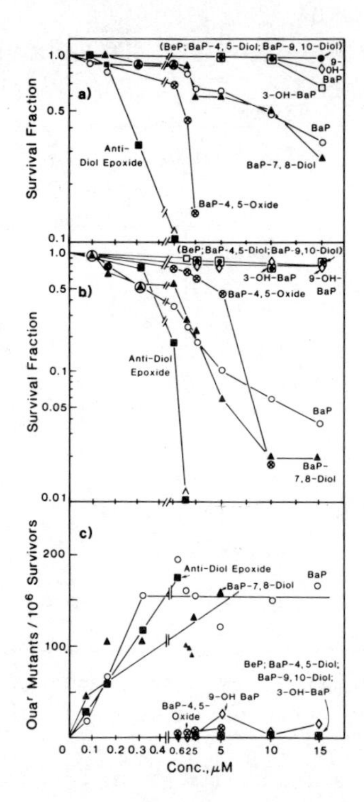

FIG. 6. Cytotoxicity and mutation produced in C3H 10T1/2 cells by BAP and derivatives. a. C3H 10T1/2 cells were treated with BAP derivatives for 48 h in the nonreseeding cytotoxicity assay. Each symbol represents the mean of values from 4 to 6 experiments, and in each experiment 5 dishes were used to determine the value at each concentration. b. C3H 10T1/2 cells were treated with BAP derivatives for 48 h in the reseeding cytotoxicity assay. Each symbol represents the mean of values from 4 to 6 experiments. For both nonreseeding and reseeding assays, a symbol with a ^ indicates that the survival fraction was less than this value. For survival values of greater than 0.1, the S.E. was less than 30%. c. C3H 10T1/2 cells were treated with BAP derivatives for 48 h and reseeded 48 h later to determine the Oua^r mutant frequency as described in "Materials and Methods." The results for BAP-4,5-oxide, 9-OH-BAP, 3-OH-BAP, BAP-9,10-diol, and benzo[e]pyrene (BEP) at each concentration (Conc.) are the averaged values from 2 experiments. For BAP-4,5-diol, one experiment was conducted. For BAP, (±)-BAP-7,8-diol, and (±)-BAP-anti-7,8-diol-9,10-epoxide (anti-diol epoxide), from 2 to 6 experiments were conducted, and the results were averaged for each concentration point. The S.E. of the experiments at each concentration of carcinogen varied up to a maximum of 40% of the averaged mean value ● includes 0.5% acetone controls.

compounds. Benedict et al. have shown that 5-AzaCR transforms C3H 10T1/2 cells (3). Jones and his collaborators demonstrated that 5-AzaCR provokes differentiation of these cells into end-stage phentotypes, such as adipocytes, chrondrocytes, and myocytes (5,6,14). We showed that 5-Aza-CR does not mutate C3H 10T1/2 cells or V79 cells to Oua^r and at best weakly mutates V79 cells to 8-azaguanine resistance (18) over concentrations that cause neoplastic transformation and induce end-stage differentiation in C3H 10T1/2 cells. Hence, we proposed that differentiation mediated by this compound is likely not mediated by simple base substitution or frameshift mutations (18). Taylor and Jones have proposed that 5-AzaCR-induced differentiation may be mediated by inhibition of DNA methylation (33). It is also likely that the transformation of C3H 10T1/2 cells may proceed via mechanisms other than by base substitution mutations.

More recently, my laboratory has been studying mechanisms of neoplastic transformation caused by carcinogenic metal salts. These compounds are known to be human carcinogens via epidemiological evidence, and for nickel and chromium salts there is evidence that these compounds are animal carcinogens as well (32). Interestingly, although arsenic has been implicated as a skin carcinogen and respiratory carcinogen in man, it has so far failed to produce tumors in animals (32). We have shown that nickel oxide and nickel subsulfide cause neoplastic transformation in C3H 10T1/2 cells but do not cause base substitution mutations to Oua^r (20,22). We have further shown that sodium arsenite, which provokes only weak transformation, functions as a promoter of neoplastic transformation initiated by 3-methylcholanthrene (22). We are currently studying the molecular biology of metal salt transformation in C3H 10T1/2 cells. In addition, in collaboration with Dr. K.S. Narayan, we have been using scanning electron microscopy to characterize chemically transformed C3H 10T1/2 cells and to determine the kinetics of expression of chemically transformed cells (26).

ACKNOWLEDGMENTS

I joined Dr. Heidelberger's laboratory in 1977 as a postdoctoral fellow and remained associated with him as Assistant Professor of Research Pathology and later as Assistant Professor of Microbiology and Pathology until his untimely and tragic death from cancer. I would like to thank Dr. Heidelberger for all the scientific training, encouragement, and support he gave to me during our association. I will always remember his scientific brilliance, his critical scientific attitudes, his humanity, and the unique training in chemical carcinogenesis I received from him. I sincerely regret his passing but remain assured that the many scientists that he trained will preserve and extend his tradition of scientific excellence.

This work was supported by Grant NP410 from the American Cancer Society, Public Health Service Grant R01 ES03341-02 from the National Institute of Environmental Health Sciences, U.S. Department of Health and Human Services; Grant R-811099-01-0 from the Environmental Protection Agency; and a grant from the R. J. Reynolds Company.

REFERENCES

1. Ames, B.N., Durston, W.D., Yamasaki, E., and Lee, F.D. (1973): Proc. Nat. Acad. Sci., USA, 70:2281-2285.
2. Barrett, J.C., and Ts'o, P.O.P. (1978): Proc. Nat. Acad. Sci., USA, 75:3297-3301.
3. Benedict, W.F., Banerjee, A., Gardner, A., and Jones, P.A. (1977): Cancer Res., 37:2202-2208.
4. Chan, G., and Little, J.B. (1978): Proc. Nat. Acad. Sci., USA, 73:3363-3366.
5. Constantinides, P.G., Jones, P.A., and Gevers, W. (1977): Nature, 267:364-366.
6. Constantinides, P.G., Taylor, S.M., and Jones, P.A. (1978): Dev. Biol. 66:57-71.
7. Gehly, E.B., Landolph, J.R., Heidelberger, C., Nagasawa, H., and Little, J.B. (1982): Cancer Res., 42:1866-1875.
8. Heidelberger, C. (1975): Annu. Rev. Biochem., 44:79-121.
9. Heidelberger, C., Landolph, J.R., Fournier, R.E.K., Fernandez, A., and Peterson, A.R. (1983): In: Progress in Nucleic Acid Research and Molecular Biology, Vol. 29, edited by W. Cohn. Academic Press, Inc., New York.
10. Huberman, E., Aspiras, L., Heidelberger, C., Grover, P.L., and Sims, P. (1971): Proc. Nat. Acad. Sci., USA, 68:3195-3199.
11. Huberman, E., Mager, R., and Sachs, L. (1976): Nature, 264:360-361.
12. Jeanette, K.W., Jeffrey, A.M., Blobstein, S.H., Beland, F.A., Harvey, R.G., and Weinstein, I.B. (1977): Biochemistry, 16:932-938.
13. Jerina, D.M., Yagi, H., Thakker, D.R., Lehr, R.E., Wood, A.W., Levin, W., and Conney, A.H. (1980): In: Microsomes, Drug Oxidations, and Chemical Carcinogenesis, Vol. 2, edited by M.J. Coon, A.H. Cooney, R.W. Estabrook, H.V. Gelboin, J.R. Gillette, and P.J. O'Brien, pp. 1041-1049. Academic Press, Inc., New York.
14. Jones, P.A., and Taylor, S.M. (1980): Cell, 20:85-93.
15. Landolph, J.R., and Heidelberger, C. (1979): Proc. Nat. Acad. Sci., USA, 76:930-934.
16. Landolph, J.R., Telfer, N., and Heidelberger, C. (1980): Mutat. Res, 72:295-310.
17. Landolph, J.R., Bhatt, R.S., Telfer, N., and Heidelberger, C. (1980): Cancer Res., 40:4581-4588.

18. Landolph, J.R., and Jones, P.A. (1982): Cancer Res., 42:817-823.
19. Landolph, J.R., and Fournier, R.E.K. (1983): Mutat. Res., 107:447-463.
20. Landolph, J.R. (1983): Proc. Am. Assoc. Cancer Res., 24:100.
21. Landolph, J.R. (1985): IARC Monographs, edited by Dr. H. Yamasaki, in press.
22. Landolph, J.R. (1985): Cancer Res., in press.
23. Link, K.H., Heidelberger, C., and Landolph, J.R. (1983): Environmental Mutagenesis, 5:33-48.
24. Maher, V.M., and Wessel, J.G. (1975): Mutat. Res., 28:279-284.
25. Mondal, S., Brankow, D.W., and Heidelberger, C. (1976): Cancer Res., 36:2254-2260.
26. Narayan, K.S., Young, R., Heidelberger, C., and Landolph, J.R. (1984): Carcinogenesis, 5:885-894.
27. Newbold, R.F., and Brookes, P. (1976): Nature, 261:52-54.
28. Reznikoff, C.A., Bertram, J.S., Brankow, D.W., and Heidelberger, C. (1973): Cancer Res., 33:3239-3249.
29. Reznikoff, C.A., Brankow, D.W., and Heidelberger, C. (1973): Cancer Res., 33:3231-3238.
30. Scribner, J.D., and Suss, R. (1978): Tumor Initiation and Promotion. International Review of Experimental Pathology, Vol. 18, edited by G.W. Richter and M.A. Epstein, pp. 138-198. Academic Press, Inc., New York.
31. Shih, C., Shilo, B.-Z., Goldfarb, M.P., Dannenberg, A., and Weinberg, R.A. (1979): Proc. Nat. Acad. Sci., USA, 76:5714-5718.
32. Sunderman, F.W. (1978): Fed. Proc., 37:40-46.
33. Taylor, S.M., and Jones, P.M. (1979): Cell, 17:771-779.
34. Tsutsui, T., Maizumi, H., McLachlan, J.A., and Barrett, J.C. (1983): Cancer Res., 43:3814-3821.
35. Weinberg, R.A. (1982): Adv. Cancer Res., 36:149-163.
36. Wood, A.W., Goode, R.L., Chang, R.L., Levin, W., Conney, A.H., Yagi, H., Dansette, P.M., and Jerina, D.M. (1975): Proc. Nat. Acad. Sci., USA, 72:3176-3180.

Carcinogenesis, Vol. 10, edited by E. Huberman
and S. H. Barr. Raven Press, New York © 1985.

Inhibition and Enhancement of Oncogenic Cell Transformation in C3H 10T½ CL8 Cells

S. Nesnow, H. Garland, and *G. Curtis

*Carcinogenesis and Metabolism Branch, Genetic Toxicology Division, Health Effects Research Laboratory, U.S. Environmental Protection Agency, Research Triangle Park, North Carolina 27711; *Environmental Health Research and Testing, Research Triangle Park, North Carolina 27709*

During the last several years, we have been studying the effects of chemicals on the inhibition and enhancement of oncogenic cell transformation in C3H 10T1/2 CL8 mouse embryo fibroblasts (C3H 10T1/2 cells). Our findings, summarized in Table 1, indicate that many diverse chemicals can affect oncogenic cell transformation. In general, their mechanisms of action seem to be at the level of the enzymes that are involved in the metabolic activation of the carcinogens used to transform the cells. Specifically, 7,8-benzoflavone, 1,2-naphthoquinone, and phenanthrene-9,10-quinone inhibit cell transformation by inhibiting cytochrome P-450-mediated oxidation (11,13). Benz[a]anthracene, 5,6-benzoflavone, phenobarbital, and pregnenolone-16-α-carbonitrile enhance cell transformation by inducing the complex of cytochrome P-450 mixed-function oxidases and increase the ability of C3H 10T1/2 cells to metabolize carcinogens (14). Cyclohexene, cyclohexene oxide, 1,2-dihydronaphthalene, styrene oxide, and 1,2,3,4-tetrahydronaphthalene-1, 2-oxide enhance cell transformation by inhibiting epoxide hydratase and thus allow increased levels of active arene oxides to accumulate in the cells (13).

The mixed-function oxidases in C3H 10T1/2 cells are efficient at metabolizing polycyclic aromatic hydrocarbons (PAH) such as 3-methylcholanthrene (3MC) (13), benzo[a]pyrene (BAP) (14), and cyclopenta(cd)-pyrene (6) and are not efficient at metabolizing 2-acetylaminofluorene (15). These results are in concert with the observation that PAH can transform C3H 10T1/2 cells while 2-acetylaminofluorene cannot (6,9,12-16).

TABLE 1. Inhibitors and enhancers of oncogenic transformation in C3H 10T1/2 cells.

Chemical	Carcinogen	Effect	Reference[a]
Benz[a]anthracene	BAP	enhancement	14
	3MC	enhancement	13
5,6-Benzoflavone	BAP	enhancement	14
7,8-Benzoflavone	3MC	inhibition	13
	BAP	inhibition	U[b]
Cyclohexene	3MC	enhancement	13
Cyclohexene oxide	3MC	enhancement	13
1,2-Dihydro-naphthalene	3MC	enhancement	13
1,2-Naphthoquinone	BAP	inhibition	11
Phenanthrene-9,10-quinone	BAP	inhibition	11
Phenobarbital	BAP	enhancement	14
Pregnenolone-16-α carbonitrile	BAP	enhancement	14
Styrene oxide	3MC	enhancement	13
1,2,3,4-Tetrahydro-naphthalene-1,2-oxide	3MC	enhancement	13

[a]Data taken from the author's publications.
[b]Unpublished data.

During our studies of cell transformation and metabolism of indirect-acting carcinogens, we observed, as had others, that C3H 10T1/2 cells could not be transformed by direct-acting alkylating agents such N-methyl-N'-nitro-N-nitrosoguanidine (MNNG) or the k-region oxides of most PAH. The only direct acting carcinogens reported to transform C3H 10T1/2 cells were cyclopenta(cd)pyrene-3,4-oxide (6) and N-hydroxy-2-acetylaminofluorene (15); however, these agents were weak in activity. In some of our experiments with enzyme inducers we noticed that when dishes containing cells were treated with carcinogens several days after seeding, more foci were produced than when the dishes were treated one day after cell seeding as described in the original procedure of Reznikoff et al. (16). We investigated this observation by treating separate groups of dishes of C3H 10T1/2 cells with BAP on various days after seeding (Fig. 1). We found that the production of foci increased with increasing time interval between seeding and treatment up to 5 days after seeding. After 5 days, foci production decreased. When this type of experiment was repeated with MNNG, similar results were obtained, except the decrease was not observed up to 8 days after seeding (data not shown). Data from this experiment were the first to show that MNNG could transform C3H 10T1/2 cells (12). Using the modified protocol of treating 5-day-old cultures of C3H 10T1/2 cells, we also obtained cell transformation with seven other carcinogens (Table 2). Several other carcinogens were not effective in transforming C3H 10T1/2 cells when this procedure was used (Table 2), probably due to the inability of C3H 10T1/2 cells to metabolically activate these specific carcinogens.

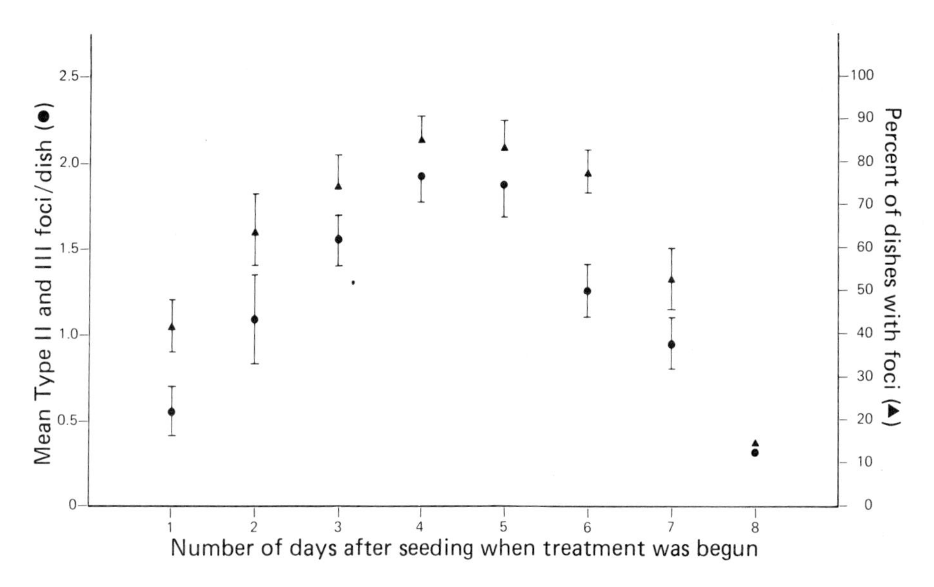

FIG. 1. Temporal relationship between day of treatment and oncogenic transformation in C3H 10T1/2 cells. Cells were seeded on day 10 and treated with 1 μg/mL BAP for 24 h on subsequent days.

TABLE 2. Carcinogens evaluated in the 5-day treatment protocol with C3H 10T1/2 cells.[a]

Chemical	Result
2-Acetylaminofluorene	no transformation
Aflatoxin B_1	transformation
BAP	transformation
N,N'-Dimethylaminoazobenzene	no transformation
7,12-Dimethylbenz[a]anthracene	transformation
Epichlorohydrin	transformation
Ethyl nitrosourea	transformation
3MC	transformation
MNNG	transformation
Nickel sulfate	no transformation
4-Nitroquinoline-N-oxide	transformation
Propane sultone	transformation
Urethane	no transformation
Vinyl carbamate	no transformation

[a]Data from the author's laboratory.

Recently we have been examining the phenomenon of inhibition of cell transformation in C3H 10T1/2 cells by chemicals that do not exert their action by affecting the microsomal mixed-function oxidases. This paper will document our findings and describe studies designed to ascertain the operable mechanisms of action.

MATERIALS AND METHODS

C3H 10T1/2 cells derived by Reznikoff et al. (17) were cultured according to previously published procedures (12,16). In general, cell transformation studies were performed by seeding 1000 cells/dish in Eagle's basal medium with Earle's salts and L-glutamine supplemented with 10% heat-inactivated fetal bovine serum. After the chemical treatment, the cells were washed and the medium replaced with fresh medium containing Garramycin sulfate, 25 μg/mL. Once the cells became confluent, the fetal bovine serum concentration was reduced to 5% (2). The medium was changed weekly. At the end of the experiments, the cells were washed with 0.85% NaCl, fixed with methanol, and stained with Giemsa. Both type II and type III foci were scored according to previously described criteria (16). Cytotoxicity was determined with 200 cells/dish, and the dishes were fixed and stained 10-12 days after treatment.

RESULTS AND DISCUSSION

We are interested in enhancing the chemical transformation of C3H 10T1/2 cells by various techniques. We reasoned that cell transformation experiments, especially those performed with C3H 10T1/2 cells, are single-exposure experiments, unlike the chronic-exposure tumorigenesis experiments. We, therefore, treated C3H 10T1/2 cells with carcinogens at several times throughout the transformation experiment. Our initial findings were that cells treated on day 1 and day 8 with BAP (each treatment at 0.5 μg/mL) produced more total type II and III foci than cells treated with 1 μg/mL BAP on day 1 (Table 3). However, C3H 10T1/2 cells treated with BAP at 0.5 μg/mL on days 1, 8, and 15 produced fewer total foci and fewer dishes with foci than dishes treated with BAP at 1.5 μg/mL on day 1. The cells treated on days 1, 8, and 15 also produced fewer foci and dishes with foci than cells treated with BAP on days 1 and 8. These results suggested that C3H 10T1/2 cells in logarithmic phase growth produced more transformation when treated with split concentrations of BAP. They also suggested that BAP-initiated cells, when at confluence, were inhibited in expressing transformation when treated again with BAP.

The time course of inhibition was explored further by treating BAP-initiated cells at various times after they reached confluence. Each treatment consisted of a 24-h exposure to 1 μg/mL BAP. C3H 10T1/2 cells treated a second time with BAP from day 14 to day 26 did not produce any transformed foci (Table 4). A small number of foci were observed in dishes treated on days 1 and 33.

To determine whether this inhibition was a temporary event or a result of delayed cytotoxicity, cells treated on day 1 or days 1 and 22 were held for longer times before scoring (Table 5). Cells treated on days 1 and 22

TABLE 3. Response of BAP-initiated C3H 10T1/2 cells to subsequent BAP treatment.

	Treatment Day(s)					
	1		1 and 8		1, 8, and 15	
BAP (μg/mL)	No. foci[a]	% Dishes with foci[a]	No. foci[a]	% Dishes with foci[a]	No. foci[a]	% Dishes with foci[a]
0	0	0	0	0	0	0
0.5	34	58	75	81	26	50
1.0	52	78	96	83		
1.5	56	83				

[a]Total of type II and type III foci per 24 dishes scored per treatment group. Treatment time was 24 h for each treatment.

TABLE 4. Inhibition of BAP-induced transformation as a function of time of second treatment.[a]

Day of second treatment[b]	Total foci[c]	% Dishes with foci[c]
14	0	0
16	1	4
22	0	0
26	0	0
33	4	3

[a]C3H 10T1/2 cells were treated on day 1 with BAP (1 μg/mL) for 24 h.
[b]Second treatment was with BAP (1 μg/mL) for 24 h.
[c]Total type II and III foci per 24 dishes scored.

normally scored at week 6 again produced no foci. Cells treated on day 1 and held 1 to 3 weeks beyond week 6 (weeks 7 to 9) produced increasing numbers of foci. However cells treated on days 1 and 22 and held to week 9 produced only 2 foci/24 dishes. The extent of inhibition did not change with the week of scoring, indicating that the change observed was not temporary.

Several experiments were performed to probe the mechanisms of this effect. Selective cytotoxic effects of BAP on transformed cells were determined by means of a reconstruction procedure. Transformed C3H 10T1/2 cells, 58-3MC cells, were co-seeded with C3H 10T1/2 cells at ratios of 8 58-3MC cells to 100,000 C3H 10T1/2 cells and 16 58-3MC cells to 100,000 C3H 10T1/2 cells. After the cells attained confluence, the dishes were treated with BAP or MNNG, 1 μg/mL, for 24 h. The dishes were fixed and stained 7 days later. Of 8 or 16 transformed cells seeded, all were recovered (Table 6). BAP killed approximately 70% of the 58-3MC cells seeded while MNNG killed approximately 50%. We next investigated the cytotoxic effects of the second BAP treatment. Cells treated on days 1 and 22 were reseeded on day 23 and the plating efficiency measured (Table 7). We observed a 92% decrease in plating efficiency compared to the plating efficiency of cells treated on day 1 and reseeded on day 23. These results suggest that many initiated cells treated on day 22 are not likely to survive the treatment. However, since it only takes one initiated

TABLE 5. Inhibition of BAP-induced transformation by BAP as a function of time of scoring.[a]

Week of scoring	No. foci/24 dishes Treatment days Day 1	 Days 1, 22	Inhibition of transformation (%)
6	34	0	100
7	68	2	97
8	84	2	98
9	82	2[b]	98

[a]BAP concentration was 1 μg/mL for each treatment. Treatment time was 24 h for each treatment.

[b]Four foci appearing in the Day 22 controls have been subtracted from this total.

TABLE 6. Determination of the cytotoxic effects of the second BAP treatment on transformed cells with a reconstruction procedure.[a]

Treatment	No. 58-3MC Cells seeded			
	8	16	8	16
	Foci/dish[b]		Survival (%)	
Acetone	8.6	14.8	100	100
BAP (1 μg/mL)			33.5	31.8
MNNG (1 μg/mL)			47.8	51.5

[a]100,000 C3H 10T1/2 cells were seeded with 8 or 16 58-3MC cells per 60 mm Petri dishes. After attaining confluence, the cultures were treated with BAP or MNNG for 24 h. The cultures were fixed and stained 7 days later and scored for 58-3MC cell foci.

[b]Mean of 104 individual determinations. The coefficient of variation was <15%.

TABLE 7. Plating efficiency of initiated C3H 10T1/2 cells treated on day 22 with BAP.[a]

Treatment group	Plating efficiency (%) Mean +/- SD[b]
Acetone (Days 1, 22)	13.3 +/- 2.5
BAP (Day 1)	15.1 +/- 2.1
BAP (Days 1, 22)	1.2 +/- 0.7

[a]Cells were treated with BAP (1 μg/mL) for 24 h on days 1 or 1 and 22 (except for the acetone controls). On day 23, the cells were trypsinized and reseeded at 200 cells/dish. The number of colonies was determined 7 days later.

[b]The number of replicates for each determination was 24.

cell to form a transformed colony and since BAP does not kill all the transformed cells, cytotoxicity cannot explain the observations of complete inhibition of transformation.

Many chemicals have been reported to inhibit cell transformation: actinomycin D (5), allylisopropylacetamide (9), antibiotics (4), antifungals (16), ascorbic acid (1), dexamethasone (B), isopropylvaleramide (9), phosphodiesterase inhibitors (3), protease inhibitors (7), retinoids (10), and serum components (2). Although the mechanisms of action of these chemicals are not known, a number of these chemicals have been examined for their ability to inhibit the mixed-function oxidases and have been found to be inactive in this regard. It has been proposed that some of these chemicals may exert their action by affecting receptors involved in the transformation process (10). We are currently evaluating this mechanism with regard to our observations on the inhibition of cell transformation of C3H 10T1/2 cells by BAP.

DEDICATION

This paper is dedicated to the memory of Charles Heidelberger: teacher, mentor, and friend.

ACKNOWLEDGMENT

The authors thank Northrup Services editorial staff for the preparation of this paper. This paper has been reviewed by the Health Effects Laboratory, U.S. Environmental Protection Agency, and approved for publication. Mention of trade names or commercial products does not constitute endorsement or recommendation for use.

REFERENCES

1. Benedict, W.F., Wheatley, W.L., and Jones, P.A. (1980): Cancer Res., 40:2796-2801.
2. Bertram, J.S. (1977): Cancer Res., 37:514-523.
3. Bertram, J.S. (1979): Cancer Res., 39:3502-3508.
4. Bertram, J.S. (1979): Cancer Lett., 7:289-298.
5. Bertram, J.S., Libby, P.R., and Merriman, R.L. (1980): J. Nat. Cancer Inst., 64:1393-1399.
6. Gold, A., Nesnow, S., Moore-Brown, M., Garland, H., Curtis, G., Howard, B., Graham, D., and Eisenstadt, E. (1980): Cancer Res., 41:1893-1897
7. Kuroki, T., and Drevon, C. (1979): Cancer Res., 39:2755-2761.
8. Kuszynski, C., and Langenbach, R. (1982): Cancer Lett., 15:215-221.
9. Kuszynski, C., Somogyi, A., Nesnow, S., and Langenbach, R. (1981): Cancer Res., 41:1893-1897.

10. Mordan, L.J., and Bertram, J.S. (1983): Cancer Res., 43:567-571.
11. Nesnow, S., Bergman, H., Garland, H., and Morris, M. (1980): J. Environ. Pathol. Toxicol., 4:17-30.
12. Nesnow, S., Garland, H., and Curtis, G. (1982): Carcinogenesis, 4:377-380.
13. Nesnow, S., and Heidelberger, C. (1976): Cancer Res., 36:1801-1808.
14. Nesnow, S., Leavitt, S., Garland, H., Vaughan, T.O., Hyatt, B., Montgomery, L., and Cudak, C. (1981): Cancer Res., 41:3071-3076.
15. Oglesby, L., Nesnow, S., Hyatt, B., Montgomery, L., and Cudak, C. (1982): Cancer Lett., 16:231-237.
16. Reznikoff, C.A., Bertram, J.S., Brankow, D., and Heidelberger, C. (1973): Cancer Res., 33:3239-3249.
17. Reznikoff, C.A., Brankow, D., and Heidelberger, C. (1973): Cancer Res., 33:3231-3238.

Carcinogenesis, Vol. 10, edited by E. Huberman and S. H. Barr. Raven Press, New York © 1985.

Genes and Membrane Signals Involved in Neoplastic Transformation

N. H. Colburn

Cell Biology Section, Laboratory of Viral Carcinogenesis, National Cancer Institute, Frederick, Maryland 21701

It is indeed a privilege to be able to contribute to this symposium volume in honor of Dr. Charles Heidelberger. Dr. Heidelberger was one of my greatest teachers during the years that I spent as a graduate student at McArdle Laboratory in the mid-sixties and an important source of continuing encouragement in the ensuing years. Table 1, from a 1975 article of Dr. Heidelberger (30), is a typical example of his clear articulation of the questions that prompted the right experiments. In particular, he focused our attention early on the somatic mutation hypothesis of carcinogenesis, its predictions, and on at least one alternative model

TABLE 1. Theories of the cellular mechanisms of chemical oncogenesis.[a]

1. Does the chemical oncogen transform cells to malignancy, or does it select somehow for pre-existing malignant cells, as proposed by Prehn.
2. Does the chemical oncogen transform cells by itself or by "switching on" an oncogenic virus that carries out the transformation?
3. If the chemical transforms cells by itself, is the mechanism a mutagenic one that involves an alteration of the primary sequence of DNA, or is the mechanism a perpetuated effect on gene expression?
4. Do chromosomal alterations produce malignancy?

[a]Adapted from C. Heidelberger (30).

for perpetuated effects on gene expression as put forth in an article by Pitot and Heidelberger in 1963 (34).

Some questions that have formed the basis for research in our laboratory are shown in Table 2. In the mouse skin initiation-promotion model, two or more apparently irreversible transitions occur during tumor promotion (4). Are these transitions determined by specific genes, and is expression of these genes inducible by tumor promoters? Further, it has been known from the pioneering work of Boutwell and colleagues (2) that one can breed mice for sensitivity or resistance to initiation-promotion carcinogenesis. It appears likely that this sensitivity is principally at the level of tumor promotion (22,25,35). What are the genes that determine sensitivity to tumor promotion and what are the gene products doing? Are the genes that are involved in induction of neoplasia also involved in maintenance of neoplasia? What relationship do they have to known retroviral oncogenes or to other transforming genes? Finally, what are the signals that regulate expression of such genes?

The model system we have used to investigate these questions is the JB6 mouse epidermal family of cell lines described in Table 3. These lines

TABLE 2. Genetic determinants of tumor promotion: Unanswered questions.

1. Are the apparently irreversible transitions that occur during tumor promotion or preneoplastic progression *in vivo* determined by specific genes ("promotogenes")? Do tumor promoters induce expression of these genes?

2. Do these genes account for sensitivity to tumor promotion in sensitive animals or people?

3. What are the activities of the corresponding gene products?

4. Are any such "promotogenes" involved not only in induction of malignancy but also in maintenance of malignancy?

5. Do "promotogene" products function to activate transforming genes (maintenance genes)? What relationship do they have to any of the known oncogenes?

6. What are the signals involved in inducing expression of "promotogenes"?

TABLE 3. JB-6 clonal cell lines and their phenotypes.

Cell line	Tumorigenicity[a]	Anchorage dependence	Sensitivity to promotion of transformation	Sensitivity to mitogenesis
$JB6_{35}$-	N	D	P^-	M^+
$JB6_{35}$+	N	D	P^+	M^+
Cl 41	N	D	P^+	M^+
Cl 22	N	D	P^+	M^+
Cl 30	N	D	P^-	M^+
Cl 25	N	D	P^-	M^+
R 219[b]	N	D	P^+	M^-
R 6141	ND	D	P^+	M^-
R 28	ND	D	P^-	M^-
RT 101	T	I	-	-
R 681	T	I	-	-
T^36274[c]	T	I	-	-
T^36271	T	I	-	-

[a]Tumorigenicity was assayed by subcutaneous injection of 1-2 x 10^6cells into nude mice and/or neonatal BALB/c mice. Assays of anchorage independence, promotion of neoplastic transformation, and mitogenesis were assayed as described (7,18,19).

[b]The R cell lines were derived from JB6-Cl 41 cells by selection for mitogenesis resistance (19).

[c]Tumor cell lines T^36274 and T^36271 were derived from P^+ JB-6 cells at passage 60 by three cycles of cloning of anchorage-independent colonies induced by 12-O-tetradecanoyl-phorbol-13-acetate (TPA) in soft agar (7).

N: nontumorigenic; T: tumorigenic; D: anchorage dependent; I: anchorage independent; $P^+/^-$: sensitive/insensitive to promotion of anchorage independence and tumorigenicity by phorbol esters and nonphorbol tumor promoters; $M^+/^-$: sensitive/insensitive to stimulation of cell division at plateau density by TPA; ND: not determined or not applicable.

were originally derived (18) from newborn BALB/C primary epidermal cultures prepared by the method of Yuspa and Harris (40). After passage 35, ($JB6_35^+$) JB6 cells acquired the property of "promotion sensitivity" or the capacity to respond to phorbol esters and other tumor promoters with induction of anchorage independence and tumorigenicity. This shift in phenotype was apparently a rare event since subculture (at six doublings per passage) of the "prepromotable" JB6 cells for up to 60 passages on two subsequent trials did not produce the promotable phenotype. At passage 50, the promotable JB6 cells were nonselectively cloned to yield promotion-sensitive (P^+) cell lines including JB6 Cl 41 and Cl 22 and promotion-insensitive (P^-) cell lines such as JB6 Cl 30 and Cl 25 (10). In addition, several JB6-derived tumorigenic cell lines were cloned from soft agar after three cycles of induction by 12-O-tetradecanoyl-phorbol-13-acetate (TPA) (7). These cell lines include the T^3 series of cell lines. The "R" series of lines were derived after a selection for resistance to mitogenic stimulation by TPA (19) and included P^+, P^-, and tumorigenic cell lines.

The P^+ cell lines turned out to be sensitive to induction of transformation not only by phorbol esters but also by several other classes of tumor promoters, including certain polypeptide hormones, oxidants, and environmental agents such as phthalate esters (27). Likewise, the P^- cell lines showed cross resistance to all of these agents. These observations suggest that the determinants of sensitivity to promotion of neoplastic transformation are likely to be on a pathway common to several types of inducers, whose initial interaction with the cell may differ.

WHAT ACCOUNTS FOR THE PROMOTION-RESISTANT PHENOTYPE?

The approach we have taken is to test systematically a series of hypotheses for required events in tumor promotion to ascertain whether lack of such an event in response to promoters may account for the resistance of the P^- cells. Table 4 shows the hypotheses that can be ruled out as explaining the resistant phenotype. These include a lack of (i) mitogenic response, (ii) hexose uptake response, (iii) phorbol diester receptors, (iv) C-kinase activity, (v) receptors for epidermal growth factor (EGF), and (vi) switched off collagen synthesis. This analysis does not rule out a requirement for these events in tumor promotion. In fact, available evidence indicates that interaction of phorbol esters, EGF, and other transformation-promoting growth factors (TGFs) with their receptors is almost certainly essential for their promoting activity. Direct proof of the sort that would come from demonstration of association between receptorless state and lack of transformation response has been obtained in the cases of EGF and certain TGFs (6,20) but not in the case of phorbol esters since in no case have phorbol ester receptorless variants been found (8,9). Nevertheless, this analysis does allow one to identify the missing or unique responses in P^- variants and study them as events likely to be causally related to promotion of transformation.

TABLE 4. What accounts for the promotion-resistant (P^-) phenotype?

Not lack of mitogenic response Mitogen-resistant, promotion-sensitive (M^-P^+) variants have been isolated (19).	Possibly, lack of ganglioside G_T synthesis response P^+, but not P^-, cells consistently show decreased G_T synthesis in response to tumor promoters (38).
Not lack of hexose uptake response Some promotion-sensitive (P^+) cells are resistant to promoter stimulation of hexose uptake (21).	Possibly, lack of superoxide dismutase (SOD) diminution P^+, but not P^-, cells consistently show to substantial SOD decrease in response to tumor promoters (13).
Not lack of phorbol diester (PDE) receptors All promotion-insensitive (P^-) variants analyzed have normal complements of PDE receptors (8,9).	Possibly, lack of P^+ gene or its expression Transfection of P^+ DNA into P^- cells transfers P^+ phenotype (16,17). P^+ (pro) genes have been cloned and are active (12).
Not lack of protein kinase C activity P^- and P^+ variants show similar levels of C-kinase activity (27).	
Not lack of EGF receptors Some promotion-sensitive (P^+) cells are EGF receptor negative (9,20).	
Not lack of switched-off collagen synthesis Both P^+ and P^- JB6 cells turn off collagen synthesis in response to tumor promoters (24).	

POSSIBLE ROLE OF DECREASED G_T SYNTHESIS IN RESPONSE TO TUMOR PROMOTERS

The first biochemical response studied that appeared to discriminate P^- from P^+ cells was a substantial decrease in net synthesis of the sialoglycosphingolipid trisialoganglioside (G_T) in response to TPA and other tumor promoters (38). Huberman, Heckman, and Langenbach (31) reported several years ago that TPA induced characteristic changes in ganglioside synthesis that accompanied induction of differentiation in human melanoma cells. This evidence, together with the findings of Hakomori and others (29), suggested that gangliosides may be involved in regulatory responses to hormones and growth factors that may be altered in neoplasia. We found that the synthesis of the major ganglioside of JB6 cells, the trisialoganglioside G_T1b, was substantially decreased after TPA exposure of P^+ cells but showed little or no decrease in four out of four P^- cells. Further evidence in support of the significance of the G_T decrease in promotion of transformation came from an independent experiment in which the uncloned JB6 35^+ population was put through a selection for resistance to the G_T response to TPA. This selection for resistance to the G_T response yielded cells that had been coselected for the P^- phenotype (12), suggesting a possible causal connection between the G_T response and promotion of transformation.

THE POSSIBLE ROLE OF OXIDANT DEFENSES

Several lines of evidence suggest that oxidant stress may be an important signal-transducing process in tumor promotion (27,32,33) and that its action may be much more specific than, for example, the production of DNA strand breaks (26). The P^- cells do differ from the P^+ cells in their decrease in levels of superoxide dismutase (SOD) following TPA exposure (13). This situation would be expected to produce oxidant stress via elevation of superoxide anion. If such elevation of superoxide anion is required for promotion of transformation, then removal of the anion by addition of SOD to cells would be expected to block promotion of transformation by TPA. This, in fact, turns out to be the case. Addition of SOD or the SOD mimetic agent Cu(II) diisopropylsalicylic acid does block promotion of transformation by TPA (33).

THE POSSIBLE ROLE OF PROMOTION SENSITIVITY GENES

Recently we have found that P^+ cells differ from P^- cells at the level of a gene that appears to specify sensitivity to promotion of neoplastic transformation in JB6 cells (16,17). As indicated in Table 5 (16,17), this difference was shown by transfection of DNA from P^+ cells into P^- cells, followed by an assay for acquisition of sensitivity to promotion of anchorage-independent transformation by TPA in 0.33% agar. The P^+ trait could

TABLE 5. Promotion sensitivity can be transferred by transfection of genomic DNA from P^+ into P^- cells.[a]

Donor DNA	Recipient cells	Phenotype of recipient cells	N	TPA-induced promotion response (agar colonies per 10^5 cells)
JB6 Cl 41	JB6 Cl 30	P^-	4	444 ± 250
JB6 Cl 41	JB6 Cl 25	P^-	3	393 ± 102
JB6 Cl 41	NIH 3T3	--[b]	4	10 ± 17
JB6 Cl 22	JB6 Cl 30	P^-	4	478 ± 164
JB6 Cl 22	JB6 Cl 25	P^-	3	356 ± 75
JB6 Cl 22	NIH 3T3	--	4	3 ± 6
JB6 Cl 30	JB6 Cl 30	P^-	4	49 ± 41
JB6 Cl 30	JB6 Cl 25	P^-	2	52 ± 30
None	JB6 Cl 30	P^-	4	40 ± 21

[a]Seven days after transfection of DNA (6 days after subculturing), the transfected JB6 P^- cells were suspended in 0.33% agar containing 10% fetal bovine serum ± TPA at 1.6×10^{-8}M. Results are given as mean number of TPA-induced agar colonies per 10^5 cells ± standard deviation for the number of independent experiments indicated by N. P^+: promotable to anchorage independence and tumorigenicity by TPA and other tumor promoters; P^-: nonpromotable.
[b]The NIH 3T3 cells failed to form colonies in agar in the absence or presence of TPA, but their phenotype may differ in several respects from JB6 P^- cells.

be transferred by DNA from either of two P^+ cell lines into either of two P^- cell lines. The P^- DNA was inactive. The NIH 3T3 cells were nonpermissive for P^+ expression, suggesting that they lack an activity expressed in JB6 P^- cells that complements or confers permissiveness for expression of the P^+ trait.

Two different promotion sensitivity (pro) genes have now been cloned (12) by following the strategy shown in Fig. 1. DNA fragments were created with a restriction enzyme shown not to affect the P^+ activity, namely Bgl II (15). A partial genomic library was constructed from the Bgl II fragments of the most active size fraction. The search routine of sib selection (5) was used to find the biologically active plasmids containing the pro genes. This cloning by activity assay contrasts with a strategy

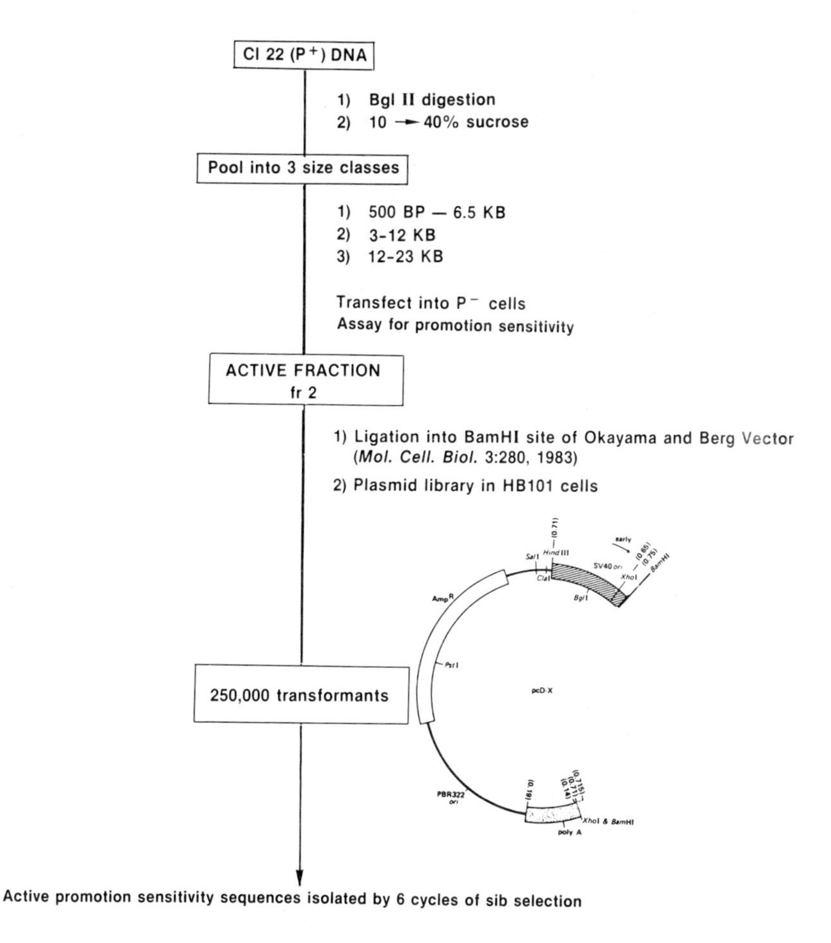

FIG. 1. Cloning of the P^+ (promotion-sensitivity) gene(s).

frequently used to clone homologues of retroviral oncogenes from the DNA of nonvirally induced tumors, namely to retrieve homologous DNA fragments by using cloned oncogene probes. The cloning yielded two active plasmids designated p26 and p40, which contain pro-1 and pro-2 genes, respectively. The details of the cloning and the characterization of the pro genes will be presented elsewhere.

One question that arises regarding the biological activity of the pro genes concerns whether a pro gene specifies the entire phenotype that is manifested by the parental P^+ cell lines. Table 6 summarizes the results of experiments designed to determine whether P^+ cell lines derived by transfection of p26 or p40 into P^- cells have acquired sensitivity not only to TPA but also to other classes of transformation promoters. The results indicate that the transfectant P^+ lines show a similar magnitude of sensitivity to induction of transformation by TPA and the phthalate ester tumor promoter di-(2-ethylhexyl)phthalate (DEHP) (39,27). The transfectants show reduced sensitivity to EGF and no sensitivity to $NaIO_4$ (37). As for inhibitors of promotion of transformation by TPA, the transfectants are sensitive to antipromotion by retinoic acid (14), the glucocorticoid fluocinolone acetonide, the ganglioside G_T (38), and superoxide dismutase (33), with the sensitivity to the gluocorticoid being diminished. Hence, by these two criteria pro-1 and pro-2 genes each appear to specify a phenotype similar to that shown by the parental P^+ cells. This finding rules out the possibility that, for example, sensitivity to EGF is specified by pro-1 while sensitivity to $NaIO_4$ is specified by pro-2.

TABLE 6. Phenotypes of parental P^+ and P^+ transfectants.[a]

	Parental P^+		P^+ Transfectants			P^-
	Cl 22	Cl 41	p26 Cl 4	p26 C19	P40 Cl 2	Cl 30
Transformation by Sensitivity to Promotion						
TPA	++++	++++	+++	+++	+++	0
EGF	++++	++++	+	+	+	0
$NAIO_4$	++	++	0	0	0	0
DEHP	+++	+++	+++	+++	ND	0
Promotion by Inhibition of TPA						
Retinoic acid	+++	++++	++++	++++	ND	
Fluocinolone acetonide	++++	++++	++	++	++	
Ganglioside G_T	++	++	++	++	++	
Superoxide dismutase	++++	++	++++	++++	++++	

[a]Activity is represented on a scale of + (low) to ++++ (high), with 0 indicating no activity and ND indicating not determined in a standard assay involving exposure to promoter at the time of suspension of 10^4 cells per dish in soft agar and scoring colonies greater than eight cells at 2 weeks. Sensitivity to antipromoters was assayed by simultaneous exposure to TPA (1.6×10^{-9}M) and antipromoters as described (14). (To be presented in greater detail elsewhere.)

Table 7 shows possible candidates for pro gene products that can be ruled out or ruled in on the basis of findings reported by our laboratory or others. Candidates consistent with the data include several molecules that may be involved in promotion-relevant signal transduction such as regulators of oxidant stress or ganglioside synthesis responses or mediators of processes activated by C-kinase.

Evidence for Transforming Genes

A final observation regarding genes in the JB6 system is that after transformation by TPA, the JB6 P^+ cells acquire a transforming activity that can be transferred by transfection into appropriate recipient cells (11) (Table 8). This transforming activity is detectable after transfection into P^+ but not into P^- cells or NIH 3T3 cells. It differs from P^+ activity also in restriction endonuclease sensitivity (11).

TABLE 7. Candidates and noncandidates for pro gene products

Candidates for a pro gene product:

1. A tumor-promoter-inducible neuraminidase that accounts for decreased net G_T synthesis (12).
2. A tumor-promoter-inducible repressor of superoxide dismutase (SOD) synthesis or inhibitor of SOD activity or of other oxidant defenses (33,27).
3. A tumor-promoter-inducible ornithine decarboxylase (3).
4. A protein kinase C substrate or a calcium binding protein linked to a phosphatidyl inositol pathway (27).
5. An adenyl cylase inhibitor (27).
6. A growth factor or growth factor receptor (not EGF, platelet-derived growth factor, transferrin, or insulin).
7. A nuclear (chromatin or nuclear matrix) protein that regulates expression of one or a few genes.

Noncandidates for a pro gene product:

1. Mediators of mitogenic response (19).
2. Phorbol diester receptors or protein kinase C (8,9,27)
3. Receptors for EGF, transferrin, or insulin (9,20,27).
4. Regulators of collagen synthesis or degradation (23,36).
5. Hexose transport molecules (21).
6. Clastogenic agents or inducers of clastogenic molecules (26).
7. Mediators of total DNA methylation (1).
8. Heat shock protein 80 (27,28).

SUMMARY AND CONCLUSIONS

Figure 2 summarizes our current understanding of the signal transduction events that appear to be involved in promotion of transformation in JB6 cells. Among the earliest promotion-relevant events (at least for phorbol esters) appear to be C-kinase activation and superoxide anion elevation. Whether such events regulate expression of pro genes needs to be elucidated.

Finally, Fig. 3 presents a model for the role of pro genes and transforming genes in inducing and maintaining neoplastic transformation in mouse JB6 cells. Our current hypothesis is that TPA interacts with its receptor (C-kinase) to trigger one or more promotion-relevant second messages, which then activate expression of pro gene(s). The product of a pro gene is postulated to then activate expression of a transforming gene so that its expression becomes constitutive in the neoplastic cell. In the P^- cells the "defect" could be at the level of a missing second messenger signal to activate pro genes or a structural change in pro genes such that they are not expressed or they lack activity even though expressed.

Recent evidence to be presented elsewhere indicates that pro genes are not limited to mouse cells but are found in certain human tumor and nontumor cells. If human pro homologues turn out to show biological activity for specifying sensitivity to neoplastic transformation, this finding will add a new dimension to testing the somatic mutation hypothesis of carcinogenesis.

TABLE 8. Transfection with DNA from epidermal tumor cells can transform P^+JB6 cell lines in the absence of TPA (11).[a]

DNA transfected	Cotransfection with gpt DNA into Clone 22 (AI colonies/ 10^4 gpt + cells)[b]	Direct transfection into Clone 22 (AI colonies/ 10^5 cells)[c]	Direct transfection into Clone 41 AI colonies/ 10^5 cells)[c]
$T^3$6274	120 ± 30	104 ± 6	99 ± 2
RT101	72 ± 21	-	52 ± 5
R681	60 ± 19	-	-
JB8	18 ± 9	-	-
Cl 30 (P^-)	17 ± 7	-	0 ± 1
None	18 ± 6	12 ± 1	-

[a]Each value represents the mean ± one half the range for two separate experiments run in duplicate.
[b]Selected for gpt after cotransfection; gpt^+ cells then tested for colony formation in 0.33% agar, 10% fetal bovine serum.
[c]Tested for colony formation in 0.33% agar, 7% fetal bovine serum. A.I.: anchorage independent.

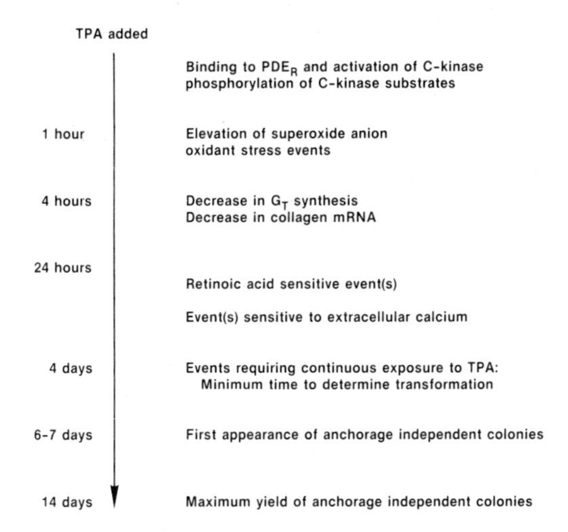

FIG. 2. Sequence of TPA-induced events relevant to promotion in JB6 P^+ cells.

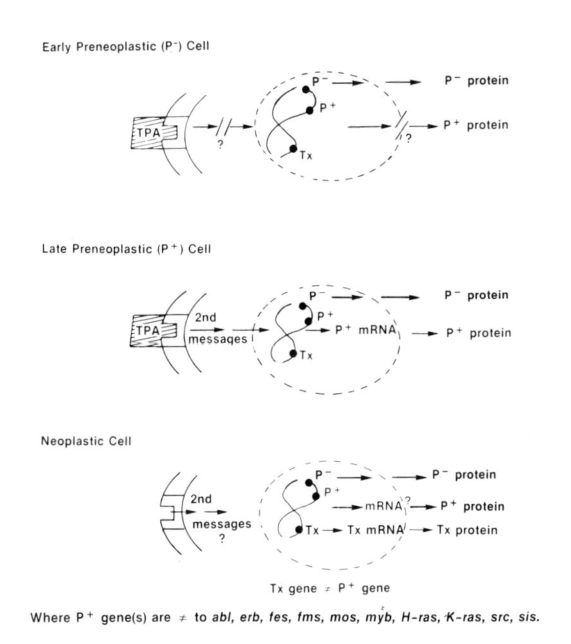

FIG. 3. Model for induction and maintenance of neoplastic transformation in JB6 mouse epidermal cells: genetic determinants.

REFERENCES

1. Bondy, G.P., and Denhardt, D.T. (1983): Carcinogenesis, 4:1599-1603.
2. Boutwell, R.K. (1964): Prog. Exp. Tumor Res., 4:207-250.
3. Boutwell, R.K. (1982): Adv. Polyamine Res., 4:127-134.
4. Burns, F.J., Vanderlaan, M., Snyder, E., and Albert, R.E. (1978): In: Carcinogenesis, Mechanisms of Tumor Promotion and Cocarcinogenesis, Vol. 2, edited by T.J. Slaga, A. Sivaka, and R.K. Boutwell, pp. 91-96. Raven Press, New York.
5. Cavalli-Sforza, L.L., and Lederberg, J. (1956): Genetics, 41:367-381.
6. Colburn, N.H., and Gindhart, T.D. (1981): Biochem. Biophys. Res. Commun., 102:799-807.
7. Colburn, N.H., Former, B.F., Nelson, K.A., and Yuspa, S. H. (1979): Nature, 282:589-591.
8. Colburn, N.H., Gindhart, T.D., Dalal, B., and Hegamyer, G.A. (1983): In: Organ and Species Specificity in Chemical Carcinogenesis, edited by R. Lagenbach, S. Nesnow, and J. Rice, pp. 189-200. Plenum Press, New York.
9. Colburn, N.H., Gindhart, T.D., Hegamyer, G.A., Blumberg, P.M., Delclos, K.B., Magun, B.E., and Lockyer, J. (1982): Cancer Res., 42:3093-3097.
10. Colburn, N.H., Kohler, B.A., and Nelson, K.J. (1980): Teratogenesis, Carcinogenesis, and Mutagenesis J., 1:87-96.
11. Colburn, N.H., Lerman, M.I., Hegamyer, G.A., and Gindhart, T.D. (1984): Mol. Cell. Biol., (in press).
12. Colburn, N.H., Lerman, M.I., Hegamyer, G.A., Wendel, E., and Gindhart, T.D. (1984): In: Genes and Cancer, edited by J M. Bishop, M. Greaves, and J.D. Rowley, pp. 137-155. Alan R. Liss, Inc., New York.
13. Colburn, N.H., Lerman, M.I., Srinivas, L., Nakamura, Y., and Gindhart, T.D. (1984): In: Cellular Interactions by Environmental Tumor Promoters, edited by H. Fujiki and T. Sugimura. Scientific Societies Press, Tokyo, in press
14. Colburn, N.H., Ozanne, S., Lichti, U., Ben, T., Yuspa, S., Wendel, E., Jardini, E., and Abruzzo, G. (1981): Ann. N.Y. Acad. Sci., 359:251-259.
15. Colburn, N.H., Srinivas, L., Hegamyer, G., Dion, L.D., Wendel, E.J., Cohen, M., and Gindhart, T.D. (1984): In: The Role of Cocarcinogens and Promoters in Human and Experimental Carcinogenesis, edited by M. Borsonyi. International Agency for Research on Cancer, Lyon, France.
16. Colburn, N.H., Talmadge, C.B., and Gindhart, T.D. (1983): In: Progress in Nucleic Acid Research and Molecular Biology, edited by W.E. Cohn, pp. 107-110. Academic Press, Inc., New York.

17. Colburn, N.H., Talmadge, C.B., and Gindhart, T.D. (1983): Mol. Cell. Biol., 3:1182-1186.
18. Colburn, N.H., Vorder-Bruegge, W.F., Bates, J.R., Gray, R.H., Rossen, J.D., Kelsey, W.H., and Shimada, T. (1978): Cancer Res., 38:624-634.
19. Colburn, N.H., Wendel, E., and Abruzzo, G. (1981): Proc. Nat. Acad. Sci., USA, 78:6912-6916.
20. Colburn, N.H., Wendel, E., and Srinivas, L. (1982): J. Cell. Biochem., 18:261-270.
21. Copley, M.P., Gindhart, T.D., and Colburn, N.H. (1983): J. Cell. Physiol., 114:173-178.
22. DiGiovanni, J., Prichett, W.P., Decina, P.C., and Diamond, L. (1984): Carcinogenesis (in press).
23. Dion, D.L., Bear, J., Bateman, J., DeLuca, L.M., and Colburn, N.H. (1982): J. Nat. Cancer Inst., 69:1147-1154.
24. Dion, L.D., Gindhart, T.D., and Colburn, N.H. (1984): Cancer Res., submitted.
25. Drinkwater, N.R., and Ginsler, J. (1984): Proc. Am. Assoc. Cancer Res., 25:126.
26. Gensler, H.L., and Bowden, G.T. (1983): Carcinogenesis, 4:1507-1511.
27. Gindhart, T.D., Nakamura, Y., Stevens, L.A., Hegamyer, G.A., West, M.W., Smith, B.M., and Colburn, N.H. (1985): In: Tumor Promotion and Enhancement in the Etiology of Human and Experimental Respiratory Tract Carcinogenesis, edited by M. Mass. Raven Press, New York.
28. Gindhart, T.D., Stevens, L., and Copley, M.P. (1984): Carcinogenesis, 5:1115-1121.
29. Hakomori, S. (1975): Biochim. Biochem. Acta, 417:56-89.
30. Heidelberger, C. (1975): Annu. Rev. Biochem., 44:79-121.
31. Huberman, E., Heckman, C., and Langenbach, R. (1979): Cancer Res., 39:2618-2624.
32. Kensler, T.W., and Trush, M.A. (1984): In: Superoxide Dismutase, edited by L.W. Oberley. CRC Press, Inc., Boca Raton, Florida, in press.
33. Nakamura, Y., Colburn, N.H., and Gindhart, T.D. (1984): Carcinogenesis, (in press).
34. Pitot, H., and Heidelberger, C. (1963): Cancer Res., 23:1694-1700.
35. Reiners, J.J., Jr., Nesnow, S., and Slaga, T.J. (1984): Carcinogenesis, 5:301-307.
36. Sobel, M.E., Dion, L.D., Vuust, J., and Colburn, N.H. (1983): Mol. Cell. Biol., 3:1527-1532.
37. Srinivas, L., and Colburn, N.H. (1984): Carcinogenesis, 5:515-519.
38. Srinivas, L., Gindhart, T.D., and Colburn, N.H. (1982): Proc. Nat. Acad. Sci., USA 79:4988-4991.

39. Ward, J.M., Rice, J.M., Creasia, D., Lynch, P., and Riggs, C. (1983): Carcinogenesis, 4:1021-1029.
40. Yuspa, S.H., and Harris, C.C. (1974): Exp. Cell. Res., 86:95.
Transformation Sensitivity Determinants

Carcinogenesis, Vol. 10, edited by E. Huberman
and S. H. Barr. Raven Press, New York © 1985.

Receptors and Endogenous Analogs for the Phorbol Ester Tumor Promoters

P. M. Blumberg, A. Y. Jeng, B. König, N. A. Sharkey, K. L. Leach, and S. Jaken

Laboratory of Cellular Carcinogenesis and Tumor Promotion, National Cancer Institute, Bethesda, Maryland 20205

For mouse skin, the phorbol esters represent the most potent class of tumor promoters. In addition to their promotional activity, the phorbol esters also have profound effects on a wide variety of cellular systems (4). These effects include (i) modulation of the activity of growth factors and other cellular effectors, (ii) stimulation of secretory cells, (iii) alteration of programs of differentiation and of differentiated cell properties, (iv) induction in normal cells of a partially transformed phenotype, and (v) enhanced expression of the transformed phenotype in cells that are already transformed. The diversity of biological responses to the phorbol esters strongly argues that the phorbol esters modulate a central regulatory pathway in cells.

Two indirect lines of evidence suggested that this modulation resulted from interaction of the phorbol esters at specific binding sites. First, the phorbol esters are highly potent, inducing biological responses at nanomolar concentrations. Second, structural modifications, which should have only limited effects on the physicochemical properties of the phorbol esters, often have dramatic effects on their biological activities. Nonetheless, early efforts to demonstrate such receptors were unsuccessful, in large part because of the high lipophilicity of the ligand, phorbol 12-myristate 13-acetate (PMA), which was used.

As an alternative approach, we synthesized [^{3}H]phorbol 12,13-dibutyrate ([^{3}H]PDBu) (Fig. 1), which we predicted should possess the optimal ratio of specific binding activity to nonspecific uptake (14). Using this derivative, we and subsequently others were able to demonstrate specific, saturable phorbol ester receptors in a wide variety of cells and in particulate preparations. Elsewhere we have reviewed in detail the results of these binding studies (6).

Briefly, [^{3}H]PDBu binds to intact cells with a dissociation constant of 7-50 nM (23 nM mean). The number of sites per cell typically ranges from

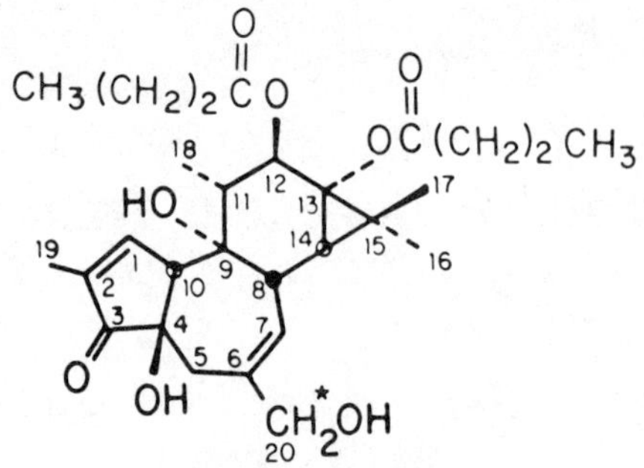

FIG. 1. Structure of [^{3}H]phorbol 12,13-dibutyrate.

1-5 x 10^5. In broken cell and tissue preparations, the binding activity is predominantly particulate. In addition, however, an apo-receptor can be demonstrated in the cytosolic fraction. This apo-receptor by itself fails to bind phorbol esters. Addition of phospholipids to it generates binding activity.

The evidence that the phorbol ester receptors mediate biological responses to the phorbol esters is quite strong (6). First, for a number of biological responses, the half-maximally effective doses of different phorbol derivatives for inducing these responses agree quantitatively with the equilibrium dissociation constants (K_ds) for receptor binding. An example is the reduction in cell surface fibronectin on chicken embryo fibroblasts treated for 3 days with phorbol esters (14). We found no more than a 3.5-fold difference between the values for binding and biological response for a series of eight derivatives spanning a range of 6 x 10^4 in biological potencies. Second, good agreement between the Kd for [^{3}H]PDBu binding and its ED_{50} has been found for additional responses in studies that did not include extensive structure-activity analysis. Third, a class of compounds structurally unrelated to the phorbol esters has been identified; these compounds are highly potent tumor promoters that also induce the variety of biological responses previously characterized for the phorbol esters (37). These compounds, teleocidin and lyngbyatoxin (Fig. 2), are indole alkaloids rather than diterpenes. Like the phorbol esters, these compounds block [^{3}H]PDBu binding with affinities corresponding to their biological potencies.

HETEROGENEITY IN PHORBOL ESTER RECOGNITION AND RESPONSE

It is important to distinguish between the conclusion that the major phorbol ester receptor mediates many of the phorbol ester responses, as

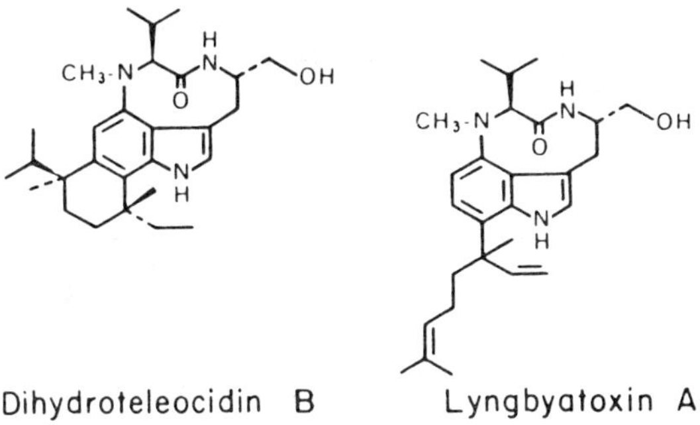

Dihydroteleocidin B Lyngbyatoxin A

FIG. 2. Structure of dihydroteleocidin B and lyngbyatoxin.

argued by the above data, and the conclusion that it mediates all responses. The reason for caution in interpretation is a considerable body of evidence for heterogeneity in phorbol ester recognition and response.

First, different structure-activity relations may exist for different biological responses. For example, tumor promotion itself has been subdivided into first and second stages (33). Mezerein and phorbol 12-retinoate 13-acetate are reported to be similar to PMA in their potency as inflammatory agents and as second-stage tumor promoters, whereas they are much less potent as complete tumor promoters. Resiniferatoxin is at least 100-fold more potent than PMA for causing acute ear reddening (28). Its irritant action differs from that of PMA, however, in being highly transient. Moreover, for a variety of other typical phorbol ester responses, resiniferatoxin is much less potent than PMA (13), and it likewise binds to the phorbol ester receptor with much lower affinity (14). Differences in pharmacokinetics may explain some of the disparities in the tumor promotion studies. These differences, however, are highly unlikely to account for the anomalous behavior of resiniferatoxin.

Second, the same cell may show different ED_{50}s for the same derivative for two different biological responses. PMA was reported to induce resistance in K562 cells to killing by natural killer cells with an ED_{50} of 50 pM (10), whereas adherence was induced with an ED_{50} of 1 nM (16). Under identical culture conditions and incubation times, 25-fold differences in ED_{50}s for mezerein were obtained for inhibition of epidermal growth factor binding and induction of prostaglandin (PGE_2), synthesis (19).

Third, a number of phorbol ester dose-response curves have been reported to be biphasic, showing stimulation at low phorbol ester concentrations and inhibition at high concentrations--which, however, are still below those concentrations at which nonspecific effects might be expected. Examples include transformation of lymphocytes by EB virus (39), induction of sister chromatid exchanges in C3H 10T1/2 cells (23), and reduction in colony-forming efficiency in human fibroblasts (21).

Fourth, instances of antagonistic combinations of phorbol esters have been observed. For example, highly soluble phorbol esters, e.g. phorbol 12,13-diacetate (PDA), have been reported to inhibit promotion by PMA (27) [this effect, however, was not found in a second mouse strain (34)]. Likewise, PDA at concentrations below its K_d for the receptor was found to inhibit macrophage chemotaxis in response to PMA (35). In other cases, however, PDA acts like a typical phorbol ester. For example, the amounts of stimulation of 2-deoxyglucose uptake in chicken embryo fibroblasts by PMA, PDA, or mixtures of the two were indistinguishable, provided concentrations of each ligand were normalized for the relative potencies of the two ligands (12).

At the more trivial level, it is obvious that the phorbol ester receptor is not the only site for phorbol ester interaction in biological systems since the compounds are degraded by esterases (32) and interact with at least one serum binding protein (31). There is no evidence, however, that these interactions directly mediate responsiveness. Second, as lipophilic molecules, the phorbol esters can interact with phospholipids and insert into lipid bilayers (18). This interaction may disrupt the properties of the bilayer and cause cell toxicity. Thus, we had observed inhibition of 2-deoxyglucose uptake and loss of protein from monolayers of chicken embryo fibroblasts by PMA with ED_{50}s of 3 and 11 μM, respectively (11). Similarly, Belman and Troll had observed lysis of erythrocytes at 4 μM PMA (2). On the other hand, as we have discussed elsewhere (29), the evidence that the phorbol esters interact simply with phospholipids to cause their high-affinity, specific effects is unconvincing.

Binding analysis supports the above biological evidence implying receptor heterogeneity. Although the usual observation is that phorbol ester binding yields linear Scatchard plots, curved plots have been obtained for intact rat embryo fibroblasts (17), mouse skin particulate preparations (15), and, in one study, lymphoid cells of myeloid origin (10). The data for the mouse skin particulate preparations are compatible with a model of three different affinity sites, of which two also possess good affinity for [^{3}H]12-deoxyphorbol 13-isobutyrate ([^{3}H]DPB) (Fig. 3).

ASSOCIATION OF KINASE ACTIVITY WITH THE PHORBOL ESTER RECEPTOR

The characterization of the phorbol ester receptor revealed marked similarities to that of an enzymatic activity being studied by Nishizuka and co-workers during this same period (Table 1). Because of these similarities, Castagna et al. examined the ability of the phorbol esters to affect protein kinase C activity (9). They observed that, under conditions of limiting Ca^{++} and phospholipid, the phorbol esters at nanomolar con-

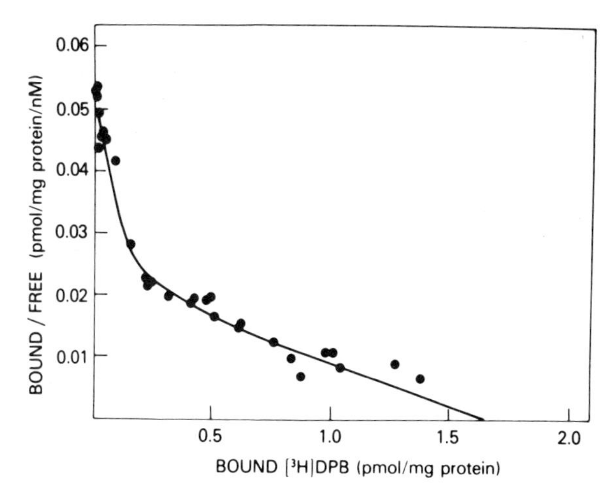

FIG. 3. Binding of [^{3}H]DPB to mouse skin particulate preparations. From Dunn and Blumberg (15).

centrations activated the kinase by reducing the Ca^{++} and phospholipid requirements. The actual identity of the cytosolic phorbol ester apo-receptor and protein kinase C was demonstrated by co-purification of the two activities in our studies (22) and in studies conducted at several other laboratories (24,26).

Although the precise relationship of the cytosolic apo-receptor and of the membrane receptor remain to be unambiguously established, the available data are consistent with the membrane receptor being a membrane-lipid-associated form of the apo-receptor. First, solubilized membrane receptors from brain co-fractionate with protein kinase C (1,20). Second, the proportion of total binding activity (7) or protein kinase C activity that is in the particulate as compared to the cytosolic fraction depends on the conditions of cell lysis. More is associated with

TABLE 1. Similarities between phorbol ester receptor and protein kinase C.

1. Tissue distribution
2. Absolute level in brain
3. Evolutionary conservation
4. High Ca^{++} sensitivity
5. Phospholipid association

the membranes in the presence of Ca^{++}, less in the presence of chelators. The outstanding issues that remain are (i) whether all or only some of the membrane receptors are the same as the apo-receptor and (ii) whether some of the membrane-associated apo-receptors have been further modified.

ROLE OF PHOSPHOLIPID IN THE PROPERTIES OF THE PHOSPHOLIPID APO-RECEPTOR COMPLEX

The essential requirement of the phorbol ester apo-receptor for phospholipids emphasizes the concept that "the receptor" is in fact a phospholipid-apo-receptor complex rather than simply the apo-receptor moiety. Association of the same apo-receptor with different lipid domains will generate different holo-receptors. We are in the process of characterizing the contributions of the phospholipids to the properties of the holo-receptor complex to assess the extent to which this mechanism may account for receptor heterogeneity.

We find that pure phospholipids and phospholipid mixtures vary markedly in whether or not they can form an active complex with the apo-receptor. The anionic phospholipids, phosphatidylserine (PS), phosphatidylinositol (PI), and phosphatidic acid (PA), are all active. The uncharged phospholipids, phosphatidylcholine (PC), phosphatidylethanolamine (PE), and sphingomyelin (SM), are inactive. Mixtures of the anionic and uncharged phospholipids, e.g. PS-PC (1:2), PS-PE (1:2), or a mixture resembling that of the red blood cell membrane, PC:PE:PS:SM:PI:DPG (30:30:13:25:1:1), are also active. Dose-response curves indicate that active phospholipids and phospholipid mixtures all fully reconstitute the apo-receptor. The mixtures differ considerably, however, in the amounts of the phospholipids required to reconstitute activity. Thus, the ED_{50} values for PS and the RBC-PL mixture were 7 and 280 μg/mL, respectively. The amount of phospholipid needed for reconstitution depends not only on the nature of the phospholipid but also on the presence or absence of Ca^{++}. For example, in the presence of excess Ca^{++}, the ED_{50}s for PS and the RBC-PL mixture drop to 0.2 and 2 μg/mL, respectively.

In addition to determining whether the holo-receptor complex will form, the phospholipids also play a role in specifying the binding properties of the complex. Control could occur at two levels: (i) the level of absolute binding affinities and (ii) the level of structure-activity relationships, i.e. in affinities for one phorbol ester relative to another. Changes are found at both levels. For example, Scatchard analysis of [^{3}H]PDBu binding to the apo-receptor complexes with PS and the RBC-PL mixture yield K_ds of 0.8 and 22 nM, respectively. Structure-activity analysis measuring competition of [^{3}H]PDBu binding by other analogs indicated a similar shift in affinities for many of the derivatives examined, including the diacylglycerol derivative diolein (see below), 4-0-methyl PMA, and

DPB. Several derivatives, however, showed considerably smaller shifts. For example, the K_d for phorbol 12,13-distearate assayed after incorporation into the liposomes showed only a twofold difference between the apo-receptor-PS and -RBC-PL complexes.

The difference in binding affinities for the apo-receptor complexed with different phospholipids suggests that heterogeneity in lipid environments should lead to curved Scatchard plots. That prediction has been confirmed. For example, the Scatchard plot of [^{3}H]PDBu binding to the apo-receptor reconstituted with a 1:1 mixture of separate liposomes of PS and of PS+ 1.8% diolein gave the curve predicted for a combination of the plots for the two individual components (7).

We conclude that the lipid domain plays an important role in determining the binding properties of the phorbol ester holo-receptor. Lipid heterogeneity within the membrane or between membranes may thus account at least in part for apparent receptor heterogeneity. Since different lipid domains can modulate binding affinities, this mechanism likewise provides a means for regulating susceptibility of kinase C to activation by both phorbol esters and endogenous ligands.

DIACYLGLYCEROLS AS ENDOGENOUS PHORBOL ESTER ANALOGS

The high evolutionary conservation of the phorbol ester binding site had strongly argued for the existence of an endogenous ligand that interacted at that site and had exerted a constraint on its divergence over evolution (5). A somewhat analogous example would be the opiate receptor, for which the endorphins and enkephalins are the endogenous ligands. Although initial efforts to demonstrate any endogenous competitive inhibitors of phorbol ester binding failed, the assays were carried out under conditions that would have failed to detect either highly unstable or highly lipophilic compounds (6).

A technically important consequence of the identification of the phorbol ester apo-receptor was that it made possible the manipulation of the lipid domain of the holo-receptor independently from the protein moiety. The effect of insoluble ligands can now be assessed by incorporating the ligands directly into the phospholipid liposomes used to complex with the apo-receptor. In vivo diacylglycerols are generated by breakdown of phosphatidylinositol as one of the earliest responses to hormone receptor occupancy by a large class of hormones and other cellular effectors (3). In vitro, diacylglycerols activate protein kinase C, and Nishizuka and coworkers have postulated that the activation of protein kinase C by this diacylglycerol constitutes one of the two effector pathways for this class of hormones (38). Diacylglycerols therefore appeared to be attractive candidates for an endogenous phorbol ester analog. Our binding studies strongly argue that this is indeed the case (30).

If incorporated into the phosphatidylserine liposomes used to reconstitute the apo-receptor, the diacylglycerol derivative 1,2-diolein reduced the [^{3}H]PDBu binding affinity without affecting the B_{max}. This inhibition could result from one of two mechanisms. The diolein could perturb the lipid domain of the holo- receptor nonspecifically and thereby shift the K_d. Alternatively, the diolein could bind at the phorbol ester binding site and function as a true competitive inhibitor. In this latter case, the equation

$$K_{app} = K_d \ (1 + I/K_I)$$

should describe the apparent binding affinity for [^{3}H]PDBu as a function of the diolein concentration where K_{app} and K_d are the apparent dissociation constants for [^{3}H]PDBu in the presence and absence of diolein respectively, I is the concentration of diolein, and K_I is the dissociation constant for the diolein. We could, in fact, demonstrate the predicted relationship over a 50-fold range in diolein concentrations (30).

A second test to distinguish the possible mechanisms of inhibition by the diacylglycerols is to determine the stoichiometry of the interaction. Were the diacylglycerol to be simply perturbing the lipid environment, inhibition should be independent of the diacylglycerol:receptor ratio. To establish the conditions for these experiments, it was first necessary to clarify the nature of the K_I for diacylglycerols.

The diolein inhibition studies described above were carried out at a fixed phosphatidylserine concentration of 960 μg/mL. Because diolein is highly hydrophobic, essentially all should be dissolved in the phosphatidylserine liposomes and none should be free in aqueous solution. The K_I for diolein, therefore, should represent the local concentration in the liposomes rather than in the aqueous volume over which the liposomes were dispersed. Experimentally, these two possibilities were distinguished by measuring the K_I for diolein as a function of the phosphatidylserine concentration in the assay (30). As expected, over a 400-fold range of phosphatidylserine concentrations, the K_I, expressed as weight percent relative to phospholipid, remained almost constant at 0.35-1.1%. In contrast, the K_I expressed in terms of concentration averaged over the aqueous assay volume varied markedly from 7-0.05 μg/mL. These findings had three consequences. First, they established conditions to probe the stoichiometry of the interaction of diolein with the receptor. Second, they permitted appropriate comparison of the relative affinities of phorbol esters and diacylglycerols for the receptor. Third, they indicated that the receptor recognizes the membrane-associated form of the ligand.

The difficulty in determining the stoichiometry of binding for an insoluble ligand is that it is not possible to separate physically the bound ligand from the free. We therefore took advantage of an alternative approach, portrayed schematically in Fig. 4. The binding of ligand to its receptor can be detected by the resultant reduction in the concentration of ligand that remains unbound. The apparent dissociation constant, K_{app}, for the ligand should, therefore, be distorted as a function of the ratio of total ligand to total receptor. The expected relationship is

$$K_{app} = K_I + \frac{n(R/2)}{1 + L/K_D}$$

where K_{app} is derived from the concentration of total ligand yielding 50% inhibition of phorbol ester binding, K_I is the dissociation constant expressed in terms of free ligand, R is the concentration of receptor, L is the concentration of phorbol ester, and K_D is the dissociation constant for the phorbol ester. The value n represents the number of ligand molecules associated with the receptor under conditions that inhibit binding. To get adequate perturbation of K_{app} for analysis, assay conditions must be adjusted so that $R >> K_I$. For diolein interacting with the phorbol ester apo-receptor-phosphatidylserine complex, this constraint pushes the measurements close to the limit of technical feasibility. Nonetheless, this condition can be met by reducing the concentration of PS sufficiently and by maximizing the holo-receptor concentration.

Under these conditions, we find that the K_{app} depends markedly on the ligand:receptor ratio (B. König, P. DiNitto, and P. M. Blumberg, manuscript submitted). The data give a good fit to the theoretical equation, moreover, and yield an experimental value for n of 1.72. These results

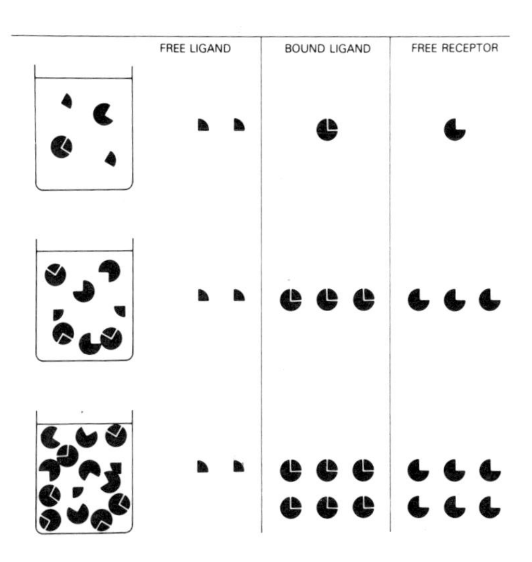

FIG. 4. Distortion of apparent affinity due to stoichiometry of binding.

strongly argue that diolein does not inhibit phorbol ester binding simply by perturbing the lipid environment. If so, no relationship between K_{app} and the inhibitor: receptor ratio should have been observed, i.e., n=0. The experiments were performed with the racemic DL-1,2-diolein. Recently, Rando and Young (25) reported that only the D-isomer (the natural isomer) is active at protein kinase C stimulation. Expressed in terms of the biologically active isomer of diolein, the measured value for n is thus 0.86, consistent with a 1:1 stoichiometry.

To determine the relative affinity of diacylglycerols and phorbol esters for the receptor, we measured the binding potencies for two homologous pairs of compounds, the dilaurate and dioleate derivatives (29). The diacylglycerols were somewhat less active. Thus, the K_I values for DL-1,2-dilaurin and DL-1,2-diolein were 17- and 78-fold greater (i.e., less potent). Expressed in terms of the active stereoisomer, the differences were 8.5- and 39-fold, respectively. For comparison, 4-O-methyl PMA was found to be 400-fold less potent than PMA for inducing fibronectin loss in chicken embryo fibroblasts (11). For nonhomologous pairs of compounds, the differences in potencies may be considerably more substantial than in the above examples. Expressed in terms of its membrane concentration, PDBu was 3-6 x 10^4-fold more potent than was DL-1,2-diolein (30).

A prediction from the above evidence that diacylglycerols are endogenous phorbol ester analogs is that treatments that elevate diacylglycerol levels _in situ_ in cells should also shift the phorbol ester receptor binding affinity in the cells and induce similar responses to the phorbol esters. These predictions were verified in collaborative experiments carried out with members of Dr. Yuspa's Section of the Laboratory of Cellular Carcinogenesis and Tumor Promotion at the National Cancer Institute. Treatment of cultured primary mouse keratinocytes with phospholipase C from _B. cereus_ caused hydrolysis of phosphatidylcholine and elevated cellular diacylglycerol levels. Depending on the phospholipase C concentration, the K_I for [^{3}H]PDBu binding increased up to fourfold (i.e., binding affinity was decreased) with no change in the number of binding sites. Similarly, four biological effects characteristic of phorbol ester treatment--altered morphology, inhibition of epidermal growth factor binding, induction of ornithine decarboxylase, and induction of transglutaminase--were observed.

Although the diacylglycerols appear to act as endogenous phorbol ester analogs, they show marked differences in other pharmacological characteristics. Diacylglycerols are very rapidly metabolized _in vivo_, whereas the phorbol esters are relatively stable. As a result, higher concentrations of soluble diacylglycerols (e.g., 1,2-dicaprylin or 1-oleoyl,2-acetylglycerol) may be required to obtain responses in whole cell experiments than would have been predicted from _in vitro_ binding studies. The phorbol esters do not behave in this manner. Likewise, whereas phorbol esters should equilibrate between cell membranes, soluble diacylglycerols that are added exogenously may be predominantly localized to the plasma membrane as a result of metabolism occurring before they can reach internal membranes. They might thus mimic some phorbol ester responses better than others (see below). For endogenously generated diacylglycerols, a second factor leading to this same result is their insolubility, which should inhibit intermembrane transfer.

Obviously, the identification of one class of endogenous ligands does not prove that other classes do not exist. Among exogenous ligands, structurally distinct classes include the phorbol esters, indole alkaloids, and polyacetates.

GENERAL DISCUSSION

The understanding of phorbol ester action provides yet another example of the tremendous utility of exogenous, bioactive natural products as probes of important physiological pathways. In only a few systems has the role of protein kinase C been documented directly. In contrast, an extensive body of knowledge about the activity of the phorbol esters has been accumulated. To the degree that the generalization holds that different phospholipid-protein kinase C complexes correspond to the phorbol ester receptor, then this knowledge of phorbol ester action provides at least preliminary identification of the physiological functions of this pathway. Because of their more favorable pharmacological characteristics of greater stability and potency, moreover, the phorbol esters constitute better probes for this pathway than do the endogenous activators, the diacylglycerols.

An important objective of the analysis of the initial actions of the phorbol esters is to identify steps suitable for intervention or monitoring. The finding that the phorbol esters act at a regulatory site on the receptor strongly implies that the development of pharmacological antagonists, i.e., compounds that bind but do not activate, should be an attainable goal. Only a limited number of exogenous compounds directly bind at the phorbol ester site. The identification of diacylglycerols as endogenous phorbol ester analogs now suggests additional, indirect mechanisms by which agents could activate the phorbol ester pathway. For example, stimulation of endogenous phospholipase C activity should generate diacylglycerol and activate the receptor. Such stimulation could occur either at the level of the enzyme *per se* or in terms of the amount or accessibility of its lipid substrate. An example of this mechanism may be that of the *src* kinase from Rous sarcoma virus, which has been reported to phosphorylate not only tyrosines on protein but also the lipid phosphatidylinositol (36). The product of this reaction, phosphatidylinositol-4,5-bisphosphate, is the substrate for the hormonally activated phospholipase C. We and investigators from Dr. Weinstein's laboratory had previously demonstrated marked overlap between the effects in chicken embryo fibroblasts of the phorbol esters and Rous sarcoma virus (4). Part of this overlap could reflect common activation of protein kinase C. An on-going problem has been the development of assays for tumor promoters. Monitoring of the in vivo levels of protein kinase C activation may provide an integrated measure of those indirectly acting promoters functioning through this pathway.

The recognition that the phorbol ester receptor contains both phospholipid and protein moieties emphasizes the complexity of the system. Heterogeneity of binding, for example, may represent a continuum of different lipid environments rather than distinct, uniform receptor subclasses. The probability that the apo-receptor can associate with different intracellular membranes suggests that intracellular localization of the receptors may be important. Localization to different internal sites should lead to different profiles of accessible substrates and thus somewhat different mixtures of biological consequences. Phorbol esters may be more efficient for activation at internal sites than are the rapidly metabolized diacylglycerols. As we have suggested elsewhere (8), differences in in vivo structure-activity relations among phorbol esters could reflect differences in the kinetics of subcellular distribution in addition to differences in their equilibrium binding affinities.

ACKNOWLEDGMENT

We thank Dr. S. Yuspa for careful reading of the manuscript and M. Bellman for expert typing.

REFERENCES

1. Ashendel, C.L., Staller, J.M., and Boutwell, R.K. (1983): Cancer Res., 43:4333-4337.
2. Belman, S., and Troll, W. (1978): In: Carcinogenesis - A Comprehensive Survey, Mechanisms of Tumor Promotion and Cocarcinogenesis, Vol. 2, edited by T.J. Slaga, A. Sivak, and R.K. Boutwell, pp. 117-134. Raven Press, New York.
3. Berridge, M.J. (1984): Biochem. J., 220:345-360.
4. Blumberg, P.M. (1981): Crit. Rev. Toxicol., 8:153-234.
5. Blumberg, P.M., Delclos, K.B., and Jaken, S. (1983): In: Organ and Species Specificity in Chemical Carcinogenesis, edited by R. Langenbach, S. Nesnow, and J.M. Rice, pp. 201-227. Plenum Press, New York.
6. Blumberg, P.M., Dunn, J.A., Jaken, S., Jeng, A.Y., Leach, K.L., Sharkey, N.A., and Yeh, E. (1984): In: Mechanisms of Tumor Promotion, Tumor Promotion and Carcinogenesis in Vitro, Vol. 3, edited by T.J. Slaga, pp. 143-184. CRC Press, Inc., Boca Raton, Florida.

7. Blumberg, P.M., Jaken, S., König, B., Sharkey, N.A., Leach, K.L., Jeng, A.Y., and Yeh, E. (1984): Biochem. Pharmacol., 33:933-940.
8. Blumberg, P.M., König, B., Sharkey, N.A., Leach, K.L., Jaken, S., and Jeng, A.Y. (1984): In: Models, Mechanisms, and Etiology of Tumor Promotion, edited by M. Borzsonyi, N.E. Day, K. Lapis, and H. Yamasaki. International Agency for Research on Cancer, Lyon, France, in press.
9. Castagna, M., Takai, Y., Kaibuchi, K., Sano, K., Kikkawa, U., and Nishizuka, Y. (1982): J. Biol. Chem., 257:7847-7851.
10. Colburn, N.H., Gindhart, T.D., Dalal, B., and Hegamyer, G.A. (1983): In: Organ and Species Specificity in Chemical Carcinogenesis, edited by R. Langenbach, S. Nesnow, and J.M. Rice, pp. 189-200. Plenum Press, New York.
11. Driedger, P.E., and Blumberg, P.M. (1979): Cancer Res., 39:714-719.
12. Driedger, P.E., and Blumberg, P.M. (1980): Cancer Res., 40:339-346.
13. Driedger, P.E., and Blumberg, P.M. (1980): Cancer Res., 40:1400-1404.
14. Driedger, P.E., and Blumberg, P.M. (1980): Proc. Nat. Acad. Sci., USA, 77:567-571.
15. Dunn, J.A., and Blumberg, P.M. (1983): Cancer Res., 43:4632-4637.
16. Fukuda, M. (1981): Cancer Res., 41:4621-4628.
17. Horowitz, A.D., Greenebaum, E., and Weinstein, I.B. (1981): Proc. Nat. Acad. Sci., USA, 78:2315-2319.
18. Jacobson, K., Wenner, C.E., Kemp, G., and Papahadjopoulos, D. (1975): Cancer Res., 35:2991-2995.
19. Jaken, S., Shupnik, M.A., Blumberg, P.M., and Tashjian, A.H., Jr. (1983): Cancer Res., 43:11-14.
20. Kikkawa, U., Takai, Y., Minakuchi, R., Inohara, S., and Nishizuka, Y. (1982): J. Biol. Chem., 257:13341-13348.
21. Kopelovitch, L., and Bias, N.E. (1980): Exp. Cell Res., 48:207-217.
22. Leach, K.L., James, M.L., and Blumberg, P.M. (1983): Proc. Nat. Acad. Sci., USA, 80:4208-4212.
23. Nagasawa, H., and Little, J.B. (1981): Carcinogenesis, 2:601-607.
24. Niedel, J.E., Kuhn, L.J., and Vandenbark, G.R. (1983): Proc. Nat. Acad. Sci., USA, 80:36-40.
25. Rando, R.R., and Young, N. (1984): Biochem. Biophys. Res. Commun., 122:818-823.
26. Sando, J.J., and Young, M.C. (1983): Proc. Nat. Acad. Sci., USA, 80:2642-2646.
27. Schmidt, R., and Hecker, E. (1982): In: Carcinogenesis - A Comprehensive Survey, Cocarcinogenesis and Biological Effects of Tumor Promoters, Vol. 7, edited by E. Hecker, N.E. Fusenig, W. Kunz, F. Marks, and H.W. Thielmann, pp. 57-63. Raven Press, New York.
28. Schmidt, R.J., and Hecker, E. (1979): Inflammation, 3:273-280.
29. Sharkey, N.A., and Blumberg, P.M. (1985): Cancer Res., in press.

30. Sharkey, N.A., Leach, K.L., and Blumberg, P.M. (1984): Proc. Nat. Acad. Sci., USA, 81:607-610.
31. Shoyab, M., and Todaro, G.J. (1982): J. Biol. Chem., 257:439-445.
32. Shoyab, M., Warren, T.C., and Todaro, G.J. (1981): J. Biol. Chem., 256:12529-12534.
33. Slaga, T.J. (1984): In: Mechanisms of Tumor Promotion, Tumor Promotion and Skin Carcinogenesis, Vol. 2, edited by T.J. Slaga, pp. 189-196. CRC Press, Inc., Boca Raton, Florida.
34. Slaga, T.J., Fischer, S.M., Nelson, K., and Gleason, G.L. (1980): Proc. Nat. Acad. Sci., U.S.A., 77:3659-3663.
35. Sturm, R.J., Smith, B.M., Lane, R.W., Laskin, D.L., Harris, L.S., and Carchman, R.A. (1983): Cancer Res., 43:4552-4556.
36. Sugimoto, Y., Whitman, M., Cantley, L.C., and Erikson, R.L. (1984): Proc. Nat. Acad. Sci., USA, 81:2117-2121.
37. Sugimura, T., and Fujiki, H. (1983): In: Gann Monograph on Cancer Research, Vol. 29, edited by A. Makita, S. Tsuiki, H. Fujiki, and L. Warren, pp. 3-15. University of Tokyo Press, Tokyo.
38. Takai, Y., Kishimoto, A., and Nishizuka, Y. (1982): In: Calcium and Cell Function, Vol. 2, edited by W.Y. Cheung, pp. 385-412. Academic Press, Inc., New York.
39. Yamamoto, N., and Zur Hausen, H. (1979): Nature, 280:244-245.

Carcinogenesis, Vol. 10, edited by E. Huberman and S. H. Barr. Raven Press, New York © 1985.

Control of Cell Differentiation and Cell Transformation *In Vitro* by Phorbol 12-Myristate 13-Acetate and 1α,25-Dihydroxyvitamin D_3

E. Huberman and C. A. Jones

Division of Biological and Medical Research, Argonne National Laboratory, Argonne, Illinois 60439

From an operational point of view, the carcinogenesis process can be divided into three sequential stages: initiation, promotion, and progression. The first two involve the steps that lead to the transformation of a normal into a malignant cell, whereas progression covers processes through which a transformed cell develops into a malignant tumor.

To understand chemical carcinogenesis, we need to know more about the basic events underlying these processes. Tumor initiation and promotion, in particular, are known to be influenced by environmental chemicals. Studies have shown that after metabolic activation many chemical carcinogens that act as tumor initiators can bind to cellular macromolecules--including DNA--and can induce mutations in various cells (1,7, 22,30,32,33,37,42). Tumor initiation appears to involve a "mutation-like event" in genes that control normal cell growth and differentiation (6,26, 39). On the other hand, chemicals that promote tumor formation usually do not bind to DNA and are devoid of mutagenic activity. These agents, however, affect a number of cellular events (12,61,68) including, in certain cells, the induction of cell differentiation (27,28,44,58). This last property led us to suggest that promotion involves the expression of altered genes, in particular growth-facilitating genes (oncogenes), which, as a result of mutation or gene rearrangements during tumor initiation, are placed under the control of genes that regulate normal cell differentiation. Tumor promotion may thus entail a process similar to gene expression during normal cell differentiation. To examine this idea further, we are studying the ability of tumor promoters and related agents to induce cell transformation and differentiation in a number of mammalian (including human) cell systems. As a tool for studies on their mode of action, we are using different cell lines that are either susceptible and resistant to induction of cell differentiation by such agents.

THE CONTROL OF CELL DIFFERENTIATION IN CULTURED HUMAN CELLS AFTER TREATMENT WITH TUMOR-PROMOTING AGENTS

In our cell differentiation studies with tumor promoters we used three different types of cells that display useful markers of cell differentiation: HO melanoma cells (29), the promyelocytic HL-60 leukemia cells (28), and T lymphoid CEM leukemia cells (59). In the HO melanoma cells phorbol 12-myristate 13-acetate (PMA), a tumor-promoting phorbol diester prototype, at doses as low as 10^{-10} to 10^{-9} M, induces a cell differentiation characterized by an inhibition of cell growth, increased synthesis of melanin, and induction of dendrite-like structures (29). Using this differentiation-susceptible cell system, we were able to demonstrate a relationship between the tumor-promoting activities of a series of phorbol diesters in vivo and the degrees to which these agents induce differentiation in vitro. We also demonstrated a similar relationship with the HL-60 cells in which phorbol diesters and teleocidin induced a macrophage cell differentiation, characterized by an inhibition of cell growth, the appearance of morphologically mature cells, an increased phagocytic capability, and changes in both the reactivity to monoclonal antibodies and activity of certain enzymes (27,28,49).

In the CEM cells, PMA causes the expression of a phenotype that resembles that of a suppressor T lymphocyte (59). The new phenotype is characterized in part by the appearance of a specific antigenic pattern. More specifically, the cells exhibit increased reactivity with OKT3 and OKT8 monoclonal antibodies, which characterize mature suppressor and cytotoxic T lymphocytes, respectively (53-55,60). PMA also reduces the reactivity of the treated cells with the OKT4 monoclonal antibody, which detects inducer/helper T lymphocytes, and with the OKT6 antibody, which detects immature "common" thymocytes (53-55). The treated CEM cells, in common with mature suppressor T lymphocytes, also suppressed [^{3}H]thymidine incorporation into phytohemaglutinin-activated peripheral blood lymphocytes (57) but, unlike cytotoxic lymphocytes, were unable to induce a cytotoxic response in a number of human and rodent target cells.

The ability of PMA to induce differentiation of CEM cells into suppressor cells raises the possibility that tumor promotion in the mouse skin may also involve suppression of the immunity against "initiator"-induced papillomas and carcinomas. Such a suppression could result from an increase in the number of mature suppressor T lymphocytes in the animal that has been given the promoter.

Despite enormous scientific effort, the mechanism by which these chemicals promote tumorigenesis and alter differentiation processes is poorly understood. To aid in analyzing the molecular events involved in these processes, we have isolated an HL-60 cell variant that, unlike the parental HL-60 cell line, is resistant to induction of cell differentiation by PMA and related agents (27,35). These resistant cells, when tested with fluorescent probes used for studies on membrane lipid dynamics, exhibit decreased fluidity of either the inner leaflet of the plasma membrane and/or of the cytosolic organellar membranes (19).

Recently it was found that phorbol diesters and teleocidin bind to cellular receptors in various types of cells including the HL-60 cell line (16,63,66). Subsequent studies have also indicated that these receptors are associated with a specific protein kinase, named protein kinase C (2,9,18,38,50). Other studies also indicated that PMA must remain bound to the surface of HL-60 cells to elicit a biological response, suggesting that a receptor-mediated transmembrane process is involved in PMA induction of differentiation (11). Since the cell surface binding of phorbol esters is not altered in the PMA resistant cells (63), it is reasonable to propose that the basis of the PMA resistance is a block in the signal transmission that occurs subsequent to receptor occupancy. A decrease in the fluidity of the membrane lipids of the resistant cells would be expected to impair such a transmission process. In accord with this hypothesis, our previous studies have demonstrated that PMA-resistant HL-60 cells are also defective in their ability to down-regulate specific ^{3}H-labeled phorbol dibutyrate (PDBu) binding (63); this membrane change, too, could result from decreased fluidity of the inner leaflet of the plasma membrane. A component of this signal may involve translocation of the PMA-receptor/ protein kinase C into the cytosol and/or transmission of the signal to the nucleus (40,52). Such a change in signal transmission(s) may, in the susceptible and resistant HL-60 cells, cause a different pattern of protein phosphorylation. In agreement with this speculation, treatment with PMA for 30 minutes resulted in specific changes in the pattern of protein phosphorylation. A number of these changes, however, were either absent or less pronounced in R-94 cells, which are derived from an HL-60 cell variant that is resistant to PMA-induced cell differentiation (unpublished results). It is thus possible that the proteins that are not phosphorylated in the R-94 cells are the ones that are associated with the PMA-induced cell differentiation in the HL-60 cell and other types of cells and perhaps also with the expression of "mutated tumor genes" during tumor promotion.

THE INDUCTION OF A MACROPHAGE CELL DIFFERENTIATION IN THE HL-60 CELLS BY THE BIOLOGICALLY ACTIVE METABOLITE OF VITAMIN D_3

It was reported that low doses of 1α,25-dihydroxyvitamin D_3 [1,25-$(OH)_2D_3$], the biologically active metabolite of vitamin D_3 (21), induced in HL-60 cells a number of differentiation markers (48,64). The biological activity of 1,25-$(OH)_2D_3$, in common with that for both PMA and teleocidin (17,25,63,66), requires binding to specific cellular receptors (10,20,46). These studies raised the possibility that 1,25-$(OH)_2D_3$ may induce differentiation processes in the HL-60 cells via a mechanism similar to that for PMA and teleocidin. To examine this possibility, we compared the ability of 1,25-$(OH)_2D_3$ and PMA to induce cell differentiation in HL-60 cells and in the R-80 cell variant that is resistant to PMA-induced cell differentiation (27,35). In addition, we investigated the ability of 1,25-$(OH)_2D_3$ to inhibit specific phorbol diester receptor binding (66) as a way of establishing whether 1,25-$(OH)_2D_3$, like teleocidin, shares similar binding sites with phorbol diesters.

We showed that treatment of HL-60 cells with 1,25-$(OH)_2D_3$ increased lysozyme and nonspecific esterase activities, increased the reactivity of the cells with the myeloid-specific OKM1 monoclonal antibody, and increased the fraction of morphologically mature cells; these markers were induced in both a time and dose dependent manner (Fig. 1). The increases in nonspecific esterase activity and in the synthesis of a number of monocyte/macrophage protein markers and the absence of a large fraction of cells with banded or segmented nuclei (characteristic of granulocytes) indicate that 1,25-$(OH)_2D_3$ treatment of HL-60 results in a cell type that more closely resembles that of a monocyte/macrophage than a granulocyte.

An interesting result was that the R-80 cells were relatively resistant to the differentiation induced by both PMA and 1,25-$(OH)_2D_3$. These, and the previous results, suggest that the two inducers may affect a similar process that leads to the monocyte/macrophage-like phenotype. However, this process involves a signal that follows the binding of these inducers to their specific but different receptors, since 1,25-$(OH)_2D_3$ did not compete for the phorbol diester binding sites (Fig. 2) nor did 1,25-$(OH)_2D_3$ alter the pattern of protein phosphorylation in the treated cells (49). Thus, other events caused by both inducers, including "down modulation" of specific binding (46,63) and alterations in phospholipid (8,24,62) and calcium metabolism (21,62,67), may perhaps be the common signal(s) that leads to the similar, although not identical, phenotype.

ENHANCEMENT OF CHEMICALLY INDUCED CELL TRANSFORMATION IN HAMSTER EMBRYO CELLS BY 1,25-$(OH)_2D_3$

In view of the similarities in the biological activities of 1,25-$(OH)_2D_3$ and PMA in HL-60 cells, we decided to determine whether 1,25-$(OH)_2D_3$, like PMA, could enhance (promote) the transformation of cells pretreated (initiated) with chemical carcinogens. We used the hamster embryo cell transformation assay (4,31) with a protocol designed to evaluate the activity of presumptive tumor-promoting agents (56).

The transformed phenotype of the Syrian hamster embryo cell is characterized by defined changes in cell organization within a colony (4,5,15,31). These changes result from alterations in the growth pattern of the transformed cells. Such cells upon isolation, propagation, and inoculation into appropriate hosts can result in tumor formation

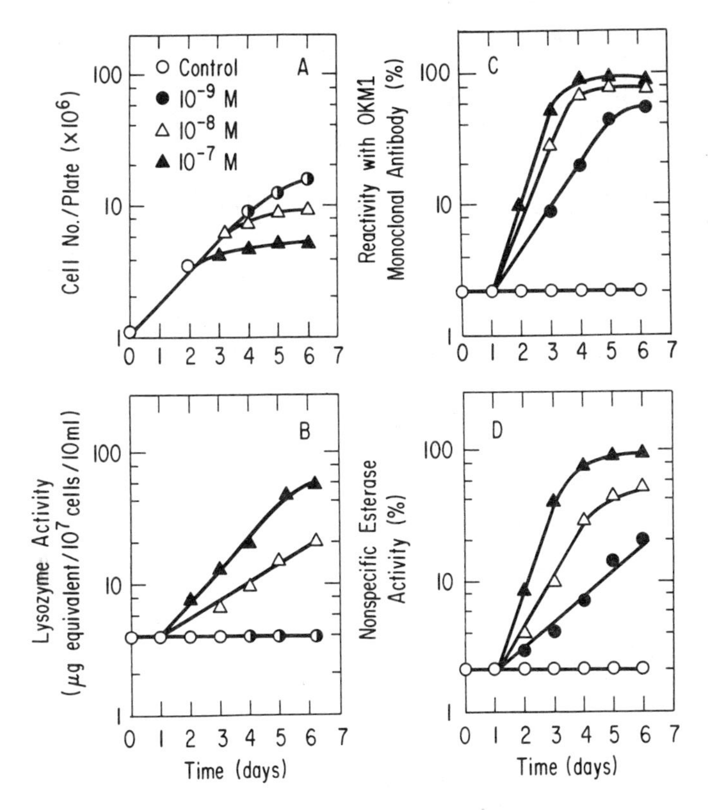

FIG. 1. Cell growth and induction of markers of cell differentiation in HL-60 cells at various times after treatment with different concentrations of 1,25-$(OH)_2D_3$: (A) cell numbers; (B) lysozyme activity; (C) reactivity with OKM1 monoclonal antibody; and (D) nonspecific esterase activity. O, control; ●, 10^{-9} M; △, 10^{-8} M; and ▲, 10^{-7} M. From Murao et al. (49).

(3,4,13,15,41). This assay is thus a useful tool for studying the fundamental mechanism(s) underlying the neoplastic process, including tumor initiation and tumor promotion. The assay is sensitive to diverse classes of carcinogens (both tumor initiators and tumor promoters), and can also be used to identify new presumptive carcinogens (14,23,51).

Our results indicate that 1,25-$(OH)_2D_3$, in common with some tumor promoters (e.g., PMA), will not effectively induce cell transformation by itself. However, when administered subsequent to known carcinogens/tumor initiators, it enhances the frequency of cell transformation. This enhancing effect has been demonstrated with the polycyclic aromatic

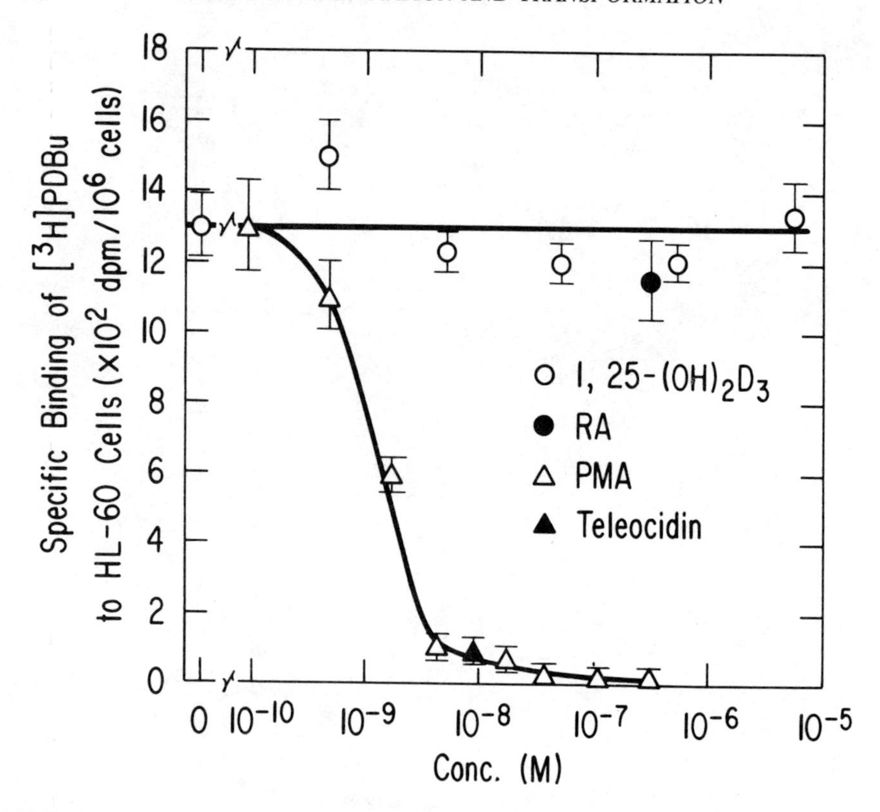

FIG. 2. Inhibition of the specific binding of (^{3}H)PDBu to intact HL-60 cells by (0) 1,25-$(OH)_2D_3$, (●) retinoic acid, (△) PMA, or (▲) teleocidin. Each point represents the mean ± S.E. for three separate experiments done in triplicate. Control value was 13.2 ± 0.8 x 10^2 dpm/10^6 cells. From Murao et al. (49).

hydrocarbons benzo[a]pyrene (BAP) and benzo[e]pyrene (BEP), which require metabolic activation (26), as well as with benzo[a]pyrene 7,8 dihydrodiol 9,10 epoxide (BPDE) (the ultimate mutagenic and carcinogenic form of BAP [34,45,61]) and N-methyl-N'-nitro-N-nitrosoguanidine (MNNG), both of which are themselves chemically reactive (Table 1). An enhancing effect of 1,25-$(OH)_2D_3$ on cell transformation induced by 3-methylcholanthrene, has also been observed in the BALB/3T3 cell system (43). We observed that the parent compound, vitamin D_3, was also effective in enhancing cell transformation in the hamster embryo cell system, although to a lesser degree than 1,25-$(OH)_2D_3$. Thus, a fraction of the embryonic hamster cells may have the capacity to convert vitamin D_3 to its reactive metabolite.

TABLE 1. Enhancement of cell transformation by 1,25-$(OH)_2D_3$ in cells pretreated with either BAP, BEP, BPDE, or MNNG.[a]

Duration of Treatment 0-3 days (Stage 1)	3-7 days (Stage 2)	Total no. colonies	Cloning effi-ciency(%)	Total no. transf. colonies	Trans. freq. (%)
-	-	1728	36	1	0.06
1,25-$(OH)_2D_3$	-	1873	39	0	<0.05
-	1,25-$(OH)_2D_3$	1920	40	0	<0.05
1,25-$(OH)_2D_3$[a]	1,25-$(OH)_2D_3$	916	40	0	<0.05
BAP (1)[b]	-	1576	33	7	0.4
-	BAP (1)	944	20	6	0.6
BAP (1)[a]	BAP (1)	1133	24	7	0.6
BAP (1)	1,25-$(OH)_2D_3$	1080	23	42	3.9
1,25-$(OH)_2D_3$	BAP (1)	1122	23	13	1.2
1,25-$(OH)_2D_3$[a] + BAP (1)	1,25-$(OH)_2D_3$ + BAP (1)	832	17	17	2.0
BEP (3)	-	1284	27	0	<0.08
BEP (3)	1,25-$(OH)_2D_3$	1459	30	5	0.3
Pyrene (3)	-	1776	31	0	<0.06
Pyrene (3)	1,25-$(OH)_2D_3$	1670	35	0	<0.06
BPDE (0.1)	-	1078	23	9	0.8
BPDE (0.1)	1,25-$(OH)_2D_3$	994	21	28	2.8
MNNG (0.3)	-	1818	38	3	0.2
MNNG (0.3)	1,25-$(OH)_2D_3$	1794	37	24	1.3

[a]Cultures treated for 7 days with the chemical without medium change.

[b]The numbers in the parentheses represent the concentrations of the chemicals in μg/mL. The concentration of 1,25-$(OH)_2D_3$ was 0.1 μg/mL. From Jones et al. (36).

Although 1,25-$(OH)_2D_3$ and PMA share similar biological properties, including enhancement (promotion) of cell transformation induced by induction of cell differentiation in human promyelocytic HL-60 leukemia cells (47,49,65), they differ in their mode of action, presumably because they do not have the same receptors for their biological activity (49). In this context, 1,25-$(OH)_2D_3$, unlike PMA, does not promote papilloma formation in the mouse. Furthermore, it diminishes the effect of PMA in inducing such tumors (69). Results from our cell transformation studies in which we used a number of cell pools, each deriving from a different pregnant hamster, also support this notion (Table 2). Thus, the high or low responsiveness of a cell pool to the enhancing (promoting) effect of 1,25-$(OH)_2D_3$ was not necessarily correlated with a similar responsiveness to PMA. These two agents, which share similar biological effects in some cell types, such as enhancing (promoting) cell transformation in rodent fibroblasts and inducing cell differentiation in human leukemia cells, may differ in their biological effects in other cell types, including mouse skin epithelial cells.

In summary, we have discussed in the present report a number of studies in golden hamster and human cell culture systems that have helped to expand to some degree our comprehension of the cellular processes involved in the induction of cell differentiation and perhaps in the promotion of tumors.

TABLE 2. Enhancement by 1,25-$(OH)_2D_3$ of BAP-induced cell transformation in embryo cell pools derived from different pregnant hamsters.

Duration of Treatment		Transformation Frequency (%)					
		Cryopreserved Cell Pools					Noncryopreserved Cell Pool
0-3 days	3-7 days	A	B	C	D	E	
-	-	<0.07	<0.10	<0.07	<0.06	<0.05	<0.06
-	1,25-$(OH)_2D_3$	<0.09	<0.12	<0.07	<0.06	<0.08	0.06
-	PMA	<0.09	<0.08	<0.09	<0.06	<0.12	<0.05
BAP	-	0.89	0.20	0.16	0.14	0.08	0.09
BAP	1,25-$(OH)_2D_3$	4.50	2.37	0.40	0.26	0.10	0.49
BAP	PMA	4.65	1.29	2.62	4.79	0.17	0.07

For each determination, 700-2200 colonies were scored. The cloning efficiency of the control cells or cells treated with either 1,25-$(OH)_2D_3$ or PMA ranged from 23-47%. BAP treatment reduced these frequencies to 10-28%. The concentration of BAP was 1 μg/mL; the concentrations of PMA and 1,25-$(OH)_2D_3$ were 0.5 and 0.1 μg/mL, respectively. From Jones et al. (36).

REFERENCES

1. Ames, B.N., Durston, W.E., Yamasaki, E., and Lee, F.D. (1973): Proc. Nat. Acad. Sci., USA, 70:2281-2285.
2. Ashendel, C.L., Staller, J.M., and Boutwell, R.K. (1983): Cancer Res., 43:4333-4337.
3. Barrett, J.C., Crawford, B.D., Mixter, L.O., Schectman, L.M., Ts'o, P.O.P., and Pollack, R. (1979): Cancer Res., 39:1504-1510.
4. Berwald, Y., and Sachs, L. (1965): J. Nat. Cancer Inst., 35:641-661.
5. Borek, C., and Sachs, L. (1966): Nature, 210:276-278.
6. Bouck, N., and DiMayorca, G. (1976): Nature, 264:360-361.
7. Brookes, P., and Lawley, P.D. (1964): Nature, 202:781-784.
8. Cabot, M.C., Welsh, C.J., Callaham, M.F., and Huberman, E. (1980): Cancer Res., 40:3674-3679.
9. Castagna, M., Takai, Y., Kaibuchi, K., Sano, K., Kikkawa, N., and Nishizuka, Y. (1982): J. Biol. Chem., 257:7847-7851.
10. Colston, K., Colston, M.J., and Feldoman, D. (1981): Endocrinology, 108:1083-1086.
11. Cooper, R.A., Braunwald, A.D., and Kuo, A.L. (1982): Proc. Nat. Acad. Sci., USA, 79:2865-2869.
12. Diamond, L., O'Brien, T.G., and Baird, W.M. (1980): Adv. Cancer Res., 32:1-63.
13. DiPaolo, J.A., Nelson, R.L., and Donovan, P.J. (1969): Science, 165:917-918.
14. DiPaolo, J.A., Nelson, R.L., and Donovan, P.J. (1969): J. Nat. Cancer Inst., 42:867-874.
15. DiPaolo, J.A., Nelson, R.L., and Donovan, P.J. (1971): Cancer Res., 31:1118-1127.
16. Driedger, P.E., and Blumberg, P.M. (1980): Proc. Nat. Acad. Sci., USA, 77:576-581.
17. Dunphy, W.G., Delclos, K.B., and Blumberg, P.M. (1980): Cancer Res., 40:3635-3641.
18. Feurstein, N., and Cooper, H.L. (1983): J. Biol. Chem., 258:10788-10793.
19. Fisher, P.B., Schachter, D., Abbott, R.E., Callaham, M.F., and Huberman, E. (1984): Cancer Res., 44:5550-5554.
20. Frampton, R.J., Suva, L.J., Eisman, J.A., Findlay, D.M., Moore, G.E., Moseley, J.M., and Martin, T.J. (1982): Cancer Res., 42:1116-1119.
21. Haussler, M.R., and McCain, T.A. (1977): N. Engl. J. Med., 297:974-983.
22. Heidelberger, C. (1970): Annu. Rev. Biochem., 44:78-121.
23. Heidelberger, C., Freeman, A.E., Pienta, R.J., Sivak, A., Bertram, J.S., Casto, B.C., Dunkel, V.C., Francis, M.W., Kakunaga, T., Little, J.B., and Schechtman, L.M. (1983): Mutat. Res., 114:283-285.
24. Hoffman, D.R., and Huberman, E. (1982): Carcinogenesis, 8:875-880.

25. Horowitz, A.D., Greenbaum, E., and Weinstein, I. B. (1981): Proc. Nat. Acad. Sci., USA, 78:2315-2319.
26. Huberman, E. (1978): J. Environ. Pathol. Toxicol., 2:29-42.
27. Huberman, E., Braslawsky, G.R., Callaham, M.F., and Fujiki, H. (1982): Carcinogenesis, 3:111-114.
28. Huberman, E., and Callaham, M.F. (1979): Proc. Nat. Acad. Sci., USA, 76:1293-1297.
29. Huberman, E., Heckman, C., and Langenbach, R. (1979): Cancer Res., 39:2618-2624.
30. Huberman, E., McKeown, C.K., Jones, C.A., Hoffman, D.R., and Murao, S.-I. (1984): Mutat. Res., 130:127-137.
31. Huberman, E., and Sachs, L. (1966): Proc. Nat. Acad. Sci., USA, 56:1123-1129.
32. Huberman, E., and Sachs, L. (1976): Proc. Nat. Acad. Sci., USA, 73:88-92.
33. Huberman, E., and Sachs, L. (1977): Int. J. Cancer, 19:122-127.
34. Huberman, E., Sachs, L., Yang, S. K., and Gelboin, H.V. (1976): Proc. Nat. Acad. Sci., USA, 73:607-611.
35. Huberman, E., Weeks, C., Herrmann, A., Callaham, M.F., and Slaga, T.J. (1981): Proc. Nat. Acad. Sci., USA, 78:1062-1066.
36. Jones, C.A., Callaham, M.F., and Huberman, E. (1984): Carcinogenesis, 5:1155-1159.
37. Jones, C.A., Marlino, P.J., Lijinsky, W., and Huberman, E. (1981): Carcinogenesis, 2:1075-1077.
38. Kikkawa, U., Takai, Y., Tanaka, Y., Miyake, R., and Nishizuka, Y. (1983): J. Biol. Chem., 258:11442-11445.
39. Knudsen, A.G., Jr., Hetchcote, H.W., and Brown, B.W. (1975): Proc. Nat. Acad. Sci., USA, 72:166.
40. Kraft, A.S., and Anderson, W.B. (1983): Nature 301:621-623.
41. Kuroki, T., and Sato, H. (1968): J. Nat. Cancer Inst., 41:53-71.
42. Kuroki, T., and Heidelberger, C. (1971): Cancer Res., 31:2168-2176.
43. Kuroki, T., Sasaki, K., Chida, K., Abe, E., and Suda, T. (1983): Gann, 74:611-614.
44. Lotem, J., and Sachs, L. (1979): Proc. Nat. Acad. Sci., USA, 76:5158-5162.
45. Mager, R., Huberman, E., Yang, S.K., Gelboin, H.V., and Sachs, L. (1977): Int. J. Cancer, 19:814-817.
46. Manolagas, S.C., and Deftos, L.J. (1980): Biochem. Biophys. Res. Commun., 95:596-602.
47. McCarthy, D., SanMiguel, J.F., Freake, H.C., Green, P.M., Zola, H., Catovsky, D., and Goldman, J.M. (1983): Leukemia Res., 7:51-55.
48. Miyaura, C., Abe, E., Kuribayashi, T., Tanaka, H., Konno, K., Nishii, Y., and Suda, T. (1981): Biochem. Biophys. Res. Commun., 102:937-943.

49. Murao, S.-I., Gemmell, M.A., Callaham, M.F., Anderson, N.L., and Huberman, E. (1983): Cancer Res., 43:4989-4996.
50. Niedel, J.E., Kuhn, L., and Vanderbark, G.R. (1983): Proc. Nat. Acad. Sci., USA, 80:36-40.
51. Pienta, R.J., Poiley, J.A., and Lebherz, W.B. (1977): Int. J. Cancer, 19:641-655.
52. Pryme, I.F., Lillehaug, J.R., Fjose, A., and Kleppe, K. (1983): FEBS Lett., 152:17-20.
53. Reinherz, E.L., Kung, P.C., Goldstein, G., and Schlossman, S.F. (1979): Proc. Nat. Acad. Sci., USA, 76:4061-4065.
54. Reinherz, E.L., Kung, P.C., Goldstein, G., Levey, R.H., and Schlossman, S.F. (1980): Proc. Nat. Acad. Sci., USA, 77:1588-1592.
55. Reinherz, E.L., and Schlossman, S.F. (1981): Cancer Res., 41:4767-4770.
56. Rivedal, E., and Sanner, T. (1982): Cancer Lett., 17:1-8.
57. Rose, N.R., and Friedman, H. (1980): Manual of Clinical Immunology. American Society for Microbiology, Washington, DC.
58. Rovera, G., Santoli, D., and Damski, C. (1979): Proc. Nat. Acad. Sci., USA, 76:2779-2783.
59. Ryffel, B., Henning, C.B., and Huberman, E. (1982): Proc. Nat. Acad. Sci., USA, 79:7336-7340.
60. Schroff, R.W., Foon, K.A., Billing, R.J., and Fahey, J.L. (1982): Blood, 59:207-215.
61. Slaga, T.J., Bracken, W.M., Vaije, A., Levin, W., Yagi, H., Jerina, D.M., and Conney, A.H. (1977): Cancer Res., 37:4130-4133.
62. Slaga, T.J., Sivak, A., and Boutwell, R.K., editors (1978): Carcinogenesis--A Comprehensive Survey. Raven Press, New York.
63. Solanki, V., Slaga, T.J., Callaham, M., and Huberman, E. (1981): Proc. Nat. Acad. Sci., USA, 78:1722-1725.
64. Tanaka, H., Abe, E., Miyaura, C., Kuribayashi, T., Konno, K., Nishii, Y., and Suda, T. (1982): Biochem. J., 204:713-719.
65. Tanaka, H., Abe, E., Miyaura, C., Shiina, Y., and Suda, T. (1983): Biochem. Biophys. Res. Commun., 117:86-92.
66. Umezawa, K., Weinstein, I.B., Horowitz, A., Fujiki, H., Matsushima, T., and Sugimura, T., 1981, Nature, 290:411-413.
67. Wasserman, R.H., Brindak, M.E., Meyer, S.A., and Fullmer, C.S. (1982): Proc. Nat. Acad. Sci., USA, 79:7939-7943.
68. Weinstein, I.B., Wigler, M., Fisher, P.B., Sisskin, E., and Pietropaolo, C. (1978): In: Mechanisms of Tumor Promotion and Cocarcinogenesis, edited by T.J. Slaga, A. Sivak, and R.K. Boutwell, pp. 313-333. Raven Press, New York.
69. Wood, A.W., Chang, R.L., Huang, M.-T., Uskokovic, M., and Conney, A.H. (1983): Biochem. Biophys. Res. Commun., 116:605-611.

Carcinogenesis, Vol. 10, edited by E. Huberman
and S. H. Barr. Raven Press, New York © 1985.

Regulation of Cell Differentiation and Tumor Promotion by 1α,25 Dihydroxyvitamin D_3

T. Kuroki, K. Chida, H. Hashiba, J. Hosoi, J. Hosomi, K. Sasaki,
*E. Abe, and *T. Suda

*Department of Cancer Cell Research, Institute of Medical Science, University of Tokyo, Shirokanedai, Minato-ku, Tokyo 108, Japan; *Department of Biochemistry, School of Dentistry, Showa University, Hatanodai, Shinagawa-ku, Tokyo 142, Japan*

Like most chemical carcinogens, vitamin D must be converted to a metabolically active form before it can function. Vitamin D_3 is produced in the skin, mainly in the malpighian layer of the epidermis, from 7-dehydrocholesterol (provitamin D_3) by a nonenzymatic photolysis reaction on exposure to sunlight. It then binds to a specific transport protein in the plasma and is translocated from the skin to the circulation. The first metabolic reaction of vitamin D_3 is 25-hydroxylation, which occurs mostly in the liver, and the resulting 25-hydroxyvitamin D_3 [25(OH)D_3] is further hydroxylated at the 1α-position to form 1α,25-dihydroxyvitamin D_3 [1α,25$(OH)_2D_3$, Fig. 1]. This 1α-hydroxylation occurs primarily in the kidney and is controlled by parathyroid hormone and some other peptide and steroid hormones (10).

1α,25$(OH)_2D_3$ is present in the plasma at a concentration of about 0.1 nM (0.04 ng/mL) and acts by a receptor-mediated mechanism. It is, therefore, considered today as a hormone rather than a vitamin. For a long time, the only known function of 1α,25$(OH)_2D_3$ was as a regulator of the blood Ca^{++} level by enhancing intestinal calcium transport and bone mineral mobilization. However, a specific cytosol receptor for 1α,25$(OH)_2D_3$ was recently found in almost all tissues examined except the liver and skeletal muscle (Table 1), and this finding raised the question of whether 1α,25$(OH)_2D_3$ has more subtle functions in a wide variety of tissues and cells other than the known target tissues associated with translocation of calcium and phosphate.

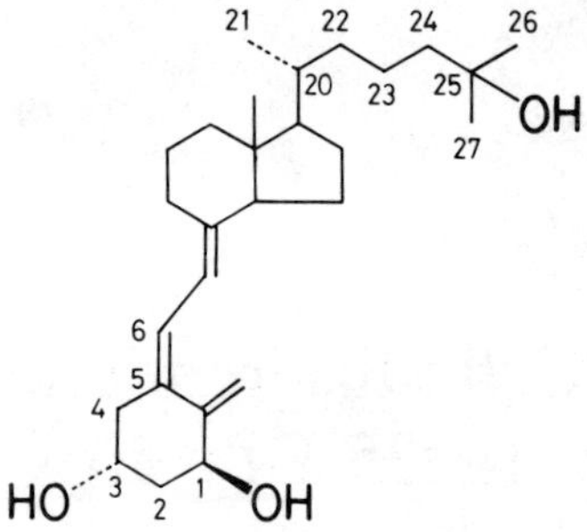

FIG. 1. Chemical structure of 1α,$25(OH)_2D_3$.

A breakthrough was made by the studies of Suda and his colleagues showing that 1α,$25(OH)_2D_3$ stimulates differentiation of myeloid leukemia cells of humans and mice (1,17). We extended this original observation to other types of cells, finding that 1α,$25(OH)_2D_3$ also regulates differentiation of epidermal keratinocytes and melanoma cells of mice. We also found that it is involved in tumor promotion, although in vivo and in vitro results were conflicting. We summarize here our recent findings on regulation of cell differentiation and tumor promotion by 1α,$25(OH)_2D_3$.

TABLE 1. Tissues and cells containing a receptor for 1α,$25(OH)_2D_3$.

Normal tissues	Cancer (cultured cells)	Cell lines
Skin	Breast	BALB 3T3
Kidney	Colon	Swiss 3T3
Intestine	Lung	CHO
Pancreas	Genito-urinary tract	LLC-PK1
Parathyroid	Bone	
Pituitary	Pancreas	
Spleen	Lymphoma	
Brain	Leukemia	
Monocytes	Melanoma	
Lymphocytes (activated)		

REGULATION OF CELL DIFFERENTIATION BY $1\alpha,25(OH)_2D_3$

Table 2 summarizes the effects of $1\alpha,25(OH)_2D_3$ on regulation of differentiation of various cells. $1\alpha,25(OH)_2D_3$ induces differentiation of myeloid leukemia cells of mice (M-1) and humans (HL-60) into macrophages (1,17,18). Olsson et al. (20) reported that it induces differentiation of human histiocytic lymphoma cells (U-937). It inhibits the differentiation of Friend erythroleukemia cells that is induced by dimethylsulfoxide (DMSO) (23). These effects of $1\alpha,25(OH)_2D_3$ on leukemia cells may be related to its physiological effects on immunoreactive cells.

TABLE 2. Cell types whose differentiation is regulated by $1\alpha,25(OH)_2D_3$.

Cells	Marker	Effect
Alveolar macrophages	Cell fusion	Stimulation
Blood monocytes	IL-2 production	Inhibition
HL-60 leukemia	Macrophage function	Stimulation
M-1 leukemia	Macrophage function	Stimulation
U-937 lymphoma	Macrophage function	Stimulation
Friend erythroleukemia	DMSO-induced differentiation	Inhibition
Epidermal keratinocytes	Terminal differentiation	Stimulation
B16 melanoma	Melanin synthesis	Stimulation

$1\alpha,25(OH)_2D_3$ as an Immunoregulatory Hormone

A specific receptor for $1\alpha,25(OH)_2D_3$ was detected on peripheral monocytes from normal humans, but not on normal B or T lymphocytes unless they were activated *in vitro* or transformed by viruses (21). More recently, Tsoukas et al. (24) reported that $1\alpha,25(OH)_2D_3$ suppresses production of interleukin-2 in human T-lymphocytes activated *in vitro*. Furthermore, Suda and his colleagues (2) found that $1\alpha,25(OH)_2D_3$ induces fusion of mouse alveolar macrophages by a direct mechanism and by a spleen cell-mediated indirect mechanism. Since bone-resorbing cells such as osteoclasts, which are often multinucleated, are thought to be derived from blood-borne mononuclear cells, it seems likely that 1α,25$(OH)_2D_3$ has physiological functions in the process of differentiation of blood monocytes into osteoclasts. All these results suggest that 1α25$(OH)_2D_3$ has a regulatory role in immune phenomena or, in other words, it is an immunoregulatory hormone.

Stimulation of Terminal Differentiation of Epidermal Keratinocytes

Besides its effects on immunocytes and leukemia cells, 1α,25$(OH)_2D_3$ also regulates differentiation of cells present in skin and skin cancer, i.e., epidermal keratinocytes and melanoma cells. These cells are known to contain a receptor for 1α,25$(OH)_2D_3$ (6-9,11,22), but the physiological function of the compound was previously unknown. During the last few years, we have been studying the responses of epidermal keratinocytes of humans and mice to initiators and promoters (4,5,13,14). In these studies, we first examined possible regulation of differentiation of epidermal cells by 1α,25$(OH)_2D_3$ (12). As seen in Fig. 2, terminal differentiation of mouse epidermal cells in primary culture was stimulated in a dose dependent manner by 1α,25$(OH)_2D_3$ at a concentration of 0.05 ng/mL or more. The number of basal cells in the treated cultures decreased sharply, and the cells underwent differentiation into squamous and enucleated cells, which became sloughed off into the medium. 1α,25$(OH)_2D_3$ markedly stimulated formation of a cornified envelope, a structure with chemically stable crosslinks formed beneath the plasma membrane.

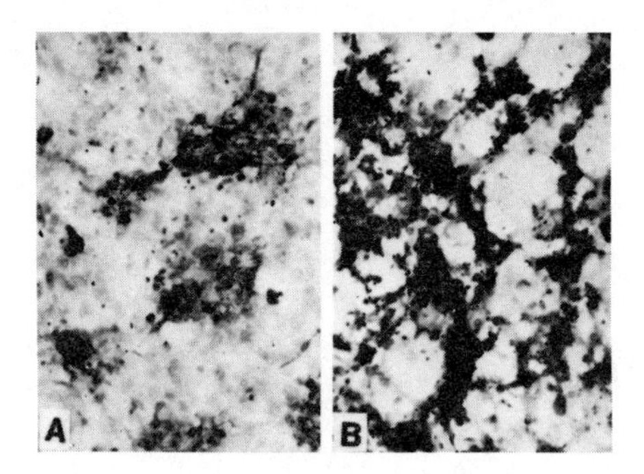

FIG. 2. Confluent sheets of primary cultures of mouse epidermal cells cultured in the absence (A) and presence (B) of 5.0 ng/mL 1α,25$(OH)_2D_3$ for 7 days (12). Focal stratification (darkly stained areas) are seen in both, but are more pronounced in the vitamin-treated culture (B).

Stimulation of Melanin Synthesis of B16 Melanoma Cells

Melanocytes are scattered singly in the sheet of keratinocytes in culture. During the above study, we noticed more melanocytes in treated cultures than in control cultures (Fig. 3), suggesting that $1\alpha,25(OH)_2D_3$ also stimulated melanin synthesis and/or growth of melanocytes. This possibility was examined by use of B16 melanoma cells (paper submitted). $1\alpha,25(OH)_2D_3$ increased the activity of tyrosinase, a key enzyme in melanin synthesis, in a dose-dependent manner in the range of 0.05 to 5.0 ng/mL. This enzyme activity was not increased by other metabolites of vitamin D_3, indicating that the effect of $1\alpha,25(OH)_2D_3$ was specific. Stimulation of melanin synthesis by $1\alpha,25(OH)_2D_3$, if it occurs in epidermal melanocytes in vivo, may act as a negative feedback mechanism in vitamin D_3 metabolism by preventing penetration of sunlight into the epidermis and thereby decreasing the conversion of provitamin D_3 to vitamin D_3.

Of particular interest is the finding that $1\alpha,25(OH)_2D_3$ and retinoic acid caused mutual interference with melanin synthesis (Fig. 4); a balance between their stimulatory and inhibitory effects was obtained at a molar ratio of 10:1, i.e. with 10 nM $1\alpha,25(OH)_2D_3$ and 1 nM retinoic acid. There may be a similar balance in other tissues and cells in regulation of cell differentiation.

REGULATION OF TUMOR PROMOTION BY $1\alpha,25(OH)_2D_3$

Enhancement of Chemically Induced Transformation of BALB 3T3 Cells

In the above studies, we noticed that $1\alpha,25(OH)_2D_3$ is very similar to 12-O-tetradecanoyl-phorbol-13-acetate (TPA) in inducing differentiation (16). This observation prompted us to examine whether $1\alpha,25(OH)_2D_3$ also

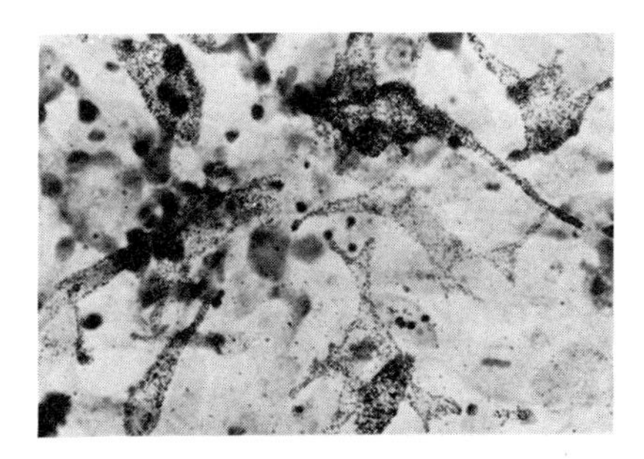

FIG. 3. Melanocytes in primary cultures of mouse epidermal cells exposed to 5.0 ng/mL $1\alpha,25(OH)_2D_3$ for 3 days (unpublished data).

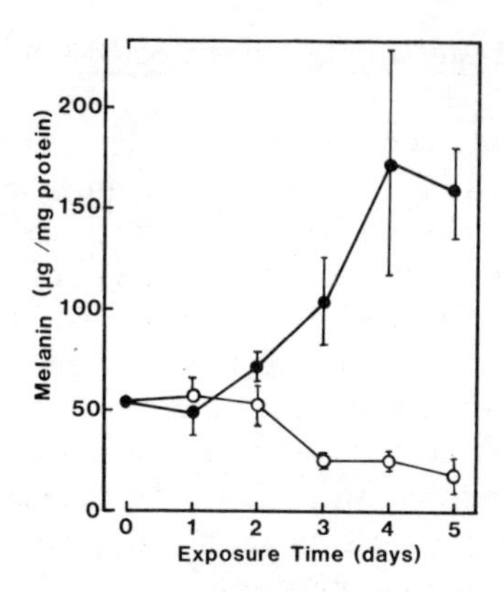

FIG. 4. Effects of 1α,25$(OH)_2D_3$ (●) and retinoic acid (○) on melanin synthesis in B16 melanoma cells. The vitamins were added at a concentration of 10 nM (4 ng/mL for 1α,25$(OH)_2D_3$ and 3 ng/mL for retinoic acid) for the indicated times (unpublished data).

has a promoting effect. First we studied this possibility in BALB 3T3 A31-1-1 cells (15). The cells were treated with a threshold dose (1 μg/mL for 72 h) of 3-methylcholanthrene (MCA) and then exposed to 1α,25$(OH)_2D_3$ at concentrations of 0.05 to 5.0 ng/mL for 2 weeks. Morphologically transformed foci were scored after cultivation for 6 weeks on a point basis with up to three points (0, 1, and 2) for transitional stages and higher points for typical transformation, for each of three parameters: basophilic staining, formation of a dense layer, and random orientation of cells at the edge of foci. Foci with scores of 4 or more points were regarded as transformed.

Treatment with MCA alone at 1 μg/mL resulted in a very low transformation frequency (0.05 ± 0.05 foci per dish in nine independent experiments), but subsequent exposure to 1α,25$(OH)_2D_3$ caused marked and dose-dependent increases in the transformation frequency (Table 3). With 1α,25$(OH)_2D_3$ at 5 ng/mL, the transformation frequency was 1.78 ± 0.59 foci per dish in seven independent experiments, which was about 35 times that in cultures treated with MCA only. 1α,25$(OH)_2D_3$ alone did not induce transformation under these conditions (Table 3).

BALB 3T3 A31-1-1 cells have a specific cytosol receptor for 1α,25$(OH)_2D_3$ with a K_d value of 28.4 pM and Nmax of 32.6 fmol/mg protein (15). This fact, together with the observation that the enhancement by 1α,25$(OH)_2D_3$ is specific, indicates a receptor-mediated mechanism. Although the enhancing effect of 1α,25$(OH)_2D_3$ on MCA-induced transformation is very similar to but rather stronger than that of TPA, other effects of 1α,25$(OH)_2D_3$ on BALB 3T3 cells are not

TABLE 3. Enhancement of MCA-induced transformation of BALB 3T3 A31-1-1 cells by $1\alpha,25(OH)_2D_3$.

1st Treatment (μg/mL)	2nd Treatment (ng/mL)	No. of Exp.[a]	Foci per dish	
			Pool data	Average ± SD
DMSO	EtOH	10	0/103	0
MCA (5)	EtOH	6	70/57	1.26 ± 0.45
MCA (1)	EtOH	9	5/87	0.05 ± 0.05
MCA (1)	$1\alpha,25(OH)_2D_3$ (5)	7	121/68	1.78 ± 0.59
DMSO	$1\alpha,25(OH)_2D_3$ (5)	2	1/18	0.06 ± 0.09
MCA (1)	TPA (1000)	4	91/36	2.03 ± 1.16

[a]Numbers of repeated experiments.

necessarily the same as those of TPA (paper submitted). Other effects of $1\alpha,25(OH)_2D_3$ on BALB 3T3 A31-1-1 cells are as follows:

1. $1\alpha,25(OH)_2D_3$ induced DNA synthesis in quiescent BALB 3T3 cells dose- and time-dependently, but its effect was less than that of TPA.
2. Unlike TPA, $1\alpha,25(OH)_2D_3$ did not interfere with binding of epidermal growth factor or phorbol dibutyrate.
3. Unlike TPA, $1\alpha,25(OH)_2D_3$ did not induce ornithine decarboxylase (ODC) activity in quiescent BALB 3T3 cells. Furthermore, $1\alpha,25(OH)_2D_3$ did not inhibit the ODC activity induced by TPA.
4. C-kinase (Ca^{++}-activated, phospholipid-dependent protein kinase) was not induced by treatment of quiescent BALB 3T3 cells with $1\alpha,25(OH)_2D_3$ (see also below).

Absence of Activation or Induction of C-Kinase

Recent studies have shown that at least some of the effects of TPA are mediated by C-kinase under physiological conditions (19). This protein kinase is activated by diacylglycerol, which is produced from inositol phospholipids in a signal-dependent manner. TPA may activate C-kinase by binding to it in place of diacylglycerol. Figure 5 shows the DEAE-cellulose chromatographic pattern of C-kinase isolated from the brain of SENCAR mice. Protein kinase was markedly activated by additions of $CaCl_2$, phosphatidylserine, and diolein to the reaction mixture, indicating the presence of C-kinase. Addition of $1\alpha,25(OH)_2D_3$ at 1 μg/mL in place of diolein did not activate the kinase, whereas addition of TPA at 1 μg/mL stimulated the activity to the same extent as diolein (Table 4). Furthermore, $1\alpha,25(OH)_2D_3$ did not inhibit the C-kinase activity stimulated by TPA. This result and the absence of induction of C-kinase in BALB 3T3

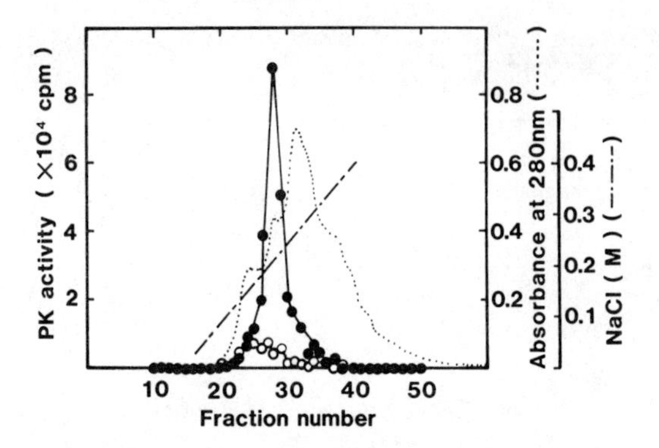

FIG. 5. Partial purification by DEAE-cellulose chromatography (DE52 column 1.2 x 5 cm) of C-kinase from mouse brain. Protein kinase (PK) activity was assayed by measuring phosphorylation of histon Hl in the absence (○) and presence (●) of 1 mM $CaCl_2$, 64 μM phosphatidylserine and 1.3 μM diolein (unpublished data).

cells by 1α,25$(OH)_2D_3$ indicate that C-kinase is not involved in 1α,25$(OH)_2D_3$-mediated events. This difference from results with TPA may be related to a difference between the receptor-mediated mechanisms of action of these two compounds. The receptor for TPA is located on the cell membrane and seems to be C-kinase itself, whereas that of 1α,25$(OH)_2D_3$, like the receptors of other steroid hormones, is found in the cytosol or nuclear fraction and a receptor-hormone complex may interact directly with DNA or RNA.

Inhibition of Tumor Promotion in Mouse Skin

The possible promoting effect of 1α,25$(OH)_2D_3$ was studied further in mouse skin. First, we examined the induction of ODC in mouse skin (3). Topical application of 1α,25$(OH)_2D_3$ did not induce ODC activity in SENCAR mice and unexpectedly inhibited ODC induction by TPA: application of 1 μg of 1α,25$(OH)_2D_3$ within 30 min before or after treatment with 10 μg of TPA resulted in 72% inhibition of ODC induction 4 h later (Fig. 6). The dose required for 50% inhibition was 0.063 μg or 0.15 nmol, which is about half the required dose of retinoic acid.

This finding suggests that 1α,25$(OH)_2D_3$ is an antipromoter in mouse skin carcinogenesis. While this study was in progress, Wood et al. (25) reported that topical application of 1α,25$(OH)_2D_3$ in the promotion stage of mouse skin carcinogenesis suppressed the formation of chemically induced tumors. Consistent with their findings, we observed that 1α,25$(OH)_2D_3$ inhibited promotion, especially Stage II of promotion (unpublished data).

TABLE 4. Effect of 1α,25$(OH)_2D_3$ and TPA on C-kinase activity.

	Kinase activity			
Additives	cpm		Ratio	
None	1297 ± 245	100		
PS + $CaCl_2$	4600 ± 173	354	100	
PS + $CaCl_2$ + 1α,25$(OH)_2D_3$	4543 ± 613	350	99	
PS + $CaCl_2$ + DG	8160 ± 485	629	177	
PS + CaCl + TPA	7690 ± 665	593	167	100
PS + CaCl + TPA + 1α,25$(OH)_2D_3$	7070 ± 910	545	154	92

Partially purified C-kinase of SENCAR mouse brain was incubated in the reaction mixture containing the additives indicated. Concentrations of 64 μM phosphatidylserine (PS), 1 mM $CaCl_2$, 9.6 μM 1α,25$(OH)_2D_3$, 6.5 μM diolein (DG), 6.5 μM TPA were added.

Thus, 1α,25$(OH)_2D_3$ enhances transformation of BALB 3T3 cells in vitro, but inhibits the promotion by TPA in mouse skin (Table 5). This apparent discrepancy may be due to a difference in cell types, i.e., mesenchymal versus epithelial cells and/or a difference in experimental conditions, i.e., in vitro versus in vivo. It is possible that 1α,25$(OH)_2D_3$ acts as a tumor promoter in certain organs or tissues other than skin.

TABLE 5. Regulation of tumor promotion in BALB 3T3 A31-1-1 cells and mouse skin by 1α,25$(OH)_2D_3$.

Activity	Effect of 1α,25$(OH)_2D_3$	
	BALB 3T3	Mouse skin
Promotion	Enhancement	Inhibition (I < II)
DNA synthesis or hyperplasia	Induction	No induction
ODC activity	No induction	No induction
TPA-induced ODC	No effect	Inhibition
TPA-induced hyperplasia	Not examined	No effect
EGF,[a] PDBu[b] binding	No interaction	Not examined
C-kinase	No induction	Not examined

[a]Epidermal growth factor.
[b]Phorbol dibutyrate.

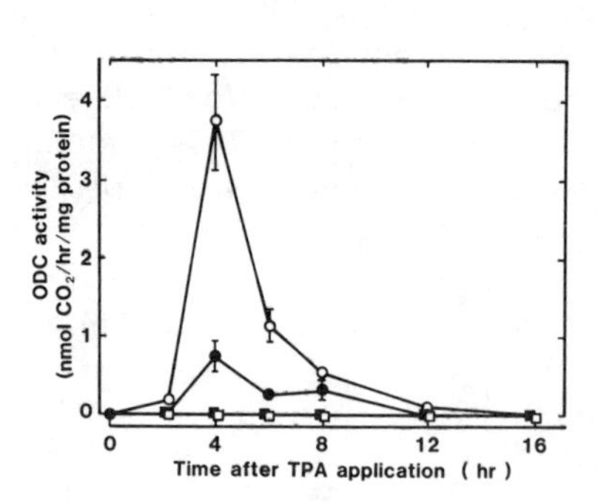

FIG. 6. Inhibition of TPA-induced ODC by 1α,25$(OH)_2D_3$ in mouse skin (3). Mice were treated with 1 μg/mL of 1α,25$(OH)_2D_3$ and 30 min later with 10 μg/mL TPA: ●, 1α,25$(OH)_2D_3$ + TPA; ○: TPA alone; ■: 1α,25$(OH)_2D_3$ alone; □: solvent (acetone) control.

ACKNOWLEDGMENT

Toshio Kuroki wishes to acknowledge his sincere gratitude to the late Professor Charles Heidelberger, first for the excellent training received during a postdoctoral fellowship under Professor Heidelberger at the McArdle Laboratory for Cancer Research at the University of Wisconsin from 1969 to 1971, and since then for continued interest and encouragement.

REFERENCES

1. Abe, E., Miyaura, C., Sakagami, H., Takeda, M., Konno, K., Yamazaki, T., Yoshiki, S., and Suda, T. (1981): Proc. Nat. Acad. Sci., USA, 78:4990-4994.
2. Abe, E., Miyaura, C., Tanaka, H., Shiina, Y., Kuribayashi, T., Suda, S., Nishii, Y., DeLuca, H.F., and Suda, T. (1983): Proc. Nat. Acad. Sci., USA, 80:5583-5587.
3. Chida, K., Hashiba, H., Suda, T., and Kuroki, T. (1984): Cancer Res., 44:1387-1391.
4. Chida, K., and Kuroki, T. (1983): Cancer Res., 43:3638-3642.
5. Chida, K., and Kuroki, T. (1984): Cancer Res., 44:875-879.
6. Clemens, T.L., Horiuchi, N., Nguyen, M., and Holick, M.F. (1981): FEBS Lett., 134:203-206.
7. Colston, K., Colston, M.J., and Feldman, D. (1981): Endocrinology, 108:1083-1086.
8. Colston, K., Colston, M.J., Fieldsteel, A.H., and Feldman, D. (1982): Cancer Res., 42:856-859.
9. Colston, K., Hirst, M., and Feldman, D. (1980): Endocrinology, 107:1916-1922.
10. DeLuca, H.F., and Schnoes, H.K. (1983): Annu. Rev. Biochem., 52:411-439.
11. Feldman, D., Chen, T., Hirst, M., Colston, K., Karasek, M., and Cone, C. (1980): J. Clin. Endocrinol. Metab., 51:1463-1465.
12. Hosomi, J., Hosoi, J., Abe, E., Suda, T., and Kuroki, T. (1983): Endocrinology, 113:1950-1957.
13. Kuroki, T., Hosomi, J., Munakata, K., Onizuka, T., Terauchi, M., and Nemoto, N. (1982): Cancer Res., 42:1859-1865.
14. Kuroki, T., Nemoto, N., and Kitano, Y. (1980): Carcinogenesis, 1:559-565.
15. Kuroki, T., Sasaki, K., Chida, K., Abe, E., and Suda, T. (1983): Gann, 74:611-614.
16. Kuroki, T., and Suda, T. (1984): In: Cellular Interactions of Environmental Tumor Promoters, edited by H. Fujiki, E. Hecker, E.E. Moore, T. Sugimura, and I.B. Weinstein, pp. 89-99. Japan Scientific Society Press, Tokyo.

17. Miyaura, C., Abe, E., Kuribayashi, T., Tanaka, H., Konno, K., Nishii, Y., and Suda, T. (1981): Biochem. Biophys. Res. Commun., 102:937-943.
18. Murao, S., Gemmell, M.A., Callaham, M.F., Anderson, N.L., and Huberman, E. (1983): Cancer Res., 43:4989-4996.
19. Nishizuka, Y. (1984): Nature, 308:693-698.
20. Olsson, I., Gullberg, U., Ivhed, I., and Nisson, K. (1983): Cancer Res., 43:5862-5867.
21. Provvedini, D.M., Tsoukas, C.D., Deftos, L.J., and Manolagas, S.C. (1983): Science, 221:1181-1182.
22. Simpson, R.U., and DeLuca, H.F. (1980): Proc. Nat. Acad. Sci., USA, 77:5822-5826.
23. Suda, S., Enomoto, S., Abe, E., and Suda, T. (1984): Biochem. Biophys. Res. Commun., 119:807-813.
24. Tsoukas, C.D., Provvedini, D.M., Manolagas, S.C. (1984): Science, 224:1438-1439.
25. Wood, A.W., Chang, R.L., Huang, M.T., Uskokovic, M., and Conney, A.H. (1983): Biochem. Biophys. Res. Commun., 116:605-611.

Carcinogenesis, Vol. 10, edited by E. Huberman
and S. H. Barr. Raven Press, New York © 1985.

HL-60 Variant Reversibly Resistant to Induction of Differentiation by Phorbol Esters

L. Diamond, B. Perussia, R. Businaro[1], and F. W. Perrella

The Wistar Institute of Anatomy and Biology, Philadelphia, Pennsylvania 19104

In acute myelogenous leukemia, cells of either the myeloid or monocytic lineage are blocked at a specific stage of the differentiation process. They maintain proliferative capacity and do not undergo further differentiation toward fully mature and functionally active monocytes or polymorphonuclear granulocytes (PMNs). Cell lines derived from leukemic patients can be maintained in culture in a proliferative state without a significant level of spontaneous differentiation. They provide useful models for studying the mechanisms involved in myelomonocytic differentiation and in the triggering of that differentiation by various agents.

One such cell line, HL-60, was established from the peripheral blood of a patient with acute promyelocytic leukemia (9,12). Many chemical agents, among which retinoic acid and dimethyl sulfoxide are the most well studied, induce differentiation of these cells along a programmed pathway that resembles that of myeloid cells (14). On the other hand, agents such as the phorbol ester tumor promoters (18,37), immune-type interferon (IFNγ) (10,19,20,33), and certain humoral factors (25,33,41,42) induce HL-60 cells to undergo differentiation along the monocytic pathway, but the differentiated cells cannot be unambiguously identified as mature cells.

The mechanisms by which the phorbol esters and IFNγ exert their activity is not understood. The mode of action of the two substances can be distinguished at least in terms of the kinetics by which they inhibit promyelocytic cell proliferation. Phorbol-ester-treated cells are first blocked in proliferation and then express some of the markers typical of monocytes (38), whereas IFNγ-treated cells are first induced to differen-

[1]Permanent address: C.N.R. Laboratory of Cell Biology, Rome, Italy 00196

tiate and then cell proliferation is arrested as a consequence of the progression of the cells along the differentiation pathway (10).

We have isolated variants of HL-60 cells that are resistant to the differentiation-inducing activity of phorbol esters and maintain their proliferative capacity in the presence of high concentrations of the promoters. However, the resistance of one of these variants to phorbol esters is reversible: if the cells are washed free of phorbol esters, they fully regain their responsiveness within 72 h and, when reexposed to phorbol esters, express new phenotypic characteristics typical of monocytes. We have found, in addition, that the reversion of the response to phorbol esters is augmented if IFNγ is present. Here we report some of the characteristics of the reversibly resistant HL-60 variant and the morphological, antigenic, biochemical, and functional changes that occur in these cells in response to treatment with phorbol esters and IFNγ.

MATERIALS AND METHODS

Cells

The human promyelocytic leukemia cell line HL-60 (9), obtained originally from Dr. R. Gallo (National Institutes of Health, Bethesda, MD), was grown in suspension in RPMI 1640 medium supplemented with 15% fetal bovine serum (Gibco Laboratories, Chagrin Falls, OH, or Armour Pharmaceutical Co., Kankakee, IL). Cultures were maintained by subculturing every 3-4 days into Falcon flasks at a density of 2×10^5 cells/mL.

Reagents

Phorbol-12,13-dibutyrate (PDBu) and 12-O-tetradecanoylphorbol-13-acetate (TPA) were obtained from Chemical Carcinogenesis (Eden Prairie, MN). [20-^{3}H]PDBu was obtained from Amersham Corporation (Arlington Heights, IL). Human recombinant IFNγ (rIFNγ) from *E. coli*, purified to homogeneity, was kindly supplied by Dr. C. G. Sevastopoulos (Genentech Inc., San Francisco, CA); it had a titer of 7×10^7 antiviral units/mg protein on HeLa cells. Recombinant type A IFNα from *E. coli*, with a titer of 10^7 antiviral units/mg protein, was kindly supplied by Hoffman LaRoche, Inc. (Nutley, NJ).

Monoclonal Antibodies

Most of the antibodies used were produced and characterized in our laboratory (34,35). B33.1 (IgG2a) detects a nonpolymorphic determinant on the HLA-DR molecule; B44.1 (IgM) and B40.9 (IgM) detect two molecules specifically present on cells of the monocytic and myeloid lineage, respectively; B43.4 (IgG2b) has specificity similar to OKM1, with which it crosscompetes, and probably detects the C3bi receptor on monocytes, granulocytes (PMNs), and NK cells; B13.4 (IgM) detects an antigen shared by monocytes and PMNs that is first expressed at the metamyelocyte stage; 5E9 (IgG1), prepared from hybrid cells obtained from the American Type Culture Collection (Rockville, MD), detects the receptor for transferrin. Antibody 906 (IgG1) (My7) (16), kindly donated by Dr. J. Griffin (Dana Farber Cancer Institute, Boston, MA), detects an antigen of apparent molecular weight 95,000 present on immature myeloid cells; the 906 antigen is lost on myeloid cells at the metamyelocyte stage but is still expressed on peripheral blood mature monocytes. Antibody 28.3.7, kindly donated by Dr. P. Andrews (Wistar Institute), detects an antigen expressed on adherent human cells (4).

Phorbol Ester Binding Assay

[^{3}H]PDBu binding was measured on intact cells in phosphate buffered saline (PBS). Approximately 10^6 cells were incubated in PBS containing [^{3}H]PDBu. Following a 2-h incubation at 0°C, ice-cold PBS containing 0.04% CHAPS, a nonionic detergent, was added, and the contents of the tubes were poured onto anion exchange glass fiber filters (5). After filtration, the filters were washed with PBS-CHAPS and placed in vials. Filter-bound radioactivity was measured by liquid scintillation counting. The amount of [^{3}H]PDBu bound in the presence of an excess of cold PDBu was used as a measure of nonspecific binding. Specific binding was calculated by subtracting nonspecific binding from total binding in the absence of cold PDBu. The total number of receptor sites and the dissociation constants of [^{3}H]PDBu binding were determined by analyzing the binding data with a nonlinear least squares curve fitting routine.

Assessment of Differentiated Characteristics

Slides of cells in suspension or of adherent cells detached by trypsinization were prepared with the aid of a Shandon-Elliott cytospin centrifuge, and the morphology of the cells was assessed after May-Grunwald-Giemsa staining.

Nonspecific α-naphthyl acetate esterase (αNAE) activity was assessed both histochemically (45) on cytocentrifuge smears and colorimetrically (36). In the latter case, Triton X-100 lysed cells were incubated for 10 min (37°C) in 40 mM Tris buffer (pH 8) containing 0.8 mM α-naphthyl acetate and the product, α-naphthol, was reacted first with fast blue RR and then with trichloroacetic acid. The dye product of the reaction (naphthyl-1-one-4-benzoyl-amino-2,5-dimethoxyaniline) was extracted with 2.5 mL ethylacetate and measured by its absorbance at 440 nm (extinction coefficient = 19.6 cm^{-1} mM^{-1}). One unit of αNAE activity corresponds to the amount of enzyme that catalyzes the production of 1 μmol of product per min.

Cathepsin B activity was measured spectrofluorimetrically with N-benzoyl-L-Phe-L-Val-L-Arg-7-amido-4-methyl coumarin used as the substrate (6). The production of 7-amino-4-methyl coumarin was determined by the fluorescence at 450 nm with an excitation wavelength of 350 nm. The total activity measured, minus the activity in replicate samples assayed in the presence of the cathepsin B inhibitor, leupeptin, corresponds to the cathepsin B activity. A standard curve of the relative fluorescence versus the amount of free 7-amino-4-methyl coumarin was used to calculate activity in terms of pmoles.

Indirect immunofluorescence staining was performed according to our published protocol (35) and analyzed on an Ortho Cytofluorograph H50 connected to a Data General MP/200 microprocessor (Ortho Instruments, Westwood, MA). Briefly, the cells were incubated in PBS, pH 7.2, supplemented with 5% human plasma and 0.1% NaN_3 (washing buffer), with the appropriate dilution of monoclonal antibody for 30 min at 4°C and, after three washes, with fluorescein-conjugated goat $F(ab')_2$ antimouse Ig (Cappel Laboratories, Cochranville, PA) preabsorbed with human Ig. Control cells were treated with the second reagent only and used to set the threshold at which 99% of the cells were negative.

For detection of Fc receptor, ox erythrocytes were sensitized with a 1:50 subagglutinating dilution of rabbit IgG anti-ox E (Cappel) (EA7S). Cells and EA7S were mixed (1:50), centrifuged (10^3 rpm for 5 min), and incubated on ice for 30 min (33). The proportion of cells forming EA7S rosettes was counted, with at least 200 cells counted by light microscopy.

Antibody-dependent cell-mediated cytotoxicity (Ab-CMC) was performed as previously described (33) in a 3-h ^{51}Cr release assay with rabbit antibody-sensitized (10^{-3} dilution) P815Y used as target cells. Different effector cell numbers were used against 10^4 target cells.

RESULTS

Isolation of TPA-Resistant Variants

To obtain TPA-resistant variants, HL-60 cells were cultured for a few passages in medium containing 0.16 nM TPA. Then, the concentration of TPA was gradually increased every four to five passages, and only the nonadherent cells were subcultured at each passage. After 3 months, a cell line had been selected that grew in medium containing 1.6 μM TPA and was, therefore, resistant to TPA-induced adherence. This cell line (R1) was cloned by limiting dilution and one clone (R1B6) was studied further. The resistant cell lines are maintained continuously in medium containing 2.4 μM PDBu, which, unlike TPA, can be washed out when required for experiments. Both R1 and R1B6 cells have a population doubling time of 20-22 h, a subtetraploid chromosome number, and a cell volume approximately twice that of HL-60 cells. Although resistant to monocyte-like differentiation induced by phorbol esters, the cells can be induced to differentiate along the myeloid pathway by treatment with dimethyl sulfoxide, dimethyl formamide, or retinoic acid (data not shown).

Both R1 and R1B6 cells, cultured in the absence of PDBu, gradually become sensitive to the adherence-inducing effect that phorbol ester tumor promoters have on wild-type HL-60 cells. Figure 1 shows the results of an experiment in which R1 cells were incubated in the absence of PDBu (without rIFNγ) for 18 h and then tested at various times thereafter for TPA-induced cell adherence to plastic surfaces. The cells began to show sensitivity to TPA 21 h after the pretreatment period (a total of 39 h in the absence of PDBu) and by 48 h $\geq$ 90% were responsive to TPA-induced adherence. The possibility that a subclone of nonresistant cells was selected can be ruled out because, during the course of the experiment, no cell death was observed and the cells grew at the same rate in the absence of PDBu as did control cells maintained in its presence. Following similar experiments with the R1B6 subclone, we concluded that the majority of the R1 and R1B6 cell populations are reversibly resistant variants of HL-60.

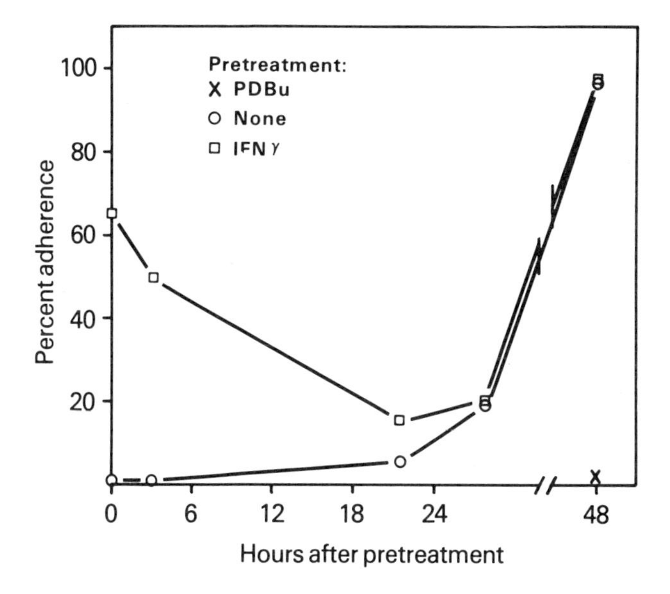

FIG. 1. Reversibility of phorbol-ester-resistant variants. R1 cells were washed free of PDBu and preincubated for 18 h in the absence of PDBu and with or without rIFNγ (10^3 U/mL). After this pretreatment period, the cells were washed three times with medium and incubated in Petri dishes with medium alone. At various times thereafter, TPA (0.016 μM) was added to the medium and the percent of cells adherent to the plastic surface was determined by microscopic examination 22 h later.

A small proportion of variant cells grown in the absence of PDBu continued to proliferate and did not adhere when phorbol esters were added back to the cultures. This nonresponsive subpopulation of R1 cells was subsequently subcultured in the absence of phorbol esters and cloned by limiting dilution; it has maintained its resistance to TPA during prolonged subcultivation in the absence of any phorbol esters. This variant, unlike R1B6, is probably similar to other HL-60 variants that are irreversibly resistant to TPA (2,17,22,23,40); its characteristics will be reported elsewhere.

Effect of rIFNγ on the TPA-Resistant Variants

Immune-type interferon induces monocytoid differentiation of HL-60 cells (10,19,20,33). We have determined its effect on the phorbol-ester-resistant variants. When R1 cells were pretreated with rIFNγ in the absence of PDBu, readdition of TPA in the presence of rIFNγ induced massive cell clumping and adherence to the plastic tissue culture dishes within a few hours. If the rIFNγ was removed before TPA was added, the cells still responded to TPA with clumping (Fig. 1, 0 time point) but sensitivity gradually decreased with time. After 27 h, the cells again showed sensitivity, similar to that of control cells preincubated in the absence of rIFNγ (Fig. 1). This potentiation of the reversible response of R1 cells to the phorbol esters is specific for immune-type interferon because it does not occur with IFNα, it requires that the cells be pretreated with IFNγ for at least 6 h (not shown), and the effect mediated by IFNγ is gradually lost during subsequent incubation in its absence.

Determination of Phorbol Ester Receptors on Parental and TPA-Resistant HL-60 Cells

The possibility that the resistance of the R1 variants to TPA could be due to a lack of phorbol ester receptors was tested in [^{3}H]PDBu binding experiments. Saturation analysis of [^{3}H]PDBu binding showed that R1B6 cells had a 42% reduction in the number of binding sites compared with HL-60 cells, when the number was corrected for cell volume differences (Fig. 2). This reduction was equivalent to a loss of 157,000 sites. The dissociation constants (K_D) for [^{3}H]PDBu binding were similar for HL-60 and R1B6 cells--26 ± 4 nM and 40 ± 2 nM, respectively.

The reversible response of these cells to phorbol esters after a period of time in the absence of phorbol esters (Fig. 1) might be explained by an increased number of phorbol ester receptors. Therefore, intact R1B6 cells were assayed for [^{3}H]PDBu binding at various periods of time after the removal of PDBu (Fig. 3). R1B6 cells incubated in the absence of PDBu showed a steady increase with time in the number of phorbol ester receptors and a net increase of 100,000 receptor sites after 72 h. These

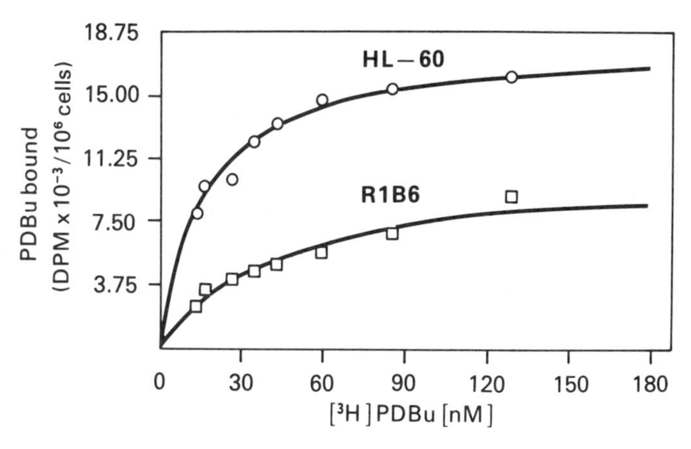

FIG. 2. Saturation analysis at equilibrium of [^{3}H]PDBu binding to parental HL-60 cells and to the phorbol ester-resistant cells, R1B6. Binding was determined by incubating 8 x 10^5 cells with [^{3}H]PDBu for 2 h at 0°C in PBS as described in Materials and Methods. The specific binding is expressed as disintegrations per minute of [^{3}H]PDBu bound/cell; the values have been normalized relative to the volume of HL-60 cells.

data suggest that R1B6 cells may be deficient in the absolute number of phorbol ester receptors or that there are two or more types of receptors with similar K_Ds, and R1B6 cells lack a particular subtype.

Expression of Enzyme Markers of Differentiation

αNAE activity in R1B6 cells treated with rIFNγ and TPA was assayed histochemically. If the cells were washed free of PDBu, pretreated with rIFNγ, and then exposed to TPA in the presence of rIFNγ for 2 days, approximately 50% stained for the enzyme and showed a high degree of activity (Table 1); more than 95% of these cells formed clumps and adhered to the plastic tissue culture dishes. If the IFNγ-pretreated cells were not treated subsequently with TPA, about 40% of the cells were stained, but with a low degree of activity, and there was no cell clumping or adherence. Treatment with TPA after preincubation in the absence of PDBu and rIFNγ resulted in a modest degree of activity in only 20-25% of the cells after 2 days.

Because the clumping of IFNγ-treated cells made it difficult to quantitate the relative αNAE activity histochemically, a biochemical assay was also used; the results (Table 2) were similar to those obtained with the histochemical analysis. Untreated R1B6 cells had a basal level of activity 45% higher than that of HL-60 cells. R1B6 cells treated with TPA, after a 22-h pretreatment period in the absence of phorbol esters,

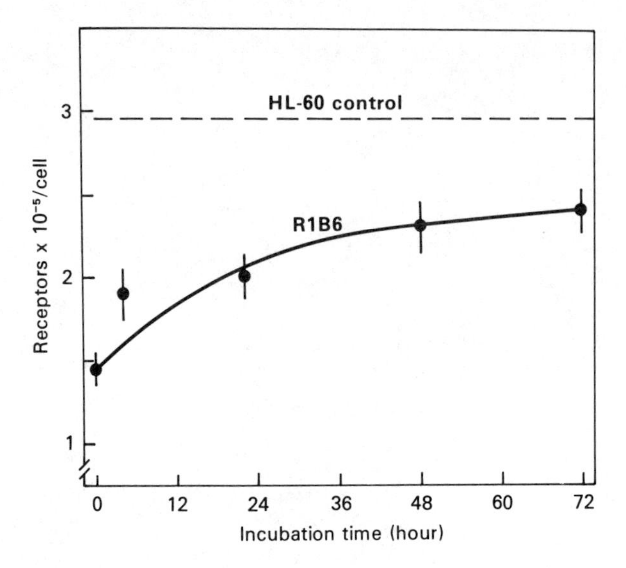

FIG. 3. Recovery of phorbol-ester-receptor binding. R1B6 cells were washed three times with PBS (Ca^{++} and Mg^{++} free) to remove PDBu. The cells were then incubated at 37°C in medium without phorbol esters for various periods of time. The specific binding of [^{3}H]PDBu was determined by incubating 10^{6} cells with 200 nM [^{3}H]PDBu for 2 h in PBS at 0°C as described in Materials and Methods. The data are expressed as the number of receptors/cell; the values have been normalized relative to the volume of HL-60 cells.

had an increased activity of about 20%, whereas pretreatment with rIFNγ and then treatment with TPA increased activity about 1.5 times. Increases in enzyme activity were also observed in HL-60 cells treated with TPA or with rIFNγ and TPA.

Cathepsin B, a cysteine proteinase, has been used as an enzyme marker for cells of the monocytoid/macrophage lineage (6,29). Measurements of cathepsin B activity in HL-60 and R1 cells are shown in Table 3. TPA treatment of HL-60 cells for 2 days resulted in a 16-fold increase in enzyme activity. R1 cells grown in the presence of PDBu had about a ninefold higher basal level than HL-60 cells, and treatment with TPA for 2 days did not increase this high level of activity.

Phenotypic Surface Characteristics of R1B6 Cells

As reported previously by us and others (17,18,34,39), a 3-day exposure of HL-60 cells to a phorbol ester tumor promoter such as PDBu induced

TABLE 1. Histochemical analysis of alpha-naphthylacetate esterase activity in R1B6 cells.[a]

Pretreatment	None				rIFNγ			
Treatment	---		TPA		---		TPA	
Days post-TPA	1	2	1	2	1	2	1	2
Cells with activity (%)	2	25	2	20	2	40	20	50
Degree of activity	±[b]	+	±	++	+	+	++	+++
Adherence (%)	0	0	25	10	0	0	>95	>95

[a]R1B6 cells were washed free of PDBu and preincubated for 18 h in the presence or absence of rIFNγ (10^3U/mL/2 x 10^5 cells). TPA (0.01 μM) was added and, 1 or 2 days later, the cells were assessed for αNAE activity and adherence to plastic, relative to control untreated cells.
[b]Maximum activity = ++++.

the appearance of some monocytic markers on HL-60 cells (Table 4); the proportion of cells expressing B13.4 antigen and C3bi receptors was increased, and approximately two-thirds of the cells also expressed the adherence antigen 28.3.7. In contrast, expression of the transferrin receptor and the 906 antigen were decreased by one-third. More changes in expression of surface markers were seen when rIFNγ was present during PDBu treatment of HL-60 cells; under those conditions, about 15% of the cells were induced to express HLA-DR and the monocyte-specific antigen B44.1.

The surface phenotypes of R1B6 cells grown in the presence and absence of PDBu were similar to each other and to that of HL-60 cells cultured in the absence of PDBu (Table 4). R1B6 cells cultured in the presence of PDBu did not express the adherence antigen but did express the transferrin receptor, the myeloid antigen detected by antibody B40.9, and the early myeloid antigen detected by 906; the latter antigen was present at significantly higher density on R1B6 than on HL-60 cells (not shown). When R1B6 cells were washed and then exposed to rIFNγ in the absence of PDBu, there was a large increase in the proportion of cells expressing HLA-DR antigens and Fc receptor detected by EA7S rosette

TABLE 2. Biochemical analysis of α NAE activity.[a]

	HL-60			R1B6		
Pretreatment	---	---	IFNγ	---	---	IFNγ
Treatment	---	TPA	TPA	---	TPA	TPA
Percent of control HL-60 cell activity	100	151	164	145	171	222
Percent adherence	0	>95	>95	0	10	>95

[a]Cells were preincubated for 22 h in the absence of PDBu and in the presence or absence of rIFNγ (225 U/mL). TPA (0.05 μM) was added, and 3 days later the cells were assayed for αNAE activity and adherence to plastic, relative to untreated cells. Control HL-60 cell enzyme activity was 22 nmol/min/mg.

TABLE 3. Cathepsin B activity in HL-60 and R1 cells.[a]

Treatment	HL-60 (pmol/min/mg protein)	R1
Control	50 (0)[b]	500 (0)
TPA (0.01 μM)	900 (>90)	478 (0)

[a]HL-60 and R1 cells were cultured with or without 0.01 μM TPA, incubated for 2 days, and cell lysates were prepared by treating the cells with 0.2% Triton X-100, 0.9% NaCl for 30 min followed by centrifugation at 500 x g for 10 min. An aliquot of the lysate (40 μg) was used to assay for cathepsin B activity. Leupeptin at a concentration of 1 μM inhibited >95% of the activity.

[b]Values in parentheses represent the percentage of cells adherent on plastic tissue culture dishes.

TABLE 4. Expression of surface markers by HL-60 and R1B6 cells treated with PDBu and rIFNγ.[a]

Target cells	Treatment		Surface markers								
	PDBu	rIFNγ	HLA-DR (B33.1)	MoAg (B44.1)	MyAg (B40.9)	MoMy Ag (B13.4)	C3bi receptor (B43.4)	Transferrin receptor (5E9)	Adherence (28.3.7)	Early MyAg (906)	FcR
Day 3											
HL-60	-	-	1.0	0.9	98.0	0.5	0.7	99.2	2.8	99.1	0.8
	+	-	2.0	2.3	92.1	30.1	45.8	57.6	66.5	58.2	7.8
	+	+	14.7	17.5	85.0	59.0	70.9	57.0	75.5	50.7	75.7
Day 5											
HL-60	-	-	1.0	0.9	74.0	0.5	0.7	97.8	2.8	77.2	0.1
	-	+	9.6	33.1	95.7	8.6	32.7	95.9	3.3	69.9	45.5
R1B6	-	-	0.3	0.6	77.4	0.4	1.4	90.3	1.5	98.5	16.3
	+	-	1.2	0.8	87.5	0.9	1.1	77.9	5.1	93.2	8.9
	-	+	63.1	4.7	87.9	3.3	5.9	83.7	5.5	95.6	87.3
	+	+	27.3	6.2	92.9	8.7	20.6	64.8	16.7	81.3	89.1

[a]HL-60 and R1B6 cells were washed twice with serum-containing medium and seeded at a density of 1.5 x 10^5 cells/mL in the presence or absence of PDBu (2.5 μM) and rIFNγ (2 x 10^3 U/mL). After 3-5 days, the proportion of cells positive for the various surface markers was determined by indirect immunofluorescence with the specific monoclonal antibodies and flow cytofluorimetry. The proportion of cells positive for Fc receptor was determined by EA7S rosette formation.

formation, whereas the transferrin receptor was not decreased and no significant induction of adherence antigen was observed.

When R1B6 cells were treated with rIFNγ in the presence of PDBu, about 20-30% of the cells expressed HLA-DR antigens, C3bi, and the adherence antigen. Fewer cells (about 5-10%) became positive for monocyte-specific (B44.1) and myelomonocytic antigens (B13.4); the percentage of cells expressing transferrin receptor and 906 antigen, and the density of these two antigens at the cell surface, decreased. Most of the cells expressed FcR and the myeloid antigen detected by B40.9. Ab-CMC was not induced to high levels in R1B6 cells by any of the treatment conditions (not shown). The changes in antigen expression in R1B6 cells treated with rIFNγ and PDBu were qualitatively similar to those seen in HL-60 cells treated similarly (Table 4) but, in general, fewer R1B6 than HL-60 cells showed the changes.

Kinetic experiments (not shown) indicated that the changes in antigen expression induced in R1B6 cells by rIFNγ were maximal after 5 days of culture. IFNα at analogous concentrations did not induce changes in antigen expression, indicating that the differentiation changes induced in R1B6 cells are a specific effect of IFNγ.

DISCUSSION

We have isolated variants of HL-60 cells that are reversibly resistant to the induction of differentiation by phorbol esters and can be maintained in the presence of high concentrations of these agents without any inhibitory effect on cell growth. Like parental HL-60 cells, these cells, termed R1 and R1B6, express surface antigens that are characteristic of the early stages of differentiation of myeloid cells (35); the B40.9 antigen is expressed at a level similar to that on HL-60 cells whereas the density of the 906 antigen is higher than that on HL-60 cells. Characteristically, the 906 antigen decreases during myeloid differentiation and is undetectable at the metamyelocytic stage, whereas detectable levels of this antigen are retained during monocytic differentiation (16). Thus, the presence of the 906 antigen on reversibly resistant R1B6 cells, together with the relatively high αNAE and cathepsin B activities of these cells, indicate that they have acquired some of the phenotypic characteristics of monocyte-like cells. Under the proper manipulation, as we describe, expression of other myelomonocytoid characteristics can be induced in these cells.

We previously reported that IFNγ induces coordinate expression of monocytic antigens on normal and leukemic myeloid cells (33), suggesting that this agent induces monocytic differentiation in these cells. These observations have been confirmed with HL-60 cells (3,10,19,20). The phorbol ester-resistant variant of HL-60, R1B6, has retained sensitivity to the differentiation-inducing effects of IFNγ; treatment with rIFNγ induces expression of HLA-DR and complement and Fc receptors. However, HLA-DR induction is partially blocked if PDBu is present. In contrast, the antigens B44.1 and B13.4, induced on HL-60 cells by treatment with a combination of PDBu and rIFNγ, are not induced on R1B6 cells. Our observation that the reversion of the R1B6 cell resistance to TPA is potentiated when IFNγ is present suggests that IFNγ may reverse, in part, the resistance of these cells to phorbol esters, perhaps by increasing the number of available phorbol ester receptors.

We found that when phorbol esters were removed from the culture medium of R1B6 cells, there was a steady increase in the number of phorbol ester receptors, with approximately 10^5 more sites being available for binding after 3 days of incubation in the absence of the promoters. Some investigators (2,13,22,23) have found a correlation between resistance of HL-60 variants to phorbol esters and the number of cells in the population with increased ploidy or other chromosome alterations. However, Mascioli and Estensen (22) found no differences in the number or affinity of phorbol ester receptors in their hyperdiploid resistant variants compared with parental HL-60 cells and suggested that, in the variants, resistance was a result of modification at a step(s) beyond phorbol ester binding. Others have also found no differences in the number or affinity of receptors for phorbol esters in cells resistant to these agents (8,15,22,40). In contrast, our reversibly resistant HL-60 variants have receptor levels that are significantly lower in density than those of the parental HL-60 cells and there is a correlation between receptor density and TPA responsiveness. That is, variants that are reversibly resistant to phorbol esters can recover the level of receptors found in parental HL-60 cells and respond to phorbol ester-induced adherence in a manner similar to the parental cells. The polyploid nature of these variant cells does not appear to affect this response.

The decreased density of phorbol ester receptors in these variants compared with the parental cells may be due to the procedure by which they were selected. Passaging the cells in medium containing phorbol esters may have resulted in the receptors being continuously down-regulated to a lower steady-state level, or in a negative feedback mechanism in which the phorbol ester turns off biosynthesis of the receptor. In either case, removal of the phorbol ester from the culture medium could result in recovery of receptors. The data suggest that there may be a threshold level of receptor required for arrest of cell growth and induction of differentiation and that phorbol ester receptor down-regulation may be involved in modulating the sensitivity of the cell to phorbol esters. It remains to be determined whether the receptor population remaining in these variants when they are grown in the presence of phorbol esters is necessary for the cells to remain viable. That is, there may be two classes of receptors (8,11,24,28,31,32,43) with similar affinities, with one class involved in the mitogenic response of the cells and the other in the differentiation response.

Several recent studies have shown that a calcium/phospholipid-dependent protein kinase C is a receptor for phorbol esters (1,7,21,26,27,30,44) and that the promoters can activate this enzyme directly (7,26). Our preliminary studies show that the increase in phorbol ester receptors in R1B6 cells following reversion to sensitivity occurs mainly in the cytosol. After partial purification of the cytosol receptor, the activity of a calcium/phospholipid-dependent protein kinase copurifies with phorbol ester binding activity. The data suggest a causal relationship between the receptor/protein kinase C of phorbol esters and the biochemical events involved in the induction of myelomonocytic differentiation by these compounds. HL-60 cells and variants that are reversibly or irreversibly resistant to the phorbol esters will be very useful in trying to understand the subsequent mechanisms that trigger these events.

ACKNOWLEDGMENTS

This work was supported, in part, by Public Health Service Grants CA 23413, CA 32898, CA 37155, and CA 10815, National Institutes of Health, U.S. Department of Health and Human Services.

REFERENCES

1. Ashendel, C.L., Staller, J.M., and Boutwell, R.K. (1983): Cancer Res., 43:4333-4337.
2. Au, W.W., Callaham, M.F., Workman, M.L., and Huberman, E. (1983): Cancer Res., 43:5873-5878.
3. Ball, E.D., Guyre, P.M., Shen, L., Glynn, J.M., Maliszewski, C.R., Baker, P.E., and Fanger, M.W. (1984): J. Clin. Invest., 73:1072-1077.
4. Blaineau, C., Avner, P., Tunnacliffe, A., and Goodfellow, P. (1983): EMBO J., 2:2007-2012.
5. Bruns, R.F., Lawson-Wendling, K., and Pugsley, T. (1983): Anal. Biochem., 132:74-81.
6. Burnett, D., Crocker, J., and Vaughan, A.T.M. (1983): J. Cell Physiol., 115:249-254.
7. Castagna, M., Takai, Y., Kaibuchi, K., Sano, K., Kikkawa, U., and Nishizuka, Y. (1982): J. Biol. Chem., 257:7847-7851.
8. Colburn, N.H., Gindhart, T.D., Dalal, B., and Hegamayer, G.A. (1983): In: Organ and Species Specificity in Chemical Carcinogenesis, edited by R. Langenbach, S. Nesnow, and J.M. Rice, pp. 189-200. Plenum Press, New York.

9. Collins, S.J., Gallo, R.C., and Gallagher, R.E. (1977): Nature, 270:347-349.
10. Dayton, E.T., Matsumoto-Kobayshi, M., Perussia, B., and Trinchieri, G. (1984): submitted.
11. Dunn, J.A., and Blumberg, P. (1983): Cancer Res., 43:4632-4637.
12. Gallagher, R., Collins, S., Trujillo, J., McCredie, K., Ahearn, M., Tsai, S., Metzgar, R., Aulakh, G., Ting, R., Ruscetti, F., and Gallo, R. (1979): Blood, 54:713-733.
13. Gallagher, R.E., Ferrari, A.C., Zulich, A.W., and Testa, J.R. (1983): Prog. Nucleic Acid Res. Mol. Biol., 29:283-286.
14. Gallo, R.C., Breitman, T.R., and Ruscetti, F.W. (1982): In: Maturation Factors and Cancer, edited by M.A.S. Moore, pp. 255-271. Raven Press, New York.
15. Goodwin, B.J., Moore, J.O., and Weinberg, J.B. (1984): Blood, 63:298-304.
16. Griffin, J.D., Ritz, J., Nadler, L.M., and Schlossman, S.F. (1981): J. Clin. Invest., 68:932-941.
17. Huberman, E., Braslawsky, G.R., Callaham, M., and Fujiki, H. (1982): Carcinogenesis, 3:111-114.
18. Huberman, E., and Callaham, M. (1979): Proc. Nat. Acad. Sci., USA, 76:1293-1297.
19. Kelley, V.E., Fiers, W., and Strom, T.B. (1984): J. Immunol., 132:240-245.
20. Koeffler, H.P., Ranyard, J., Yelton, L., Billing, R., and Bohman, R. (1984): Proc. Nat. Acad. Sci., USA, 81:4080-4084.
21. Kraft, A.S., Anderson, W.B., Cooper, H.L., and Sando, J.J. (1982): J. Biol. Chem., 257:13193-13196.
22. Mascioli, D.W., and Estensen, R.D. (1984): Cancer Res., 44:3280-3285.
23. Mendelsohn, N., Calderon, T., Acs, G., and Christman, J.K. (1983): Exp. Cell Res., 148:514-519.
24. Mendelsohn, N., Gilbert, H.S., Christman, J.K., and Acs, G. (1980): Cancer Res., 40:1469-1474.
25. Metcalf, D. (1983): Leukemia Res., 7:117-132.
26. Niedel, J.E., Kuhn, L.J., and Vandenbark, G.R. (1983): Proc. Nat. Acad. Sci., USA, 80:36-40.
27. Nishizuka, Y. (1984): Nature, 308:693-698.
28. Novogrodsky, A., Patya, M., Fujiki, H., Rubin, A.L., and Stenzel, K.H. (1984): J. Cell. Physiol., 120:36-40.
29. Ostensen, M., Morland, B., Husby, G., and Rekvig, O.P. (1983): Clin. Exp. Immunol., 54:397-404.
30. Parker, P.J., Stabel, S., and Waterfield, M.D. (1984): EMBO J., 3:953-959.
31. Perrella, F.W., Ashendel, C.L., and Boutwell, R.K. (1982): Cancer Res., 42:3496-3501.

32. Perrella, F.W., Bussell, P.A., and Boutwell, R.K. (1982): Biochem. Biophys. Res. Commun., 108:1722-1727.
33. Perussia, B., Dayton, E.T., Fanning, V., Thiagarajan, P., Hoxie, J., and Trinchieri, G. (1983): J. Exp. Med., 158:2058-2080.
34. Perussia, B., Lebman, D., Ip, S.H., Rovera, G., and Trinchieri, G. (1981): Blood, 58:836-843.
35. Perussia, B., Trinchieri, G., Lebman, D., Jankiewicz, J., Lange, B., and Rovera, G. (1982): Blood, 59:382-392.
36. Robbi, M., and Beaufay, H. (1983): Eur. J. Biochem., 137:293-301.
37. Rovera, G., O'Brien, T., and Diamond, L. (1979): Science, 204:868-870.
38. Rovera, G., Olashaw, N., and Meo, P. (1980): Nature, 284:69-70.
39. Rovera, G., Santoli, D., and Damsky, C. (1979): Proc. Nat. Acad. Sci., USA, 76:2779-2783.
40. Solanki, V., Slaga, T.J., Callaham, M., and Huberman, E. (1981): Proc. Nat. Acad. Sci., USA, 78:1722-1725.
41. Tanaka, H., Abe, E., Miyaura, C., Shiina, Y., and Suda, T. (1983): Biochem. Biophys. Res. Commun., 117:86-92.
42. Tomida, M., Yamamato, Y., and Hozumi, M. (1982): Biochem. Biophys. Res. Commun., 104:30-37.
43. Tran, P.L., Castagna, M., Sala, M., Vassent, G., Horowitz, A.D., Schachter, D., and Weinstein, I.B. (1983): J. Biochem., 130:155-160.
44. Vandenbark, G.R., Kuhn, L.J., and Niedel, J.E. (1984): J. Clin. Invest., 73:448-457.
45. Yam, L.T., Li, C.Y., and Crosby, W.H. (1971): Am. J. Clin. Pathol., 55:283-290.

Carcinogenesis, Vol. 10, edited by E. Huberman
and S. H. Barr. Raven Press, New York © 1985.

Oncogenes and Cellular Controls in Radiogenic Transformation of Rodent and Human Cells

C. Borek

Radiological Research Laboratory, and Department of Pathology, Columbia University College of Physicians and Surgeons, New York, New York 10032

Radiation is the most ubiquitous oncogenic agent. Although weak compared to some chemicals, it is far more pervasive and measurable at low doses. The interaction of radiation with cells occurs within a fraction of a second. Yet radiation can elicit neoplastic processes whose phenotypic expressions are akin to those induced by much lengthier exposure to chemicals or by viruses introducing new genetic information (6). Although x-rays were discovered in 1895, we still know relatively little about the mechanisms of their oncogenic action. Even though epidemiological and animal data have yielded much information on the carcinogenic action of radiation (3,43), they are limited in studies at low doses. Furthermore, the complex homeostatic mechanisms that prevail *in vivo* limit the exploration of cellular and molecular mechanisms underlying radiation-induced malignancies (6).

RADIOGENIC *IN VITRO* CELL TRANSFORMATION

Cell culture systems afford the opportunity to study the oncogenic action of radiation on cells under defined conditions free from host-mediated effects (6). Among the systems that have been most used in radiation studies is the hamster embryo cell system (6,16,23,24) in which primary diploid cells are exposed to radiation. Transformation is scored by counting the presence of colonies that differ morphologically from normal (Fig. 1) and subsequently grow in agar and give rise to tumors (2,6). Another system is the C3H 10T1/2 established mouse cell line developed in Heidelberger's laboratory (40,42), in which transformation is scored by counting foci (Fig. 1). The principal differences between these systems have been reviewed (6,10).

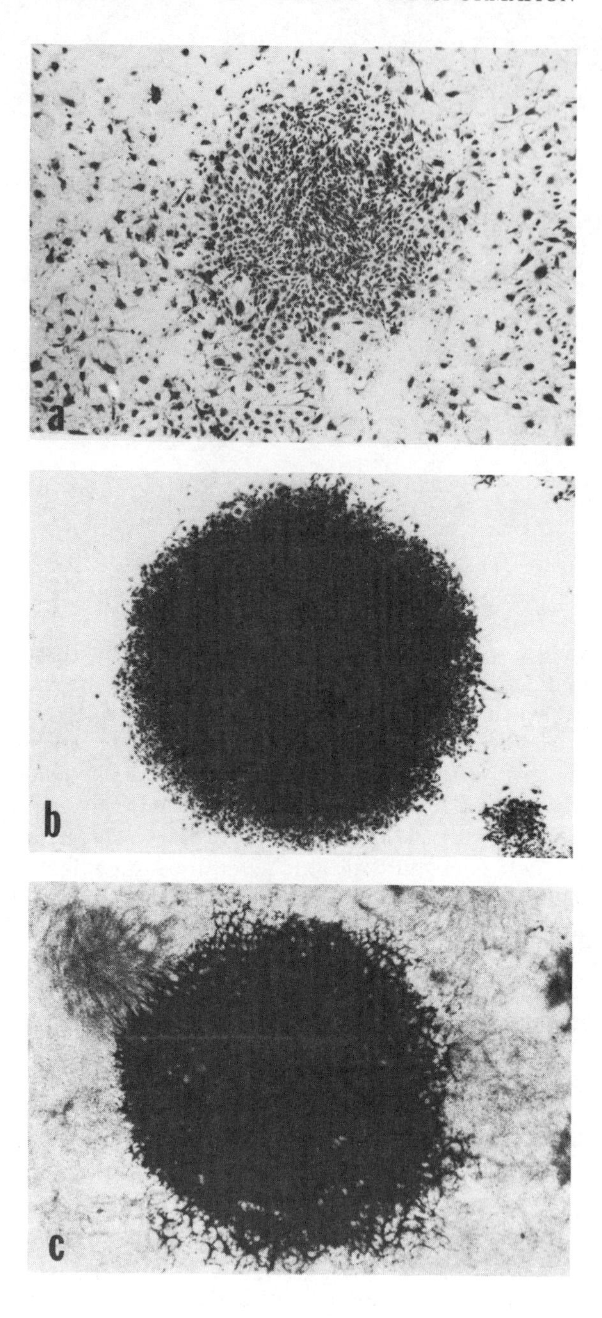

FIG. 1. (a) Normal hamster embryo colonies, (b) radiation-transformed hamster embryo colonies, and (c) a focus of C3H 10T1/2 cells transformed by radiation.

Radiogenic transformation is a multistep process (6). Cell division shortly after exposure to radiation is essential for fixation of transformation (6,23,24), which indicates DNA as a target of radiation. Additional replications are required for expression of the transformed state (6,23, 24). DNA-damaging agents that act in concert with radiation at exposure times serve as cocarcinogens (18,28), whereas those that affect the neoplastic process at later stages and enhance the frequency of transformation may serve as a tumor promoters (6).

Human diploid cells can also be transformed in vitro by radiation, both by x-rays (5) and ultraviolet (UV) light (11,37,41). The steps involved in human cell transformation by x-rays are similar to those previously observed in rodents (6). However, the course and the frequency of human radiogenic cell transformations are somewhat different (5,11). This differential pattern between human and rodent cells for course and frequency of transformation has also been observed with chemically transformed cells (34,37,38,47) and may vary also with the origin of the human cells (11,37). One of the most striking differences between human and hamster radiogenic cell transformation is that the growth of the human cells as transformed foci in culture (Fig. 2) occurs at the same time that these cells acquire the ability to grow in agar or on agar (5). Their lifespan may be limited and their karyotype may remain diploid or close-to-diploid despite their ability to induce tumors (5).

FIG. 2. Human embryo cells transformed by 200 rad of x-ray.

PERMISSIVE AND PROTECTIVE FACTORS IN TRANSFORMATION

Permissive Factors

Neoplastic cell transformation can be regarded as an abnormal expression of cellular genes. The initial event may be one or a combination of events that disrupt the genetic apparatus. These events may comprise a variety of DNA alterations including the activation of transforming genes (oncogenes) (4,36,39). Events associated with initiation and promotion and their control may overlap to some degree (14), but for them to occur, permissive cellular physiological conditions must prevail (9). These permissive conditions will confer upon a genetically susceptible cell a physiological state conducive to transformation (9,24).

If this permissive state does not prevail and the cell is adequately protected by inherent or external conditions (Table 1), initiation, as well as its subsequent expression, will be prevented and oncogenic transformation will not be observed.

Among permissive factors are stages in the cell cycle (6,9, 38,47). Synchronization of cellular growth in human cells results in higher transformation when they are exposed to radiation or to chemicals (5,38) at the G_1/S phase of the cycle.

Genetic Factors

The cells from xeroderma pigmentosum patients (11,37) and cells from patients with Bloom's syndrome (11) show higher transformation rates than normal human cells implying that specific genetic defects may confer upon the cells a higher susceptibility to transformation. Genetic factors are also associated with species susceptibility, and these factors may be compounded by age differences (6).

Hormonal Effects

Hormones may serve as permissive and potentiating factors in transformation (5,6,15,30,31). The hormone β estradiol potentiates human radiogenic transformation (5) and thyroid hormones serve as critical factors for initiation of radiation- (30,31), chemical- (15), and retroviral-induced transformation (22). In the absence of thryoid hormones, a condition obtained by removing the hormones from the serum (15,30,31), no radiogenic or chemically induced transformations are observed in hamster or C3H 10T1/2 cells (15,30,31) and a twofold decrease in viral transformation can be observed under the same conditions in NRK cells (22). When the hormone is added at physiological levels, transformation is hormone dose dependent in all cases. Time course studies indicate that thyroid hormones exert their permissive action when added prior to the oncogenic

TABLE 1. Permissive and protective factors in transformation in vitro.

Factors	Parameter Observed
Permissive factors	
High genetic susceptibility	XP[a] cells and Bloom cells transform at a higher rate than normal following UV exposure
Species difference in susceptibility	Hamster > mouse > man
Thyroid hormones	Altered oxidative state and/or induction of "cellular transformation-associated proteins"
Stage in cell cycle	Some stages (e.g., G_1/S in human cells) are more permissive
Protective factors	
Dietary and cellular antioxidants	Prevent the deleterious effects of reactive oxygen species
Selenium	Induces catalase and GSHpX; doubles cellular capacity to destroy peroxides
Superoxide dismutase	Scavenges superoxides formed by radiation and chemicals
NPSH (nonprotein thiols)	Induced by selenium and scavenges reactive oxygen species
Retinoids	Various actions, some on cell membrane, e.g., induces Na/K ATPase; some derivatives scavenge free radicals; affect oncogene expression (?)

[a]Xeroderma pigmentosum.

agents and that maximum transformation is observed when the hormone is present at least 12 h before the cells are exposed to any of the agents (15,22,30,31). The data suggest that one possible mechanism of thyroid hormone action could be mediated via cellular thyroid-hormone-dependent "transforming proteins" (15,31).

The role of thyroid hormones in later stages of transformation (promotion and progression) is still unclear. Our experiments do suggest that in the presence of tumor promoters such as the phorbol ester 12-O-tetradecanoylphorbol-13-acetate (TPA) or telecidin B, thyroid hormones serve as potentiators and may be involved in the onset of biochemical events initiated by the promoters (14), such as the production of free oxygen species. Supportive evidence for this possibility comes from the fact that superoxide dismutase, a scavenger of superoxide ($O_2^{\cdot}$) inhibits the action of the thyroid hormone T_3 in transformation (14), thus suggesting that the hormone also may be influencing the oxidative state of the cells.

Protection by Free Radical Scavengers and Antioxidants

Both ionizing and ultraviolet radiation, as well as a variety of chemicals, initiate the production of free radical species upon interaction with aqueous environments (6). The process results in the production of free oxygen species and peroxides, which are toxic to cells and can produce crosslinking of a variety of molecules, as well as initiate further cascade reactions (13). These free radicals appear to play important roles in oncogenic transformation and tumor promotion (12,13,25). Antioxidants and enzymes that scavenge free radicals can serve as protective factors (1,12,13,25,46), reduce the oncogenic effects of radiation (Fig. 3) and chemicals (12,13,25), and inhibit the action of tumor promoters (25).

Inherent cellular protective agents may vary with species and tissue (25) and in part account for species variation in transformation. Agents added externally such as dietary antioxidants or enzymatic factors can confer protection (12,13,25). Superoxide dismutase inhibits both radiogenic transformation and its enhancement by the tumor promotor TPA (8,25). Retinoids, which have a variety of actions including scavenging of free radicals by some derivatives, also inhibit radiogenic transformation and the effect of TPA (7). More recently, we have found that cellular pretreatment with seleninum, a component of glutathione peroxidase, confers protection on cells and inhibits transformation by doubling cellular capacity to destroy peroxide (12,13) (Table 2). This action is mediated via the induction of glutathione peroxidase, catalase, and nonprotein thiols, the important one among them being glutathione (12,13).

The inhibitory action of antioxidants and scavenging agents, some of them dietary factors (Vitamin A, selenium), implicates free radical processes in neoplastic transformation. The protective action of these

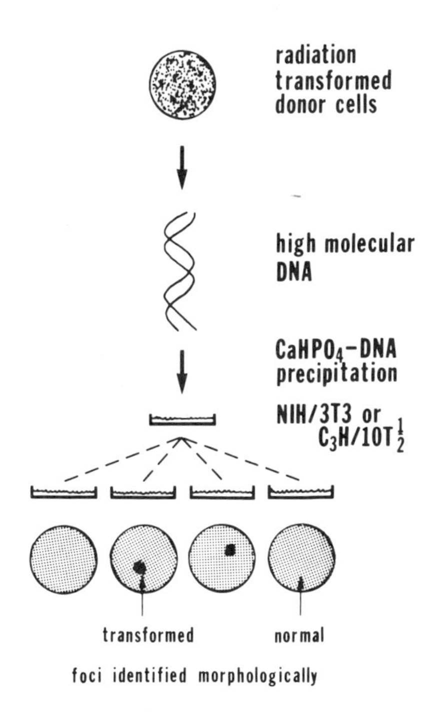

FIG. 3. Protocol for DNA-mediated gene transfer in which the calcium phosphate precipitation method is used.

compounds underscores the interplay between genetic factors and life-style in determining the onset and progression of the neoplastic process following exposure to oncogenic agents.

POLY(ADP-RIBOSE)

The relationship between DNA repair following radiation damage and radiogenic transformation is unclear. Similarly, the relationship between radiation-induced mutation and transformation is obscure. Our studies indicate that some critical biochemical events controlling mutagenesis may differ from those involved in transformation (17). For example, 3-amino benzamide, an inhibitor of the regulatory molecule poly-(ADP-ribose), inhibits radiogenic transformation but enhances sister chromatid exchange and mutagenesis and has no effect on the ligation of induced lesions in DNA.

TABLE 2. Levels of transformation, glutathione peroxidase, catalase, and nonprotein thiols in selenium-pretreated and untreated C3H 10T1/2 cells.

Parameters	Untreated	Selenium treated (0.1 M Na_2SeO_3)
Transformation by 400 rad x-ray	1.2×10^{-3}	6.1×10^{-4}
Transformation by 1.2 g/mL BAP	1.1×10^{-3}	2.2×10^{-4}
Glutathione peroxidase[a]	5.2	10.0
Catalase[a]	4.3	6.0
Nonprotein thiols[b]	1.0	2.1

[a]N moles H_2O_2 reduced/min/mg protein.
[b]N moles/mg protein.

ONCOGENES IN RADIOGENIC TRANSFORMATION *IN VITRO*

The role of DNA as a target in radiation transformation had been suggested by the requirement for DNA metabolism for fixation of the transformed state. Yet, no direct proof was available. Recent developments enabling DNA-mediated gene transfer (transfection) (45) have made it possible to address this question.

Hamster cells and C3H 10T1/2 cells exposed to x-rays and transformed *in vitro* yield populations of transformed cells from which high-molecular-weight DNA can be purified with established methods (36,39,45). The high-molecular-weight DNA is added to NIH/3T3 cells or the C3H 10T1/2 cells by means of a modified calcium phosphate precipitation method (45) in the presence or absence of a selective marker such as the Eco-*gpt* (36) (Fig. 3). The ability of DNA from *in vitro* x-ray-transformed cells to confer the transformed phenotype on normal 3T3 or C3H 10T1/2 cells by producing transformed foci (Table 3) that grow in agar (Fig. 4) (19-21) indicates that DNA codes for the radiation-transformation phenotype following direct *in vitro* exposure to radiation. It also indicates that specific cellular oncogenes (27,39) are activated (20). Restriction enzyme analysis indicates that specific segments of DNA are encoding the transformation phenotype and that these may differ in the *in vitro* transformed

TABLE 3. Transfection of DNA from in vitro radiation transformed C3H 10T1/2 and hamster embryo cells onto NIH/3T3 cells.

DNA Source	No. foci/μg DNA
Normal hamster embryo	0
X-ray-transformed hamster embryo	0.27
Normal C3H 10T1/2	0
X-ray-transformed C3H 10T1/2	0.25
NIH 3T3	0

C3H 10T1/2 and the hamster cells (20). DNA isolated from the transfected foci can retransfect the 3T3 and 10T1/2 cells in a secondary and tertiary round of transfection (20).

Thus far, we have used three approaches to carry out molecular analysis of the events involved in oncogene activation and expression following irradiation (20).

Quick blot analysis (26) involves the selective immobilization of mRNA on nitrocellulose filters from NaI-dissolved cells that have been molecularly hybridized with radioactive DNA. The quick blot technique was used to identify elevated transcripts of known oncogenes in the in vitro radiation-transformed hamster and C3H 10T1/2 cells. The C3H 10T1/2 transformed cells as compared to the controls showed elevated expression of three oncogenes (20,21), the c-abl, c-fms, and B-lym-1. None of the ras family gene transcripts appeared to be significantly elevated (19-21). There was no elevated expression of the Ki-ras gene, the cellular counterpart of the viral oncogene in Kirsten murine sarcoma virus (35), of the Harvey ras gene, a counterpart of the Harvey sarcoma virus transforming gene (33), or of the N-ras gene.

A Southern blot analysis carried out (32) on DNA from normal NIH/3T3 and from NIH/3T3 cells transformed by DNA from C3H 10T1/2 or hamster cells indicated that no amplification or extra copies of any of the ras oncogenes were present in the transformants as compared to the controls. It appears that as observed at the RNA level with the quick blot (21,26) and the cytoblot methods (21,44), the activated oncogenes in the radiation in vitro transformed cells do not belong in the ras family of genes (19,21). This situation differs from that observed in vivo by Guerrero et al. (32), who saw an amplification of the Kirsten ras gene (Ki-ras) in thymic lymphomas induced by gamma radiation. Although one must take into account

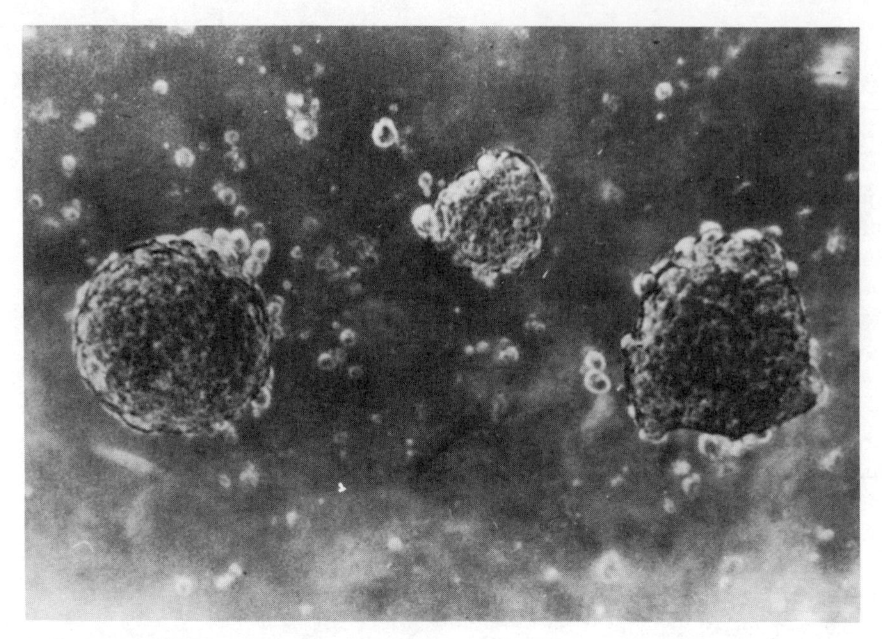

FIG. 4. NIH/3T3 cells transfected by DNA from radiation-transformed hamster cells (19,20) growing in agar.

the different types of cells involved, the difference in the results underscores the possibility that the activation of the ras oncogenes occurs late in the progression of the neoplastic process when cells have been established as progressive tumors.

A characterization of the 21,000-dalton transforming proteins (p21) encoded by the cellular ras gene family was carried out with monoclonal antibodies of various specifications (29). These antibodies immunoprecipitate the predominant p21 species from transformed and untransformed cells. The secondary tranfectants of 3T3 cells originally transfected with DNA from the x-ray-transformed C3H 10T1/2 cells or x-ray-transformed hamster embryo cells were analyzed.

Preliminary results indicate that none of the p21 ras products were present in cells transformed by hamster DNA. Extracts from 3T3 cells transformed by the C3H 10T1/2 DNA showed no immunoprecipitation with the antibodies directed against the Harvey ras or N-ras although, as suggested, some elevated expression of p21 encoded by the Kirsten ras gene was observed.

Clearly the results are too preliminary to rule out or in part support the involvement of the ras family of cellular oncogenes in radiogenic transformation *in vitro* of the C3H 10T1/2 cells.

CONCLUSIONS

Cellular transformation in vitro affords us the opportunity to probe into early events in the neoplastic process that occur following exposure to radiation.

Ionizing radiation induces DNA change, which includes strand breaks, release of bases from the DNA backbone, and modification of the bases themselves. Such DNA alterations or subsequent DNA repair have been implicated in radiogenic transformation, althrough molecular approaches to prove these changes have become available only recently with the advent of DNA-mediated gene transfer and recombinant DNA technologies. Using transfection techniques, we found that indeed the radiation-transformed phenotype induced following exposures of diploid or heteroploid cells to radiation is encoded in the DNA of the transformed cells. The oncogenes activated in the in vitro radiation-transformed hamster and C3H 10T1/2 cells do not appear to be of the ras family since no alterations or amplifications of these genes were observed in the transformants as compared to the normal cells. In the C3H 10T1/2 cells, the data are not conclusive because there is some suggestive evidence showing a small enhancement in the levels of the 21,000 dalton protein (p21) encoded by the Kirsten gene.

The activation of other genes in the radiation-transformed cells is suggested by elevated transcripts of c-abl, c-fms, and B-lym-1 in the transformed C3H 10T1/2 cells but not the hamster cells. This situation is currently being evaluated.

The lack of activation of the Ki-ras gene in vitro differs from the situation in vivo where the Ki-ras is activated in radiation-induced lymphomas. This difference perhaps underscores the suggestive evidence in a number of systems that the ras oncogenes are activated at a later stage in the neoplastic process--such as in the bona fide tumor developing in vivo following exposure of the animal to radiation.

Although we are ignorant of the exact events that trigger the initiation of radiation-induced gene alteration, we know that permissive and protective factors can modulate the outcome of the events. We find that thyroid hormones are required for the initiating events to take place, indicating that particular physiological conditions are essential for transformation. We also find that free-radical-mediated processes play an important role in the radiation-induced neoplastic process and that specific scavenging enzymes can be induced by additions to the diet, thereby conferring protection from radiation and chemically induced transformation. This finding suggests that nongenetic events play a role in the transcriptional induction of cellular molecules, which may be protective and influence the neoplastic process.

ACKNOWLEDGMENT

This research was supported by Public Health Service Grant CA-12536 from the National Cancer Institute, U. S. Department of Health and Human Services, and by a contract from the National Foundation for Cancer Research. I thank Drs. Guerrero and Pellicer for help with the Southern blots and Dr. M. Furth for help with the p21 analysis.

REFERENCES

1. Ames, B.N. (1983): Science, 221:1256-1264.
2. Barret, J.C., and Ts'o, P.0.P. (1978): Proc. Nat. Acad. Sci., USA, 71:3761-3765.
3. BEIR III (Committee on Biological Effects of Ionizing Radiation) (1972): The Effects on Populations of Exposure to Low Levels of Ionizing Radiation. Nat. Res. Council, Nat. Acad. Sci., Washington, DC.
4. Bishop, M. (1982): Adv. Cancer Res., 37:1-29.
5. Borek, C. (1980): Nature, 283:776-778.
6. Borek, C. (1982): Adv. Cancer Res., 37:159-232.
7. Borek, C. (1983): In: Molecular Interrelations of Nutrition and Cancer, edited by M.S. Arnot, J. Van Eyes, and Y-M. Wang, pp. 337-350. Raven Press, New York.
8. Borek, C. (1984): In: Methods in Enzymology, Vol. 105: Oxygen Radicals in Biological Systems, edited by C.P. Colowick, N.O. Kaplan, and L. Packer, pp. 465-479. Academic Press, New York.
9. Borek, C. (1984): In: The Biochemical Basis of Chemical Carcinogenesis, edited by H. Greim, R. Juna, M. Kraemer, H. Marquardt, and F. Oesch, pp. 175-188. Raven Press, New York.
10. Borek, C. (1984): J. Pharmacol. Exp. Ther., (in press).
11. Borek, C., and Andrews, A. (1983): In: Human Carcinogenesis, edited by C.C. Harris and H. Antrup, pp. 519-541. Academic Press, New York.
12. Borek, C., and Biaglow, J.E. (1984): Proc. Am. Assoc. Cancer Res., 25:125.
13. Borek, C., Ong, A., Donahue, L., and Biaglow, J.E. (1984): Proc. Nat. Acad. Sci., USA, submitted.
14. Borek, C., Cleaver, J.E., and Fujiki, H. (1984): In: Cellular Interaction by Environmental Tumor Promoters and Relevance to Human Cancer, Proc. of the 14th International Symposium of the Princess Takamatsu Cancer Research Fund, pp. 195-206.
15. Borek, C., Guernsey, D.L., Ong, A., and Edelman, I.S. (1983): Proc. Nat. Acad. Sci., USA, 80:5749-5752.
16. Borek, C., and Hall, E.J. (1973): Nature, 243:450-453.

17. Borek, C., Morgan, W.R., Ong, A., and Cleaver, J.E. (1984): Proc. Nat. Acad. Sci., USA, 81:243-247.
18. Borek, C., and Ong, A. (1981): Cancer Lett., 12:61-66.
19. Borek, C., and Ong, A. (1984): Radiat. Res., Abstracts, Radiat. Res. Soc. Meeting, San Antonio, Texas, February 1983.
20. Borek, C., and Ong, A. (1984): in preparation.
21. Borek, C., Ong, A., Bresser, J., and Gillespie, D. (1984): Proc. Am. Assoc. Cancer Res., 25:100.
22. Borek, C., Ong, A., and Rhim, J.S. (1985): Cancer Res., in press.
23. Borek, C., and Sachs, L. (1966): Nature, 210:276-278.
24. Borek, C., and Sachs, L. (1967): Proc. Nat. Acad. Sci., USA, 57:1522-1527.
25. Borek, C., and Troll, W. (1983): Proc. Nat. Acad. Sci., USA, 80:1304-1307.
26. Bresser, J., Hubbel, H., and Gillespie, D. (1983): Proc. Nat. Acad. Sci., USA, 80:6523-6527.
27. Cooper, G.M. (1982): Science, 218:801-806.
28. DiPaolo, J.A., Donovan, P.J., and Popescu, N.C. (1976): Radiat. Res., 66:310-325.
29. Furth, M.F., Davis, L.J., Fleurdelys, B., and Scolnick, E.M. (1982): J. Virol., 43:294-304.
30. Guernsey, D.L., Ong, A., and Borek, C. (1980): Nature, 288:591-592.
31. Guernsey, D.L., Borek, C., and Edelman, I.S. (1981): Proc. Nat. Acad. Sci., USA, 78:5708-5711.
32. Guerrero, I., Calzada, P., Mayer, A., and Pellicer, A. (1984): Proc. Nat. Acad. Sci., USA, 81:202-205.
33. Harvey, J.J. (1964): Nature, 204:1104-1105.
34. Kakunaga, T. (1978): Proc. Nat. Acad. Sci., USA, 75:1334-1338.
35. Kirsten, W.H., and Mayer, L.A. (1967): J. Nat. Cancer Inst., 39:311-335.
36. Land, H., Parada, L.F., and Weinberg, R.A. (1983): Nature, 304:596-602.
37. McCormick, J., and Maher, V.M. (1983): In: Human Carcinogenesis, edited by C.C. Harris and H. Antrup, pp. 401-429. Academic Press, New York.
38. Milo, G.E., and DiPaolo, J.A. (1978): Nature, 275:130-132.
39. Perucho, M., Goldfarb, M., Shimizuk, M., Fogh, J., and Wigler, M. (1981): Cells, 27:467-476.
40. Reznikoff, C.A., Bertram, J.S., Brankow, D.W., and Heidelberger, C. (1973): Cancer Res., 33:3239-3249.
41. Sutherland, B.M., Delihas, N.C., Oliver, R.P., and Sutherland, J.C. (1981): Cancer Res., 41:2211-2214.
42. Terzaghi, M., and Little, J.B. (1976): Cancer Res., 36:1367-1374.
43. Upton, A.C., Randolph, M.L., and Conklin, J.W. (1970): Radiat. Res., 41:467-491.

44. White, B.A., and Bancroft, F.C. (1982): J. Biol. Chem., 257:8567-8572.
45. Wigler, M., Pellicer, A., Silverstein, S., and Axel, R. (1979): Cells, 14:725-731.
46. Zimmerman, R.J., and Cerutti, P.A. (1984): Proc. Nat. Acad. Sci., USA, 81:2085-2087.
47. Zimmerman, R.J., and Little, J.B. (1983): Cancer Res., 43:2183-2189.

Carcinogenesis, Vol. 10, edited by E. Huberman
and S. H. Barr. Raven Press, New York © 1985.

Repair and Misrepair in Radiation-Induced Neoplastic Transformation

M. M. Elkind, *C. K. Hill, and †A. Han[1]

*Department of Radiology and Radiation Biology, Colorado State University, Fort Collins, Colorado 80523; *Division of Biological and Medical Research, Argonne National Laboratory, Argonne, Illinois 60439; †Cancer Research Laboratory, Southern California Cancer Center, University of Southern California School of Medicine, Los Angeles, California 90015*

Ionizing radiation is an indisputed oncogenic agent as made evident, for example, by the excess tumors appearing among the survivors at Hiroshima and Nagasaki. For this reason, some six years ago we undertook to extend studies of mechanisms of action in vitro, which were started by C. Borek and her collaborators with Syrian hamster embryo cells (e.g., 3-5), and by J. B. Little and his collaborators with C3H mouse embryo cells (e.g., 25,38). Because of what we judged to be coherence of the data in the initial report by Terzaghi and Little (38) and the relative simplicity of their experimental system, we were attracted to the cells they used. Hence, we approached Charlie Heidelberger, and through his kindness and generosity we were able to establish the C3H mouse 10T1/2 cell system (31,32) at Argonne National Laboratory. The experiments that were initiated there have continued to the present and have been more recently extended at our respective institutions.

In this review, we describe features of the process of neoplastic transformation in vitro that we believe to be characteristic of the inducing agent, ionizing radiation. Because these features have their

[1]Deceased, 7 May 1984.

counterparts in tumor induction in rodents by radiation, and because the biophysics of the induction process by radiation can be expected to differ in several fundamental ways from that connected with transformation induced by chemicals or viruses, radiation studies offer the prospect of developing insights of mechanism that may be unique.

TRANSFORMATION ASSAY SYSTEM

The feature of 10T1/2 cells that facilitates their use for transformation studies is the colonial morphological change that correlates with the induction of fibrosarcomas in C3H mice. Reznikoff et al. (31) demonstrated this feature for chemical induction; Terzaghi and Little (38) showed that x-rays produce the same effect; and Han and Elkind (14) verified that the colonial transformants induced by fission-spectrum neutrons were also able to give rise to fibrosarcomas in C3H mice. Thus, as far as ionizing radiation is concerned, a sparsely ionizing radiation like x- or γ-rays, and a densely ionizing radiation like fission-spectrum neutrons (average energy 0.85 MeV), are both potent transforming agents although, as we will show, these two kinds of radiation differ significantly in other respects. [Nonionizing radiations, like the bacteriocidal 254 nm line of a low-pressure Hg discharge, or near-ultraviolet radiations, which characterize solar radiation on the surface of the earth, are also quite effective in transforming 10T1/2 cells (35).]

To identify cells whose "societal" properties have been changed, transformed colonies are scored as densely staining areas of cells growing in an unregulated way on top of a contact or density inhibited layer of cells (31,32) as sketched in Fig. 1 (9).

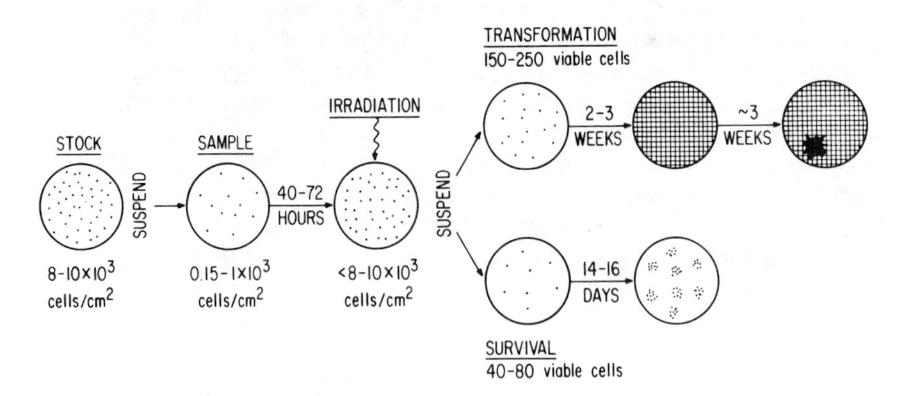

FIG. 1. Flow diagram of the use of C3H mouse 10T1/2 cells for the assessment of neoplastic transformation (upper track) and cell survival by colony formation (lower track). From Elkind et al. (9).

Several important features of our protocol should be noted.

1. The stock cultures that are used to start an experiment consist of actively growing, substrate-attached cells that are not more than about 50% confluent when suspended.

2. Cells to be used in an experiment are diluted, plated, and incubated for 2-3 days before treatment is started.

3. At the end of a radiation treatment, which could take a fraction of a minute to several days, cells are again suspended, diluted as required, and inoculated into fresh dishes according to whether transformation (150-250 viable cells per 100 mm dish) or survival (40-80 viable cells per 100 mm dish) is to be assessed. In none of the experiments to be described were the conditions and/or lengths of the irradiations such that cells became confluent during treatment and, indeed, in the case of a multifractionation irradiation protocol in which cell growth extended over 7 days, tests were made (21) to insure that no loss of transformants occurred that might be due to the onset of confluency (37).

4. The inoculum of 150-250 viable cells (upper track, Fig. 1) is dictated by the observation that both for fission-spectrum neutrons (14) and x-rays (14,18) the transformation frequency per viable cell (or per surviving cell) is independent of the inoculum size from about 20-300 viable cells per 100 mm dish. Although similar to what had been reported first by Terzaghi and Little (38), the foregoing observation disagrees with the report of Kennedy et al. (24) as we have noted (18) and will discuss again in this paper.

5. The protocol as developed in Heidelberger's laboratory (31,32) calls for weekly refeedings, for as long as 6 weeks, for the expression of transformation (upper track, Fig. 1). This procedure can overestimate the frequency of transformation because of the possible mechanical production of satellite colonies. To avoid this problem, we introduced a null method for estimating the average number of transformants per dish (14). For a given treatment, the frequency of dishes with no transformed colonies, f, was used to estimate the average number of transformed colonies per dish, which from Poisson statistics, is

$$\lambda = -\ln f$$

The use of this method eliminates overestimation of transformation frequencies but it also means that on the average only one, or fewer, transformed colonies per ~ 200 viable cells is obtained per dish. Consequently, many dishes must be used to estimate rates of transformation with moderate uncertainties when the frequencies are low.

AGE-RESPONSE DEPENDENCIES

In 1963, Terasima and Tolmach (36) demonstrated that, following a fixed x-ray dose, the survival of actively growing HeLa cells varied with age (or position) in their growth cycle. Radiobiologists have since confirmed this report and have been sensitive to the need to determine age-response patterns for the induction of different end points as a means of identifying the timing of critical events in the growth cycle of cells and relating such events to cyclic metabolic/molecular processes.

We have measured the age-survival dependence using 10T1/2 cells and the technique of harvesting mitotic cells to obtain synchrony (e.g., 36). We have found it to be typical of cells in culture that have an appreciable G_1 period relative to the rest of the cell cycle (unpublished data). That is, the age-intervals of resistance are early in the G_1-phase and late in the S-phase; sensitive periods are G_2-mitosis and the G_1-S border. Although the synchronization of cells requires special techniques, the measurement of an age-survival dependence is quite feasible because survival is essentially a high-frequency event.

We have not yet undertaken an analogous measurement of the dependence on age of transformation because, as will be evident from the figures to follow, this end point is induced with low frequency. Still, it is likely that the susceptibility to induction varies with cell age as has been reported in respect to chemical induction by a methylating agent (1). For this reason, and to explain further the timing of the treatment schedule in Fig. 1, we show in Fig. 2 the dependence on time after replating of the proportion of cells in the S-phase, as well as the transformation frequency induced by a fixed dose of x-rays delivered at the times indicated. In a population undergoing steady-state exponential growth, about 40% of the cells are in the S-phase. Even though stock cultures of actively growing cells were used to initiate the measurements, 1-2 days were required for the cells to recover from the perturbations that are induced by the suspension, dilution, and replating of cells; indeed, cell division did not commence until some 12 h after replating (18). Similar fluctuations with 10T1/2 cells have been reported by Bertram and Heidelberger (1) and by Radner et al. (30). [Perturbations in growth and in macromolecular synthesis, as illustrated, were first reported by Salzman in 1959 (33).]

Associated with the fluctuations in the DNA labeling index are fluctuations in the frequency of transformation induced by an x-ray dose of 350 rad. It would be premature to identify maxima and minima in the age-transformation dependency from the results in Fig. 2; nonetheless, they suggest first that the frequency depends upon cell age, and second that S-phase may be a period of relatively high frequency as was reported for chemical induction (1).

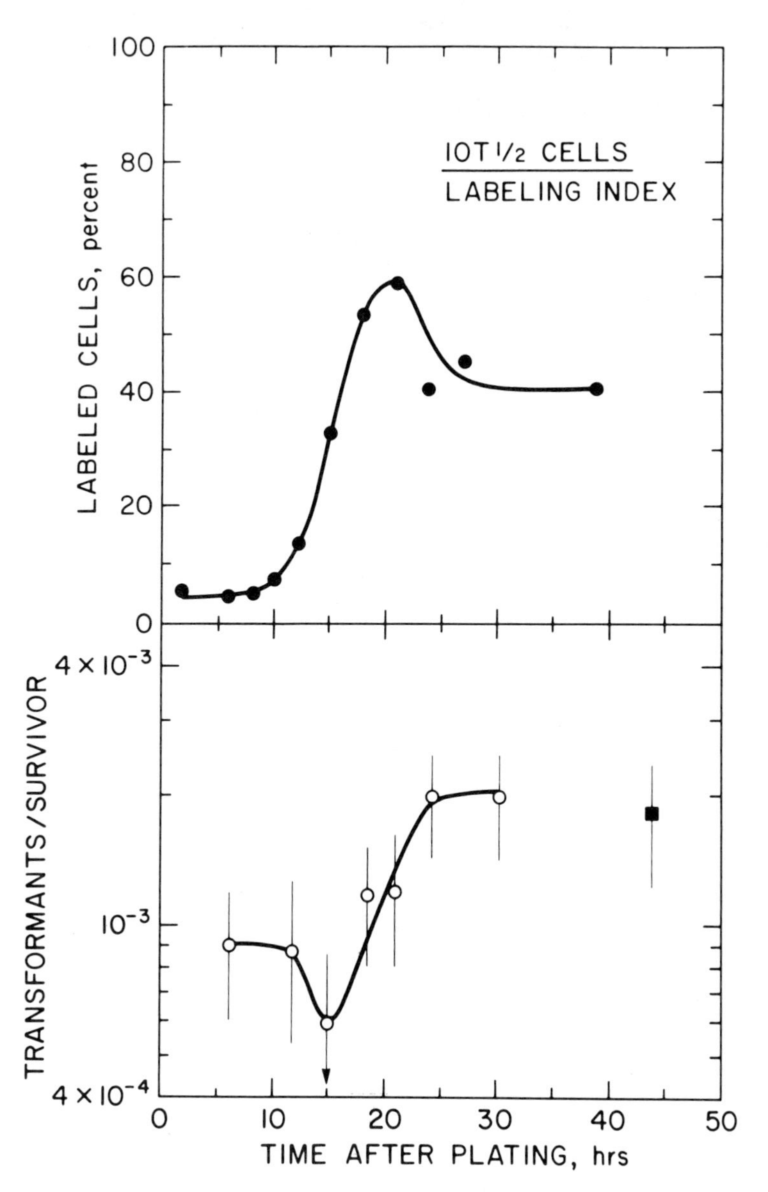

FIG. 2. Variations in the DNA labeling index (unirradiated cells) and in transformation (350 rad, 50 kV x-rays, ref. 14) following the suspension, dilution, and inoculation of cells derived from an actively growing, non-confluent stock culture. The closed square was obtained in a separate experiment. Uncertainties are standard errors. From Hill et al. (21).

TRANSFORMATION DEPENDENCE ON SEEDING DENSITY

The results that we have obtained using our methods are summarized in Figs. 3 and 4 relative to the dependence of transformation frequency on the seeding density of viable cells (in 100 mm dishes). Figure 3 shows the data for a 65% survival dose, 250 rad of x-rays (18). The dashed curve in Fig. 3, as well as in Fig. 4, is shown for comparison (see ref. 14 for data); an x-ray dose of 700 rad corresponds to a survival of 5.5% and also to a maximum transformation frequency per surviving cell (14). Thus, in agreement with Terzaghi and Little (38), we find that a minimum of an average of ~ 12 divisions is required to yield a frequency independent of seeding density. Twelve divisions is also the number for the expression of a maximum frequency in the studies of Mordan et al. (29).

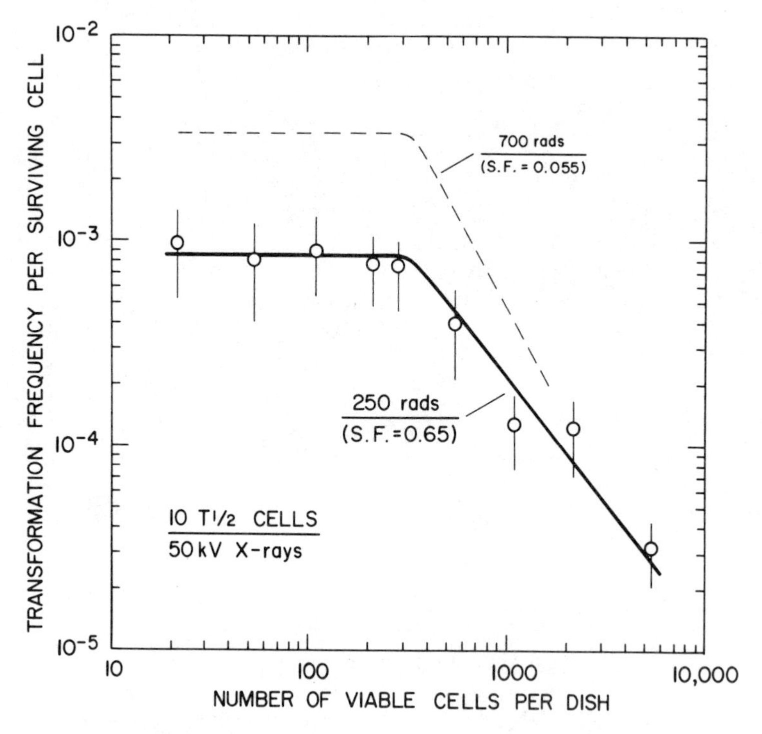

FIG. 3. The dependence of transformation frequency per viable (or surviving) cell on seeding density (100 mm dishes) for cells exposed to 250 rad of x-rays (50 kV, 1026 rad/min). S.F. = surviving fraction. From Han and Elkind (14) (dashed curve) and Han et al. (18). Uncertainties are standard errors.

Kennedy et al. (24) reported that the transformation frequency of irradiated 10T1/2 cells progressively decreases throughout the range of inocula shown in Fig. 3. A suggestion of the basis for the discrepancy between the results of Terzaghi and Little and our own (Figs. 3 and 4), on the one hand, and those of Kennedy and co-workers, on the other, comes from data that show a change in the seeding-density curve that results from the addition of 12-O-tetradecanoylphorbol-13-acetate (TPA) 2 days after exposure. TPA is reported to interfere with cell-to-cell communication (41). Although this promoter does not affect the viability or the growth rate of unirradiated 10T1/2 cells, or the survival of cells (15), it does result in twice the density at confluency, which is an indication of a reduction of intercellular controls. The data points in Fig. 4 suggest a curve made up of, say, two straight lines as opposed to only one as reported by Kennedy et al. (24). The effect of TPA is to shift our seeding-density results for experiments without TPA toward those reported by Kennedy et al. also for experiments without TPA (24). The data in Fig. 4 suggest, therefore, the possibility that a TPA-like material may have been present in the medium used by this group, perhaps introduced by the serum that they used.

Although we agree with Mordan et al. (29), and therefore with Haber et al. (13), that a full yield of transformants requires a minimum average colony size (which happens to correspond to an average of about 12 divisions at an inoculum of ~ 300 cells) the plateaus (without TPA) from about 20-300 cells in Figs. 3 and 4 suggest to us that another factor becomes important when more than 12 divisions occur (inocula of fewer than 300 viable cells). The "flat" character of colonies of nontransformed cells means that as colonies form, division is progressively confined to the periphery of the colonies. Consequently, with division, the probability that a transformed phenotype will appear (i.e., a cell whose growth is noncontact inhibited) decreases because a progressively smaller proportion of the total population continues to divide. To some extent, TPA relieves this condition and, hence, the frequencies are greater from 20-300 viable cells per dish than otherwise.

Mordan and coauthors (29) concluded that transformation frequency depends on colony size at confluency and is a maximum for at least about 12 divisions. We agree with this conclusion but for somewhat different reasons. Our results without TPA suggest a fading of the transformation signal with division (15). On the other hand, our data, as well as those of Mordan et al., support the proposal of Reznikoff and coauthors as originally published (32) that frequencies should be quantitated relative to the number of cells that survive the inducing treatment rather than to the number of foci per dish (11,24). To this end, we have been careful to use viable cell inocula between approximately 150-250 cells, as noted, even in the instances where TPA is added as a promoter, in order to minimize the influence of any factor(s) that might alter the quantitative expression of the end point.

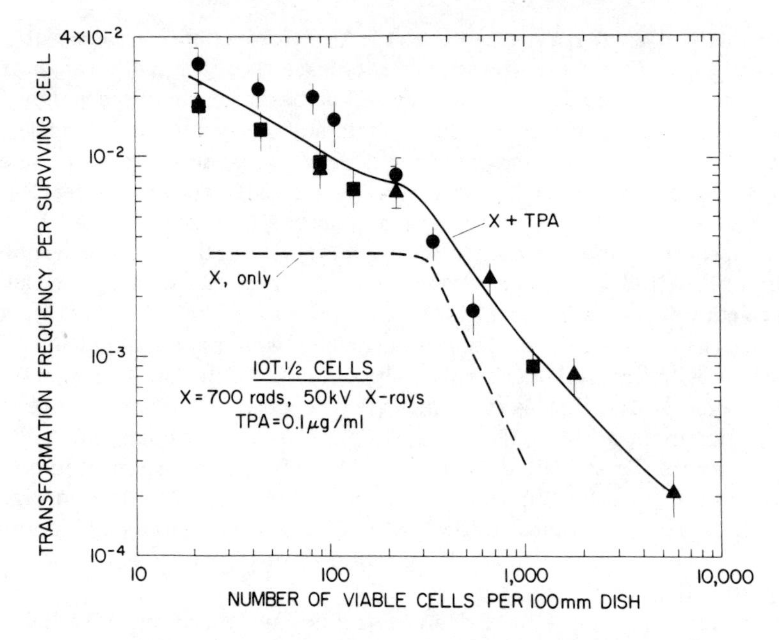

FIG. 4. The change in the dependence of transformation frequency per surviving cell for cells given an x-ray dose of 700 rad (50 kV, 1500 rad/min) followed 48 h later by the addition of 0.1 μg/mL of the phorbol ester 12-O-tetradecanoylphorbol-13-acetate (TPA). Other details as given in Figs. 2 and 3. From Han and Elkind (15).

TRANSFORMATION AND SURVIVAL

The essential features of the single-dose induction of neoplastic transformation for a sparsely ionizing radiation like ^{60}Co γ-rays are shown in Fig. 5. The survival curve is conventional in shape and, in view of its broad shoulder, is characteristic of cells that are able to repair an appreciable quantity of sublethal damage (10). The transformation curve rises from a spontaneous level of 1.1×10^{-5} to a maximum at ~ 1.2×10^{-3} and then drops off with dose at a rate that is equal to the rate of survival loss. Similar results have been obtained with fission-spectrum neutrons, except that both curves are displaced toward the left, because of the greater biological effectiveness of neutrons. When the transformation curve is plotted on linear-linear coordinates, a typical bell-shaped curve is obtained, which starts from essentially zero, rises to a maximum, and then approaches zero again with increasing dose.

An important biological statement is contained in Fig. 5, i.e., cells destined to express the transformed phenotype are neither more nor less

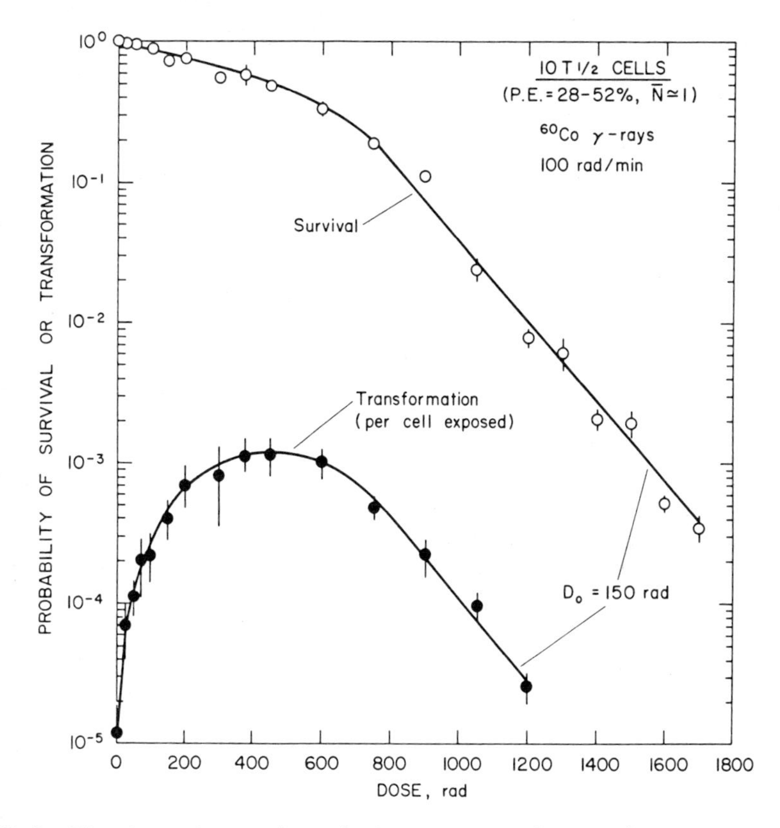

FIG. 5. The dependence of survival and neoplastic transformation on the dose of γ-rays. Transformation is expressed per cell exposed. D_0 is the dose that reduces the probability of survival or transformation to 1/$\underline{e}$ of its value along the straight line portion of a curve. Uncertainties are standard errors. PE = plating efficiency, N = multiplicity. From Han et al. (16).

sensitive than untransformed cells in view of the equal D_0s of the terminal portions of the curves. A qualification to the foregoing is required by the fact that we do not as yet have a detailed picture of the age dependence of transformation and therefore how it corresponds to the age dependence of survival.

When the transformation data in Fig. 5 are normalized by surviving fraction to derive the frequency per surviving cell as a function of dose, then as first reported by Terzaghi and Little (38) an induction curve is obtained that rises to a plateau at 3×10^{-3} (see ref. 14 and Figs. 3 and 4). A curve of similar shape is obtained with fission-spectrum neutrons although the plateau is some twofold higher (14). Plateaus of this type probably result from a combination of a limited age interval(s) sensitive to transformation induction plus less than a 100% probability of transformation within that interval(s) due, possibly, to a saturation process or to the inactivation of a target essential to the expression of transformation. The internal consistency suggested by the results in Fig. 5, as well as those in the figures to follow, speaks well for a transformation end point related to surviving cells (32) and the effectiveness of the seeding-density precaution discussed in reference to Figs. 3 and 4.

REPAIR OF SUBLETHAL AND SUBEFFECTIVE TRANSFORMATION DAMAGE BY γ-RAYS

Using the assay method as described, we have examined the modulations of the transformation end point that result from an inherently radiobiological parameter, dose protraction. In an initial study, we found that separating two fairly large doses of x-rays (i.e., 350 rad each) by various intervals up to 16 h led to a significant reduction in frequency, as well as an increase in survival (14), in agreement with reports by Terzaghi and Little (39) and Borek (2). To explore more fully the influence of intracellular processes on the expression of subeffective transformation damage, γ-ray exposures were protracted by the use of low dose rates, or of multidose fractions in which each fraction was delivered at a high dose rate.

Figure 6 shows the enhanced survival that results from the repair of sublethal damage during irradiation when the dose rate of γ-rays is reduced. The yield of transformants per surviving cell was found to decrease progressively with decreasing dose rate (16) consistent with the reduction in the frequency when two-dose fractionation is used (14).

To examine the influence of a reduction in the γ-ray dose rate in the low dose region, we concentrated on the initial parts of the 100 and 0.1 rad/min induction curves. The results are shown in Fig. 7 (16,17) along with data for five high-dose-rate fractions, each separated by 24 h (17,21). Very little cell killing is associated with any of the doses in this figure and, hence, substantially the same lines would result if the ordinate were the frequency per cell exposed instead of per surviving cell.

The points that are evident in Fig. 7 are that each set of data is fitted quite well by a straight line and that evidently intracellular repair processes reduce the rate of transformation by at least twofold, for 0.1 compared with 100 rad/min, and by threefold for the five-fraction sequence compared with single fractions. The slopes of the lines for the protracted exposures differ significantly from that for single short exposures (20). Separating each of five fractions by 2 days leads to further reductions in the frequency of transformation (21). Our results with x-rays (14) and with γ-rays (16,17,21) clearly show, therefore, that

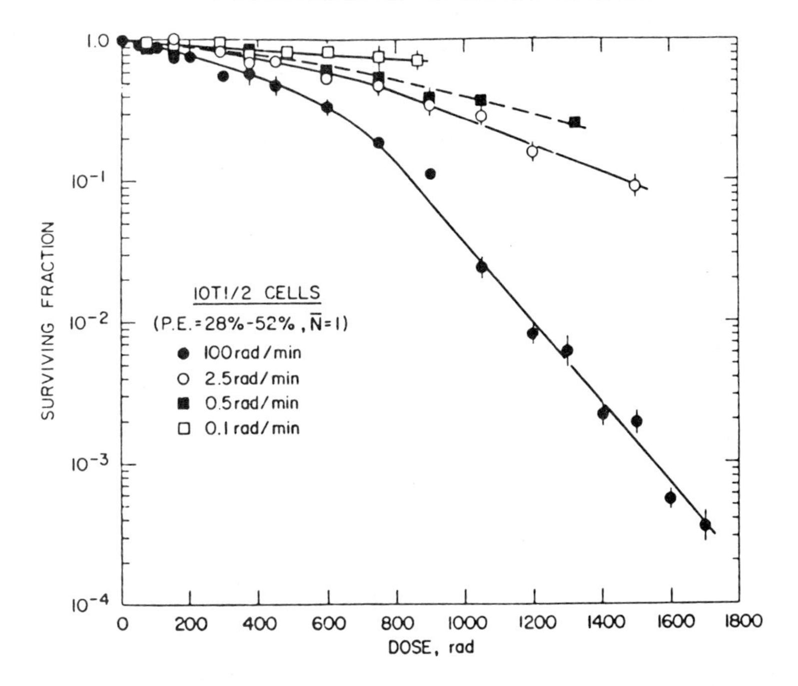

FIG. 6. Enhanced survival resulting from γ-irradiation at reduced dose rates. Cells were incubated in growth medium at 37°C during all of the exposures except for the irradiations at 100 rad/min, which were performed at room temperature. Other details as in Fig. 5. From Han et al. (16).

intracellular processes progressively reduce the frequency of transformation with protraction of the dose and that this progressively reduced effectiveness, while accompanied by a corresponding increase in survival, is nonetheless a real reduction in the rate of expression of the transformation end point since the reductions are clearly evident even in the small dose region where little cell killing is involved.

REPAIR/MISREPAIR IN TRANSFORMATION DUE TO FISSION-SPECTRUM NEUTRONS

Biophysically, neutrons can be an interesting radiation for biological studies. Like x- or γ-rays, fast neutrons are penetrating because they are neutral particles and, consequently, do not ionize by themselves. However, x- and γ-rays lose energy by setting electrons into motion whereas fast neutrons energize primarily protons. Since both particles carry a single charge, it is the some 1800-fold difference in mass that is responsible for

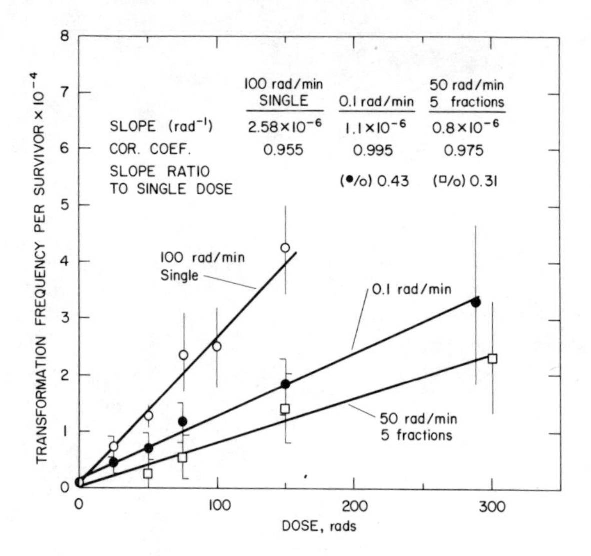

FIG. 7. The reduced induction of neoplastic transformation of 10T1/2 cells due to a reduction in dose rate or the fractionation of high-dose-rate exposures. During irradiation at 0.1 rad/min, cells were under active growth conditions (37°C) and the five high-dose fractions were 24 h apart. Other details as for Fig. 5. From Han et al. (17) with minor changes.

the difference in radiobiological effectiveness between an electron and a neutron. X- and γ-rays have about equal effects per unit dose; both are considered to be low linear energy transfer (LET) radiations, and both produce ionizations uniformly distributed in volume. The recoil protons (plus some recoil nuclei) set into motion by fast neutrons are densely ionizing relative to x- and γ-rays. Ionization is distributed nonuniformly in volume by the recoil protons and is densely distributed along their, tracks which are relatively straight. In addition to being quantitatively more effective than low-LET radiations, in a number of contexts a high-LET radiation like fast neutrons produce qualitatively different effects as we will now proceed to illustrate.

Figure 8 shows the survival results for 10T1/2 cells irradiated with fission-spectrum neutrons at both high and at low dose rates. In contrast to the result with γ-rays (Fig. 6), essentially no survival sparing accompanied the low-dose-rate exposures. This result indicates that relative to this high-LET radiation, no significant repair of sublethal damage occurred.

A result consistent with the foregoing is contained in Fig. 9, which also contains information that another kind of repair of sublethal lesions takes place during neutron exposure. To begin with, the single-dose survival curves for x-rays (●) and for fission-spectrum neutrons (■) illustrate

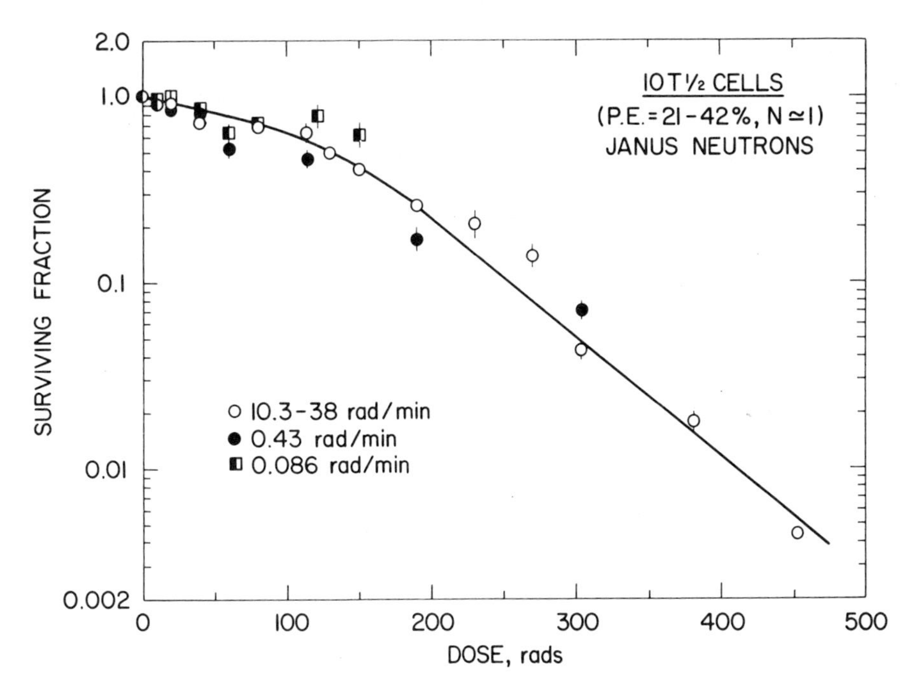

FIG. 8. The survival curve for cells exposed at both high and low dose rates with fission-spectrum neutrons produced by the JANUS Reactor at Argonne National Laboratory. During the low-dose-rate exposures, cells in T-25 flasks were incubated with medium at 37°C. Other details as for Fig. 5. From Hill et al. (19).

the qualitative (reduced shoulder width), as well as the quantitative, differences in the effectiveness of this high-LET radiation compared with x-rays. The square (◩) at 175 rad shows that the same survival is obtained from neutron exposures at both a low dose rate (exposure time 6.8 h) and a high dose rate (exposure time 10.9 min) consistent with Fig. 8. Thus, repair of sublethal damage did not occur at the reduced dose rate.

However, the remaining two curves in Fig. 8 indicate that, even so, a repair process did in fact take place. When a 175-rad neutron dose delivered at a high dose rate is followed promptly by graded x-ray doses, also delivered at a high dose rate, the survival curve (◧) of the 9% of the cells that survived the neutron exposure has a reduced shoulder compared to the normal x-ray survival curve (●). This result indicates that the neutron dose registered sublethal x-ray damage in these cells because part of the shoulder of the x-ray survival curve was removed. But when 175 rad delivered at a low dose rate was followed by graded x-ray doses at a high dose rate, the survival curve (◑) has an increased shoulder. Consequently,

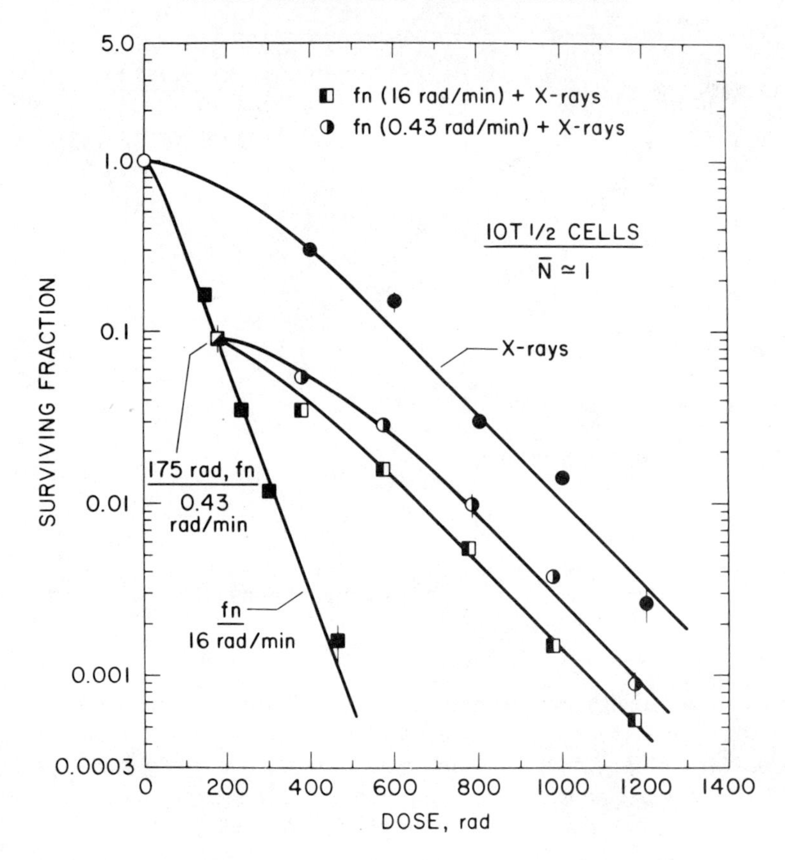

FIG. 9. Survival curves for cells exposed to single doses of fission-spectrum neutrons (■) or to x-rays (●; 50 kV, 1020 rad/min), and to combination doses as follows: 175 rad of fission-spectrum neutrons (16 rad/min) plus graded x-ray doses (1020 rad/min, ◧); or 175 rad of fission-spectrum neutrons (0.43 rad/min) plus graded x-ray doses (1020 rad/min, ◑). Other details as for Figs. 5 and 8. From Han et al. (18).

even though sublethal neutron damage is not repaired during a protracted neutron exposure, neutron-induced sublethal x-ray damage is repaired.

In discussing the influence that dose protraction has in the case of neutron exposures, the preceding two figures must be kept in mind. Figure 10 is a composite of our high, low, and multifractionation results for fission-spectrum neutrons and our high and multifractionation data for γ-rays. To begin with, here too the high-dose-rate curves indicate a quantitative difference between neutrons (□) vs. γ-rays (△). In addition, the data, particularly in the inset, show that neutrons at a low dose rate are more effective than at a high dose rate (22), and that the results with multiple high-dose-rate exposures agree with those at a low dose rate (19,20).

In contrast with what we have found with low-LET radiations (Fig. 7; 14,16,21), protracted exposures of neutrons are more effective than protracted exposures of x- or γ-rays. For x-rays, repair of sublethal damage is accompanied by the repair of subeffective transformation damage. For neutrons, repair of sublethal damage does not occur but, as shown in Fig. 9, during such an exposure a repair process connected with x-ray survival can proceed. Whether or not neutron-induced sublethal x-ray damage affects the intracellular processes that modulate the induction of transformation, it is clear that protracted exposure to neutrons results in the enhanced expression of transformation; hence, for this radiation, subeffective transformation damage is misrepaired.

DISCUSSION

To pursue mechanisms of transformation with an in vitro system, one has to be confident that the system is capable of reproducing at least qualitative aspects of oncogenesis in vivo. Because 10T1/2 cells are presumably immortal, they may not be able to mimic in all regards the multiple steps in oncogenesis. Also, because we have used actively growing cells, our results may not be entirely applicable to those tissues where the target cells may be quiescent.

In spite of the foregoing reservations, our data with both low- and high-LET radiations are qualitatively similar to a number of observations from in vivo systems. The results of Burns and Vanderlaan (6) show that the fractionation of the dose to the hair follicles of rat skin leads to reductions in the frequency of tumors, and to a flattening of the bell-shaped induction curve due to single exposures, in a manner predicted by our observations with protracted doses of x- or γ-rays (14,16). Similarly, our data with protracted low-LET radiations are qualitatively similar to those of Upton et al. (44), Ullrich and Storer (43), and Mole and coauthors (27,28) relative to tumor induction, and to those of Thomson et al. (40) in connection with life-shortening in the mouse, an end point to which tumorigenesis makes a major contribution. A corresponding qualitative agreement exists between our results with protracted compared to single short exposures of fission-spectrum neutrons and the observations of enhanced tumorigenesis in the mammary gland (42) and in the Harderian gland (12), and of enhanced life-shortening in the mouse (40).

In addition, the considerable importance of cell survival when low-LET radiations are used is also evident in an inversion in the frequencies of the induction of reticulum cell sarcoma by low versus high dose rates of γ-rays (43). As has been noted (9), such an inversion is consistent with a sparing of the killing of the precursor cells of such tumors when the rate of exposure is low, a prediction from our results, which is supported by the data of Metalli et al. (26). Thus, our studies lead to the inference that the cells responsible for the rate of spontaneous reticulum cell sarcoma were present at the time of exposure, that is, early in the lifetime of the host (9,26).

The considerable concordance, as just outlined, between our results in vitro and a signficant number of observations in vivo not only supports the use of 10T1/2 cells for studies of mechanism, but also validates the quantitative techniques that we adopted for the assessment of transformation as described in reference to Figs. 1-4. It would appear to be useful and justified, therefore, to consider working hypotheses based upon the observations that we have made thus far.

Kennedy and coauthors (23,24) proposed a two-event process for the induction of transformation by ionizing radiation. The first, due to the radiation, is frequent and is registered in a large fraction of the surviving cells. The second is very infrequent and occurs randomly between the time of exposure and the time when single cells grow into a confluent population (23).

We would modify and add to the foregoing as follows. The first event depends upon (i) radiation quality because the induction curve for fission-spectrum neutrons is shifted toward smaller doses compared with that due to x- or γ-rays (14) and (ii) the effective magnitude of the radiation-induced signal because the temporal aspects of the exposure, as well as the quality of the radiation, signficantly influence the outcome due to repair (Fig. 7) or to misrepair (Fig. 10). In view of the preceding, it is evident that radiation deposits damage that may be only subeffective and therefore modifiable by cell processes during the overall exposure.

In reference to the second event proposed by Kennedy and co-workers, the results described here, as well as data as yet unpublished, suggest that the net frequency of transformation is subject to considerable enhancement by TPA or to suppression by the antipromoter antipain, depending on the temporal course of the exposures, as well as the quality of the radiation. Thus, potentially effective damage, in which the expression is modifiable by chemical means, is registered in addition to subeffective damage.

Last, we consider the possibility that a point mutation rather than a chromosomal alteration is the more likely starting point on the assumption that transformation results from a heritable genomic change. Experimental evidence indicates that x-ray-induced single-strand lesions in DNA are completely repaired in surviving cells (8). Still, point mutations are not induced by x-rays in genes that affect essential cellular functions whereas mutants connected with nonessential functions can be induced (34). In the instance of resistance to 6-thioguanine, for example, cytogenetic evidence supports the inference that deletions or translocations of the chromosome arm containing the hypoxanthine-guanine phosphoribosyl

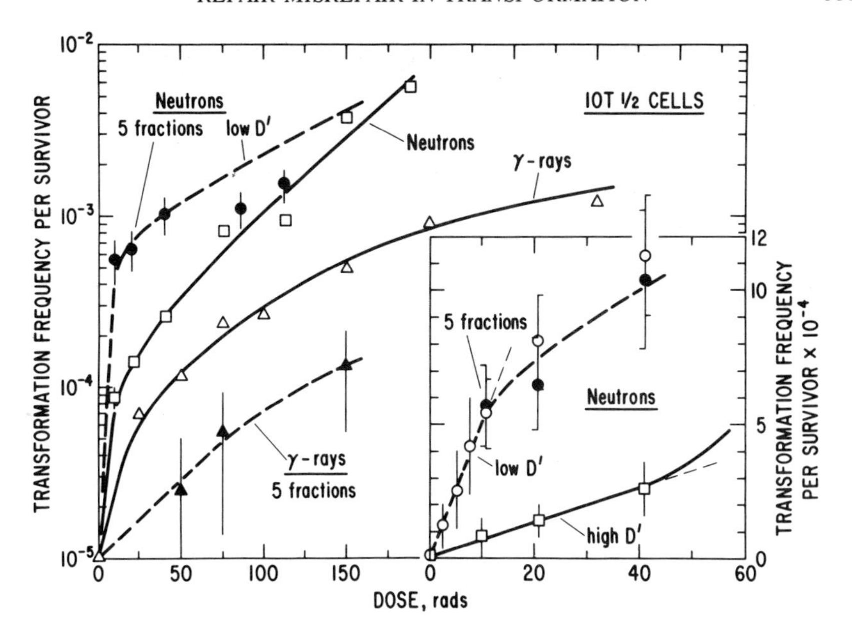

FIG. 10. Composite data for the effect of dose protraction on the induction of transformation due to fission-spectrum neutrons produced by the JANUS reactor. For orientation, the induction curves for high dose rates of γ-rays (△) and neutrons (□) are shown (see ref. 14). The dashed curve for neutrons, both in the inset as well as in the main figure, is for low-dose-rate (low D') exposures (see ref. 17 for doses greater than 10 rad). The data on the dashed curve for fractionated γ-rays (▲) come from Fig. 7. On linear-linear coordinates, the inset shows the results at 0.083 rad/min (○) and at a high-dose-rate (□, see also ref. 20). Frequencies induced by five fractions of high-dose-rate neutrons (●) or γ-rays (▲), 24 h apart, are shown in both parts of the figure (see ref. 20 and 21). The slopes of the initial, linear parts of the two curves in the inset are in the ratio 8.1:1, low to high dose rates (22). Other details as for Fig. 7.

transferase locus are responsible for the change (7). These observations suggest that, relative to cell killing, a point mutation represents too small a target and, hence, genetic change in cells that are able to survive requires a target comparable to that connected with lethality.

Therefore, we believe it is unlikely that the frequent event proposed by Kennedy and coauthors (23) will be a point mutation in, say, a proto-oncogen or a regulating genetic element of such a gene. Rather, from considerations of relative target sizes, we would propose that gross chromosomal lesions are the starting point of the process which, in view of the well-known clastogenic properties of radiation, give rise to chromosomal rearrangements and a loss of growth control.

Thus, the essentials of the inductive process of the transformation of cells by radiation are probably qualitatively different from those associated with chemicals and viruses. In addition to an absence of a question of penetration or a need to activate the carcinogen in question, the properties of ionizing radiation appear to be sufficiently unique to suggest an important role for radiobiology in studies of the etiology of cancer.

ACKNOWLEDGMENTS

We are indebted to G. Holmblad, J. Trier, G. Fox, and A. Shirvin for their assistance and to support from the U.S. Department of Energy, Contract No. W-31-109 ENG-38, and Public Health Service Grant CA 29940, the National Cancer Institute, U.S. Department of Health and Human Services.

REFERENCES

1. Bertram, J.S., and Heidelberger, C. (1974): Cancer Res., 34:526-537.
2. Borek, C. (1979): Br. J. Radiol., 50:845-846.
3. Borek, C., and Hall, E. (1973): Nature, 243:450-453.
4. Borek, C., and Hall, E. (1974): Nature, 252:499-501.
5. Borek, C., and Sachs, L. (1966): Nature, 210:276-278.
6. Burns, F.J., and Vanderlaan, M. (1977): Int. J. Radiat. Biol., 32:135-144.
7. Cox, R., and Masson, W. R. (1978): Nature, 276:629-630.
8. Elkind, M.M. (1984): Radiat. Res., in press.
9. Elkind, M.M., Han, A., Hill, C.K., and Buonaguro, F. (1983): In: Radiation Research, edited by J.J. Broerse and G.W. Barendsen, pp. 33-42. Martinus Nijhoff, Amsterdam.
10. Elkind, M.M., and Sutton, H.A. (1959): Nature, 184:1293-1295.
11. Fernandez, A., Mondal, S., and Heidelberger, C. (1980): Proc. Nat. Acad. Sci., USA., 77:7272-7276.
12. Fry, R.J.M. (1977): Int. J. Radiat. Oncol. Biol. Phys., 3:219-226.
13. Haber, D.A., Fox, D.A., Dynan, W.S., and Thilly, W.G. (1977): Cancer Res., 37:1644-1648.
14. Han, A., and Elkind, M.M. (1979): Cancer Res., 39:123-130.
15. Han, A., and Elkind, M.M. (1982): Cancer Res., 42:477-483.
16. Han, A., Hill, C.K., and Elkind, M.M. (1980): Cancer Res., 40:3328-3332.
17. Han, A., Hill, C.K., and Elkind, M.M. (1984): Br. J. Cancer, 49: 91-96.
18. Han, A., Hill, C.K., and Elkind, M.M. (1984): Radiat. Res., 99:249-261.

19. Hill, C.K., Buonaguro, F.M., Myers, C.P., Han, A., and Elkind, M.M. (1982): Nature, 298:67-69.
20. Hill, C.K., Carnes, B., Han, A., and Elkind, M.M. (1984): Radiat. Res., submitted.
21. Hill, C.K., Han, A., Buonaguro, F., and Elkind, M.M. (1984): Carcinogenesis, 5:193-197.
22. Hill, C.K., Han, A., and Elkind, M.M. (1984): Int. J. Radiat. Biol., 46:11-16.
23. Kennedy, A.R., Cairns, J., and Little, J.B. (1984): Nature, 307:85-86.
24. Kennedy, A.R., Fox, M., Murphy, G., and Little, J.B. (1980): Proc. Nat. Acad. Sci., USA., 77:7262-7266.
25. Kennedy, A.R., and Little, J.B. (1980): Cancer Res. 40:1915-1920.
26. Metalli, P., Covelli, V., and Silini, G. (1978): In: Late Biological Effects of Ionizing Radiation, pp. 341-349. International Atomic Energy Agency, Vienna.
27. Mole, R.H. (1983): In: Radiation Carcinogenesis, edited by J. Boice and J.F. Fraumeni. Raven Press, New York.
28. Mole, R.H., Papworth, D.G., and Corp, M.J. (1983): Br. J. Cancer, 47:285-291.
29. Mordan, L.J., Martner, J.E., and Bertram, J.S. (1983): Cancer Res., 43:4062-4067.
30. Radner, B.S., Abersold, P.M., and Kennedy, A.R. (1982): Mutat. Res., 94:397-403.
31. Reznikoff, C.A., Bertram, J.S., Brankow, D.W., and Heidelberger, C. (1973): Cancer Res., 33:3239-3249.
32. Reznikoff, C.A., Brankow, D.W., and Heidelberger, C. (1973): Cancer Res., 33:3231-3238.
33. Salzman, N.P. (1959): Biochem. Biophys. Acta, 31:158-163.
34. Singh, B., and Gupta, R.S. (1982): Environ. Mutagen., 4:543-551.
35. Suzuki, F., Han, A., Lankas, G.R., Utsumi, H., and Elkind, M.M. (1981): Cancer Res., 41:4916-4924.
36. Terasima, T., and Tolmach, L.J. (1963): Nature, 190:1110-1211.
37. Terzaghi, M., and Little, J.B. (1975): Nature, 253:548-549.
38. Terzaghi, M., and Little, J.B. (1976): Cancer Res., 36:1367-1374.
39. Terzaghi, M., and Little, J.B. (1976): Int. J. Radiat. Biol., 29:583-587.
40. Thomson, J.F., Williamson, F.S., Grahn, D., and Ainsworth, E.J. (1981): Radiat. Res., 86:559-572.
41. Thomson, J.F., Williamson, F.S., Grahn, D., and Ainsworth, E.J. (1981): Radiat. Res., 86:572-588.
42. Trosko, J.E., Yoti, C.P., Warren, S.T., Tsushimoto, G., and Chang, C.C. (1981): In: Carcinogenesis: the Biological Effects of Tumor Promoters, edited by E. Hecker. Raven Press, New York.
43. Ullrich, R.L. (1984): Radiat. Res., 97:587-597.
44. Ullrich, R.L., and Storer, J.B. (1979): Radiat. Res., 80:325-342.

45. Upton, A.C., Randolph, M.L., and Conklin, J.W. (1970): Radiat. Res., 41:467-491.

Carcinogenesis, Vol. 10, edited by E. Huberman
and S. H. Barr. Raven Press, New York © 1985.

Mechanisms of Malignant Transformation of Human Diploid Cells

J. B. Little

Department of Cancer Biology, Harvard School of Public Health, Boston, Massachusetts 02115

Most studies of the induction of malignant transformation by radiation have been carried out in rodent cell systems. Rodent cells grown and maintained under appropriate conditions have several clear advantages for such studies. First, malignant transformation can be readily induced by physical and chemical carcinogens despite a very low background frequency of spontaneous transformants. Second, transformation can be easily scored by either colony morphology or focus assays. Third, such morphological transformation correlates well with the ability of these cells to form progressively growing tumors in syngeneic hosts. Tumorigenicity is usually considered the ultimate criterion for complete transformation in rodent systems. It has been shown, however, that transformation of rodent cells is a multistage process whereby the cells progressively acquire the various phenotypic characteristics of transformed cells, characteristics such as changes in growth pattern and morphology, the ability to grow under anchorage-independent conditions, and finally tumorigenicity.

One other characteristic of the process of transformation in rodent cells is sometimes overlooked--the process of immortalization. Immortality is one of the prime characteristics that differentiates tumor cells from normal diploid cells, which have a limited proliferative capacity in vitro. The process of immortalization appears to be an early and frequent event in many rodent cells, occurring either spontaneously or as the result of carcinogen treatment. Immortalization appears to be a very early step in the transformation of primary Syrian hamster embryo fibroblasts. Established mouse embryo cell lines such as BALB/3T3 and C3H 10T1/2 are already immortal and aneuploid, this process having occurred spontaneously when these cell lines were established in culture.

It has become clear during the past decade that the complete transformation of human diploid fibroblasts occurs rarely and is very difficult to induce in vitro, probably because the process of immortalization is an extremely rare event in human diploid cells. To my knowledge, only two laboratories have succeeded in inducing complete transformation of human cells in vitro, and those only in rare instances (3,7). Recently, several investigators have shown that certain characteristics of transformation may be more readily induced in human fibroblasts (1,6,8-10). These characteristics include the ability to grow under anchorage-independent conditions, an ability occasionally associated with the appearance of changes in growth pattern and morphology resembling focus formation. Cells isolated from anchorage-independent colonies may form small tumor nodules when injected into nude mice, although these nodules rarely grow to more than 1 cm in diameter and usually regress (8,11). Moreover, the environmental conditions in which the cells are grown and assayed must be very carefully controlled to maintain a low frequency of spontaneous anchorage-independent growth, as well as a measurable induced frequency.

The present investigation was undertaken to gain further information concerning the mechanisms for the transformation of human diploid fibroblasts, in particular those factors that might be related to the process of immortalization. Chromosomal rearrangements are associated with most human and rodent tumors. The low efficiency of complete transformation of cultured human cells has been ascribed to the stability of human chromosomes. We have therefore focused on chromosomal aberrations and rearrangements induced by x-irradiation. The experimental approach has been to irradiate cultures at early passage and to follow them throughout their normal life-span by subculturing them regularly at 1:4 dilutions. Chromosome rearrangements have been studied by obtaining G-banded karyotypes at regular intervals. The results described below relate to (i) the induction and persistence of chromosomal rearrangements in x-irradiated cells, (ii) the emergence of abnormal clones of cells during proliferation, and (iii) changes in the life-span of irradiated cultures. Our eventual goal is to relate specific chromosomal rearrangements to characteristics of transformation, in particular increased longevity and immortalization, and to investigate the role of oncogene activation in these processes.

RESULTS

Morphologic Transformation

Under carefully controlled environmental conditions, the treatment of normal human diploid fibroblasts with physical or chemical carcinogens can induce in a fraction of them the capacity to grow under anchorage-independent conditions. This effect can be dose dependent. The results of such an experiment in which the cells were treated with the chemical carcinogen N-acetoxy-acetylaminofluorene (N-Ac-AAF) are shown in Fig. 1. The frequency of anchorage-independent colonies clearly increased as a function of dose over the range of 1 to 10 μM N-Ac-AAF. The spontaneous frequency of anchorage-independent colonies in this experiment was approximately 10^{-4}. As can be seen in Fig. 2, the amount of anchorage-independent growth in this assay was independent of the number of treated cells seeded over initial cell densities that varied by two orders of magnitude. Approximately 10-15 population doublings following treatment and prior to seeding were required for maximal expression of the anchorage-independent trait.

A photomicrograph of an anchorage-independent colony is shown in Fig. 3. When such colonies were isolated and the cells grown in sufficient numbers and injected subcutaneously into nude mice, nodules developed at the site of injection in about 40% of the cases. Usually, however, these nodules regressed quite rapidly. When they were excised and examined, they showed several pertinent characteristics. First, histopathologic sections were characterized as coming from a poorly differentiated malignant tumor composed of primitive cells. One of these sections is shown in Fig. 4. Second, metaphase spreads obtained from cells in the tumor nodules clearly indicated they were of human origin. Finally, when the cells were dispersed and cultivated in vitro they soon resumed a normal fibroblastic morphology, showed a diploid karyotype, and became senescent. Thus, although these cells apparently gained certain phenotypic characteristics of transformation including the ability for limited growth in nude mice, they did not exhibit the aneuploidy or immortality associated with complete transformation.

In only one instance have we observed a tumor nodule to grow to greater than 1 cm in diameter. A photograph of the mouse bearing this nodule is shown in Fig. 5. The large locally invasive tumor grew to approximately 1.5 x 3 cm before the mouse was sacrificed. The histopathologic features of this tumor were identical to those seen in the smaller, regressing nodules shown in Fig. 4. Unfortunately, the cell line was lost, and we have no information as to the eventual life-span of cells isolated from this progressively growing tumor.

Silinskas et al. (8) have shown that the induction of anchorage-independent growth in human diploid fibroblasts correlates very closely with the induction of mutations. The results described above suggest that the induction of anchorage-independent growth may be an early carcinogen-induced change in normal human fibroblasts that might result from a mutation controlling certain membrane functions or perhaps a change in gene expression. It does not, however, appear to be associated with a number of the other criteria of transformation such as aneuploidy and immortality. These results also bring into question the value of the nude mouse assay in assessing the complete transformation of human cells.

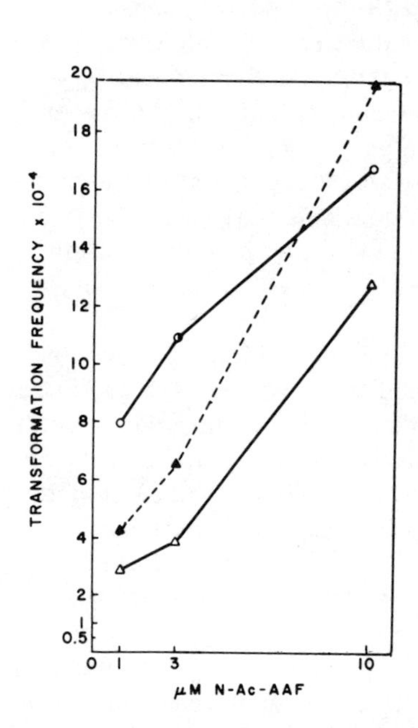

FIG. 1. Dose-response curves for induction of anchorage-independent growth in normal diploid fibroblasts treated for 15 min with N-Ac-AAF and grown under differing environmental conditions. Data taken from Zimmerman and Little (10) where details of these conditions are described.

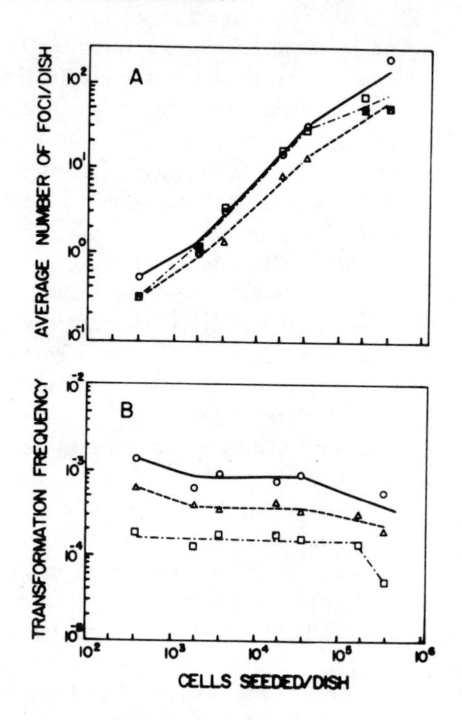

FIG. 2. Dependence of transformation to anchorage-independent growth on the number of cells initially seeded. Cells were treated with 3 μM N-Ac-AAF and allowed to proliferate in vitro for 14 population doublings prior to seeding at the indicated densities under anchorage-independent conditions. From Zimmerman and Little (10).

The capacity for anchorage-independent growth is usually considered to be the highest correlate to tumorigenicity in rodent cell systems. Perhaps it reflects primarily characteristics associated with morphologic transformation as distinct from immortalization. As immortalization is an early step in the transformation of rodent cells, this process is usually complete before the final morphologic changes that allow the cells to grow under anchorage-independent conditions have occurred.

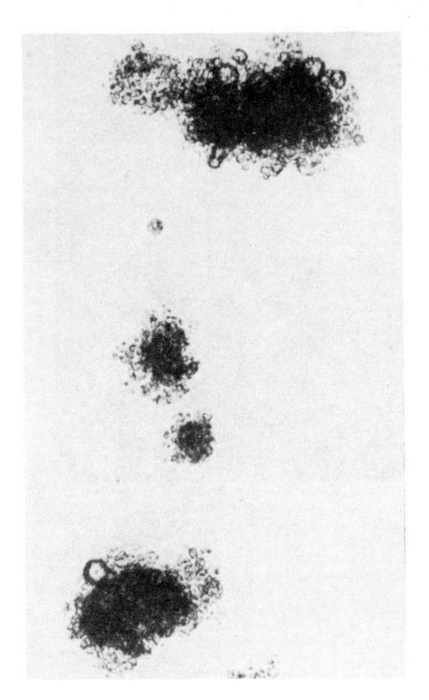

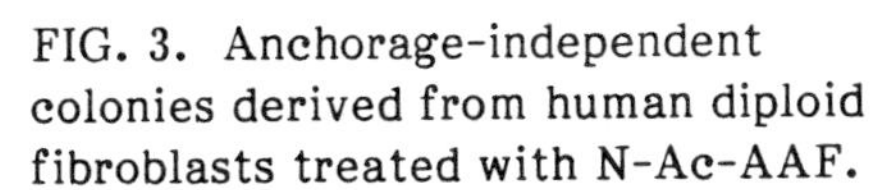

FIG. 3. Anchorage-independent colonies derived from human diploid fibroblasts treated with N-Ac-AAF.

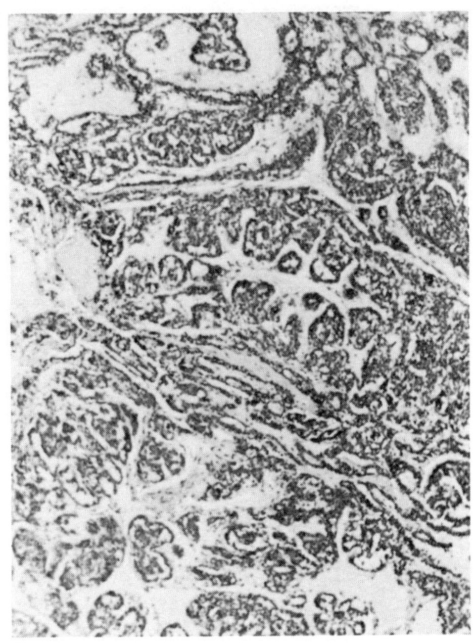

FIG. 4. Histopathological section (x100) of tumor nodule that developed at the site of injection of 10^6 cells isolated and grown from an anchorage-independent colony. From Zimmerman and Little (11).

X-ray-induced chromosomal rearrangements in long-term cultures

Although the induction of gross chromosomal aberrations in mammalian cells by ionizing radiation has been widely studied, surprisingly little information is available concerning the frequency of chromosomal rearrangements in x-irradiated human diploid cells. Yet, the occurrence of chromosomal rearrangements in cancer cells is well known. In this series of experiments, we examined the change in the frequencies of x-ray-induced chromosomal rearrangements in human diploid fibroblasts as a function of subculture time (4). The cells were followed throughout their life-span *in vitro*; metaphase spreads and G-banded karyotypes were

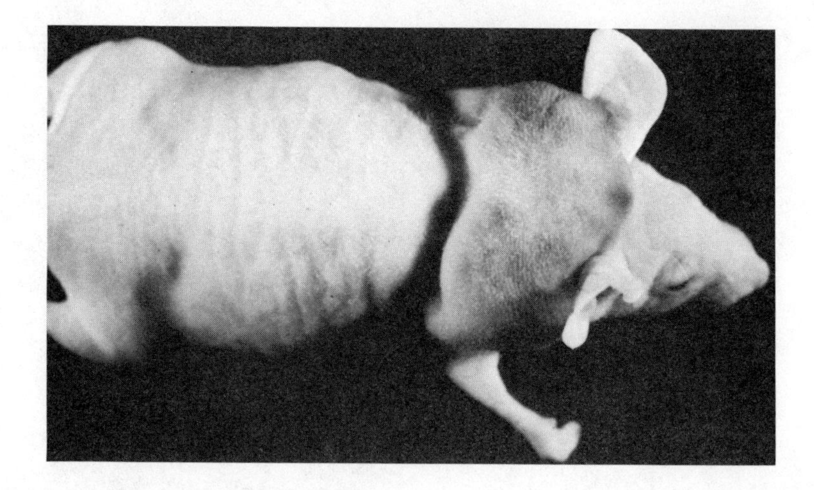

FIG. 5. Nude mouse bearing large, invasive tumor, which resulted from injection of 10^6 cells from an anchorage-independent colony derived from human diploid fibroblasts treated with N-Ac-AAF.

examined at regular intervals. Evidence was sought not only for induction of chromosome rearrangements but also for the stability of these rearrangements in cultures allowed to proliferate for long periods after irradiation.

The frequency of gross chromosomal aberrations was examined in metaphase spreads in cells irradiated with 400 rad of x-rays while in the confluent, density-inhibited phase of growth. A series of dishes was regularly subcultured at a 1:4 dilution, while cells in other dishes were maintained in confluence without subculturing for up to 43 days. Chromosome-type aberrations including dicentrics, rings, and fragments were measured at various times after irradiation. As can be seen in Fig. 6, the frequency of chromosome aberrations declined rapidly in cells allowed to proliferate after radiation exposure. Few aberrations were observed at the second subculture, and none were present at subculture 5. This rapid decline in chromosomal aberrations with subcultivation suggests that the presence of such aberrations at mitosis is lethal to the cells; the cells containing aberrations are thus lost from the population. In contrast, in cells not subcultured after irradiation but maintained under density-inhibited conditions (no proliferation), the frequency of aberrations in first division metaphases declined rapidly with holding periods of 4-24 h but thereafter remained stable at a level of 30-40% of the initial frequency up to 43 days after irradiation. The kinetics of this decline in chromosomal aberrations correspond exactly to those previously described for the repair of potentially lethal damage in x-irradiated cells (2). These results suggest that a certain amount of the damage that results in chromosomal

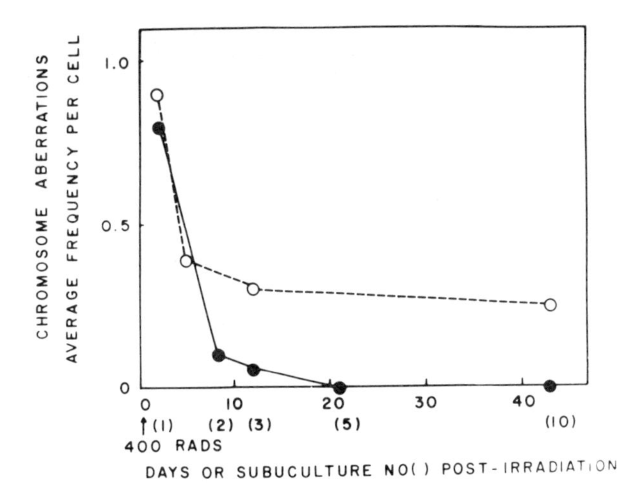

FIG. 6. Change in the frequency of chromosome type aberrations (rings, dicentrics, fragments) with subculture number and/or time (days) after treatment with 400 rad of x-rays. O----O: Nonproliferating cells (cells held in confluent growth with no subculture). ●----●: continually subcultured (proliferating) cells. From Kano and Little (4).

aberrations at the first mitosis is rapidly repaired following radiation exposure but that a significant fraction of this damage persists for long periods of time in nonproliferating cells. When the cells are allowed to proliferate, the presence of the aberrations that occur at mitosis are lethal to the cells, which are thus lost from the population.

The changes in the frequency of various chromosomal rearrangements during the growth of irradiated cells examined on G-banded karyotypes are shown in Table 1. Data are presented for translocations, dicentrics and rings, fragments or deletions, and inversions in cells irradiated with 400 rad and examined at the first, fifth, and tenth subcultivation after irradiation. As was seen in Fig. 6, the frequency of gross chromosomal aberrations (dicentrics and rings) declined to 0 after the first subcultivation. The frequency of translocations, however, remained at about 50% of the initial value at subcultures 5 and 10. Similar results were seen in cultures irradiated with 600 rad (Table 2), although the frequency of these chromosomal changes was increased about twofold.

A certain number of the cells that contained translocations may also have contained gross chromosomal aberrations that were lethal to the cells. Data on the persistence of translocations in cells exposed to 400 rad that contained no other visible chromosomal damage are shown in Fig. 7. The open circles represent the overall frequency of translocations at subculture 1, 5, and 10 after irradiation. As was evident in Table 1, this frequency declined considerably at the later subcultivations. How-

TABLE 1. Type and frequency of chromosome rearrangements at various subculture times following 400-rad x-irradiation.[a]

Dose (rad)	Subculture No. after Irradiation[b]	Passage No.	No. Cells Analyzed	Type of Chromosome Rearrangements[c]: Trans-locations (per cell)	Dicentrics & Rings (per cell)	Fragments or Deletion (per cell)	Inversions (per cell)
0	-	13	25	0	0	0	0
0	-	28	25	0	0	1 (0.04)	0
400	1	9	25	12 (0.48)	8 (0.32)	14 (0.56)	1 (0.04)
400	5	13	25	4 (0.16)	0	5 (0.20)	1 (0.04)
400	10	18	25	6 (0.24)	0	3 (0.12)	0

[a]Reproduced from Kano and Little (4).
[b]The first mitosis after the first, 4th or 10th subcultures (1:4) following 400-rad x-irradiation.
[c]Total number of chromosome-type aberrations or translocations present in 25 cells analyzed; average frequency per cell in parentheses.

TABLE 2. Type and frequency of chromosome rearrangements at various subculture times following 600-rad x-irradiation.[a]

Dose (rad)	Subculture No. after Irradiation[b]	Passage No.	No. Cells Analyzed	Type of Chromosome Rearrangements[c]: Trans-locations (per cell)	Dicentrics & Rings (per cell)	Fragments or Deletion (per cell)	Inversions (per cell)
600	1	6	25	22 (0.88)	23 (0.92)	28 (1.12)	2 (0.08)
600	5	10	25	11 (0.44)	0	1 (0.04)	1 (0.04)
600	10	15	25	14 (0.56)	0	3 (0.12)	2 (0.08)

[a]Reproduced from Kano and Little (4). Data presented as in Table 1.

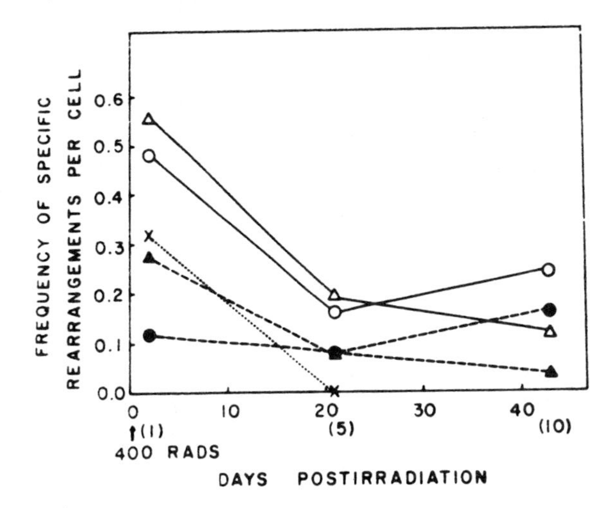

FIG. 7. The change in the frequency of chromosomal rearrangements per cell with subculture number and time (days) after 400 rad x-irradiation in proliferating (continually subcultured) cells. ○, translocation frequency, △, deletion or fragment frequency. ×, dicentric frequency. The cells with translocations (○) include those with dicentrics, inversions, or deletions (fragments) as well; the cells with deletions (△) include those dicentrics, inversions, or translocations as well; the cells with dicentrics (×) include those with translocations and deletions as well. ●, the frequency of cells that contain only translocations. ▲, frequency of cells that contain only deletions or fragments. From Kano and Little (4).

ever, cells with translocations also included those with dicentrics, inversions, or deletions as well. The solid circles represent the frequencies of cells with translocations only. As can be seen in Fig. 7, the frequency of such cells did not change significantly from the first through the tenth subcultivation. Similar results were seen in cells initially irradiated with 600 rad. These results suggest that in cells containing x-ray-induced translocations alone, translocations persist over many cell generations. Such chromosomal rearrangements thus appear to be very stable. The apparent decline in the frequency of translocations with subculture observed in Tables 1 and 2 reflects the death of cells that also contained deletions and dicentrics.

Emergence of abnormal clones

Nonrandom chromosome changes, particularly specific translocations, have been recognized in cells from a number of human cancers. Cells with specific chromosomal rearrangements form abnormal clones, and it has been suggested that the formation of such clones may be an important factor in the genesis of cancer. The findings that stable translocations are induced by irradiation in cultured human diploid fibroblasts and that these translocations persist over many generations of replication, led us to examine whether certain of these cells might gain a selective growth advantage such that clonal populations of cells containing specific marker chromosomes emerged, as apparently occurs in human tumors. To examine this question, we exposed cultured human diploid fibroblasts to single or multiple radiation doses, subcultivated them thereafter at regular intervals at a 1:4 dilution, and followed them throughout their life-span. At several intervals, G-banded karyotypes were prepared on 25-50 metaphase spreads; these karyotypes were examined for the presence of marker chromosomes suggesting the presence of abnormal clones. Such clones contained stable chromosomal rearrangements including translocations, deletions, or inversions. The term "clone" was used only when the following conditions were satisfied: (i) at least three cells containing identical structural chromosomal rearrangements were found and (ii) one cell type was observed in at least two different passages.

No abnormal clones were observed in four nonirradiated control cultures examined at several intervals throughout their life-span by G-banded karyotypes. However, abnormal clones did emerge in two out of eight cultures exposed to single doses of 400 or 600 rad of x-rays. Because the initial frequency of chromosomal rearrangements is highly dependent upon radiation dose, several cultures were exposed to multiple doses of radiation at successive subcultivations. Abnormal clones developed in five out of six cultures exposed to multiple radiation doses. Typically, these clones appeared between the 30th and 50th mean population doubling (MPD). Earlier-appearing clones disappeared from the population at later times as the clonal cells apparently became senescent. Some later-appearing clones persisted in the mass population until complete senescence. At certain times during the population's life-span, 80% of more of the cells might belong to a single abnormal clone. In two cases, the terminal cell population consisted entirely of a single clone of cells.

In all but one culture treated with multiple radiation doses, several different abnormal clones were observed. Evidence of clonal succession, that is the appearance and disappearance of successive clones of cells over the life-span of the culture, was also observed in several of these cultures. Such clonal succession probably represents the sequential emergence and attenuation of clones that initially attain a selective growth

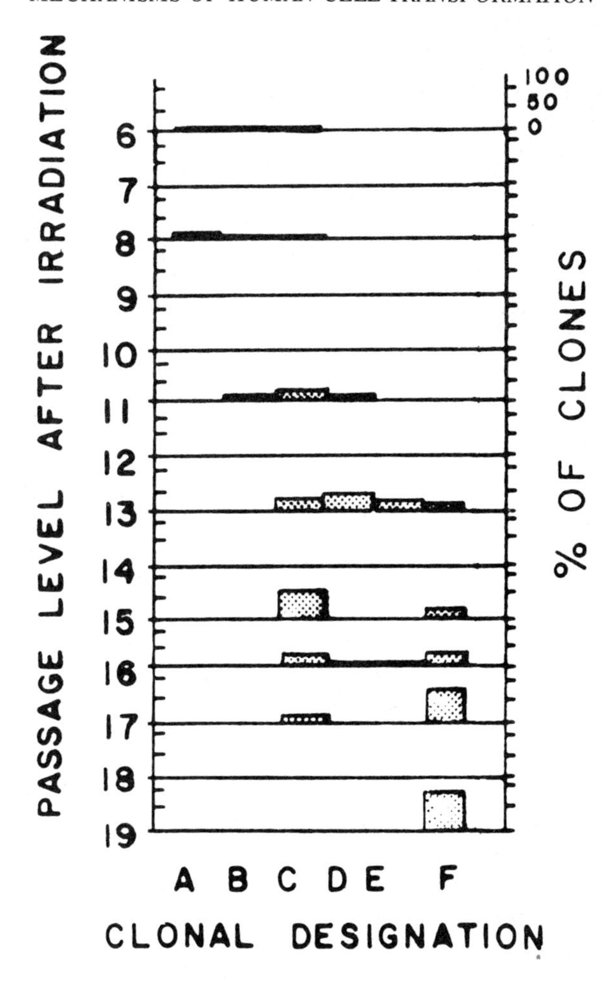

FIG. 8. Development of abnormal clones (clones of cells bearing identical structural chromosomal rearrangements) in a culture of normal human diploid fibroblasts treated with three doses of 600 rad each of x-rays at successive subcultures beginning five passages (10 MPD) after establishment of the cultures. Following irradiation, cells were subcultured continually when they became confluent at a 1:4 dilution. Vertical axis at right indicates percent of entire population of cells that arose from the particular clone at postirradiation passage level indicated on the left (Kano and Little, unpublished data).

advantage and then undergo early senescence. An example of clonal succession in such a culture is shown in Fig. 8. Three different clones possessing unique chromosomal rearrangements were first observed at MPD 22. One of these (a chromosome 12:15 translocation) became predominant in the culture between MPD 31-43, constituting over 50% of the population at MPD 37. This clone disappeared from the population after MPD 43. Three other clones also emerged during these later times. One of these, a chromosome 1:13 translocation, became predominant from MPD 43 until the culture senesced at MPD 48. By this time, the population consisted primarily of cells derived from this single clone.

The results in Table 1 and Fig. 7 above indicate that x-irradiation of human diploid cells induces stable chromosome rearrangements, which may persist throughout the life-span of the cell. These results indicate that certain of these cells apparently gain a selective growth advantage permitting clonal expansion to the point where single clones of cells bearing specific rearrangements can be recognized as constituting a significant fraction of the mass population. Presumably, the greater frequency of abnormal clones in cultures exposed to multiple radiation doses is related to the fact that surviving cells in cultures irradiated with high doses of x-rays will possess a much greater frequency and variety of chromosomal rearrangements than cells exposed to lower doses. Thus, the chance of clones emerging with specific rearrangements is enhanced. It is not known whether an increased probability of clonal expansion is related to certain specific rearrangements that confer a selective growth advantage to the cells or whether clonal expansion is a purely random phenomenon independent of the rearrangements present. To date, however, no specific pattern of rearrangements has evolved among the various abnormal clones we have observed in different cultures. Should the emergence of abnormal clones with specific chromosomal rearrangements, perhaps involving critical oncogenes, be a necessary step in the development of the transformed phenotype, the results described above indicate that carcinogen exposure can facilitate the development and expansion of such clones.

Emergence of clones with prolonged life-span

These observations on the emergence of abnormal clones in irradiated cultures led us to examine whether such cells might show any changes in life-span *in vitro*. In these experiments, cultures were exposed to single or multiple doses of irradiation and followed throughout their life-span with regular subcultivation at a 1:4 dilution (5). The life-spans (MPD) of 46 irradiated cultures were compared with those of nine nonirradiated control cultures. The results are shown in Table 3. As a group, the mean life-span of 44 irradiated cultures was slightly but significantly prolonged as compared with nine control cultures (58.4 vs. 53.0 MPD, $p<0.05$). As the cytotoxic effect of radiation was not included in the calculation of

TABLE 3. Life-span of control and irradiated human diploid fibroblasts (strain AG1522) measured as mean population doublings.[a]

Cells	No. of Cultures	Life-span (MPD)[b] Mean	Range
Controls	9	53.0	48-57
Irradiated, single doses	26	57.3	51-66
Irradiated, multiple doses	18	59.7	51-67
Irradiated, prolonged life-span	2	76[c] 82[c]	-

[a]Data from Kano and Little (5).
[b]The increased number of population doublings surviving cells must undergo in irradiated cultures to repopulate the dish not included in calculations of life-span. Thus, the actual life-span measured in terms of MPD of surviving irradiated cells will be greater than the figure shown. Strain AG1522 used in all experiments.
[c]Abnormal clones. Actual MPD thus greater by about 20 (see text).

MPD, the actual life-span of surviving irradiated cells was significantly greater than 58.3.

In the two remaining irradiated cultures, cell strains emerged with a considerably prolonged life-span. Data on one of these cultures is shown in Fig. 9. Several abnormal clones emerged at earlier passages and eventually disappeared, but one that appeared at MPD 44 expanded to include the entire population by MPD 57. Subsequent to MPD 57, this culture has consisted entirely of monoclonal cells; the mass culture senesced at MPD 47. This culture is currently surviving at MPD 82. However, this clone of cells must have undergone another 20 MPD during clonal expansion from a single cell to include the entire population ($2x10^6$ cells/dish). Furthermore, as a survivor of the radiation exposures, it will have undergone at least 10 additional MPD during the repopulation of cultures following irradiation (each dose of 600 rad kills more than 95% of the cells). Thus, the cells in this clone have undergone at least 110 MPD, more than twice

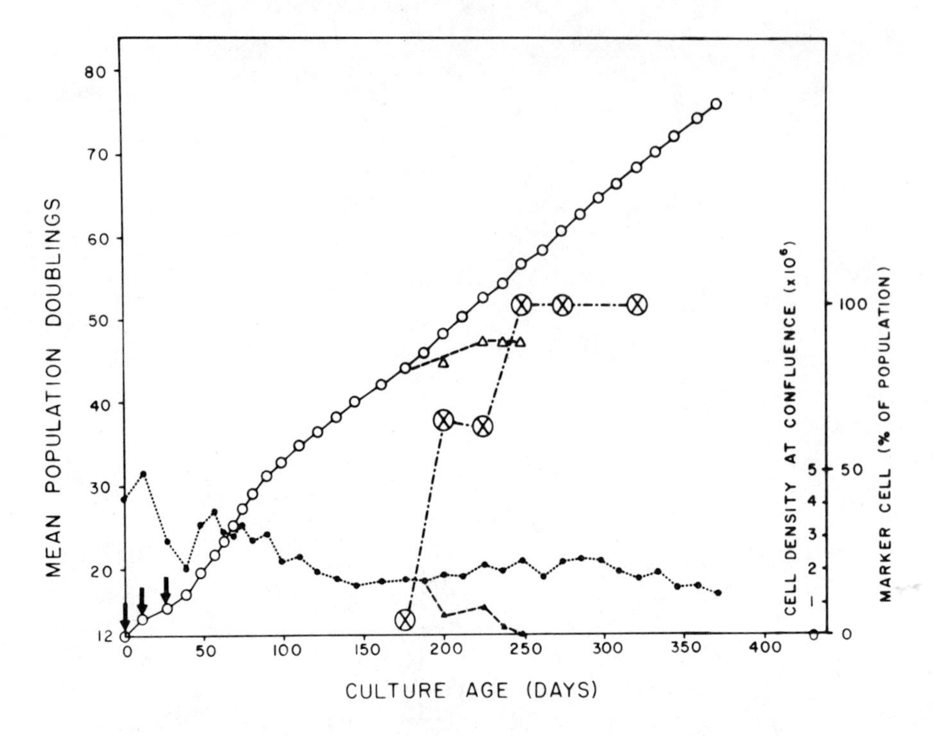

FIG. 9. Growth of a culture of human diploid fibroblasts treated with three doses of 600 rad of x-rays and subcultured continually when they became confluent at a 1:4 dilution. An abnormal clone was first observed at MPD 44. By MPD 57, the clone had expanded such that all cells in the culture were derived from it. This clone of cells is still surviving at MPD 82. Data from Kano and Little (5).

the mean life-span of control cells or of the other cells in the mass culture.

The karyotype of these monoclonal cells was highly abnormal, showing a variety of stable chromosomal rearrangements, including a deletion in the short arm of chromosome 1 (pp22, p32), and a translocation between chromosomes 11:12 (5). The human N-ras and B-lym oncogenes have been mapped to the short arm of chromosome 1 in the vicinity of the breakpoints of the deletions. The karyotype of the other clone of cells with a prolonged life-span (Table 3) showed two translocations involving chromosome 22 (1;22 and 6;22). The sis oncogene has been mapped to chromosome 22 (22q12-q13), the region of the breakpoint in the chromosome 6;22 translocation. These observations have led us to hypothesize that the prolonged life-span of this clone of cells might be related to oncogene activation. Interestingly, the morphology of the cells in both of these cultures was normal, and they did not grow in soft agar. Thus, the prolonged life-span was not associated with the acquisition of characteristics of morphologic transformation.

CONCLUSIONS

It is clear that exposure to chemical or physical carcinogens can induce certain alterations in the morphology and growth of human diploid fibroblasts of the type usually associated with the transformed phenotype. If the cultural conditions are controlled very carefully, for example, carcinogen exposure will produce a dose-dependent increase in the capacity of these cells to grow under anchorage-independent conditions. Occasionally, morphologic changes resembling focus formation in rodent fibroblasts can be observed in treated cultures. Such cells may show karyotypic changes, although they remain diploid or pseudodiploid. Finally, cells isolated from anchorage-independent colonies may form small tumor nodules in nude mice. Rarely, however, do such nodules grow progressively and become locally invasive; rather, they usually regress. Cells isolated from such nodules and grown in vitro retain a diploid karyotype and eventually senesce. Thus, although these cells show certain characteristics of morphologic transformation including the ability for limited growth in nude mice, they have not gained immortality.

On the basis of these findings, we hypothesize that morphologic transformation and immortalization are separate and distinct steps in the process of transformation of human cells. Immortalization is a very rare event that is induced at extremely low frequencies in human diploid fibroblasts, whereas characteristics of morphologic transformation may be induced quite easily. This finding is in contrast to the situation in rodent cells, in which immortalization is an early and frequent occurrence. Indeed, the mouse 3T3 and 10T1/2 cell lines are examples of cells that have undergone the immortalization step but show none of the morphologic characteristics of the transformed phenotype. Thus, cells exist that have undergone either of these two steps, but not both, suggesting that they are indeed two independent events.

The low frequency of immortalization of human fibroblasts as compared with rodent cells has been ascribed to the chromosomal stability of human cells. Irradiation of normal human diploid cells in vitro induces stable chromosomal rearrangements including deletions and translocations that persist throughout the life-span of the cell. Occasionally, cells containing specific rearrangements may gain a selective growth advantage and be recognized within the population as an abnormal clone. Such clones may expand to include the majority of the population, although they usually senesce and disappear. The life history of heavily irradiated cultures of human diploid fibroblasts therefore appears to be characterized by clonal succession, that is, the successive emergence and disappearance of several abnormal clones of cells. Occasionally, such cells have a significantly prolonged in vitro life-span. Under these conditions, the clone expands until it includes 100% of the population. Whether this increased life-span is related to the specific chromosomal rearrangements characteristics of the clone or whether it is related to some other factor independent of these rearrangements is not, at present, evident.

On the basis of these observations, we propose the hypothesis (5) that immortality in human fibroblasts develops during the growth of a population of damaged cells that contain stable chromosomal rearrangements. Among this population, abnormal clones of cells emerge that can expand to take over the entire population and show a considerably prolonged life-span. This phenomenon may be associated with the activation of cellular oncogenes as a result of specific chromosomal rearrangements. The prolonged life-span allows additional time for the process of immortalization to occur. Immortalization may result from a mutational change that occurs at random during cellular proliferation or from another as yet unidentified mechanism. At any rate, this process is a very rare one in human diploid fibroblasts grown in vitro, whereas the induction of phenotypic characteristics such as anchorage-independent growth usually associated with transformation may occur relatively frequently in rodent cells. It will be interesting to determine whether these two steps in the complete transformation of human diploid fibroblasts are associated with the activation of different cellular oncogenes.

ACKNOWLEDGMENTS

This research was supported by Public Health Service Grant CA-11751 from the National Cancer Institute and Center Grant ES-00002 from the National Institute of Environmental Health Sciences, U.S. Department of Health and Human Services.

REFERENCES

1. Borek, C. (1980): Nature, 283:776-778.
2. Fornace, A.J., Jr., Nagasawa, H., and Little, J.B. (1980): Mutat. Res., 70:323-336.
3. Kakunaga, T. (1979): Proc. Nat. Acad. Sci., USA, 75:1334-1338.
4. Kano, Y., and Little, J.B. (1984): Cancer Res., 44:3706-3711.
5. Kano, Y., and Little, J.B. (submitted).
6. Milo, G., and DiPaolo, J.A. (1978): Nature, 275:130-132.
7. Namba, M., Nishitani, K., and Kimoto, T. (1978): Japan J. Exp. Med., 48:303-311.
8. Silinskas, K.C., Kateley, S.A., Tower, J.E., Maher, V.M., and McCormick, J. (1981): Cancer Res., 41:1620-1627.

9. Sutherland, B.M., Cimino, J.S., Delihas, N., Shih, A.G., and Oliver, R.P. (1980): Cancer Res., 40:1934-1939.
10. Zimmerman, R.J., and Little, J.B. (1983): Cancer Res., 43:2176-2182.
11. Zimmerman, R.J., and Little, J.B. (1983): Cancer Res., 43:2183-2189.

Carcinogenesis, Vol. 10, edited by E. Huberman
and S. H. Barr. Raven Press, New York © 1985.

Cancer in Ataxia-Telangiectasia

R. B. Painter

Laboratory of Radiobiology and Environmental Health, University of California, San Francisco, California 94143

Ataxia-telangiectasia (A-T) is a human autosomal recessive genetic disease of childhood that affects about one in 40,000 people in the United States. The ataxia is progressive, starting at the age of six months to six years, with death occurring most often in the late teens or early twenties. Most patients die from pulmonary infections. Although almost all A-T patients show some form of immunodeficiency, their infections have the same microbial etiology as in normal humans, rather than that containing the "opportunistic" organisms seen in infections of severely immunodeficient patients. For this reason, some investigators believe the infections come about as a result of complications of the ataxia, rather than from the immunological defect. Of particular interest for this volume, however, is the high incidence of cancer in A-T patients; between 10 and 20% of them display neoplastic disease, most of which is in the lymphoreticular system. A-T patients are also uniformly hypersensitive to ionizing radiation. Therefore, when the defective gene or genes in A-T are identified and characterized, it may be possible not only to clarify the relationship between radiation sensitivity and susceptibility to cancer in A-T patients, but also to identify the processes leading to radiation-induced cancer in normal humans.

Our laboratory and a few others throughout the world are in the process of attempting to identify and clone one of the genes in the A-T phenotype. That several genes are involved is indicated by complementation studies with cells from different A-T patients; at least four and probably five complementation groups have already been identified (9, 11). Until someone successfully identifies one or more of the A-T genes and determines the functions of its products, we cannot be certain whether the sensitivity of A-T patients to ionizing radiation causes their cancer susceptibility. However, some features of the cells from A-T homozygotes suggest there is such a relationship.

First, there is in A-T fibroblasts an increased incidence of spontaneous and ionizing-radiation-induced chromosomal aberrations (7,18,25,26). One of the principal cellular changes that precede or accompany acquisition of the cancerous state is chromosomal instability, i.e., the appearance of aneuploidy, often accompanied by chromosomal rearrangements. In A-T cells, the inherent chromosomal instability is aggravated by exposure to ionizing radiation, equal doses of which induce many more aberrations in A-T cells than in normal cells. This result suggests that the same kind of damage may be the basis for both the spontaneous and radiation-induced chromosomal aberrations. Although the molecular steps leading to chromosomal aberrations are not known, the preponderance of the evidence suggests that the lesions that lead to radiation-induced chromosomal aberrations are unrepaired or misrepaired double-strand breaks in the DNA (17,19,20).

A second characteristic of A-T cells studied so far is radioresistant DNA synthesis. This characteristic (Fig. 1), first reported by Houldsworth and Lavin (8) for lymphoblastoid A-T cells and independently by Painter and Young (16) and de Wit et al. (4) for A-T fibroblasts, had never previously been observed in any kind of mammalian cells. (A recent report from the Netherlands shows that the cells from two brothers with symptoms completely different from A-T demonstrate radioresistant DNA synthesis, enhanced cellular radiosensitivity, and increased chromosomal aberrations; this condition has been named the Nijmegen breakage syndrome [24].) It was both surprising and revealing to find that the principal basis for the radioresistant DNA synthesis in A-T cells is radioresistant chain elongation (5,12,16). It was surprising because the steep, low-dose component of inhibition in normal cells is primarily due to a block to initiation (Fig. 2A). Since this component of inhibition is missing in A-T cells (Fig. 1), one would assume that replicon initiation is completely resistant to radiation. Instead, analysis by alkaline sucrose gradient sedimentation shows that DNA chain elongation in A-T cells is completely resistant to high doses of x-rays (Fig. 2B), despite the fact that DNA damage is the same in normal and A-T cells (14). The inhibition of DNA synthesis that does occur in irradiated A-T cells is due to inhibition of replicon initiation. However, because the slope of the curve for radiation-induced inhibition of DNA synthesis in A-T cells is parallel to the slope for the shallow, high-dose component of normal cells, it seems that an effective "hit" in the DNA of A-T cells blocks the initiation of only that replicon in which the hit occurs, rather than the initiation of a whole cluster of replicons, as in normal cells (15).

The lack of inhibition of DNA chain elongation in A-T cells is revealing because it clearly demonstrates that there is a factor in normal cells that mediates between the radiation-induced lesions in DNA and the advancing DNA growing point, which is consequently blocked. This factor fails to act in A-T cells because it is either missing or defective; I have called this a "damage-recognition factor." It may be the same factor that is involved in blocking initiation of whole clusters of replicons in normal

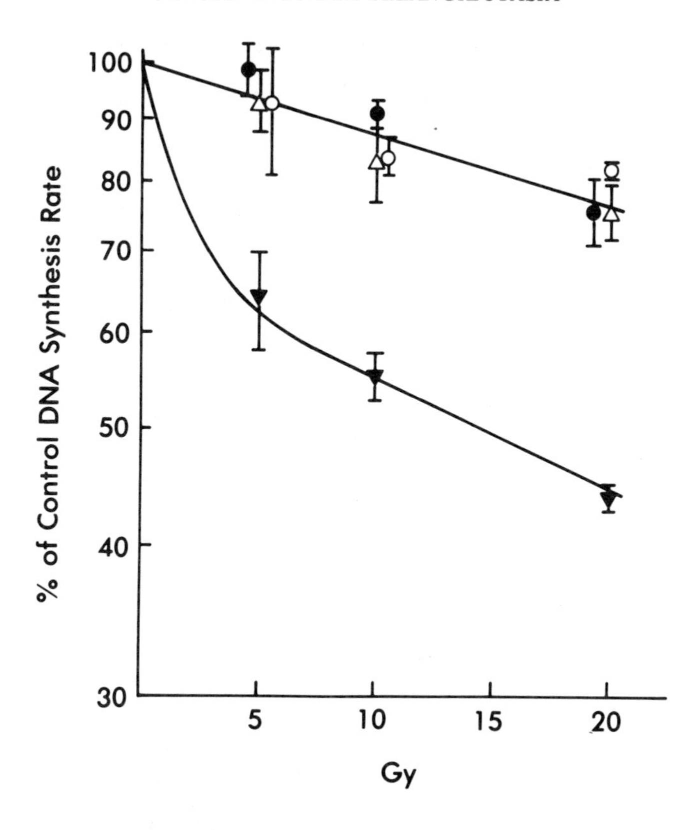

FIG. 1. Effect of radiation dose on rate of DNA synthesis in normal human cells (HS-27) (▼), and in three A-T fibroblast cell lines: AT5BI (●), GS-BRO (△), and GS-TJR (○). GS-BRO and GS-TJR were gifts from Dr. Richard Gatti, University of California at Los Angeles.

cells. It may also be responsible for radiation-induced G_2 delay, because Scott and Zampetti-Bosseler (21,27) have shown that A-T fibroblasts have a shorter mitotic delay than do normal cells.

I have previously proposed that, as a consequence of the failure of this damage-recognition factor, irradiated A-T cells progress precociously, either through S-phase, thereby replicating DNA lesions before they are repaired, and/or from G_2 into mitosis, thereby expressing chromosome damage before it is repaired. However, there are certain data that suggest this concept is too simple. Caffeine has little or no sensitizing effect on irradiated S-phase cells, even though caffeine reverses the radiation-induced inhibition of DNA synthesis (2), and the premature chromosome condensation (PCC) technique shows increased frequencies of unrepaired chromosome fragments in irradiated G_1 phase A-T cells (3).

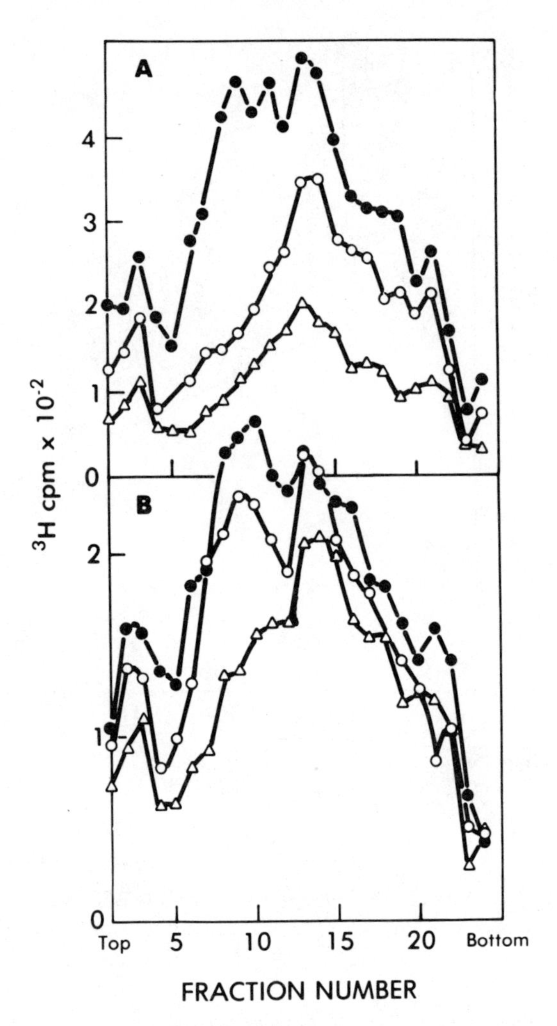

FIG. 2. Alkaline sucrose gradient profiles of DNA from (A) normal diploid HS-27 cells and (B) AT5BI cells irradiated with 0 (●), 5 (○), or 20 (△) Gy and pulse-labeled with [^{3}H] thymidine 30 min later. Sedimentation is from left to right. A dose-dependent inhibition of initiation of replicons is deduced from reduced radioactivity in low molecular weight DNA (fractions 4-11). A dose-dependent inhibition of chain elongation is deduced from reduced radioactivity in high molecular weight DNA (fractions 12-20). Note that A-T cells show less inhibition of replicon initiation and almost no inhibition of chain elongation. (Reproduced from ref. 13 with permission of Elsevier/North Holland.)

Nevertheless, the defective damage-recognition factor may still be responsible for the increased chromosome aberration frequency in A-T by, for instance, allowing a late step in chromosomal reconstitution to occur before necessary DNA repair steps have been completed (Fig. 3). Irrespective of the exact mechanism involved, this defect in regulation of DNA synthesis in A-T cells is so intrinsic to chromosomal replication that one must assume it is involved in the increased chromosomal fragility observed in these cells.

If A-T cells are killed more rapidly than normal cells by ionizing radiation and other agents that generate DNA strand breaks via free radicals (22), why is the incidence of cancer in A-T patients so high? It is not because A-T cells are hypermutable by x-ray-type damage; the evidence suggests that, if anything, they are hypomutable (1,23). The most likely explanation is the one put forth by Gatti and Hall (6). Chromosomal aberrations, induced by endogenous OH-radical generators, occur more frequently in the cells of A-T patients than in cells from normal humans because the damage-recognition factor fails (Fig. 3). The great majority of these aberrations lead to cell death, but normal cell renewal systems are capable of replenishing the cell system so affected. Occasionally (but much more often than in normal humans) a stable chromosomal rearrangement occurs so that the cell survives. Less frequently, but still much more often than in normal humans, the stable rearrangement is one that allows expression of an endogenous oncogene. This event still may not be sufficient for formation of a cancerous cell, because recent evidence indicates that there must be at least two such events (one to induce cellular immortality and one to induce invasiveness) to transform a normal cell fully (10). The faulty damage-recognition factor in A-T cells causes so many aberrations that eventually a cell is generated with the proper chromosomal rearrangements to be expressed as a cancerous cell. The immunodeficiency accompanying the A-T syndrome may also play a role by failing to suppress these transformed cells once they are formed.

In normal humans exposed to radiation, the increased aberration frequency, which in A-T cells is caused by the defective damage-recognition system, is instead caused by the abrupt appearance of many preaberrational lesions in the cell. It is instructive to consider the difference between the damage to human DNA caused by background radiation and that caused by "low-dose" exposure. Background radiation is about 150 mrem per year. Consequently, during the entire year each cell in the human body experiences only about 1-2 single-strand breaks and 1-2 damaged bases, and only about 10% of cells experience a double-strand break in their DNA. Moreover, only a tiny fraction of the cells are exposed at any one time. But when an individual is exposed to a "low" dose, such as a dose of 1 rad from a CT scan to the upper body, every cell in the field of exposure suddenly and simultaneously acquires about 10 single-strand breaks, 10 damaged bases, and one double-strand break in

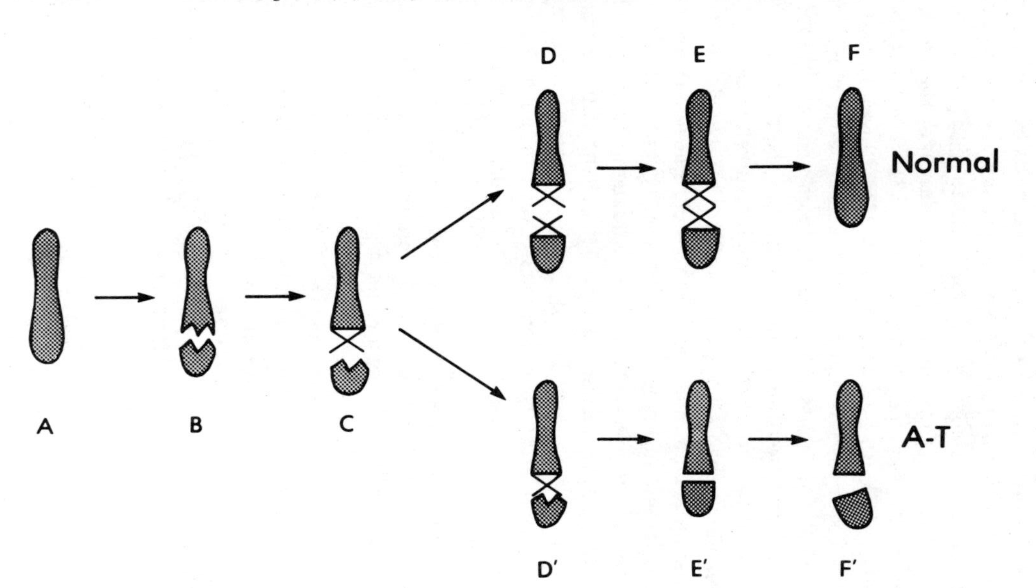

FIG. 3. Model for a mechanism by which the defective damage-recognition factor causes increased frequencies of chromosome aberrations. Radiation disrupts the linear continuity of a G_1 chromosome (shown in its mitotic form, A), to form a break (B). Restitution begins by the onset of DNA repair (C), which continues normally in D but is interrupted by a precocious attempt at rejoining (D') in A-T cells. DNA repair in normal cells terminates (E) and restitution is completed (F). In A-T cells, however, the attempt at early restitution (E') fails, resulting in a deletion (F').

its genome. Because billions of cells are so exposed, one would expect, on the basis of chance alone, to find a few that form aberrations because the damage-recognition factor fails. With increasing dose, of course, the aberration frequency increases and so does the probability for the occurrence of the complex rearrangements leading to the precancerous cell.

ACKNOWLEDGMENT

This work was supported by the U.S. Department of Energy.

REFERENCES

1. Arlett, C.F. and Harcourt, S.A. (1978): In: DNA Repair Mechanisms, edited by P.C. Hanawalt, E.C. Friedberg, and C.F. Fox, pp. 633-636. Academic Press, New York.
2. Busse, P.M., Bose, S.K., Jones,R.W., and Tolmach, L.J. (1977): Radiat. Res., 71:666-677.
3. Cornforth, M.N., and Bedford, J.S. (1984): Proc. of the 32nd Annual Meeting of the Radiation Research Society, p. 23, March 25-29, 1984, Orlando, Florida.
4. de Wit, J., Jaspers, N.G.J., and Bootsma, D. (1981): Mutat. Res., 80:221-226.
5. Ford, M.D., and Lavin, M.F. (1981): Nucleic Acids Res., 9:1395-1404.
6. Gatti, R.A., and Hall, K. (1983): In: Chromosome Mutation and Neoplasia, edited by J. German, pp. 23-41. Alan R. Liss, Inc., New York.
7. Higurashi, M., and Conen, P.E. (1973): Cancer, 32:380-383.
8. Houldsworth, J., and Lavin, M.F. (1980): Nucleic Acids Res., 8:3709-3720.
9. Jaspers, N.G.J., and Bootsma, D. (1982): Proc. Nat. Acad. Sci., USA, 79:2641-2644.
10. Land, H., Parada, L.F., and Weinberg, R.A. (1983): Nature, 304:596-602.
11. Murnane, J.P., and Painter, R.B. (1982): Proc. Nat. Acad. Sci., USA, 79:1960-1963.
12. Ockey, C.H. (1983): Radiat. Res., 94:427-438.
13. Painter, R.B. (1981): Mutat. Res., 84:183-190.
14. Painter, R.B. (1983): Radiat. Res., 95:421-426.
15. Painter, R.B., and Young, B.R. (1976): Biochim. Biophys. Acta, 418:146-153.
16. Painter, R.B., and Young, B.R. (1980): Proc. Nat. Acad. Sci., USA, 77:7315-7317.
17. Painter, R.B., Young, B.R., and Burki, H.J. (1974): Proc. Nat. Acad. Sci., USA, 71:4836-4838.

18. Rary, J.M., Bender, M.A., and Kelly, T.E. (1974): Amer. J. Hum. Genet., 26:70A.
19. Ritter, M.A., Cleaver, J.E., and Tobias, C.A. (1977): Nature, 266:653-655.
20. Sakai, K., and Okada, S. (1984): Radiat. Res., 98:479-490.
21. Scott, D., and Zampetti-Bosseler, F. (1982): Int. J. Radiat. Biol., 42:679-683.
22. Shiloh, Y., Tabor, E., and Becker, Y. (1982) Cancer Res., 42:2247-2249.
23. Simons, J.W.I.M. (1982): In: Ataxia-Telangiectasia: A Cellular and Molecular Link Between Cancer, Neuropathology, and Immune Deficiency, edited by B.A. Bridges and D.G. Harnden, pp. 155-167. Wiley, Chichester.
24. Taalman, R.D.F.M., Jaspers, N.G.J., Scheres, J.M.J.C., de Wit, J., and Hustinx, T.W.J. (1983): Mutat. Res., 112:23-32.
25. Taylor, A.M.R. (1982): In: Ataxia-Telangiectasia: A Cellular and Molecular Link Between Cancer, Neuropathology, and Immune Deficiency, edited by B.A. Bridges and D.G. Harnden, pp. 53-81. Wiley, Chichester.
26. Taylor, A.M.R., Metcalfe, J.A., Oxford, J.M., and Harnden, D.G. (1976): Nature, 260:441-443.
27. Zampetti-Bosseler, F., and Scott, D. (1981): Int. J. Radiat. Biol., 39:547-558.

Carcinogenesis, Vol. 10, edited by E. Huberman and S. H. Barr. Raven Press, New York © 1985.

Cellular Responses in Chronic Radiation Leukemogenesis

T. M. Seed, L. V. Kaspar, T. E. Fritz, and D. V. Tolle

Division of Biological and Medical Research, Argonne National Laboratory, Argonne, Illinois 60439

Leukemia, in particular the nonlymphocytic subtype, is a well-recognized late pathological consequence of ionizing radiation; its occurrence has been documented in a wide variety of mammalian species (6,9,26), including man (1,10,23). It is also well recognized that the frequency of leukemia induction varies widely with the type and course of radiation exposure (10). Acute, single, whole-body exposure to low linear energy transfer (LET) irradiation is particularly effective and in the sublethal range induces leukemia generally in proportion to radiation dose. In contrast, when the radiation exposure is protracted, the proportionality between leukemia frequency and total radiation dose becomes much less obvious. Peak leukemia frequencies are substantially reduced but extended over a greater range of accumulated radiation doses, i.e., dose-response curves tend to be flatter and broader. This effect of radiation protraction has been observed in man through epidemiological evaluation of x-irradiated ankylosing spondylitic patients and their associated excess risk to leukemia (23). Further, the effect has been verified experimentally with murine leukemia RF and CBA/H models (11,27).

The observation of different dose-response patterns (i.e., "linear" vs. "flat" dose responses) suggests inherent differences in leukemogenic processes as a consequence of acute vs. chronic radiation exposure. The instantaneous dose rate of irradiation appears not to be the critical difference, however. Several thousandfold differences in the dose rate produce little change in the overall low leukemia frequency seen under protracted exposure regimens (11,27).

Clearly, the critical factor in determining the frequency of leukemia induction under fractionated radiation regimens must be related to the unique demands and stresses placed on the responding hematopoietic system under extended radiation exposure. It would appear that the larger the daily dose rate and cumulative dose, the more important this factor becomes. As we will demonstrate later, the "critical factor" appears to involve, in part, the reparative capacity of the hematopoietic system under protracted irradiation, which, if properly invoked, "promotes" the leukemogenic process.

METHODS

Animals

For survival data and the determination of leukemia incidence, we used a total of 1025 purebred beagles: 735 and 290 dogs in experimental and control groups, respectively. This study was part of a long-range project with the overall aim of evaluating morbidity and mortality rates as a consequence of protracted whole-body gamma irradiation in a relatively large, long-lived mammalian species (2,12). For the preclinical phase studies, 72 dogs were used. At the start of specific experiments, all dogs were approximately 400 days old, anatomically normal, and in good health. Both male and female animals, in approximately equal ratios, were used in the various groups. Prior to and during the experiment, all dogs were monitored clinically and hematologically on a regular basis.

Irradiation of Animals

Experimental dogs were continuously irradiated under either "duration-of-life" or "fraction-of-life" exposure regimens at rates ranging from 0.4 to 35 R per 22-hour day. Average absorbed radiation doses were calculated to be 75% of the dose in air. Dosimetric methods and calculations are outlined in detail elsewhere (22,28). Control animals were maintained in shielded, adjoining anterooms.

Hematology and Pathology

Hemograms were performed periodically by standard methods on each irradiated and control animal (24,25).

For morphological and functional analyses of bone marrow with time and cumulative radiation dose, marrow biopsies and aspirates were performed periodically (~100-day intervals). The collected tissue specimens were assessed morphologically by light and electron microscopic methods and functionally by determining absolute marrow cellularity and granulocyte reserves (15), as well as the concentration and radiosensitivity of hematopoietic progenitors (GM-CFUa) (17).

RESULTS AND DISCUSSION

Survival and Pathological Response Patterns

Distinct time, radiation dose, and dose-rate relationships exist in the induction by chronic irradiation of nonlymphocytic leukemia and its opposite hematopoietic response, aplastic anemia (16). The induction of these two contrasting pathological entities is directly associated with survival time. As shown in Fig. 1, survival is progressively extended (i.e., curves displaced to the right) as the daily dose rate of irradiation decreases. For example, at 35 R/day the mean survival time is 57 days, whereas at 5 R/day it is 1830 days. Similarly, curves for survival responses based on cumulative dose rather than time are widely displaced at the extreme dose rates; the LD_{50} value of 1855 rad at 35 R/day is increased by severalfold to 9700 rad at 5 R/day (Fig. 2). At the intermediate rates of exposure, i.e., 17 and 10 R, however, the survival curves overlap through the 50% survival level and separate only below this level (Fig. 2). These transitions in survival rates occur at cumulative doses of about 2500 rad and at survival levels of 40 and 25% at the 10 and 17 R/day dose rates, respectively (Fig. 3). At cumulative radiation doses below (left of) the "transition" point, there is a predominance of aplastic anemias and septi-

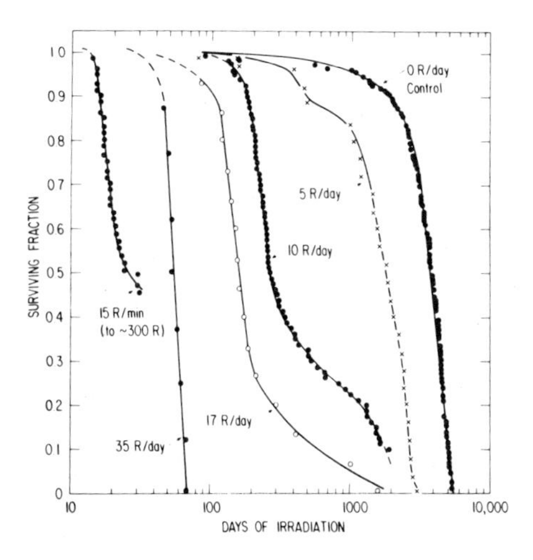

FIG. 1. Survival curves based on days of irradiation for groups of dogs exposed to either acute single doses of whole body gamma irradiation (15 R/min; 300 R doses) or continuous low daily doses of whole body gamma irradiation for duration of life (5 to 35 R/22 h day). From Seed et al. (16).

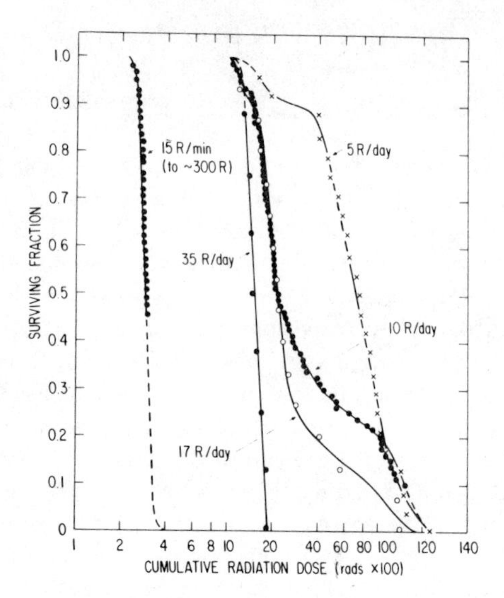

FIG. 2. Survival curves based on cumulative dose for groups of dogs exposed to either acute single doses of whole body gamma irradiation (15 R/min; 300 R doses) or continuous low daily doses of whole body gamma irradiation for duration of life (5 to 35 R/22 h day). From Seed et al. (16).

cemias, whereas above the transition point (to the right) the nonlymphocytic leukemias (NLL) predominate (Fig. 3).

The abrupt transitions in survival under continuous gamma irradiation indicate a marked heterogeneity of the population relative to radiation sensitivity. In terms of pathological predisposition and survival, two distinct subgroups are clearly identified: a radiosensitive subgroup composed of short-lived, aplasia-prone individuals and a radioresistant subgroup of long-lived, leukemia-prone individuals. The major factor responsible for these distinct subgroups lies in the differential reparative capacity of the hematopoietic system under chronic irradiation.

Leukemia Subtypes, Incidences, and Dose-Rate Effects

To date there have been a total of 46 cases of NLL in a total of 735 dogs under chronic irradiation, representing an overall leukemia incidence

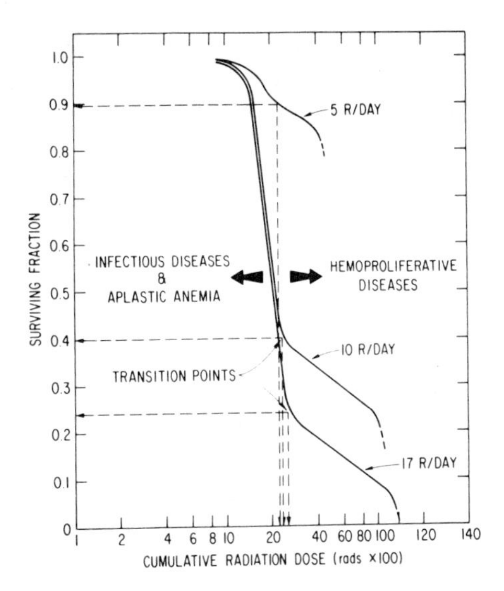

FIG. 3. Survival curves based on cumulative radiation dose show lethality rate transitions occurring at 90, 40, and 30% survival levels at 5, 10, and 17 R/day dose rates, respectively. These transitions reflect a switch from the ablative lymphohematopoietic syndromes (e.g., infectious diseases and aplastic anemia) to the hemoproliferative diseases (e.g., nonlymphocytic leukemia).

of 6.3%[1]. Seventy-eight percent of these NLL cases were either myelogenous, monocytic, or myelomonocytic leukemias; 20% were either erythroleukemias or erythremic myeloses; the remaining 2% were megakaryocytic leukemias. In addition to the NLL cases, there were two cases of lymphocytic leukemia (LL), representing an overall incidence of 0.3%.

Within the unirradiated control population, the observed incidences of NLL and LL are 0 and 1%, respectively.

---

[1]Incidence(s) based on total number of leukemic cases per total number irradiated dogs (living and dead). These listed incidences should be considered relative and not absolute. Substantial numbers of animals are still alive in the lower dose-rate groups and may still develop leukemia. However, due to recognized constraints of exposure time, radiation dose, and certain biological factors, the majority of these animals are at "low risk" and will not, most likely, significantly modify the incidence.

Relative to specific daily dose rates of exposure, the highest incidence of NLL (31.4%) occurred at a dose of 5 R (3.8 rad) per day, with appreciably lower incidences at both higher and lower dose rates (Fig. 4, top panel).

In contrast to the continuous duration-of-life exposures, termination of continuous irradiation following the accumulation of preset radiation doses (e.g., 450-3000 rad) greatly reduced the overall incidence of NLL (Fig. 4, bottom panel). Cumulative doses less than 1500 rad (1050 and 450 rad) failed to elicit NLL; cumulative doses of 1500 or 3000 rad resulted in NLL induction, but at low frequencies (Fig. 4, bottom panel). All NLL occurred within a broad but definite time interval, ranging from 300-2000 days (4).

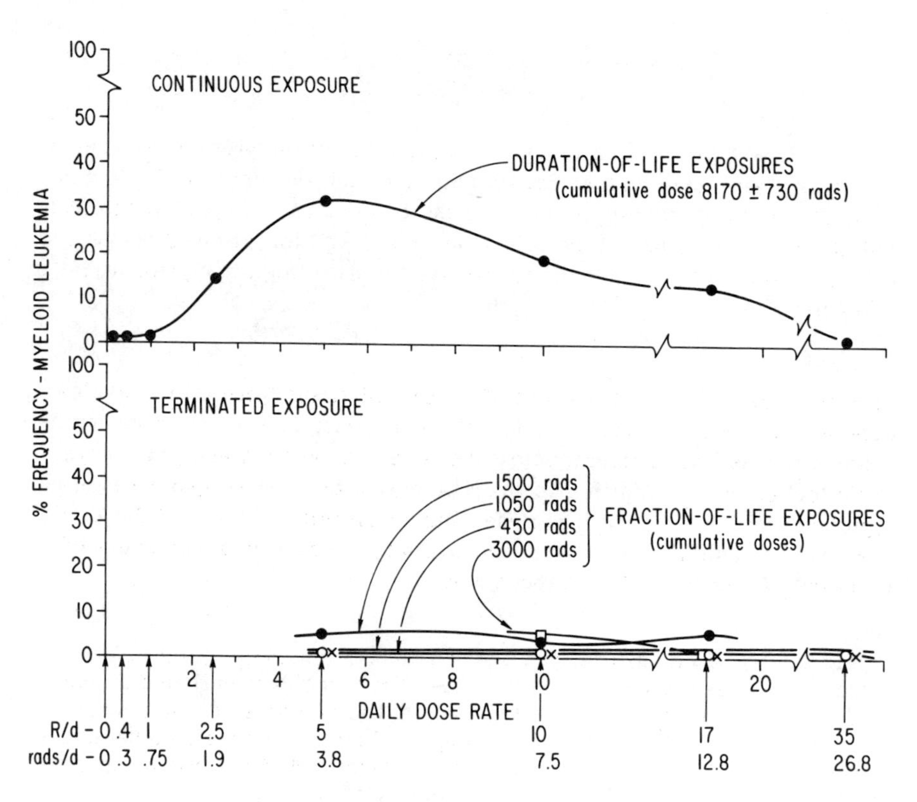

FIG. 4. Frequency of NLL (myeloid) in continuously irradiated dogs as a function of daily dose rate of exposure. Duration-of-life exposure regimens (top panel) exhibit higher leukemogenic frequencies than the initially continuous, terminated exposure regimens (lower panel).

Preclinical Responses of the Aplasia-prone and Leukemia-prone Subgroups

Blood responses.

The response of the "radiosensitive" subgroup to continuous irradiation was a progressive decay in hematopoietic function. This decay resulted in short-term survival and death from aplastic anemia (Fig. 5, panel A). In contrast, the "radioresistant" subgroup displayed a multiphasic response (Fig. 5, panel B). On the basis of characteristic blood responses, we have identified several distinct preclinical phases: I, the initial radiotoxic period characterized by declining blood cell values; II, partial recovery,

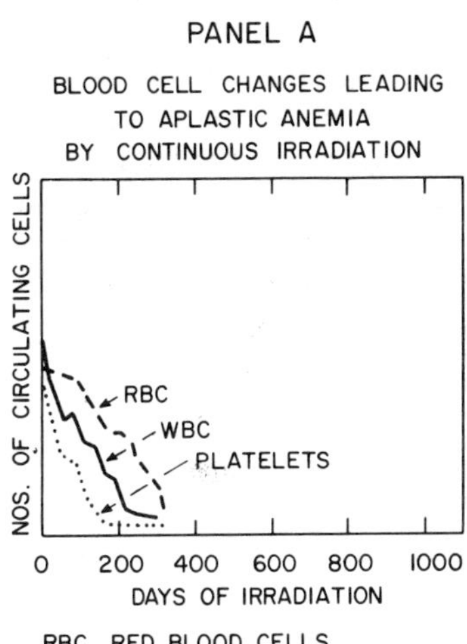

RBC RED BLOOD CELLS
WBC WHITE BLOOD CELLS

THERE IS PROGRESSIVE DEPRESSION OF CIRCULATING CELL NUMBERS CONTRASTED TO THE ALTERNATIVE RESPONSE IN PANEL B.

PANEL B

BLOOD CELL CHANGES LEADING TO LEUKEMIA INDUCTION BY CONTINUOUS IRRADIATION

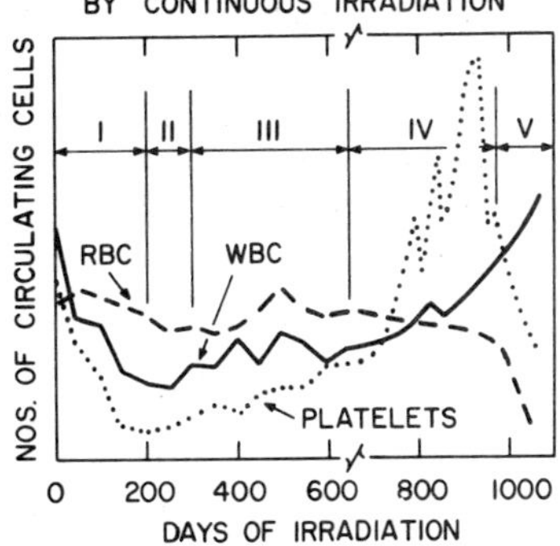

0 200 400 600 800 1000
DAYS OF IRRADIATION

RBC RED BLOOD CELLS
WBC WHITE BLOOD CELLS

PHASE I INITIAL RADIOTOXIC PHASE
PHASE II PARTIAL RECOVERY OF BLOOD-FORMING SYSTEM
PHASE III SUBNORMAL FUNCTIONING
PHASE IV "PRELEUKEMIA"
PHASE V LEUKEMIA

FIG. 5. Blood responses of the aplasia-prone subgroup (panel A) and the myeloid-leukemia-prone subgroup (panel B). Panel A for the aplasia-prone animals shows the single phase of progressive decline in peripheral blood values (erythrocytes, leukocytes, platelets) with time of radiation exposure. Panel B for the more radioresistant, leukemia-prone animals shows a multiphasic response pattern for the dogs under continuous irradiation. Phase II, partial recovery, bridges the initial radiotoxic phase with the late preclinical phases, augmenting pathological progression to overt leukemia. Redrawn from Seed et al. (21).

the bottoming out of the initial decline and the start of regeneration; III, accommodation, fluctuation of circulating blood cell values at 50-75% of normal; IV, cytologically defined preleukemias, characterized by refractory anemia with red cell abnormalities, thrombocytosis with exaggerated oscillations in platelet number, and leukoerythroblastosis with associated cytoplasmic/nuclear abnormalities; and V, overt leukemias exhibiting all the standard hallmarks. The features of these preclinical phases are described in detail elsewhere (20,21,24,25).

Bone-marrow responses.

The blood responses directly manifested the induced alterations of marrow function during the preclinical phases. Using bone marrow cellularity as an indicator of overall hematopoietic function, we found that individuals of both subgroups responded during the initial exposure periods (phase I) with progressively severe marrow hypoplasia (Fig. 6). In the radiosensitive subgroup, this condition quickly progressed (following ~150-250 days of exposure) to terminal aplastic anemia. In contrast, during the same period of exposure, the radioresistant subgroup exhibited signs of improved marrow function, as evidenced by the rebound in marrow cellularity and related parameters (as listed below). Subsequent to recovery (phase II) in late preclinical phases (phases III-IV), secondary increases in cellularity occurred (Fig. 6).

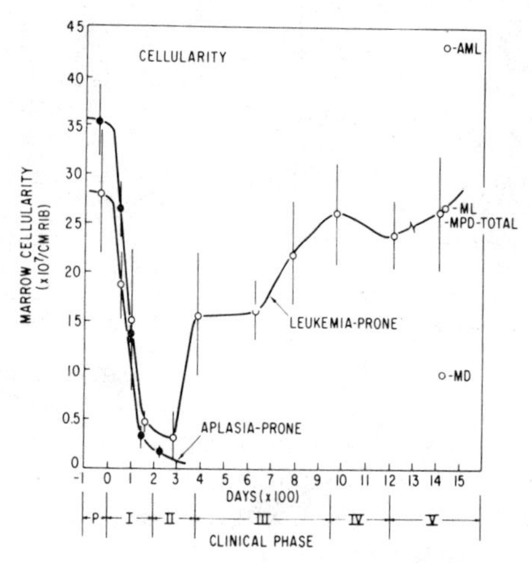

FIG. 6. Sequential phase-related changes in absolute bone marrow cellularity for the aplasia-prone subgroup and the leukemia-prone subgroup of dogs under continuous gamma irradiation.

Both the maturational and the proliferative cell compartments of the marrow exhibited similar, subgroup-specific, phase-related changes. In the aplasia-prone animals, granulocyte reserves were rapidly and irreversibly depleted; in leukemia-prone animals, following the initial depletion phase, there were primary and secondary restorative responses (Fig. 7).

Each of the major marrow cell lineages (i.e., erythrocytic, granulocytic/monocytic, megakaryocytic), were progressively, but variably, suppressed during the initial period of exposure. Aplasia-prone dogs exhibited a marked and progressive suppression of all three cell lineages that resulted in collapse of functional hematopoiesis. In contrast, the leukemia-prone animals exhibited more selective differential responses in both early and late preclinical phases (Fig. 8). During the early preclinical period, the erythropoietic elements were the least suppressed and recovered more quickly and strongly than the other cell lineages. In contrast, the megakaryopoietic elements were the most sensitive, as evidenced by the rapid decline and slow rate of recovery. The granulopoietic and monopoietic lineages responded in a somewhat intermediate fashion. During the late preclinical phases, aberrantly intense proliferation of the granulocytic/monocytic elements occurred and was accompanied by secondary suppression of both the erythrocytic and megakaryocytic elements (3,14,24,25). Alternative hemoproliferative responses during the late preclinical phases were observed and are shown in Fig. 8.

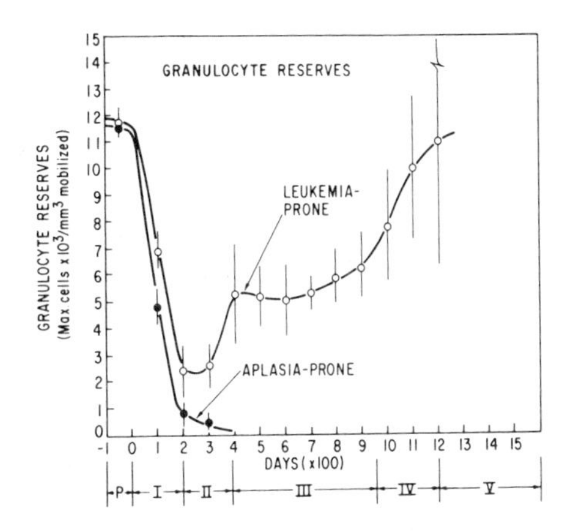

FIG. 7. Sequential phase-related changes in granulocyte reserves of the bone marrow, assessed by an endotoxin stress assay, are shown for aplasia-prone and leukemia-prone subgroups.

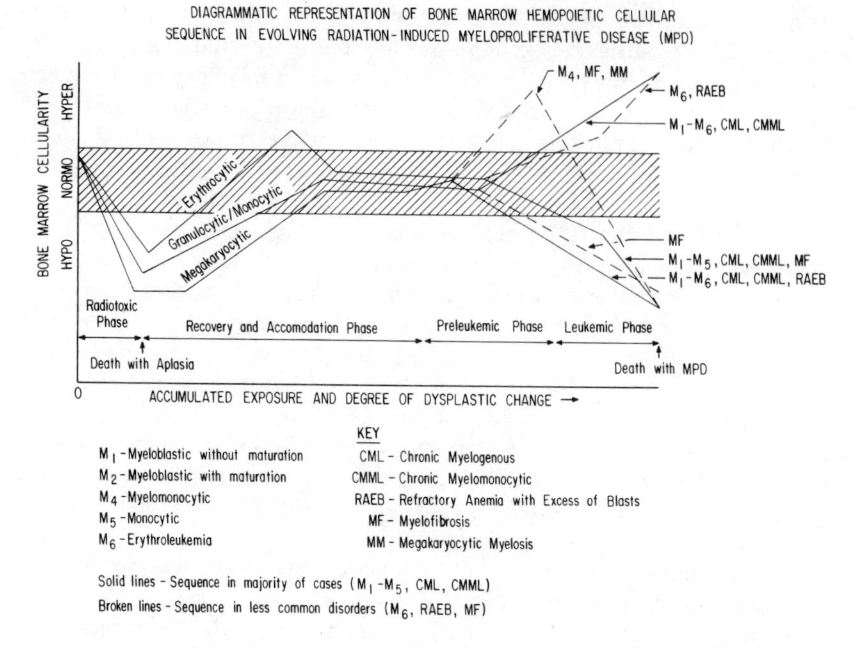

FIG. 8. Sequential cellular responses of the three major marrow cell lineages, i.e., erythrocytic, granulocytic, and megakaryocytic cell lines, with time of exposure and progression of preclinical phases. Dominant early and late preclinical responses are shown, along with specific pathologic myeloproliferative (MPD) variants (i.e., predominantly the NLL) seen during the terminal clinical phase. From Fritz et al. (16).

Hematopoietic stem cell responses.

A close correspondence was observed between the overall phase-related change in hematopoietic function and the response of the committed stem cell compartment (committed to granulo- and monopoiesis) (Fig. 9). Within the aplasia-prone subgroup, the committed stem cell compartment was progressively and irreversibly depleted, concomitant with an overall decay in hematopoietic function. The leukemia-prone subgroup responded similarly during the initial phase of exposure with a marked compartmental depletion and associated severe hematopoietic suppression. However, unique to the leukemia-prone subgroup was the partial restoration (phase II) and subsequent accommodative, but cyclic, maintenance (phase III) of the compartment's size and function (Fig. 9). Progression to the late preclinical phases (late phase III to preleukemia phase IV) resulted in an apparent secondary depletion of the stem cell compartment that temporally coincided with an aberrantly intense granulomonopoietic proliferation within the marrow (Fig. 9). These

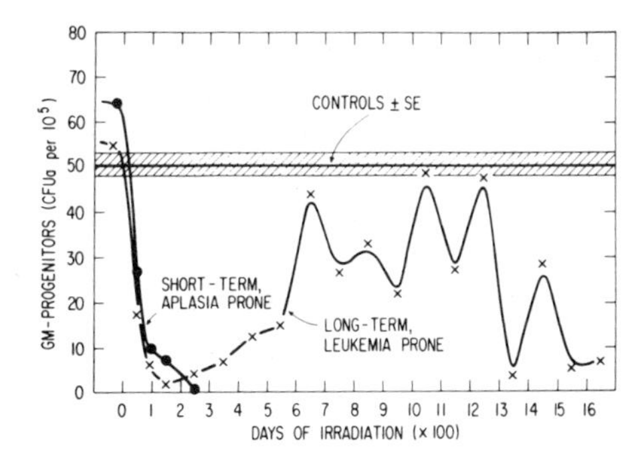

FIG. 9. The differential responses of hematopoietic progenitors reflect the overall blood and marrow responses of the aplasia-prone and leukemia-prone subgroups. From Seed et al. (17).

disparate cellular responses appeared to result from the clonal outgrowth of "preleukemic" progenitors with radically altered clonogenic properties. The dominant clonal aberrancy, as detected by in vitro cloning assays, appeared to involve a restriction of proliferative capacity of early progeny of clonogenically hyperactive cells (18).

Cellular Mechanisms of Hematopoietic Recovery

It is clear from the previous observations that repair and recovery of the hematopoietic system under continuous irradiation functions to prolong survival and, in turn, promotes leukemogenesis by bridging the initial and late preclinical phases. The recovery phase (II) is therefore, by definition, an obligatory early leukemogenic event.

What is the nature of this recovery process?

Acquired radioresistance of hematopoietic elements as a primary mechanism of recovery and accommodation.

A selective alteration in marrow cell-cycle kinetics has been stressed as the major mechanism of hematopoietic accommodation under protracted/continuous radiation exposure (5,7,8,29). Without question, such factors as enhanced progenitor-cell cycling, shortened cycle times, broadened proliferative zones, and altered transit patterns play an important and critical role in accommodation. These accommodative processes, however, cannot be invoked for prolonged periods at relatively high rates of exposure without eventually depleting vital cell stem compartments below critical threshold levels necessary to maintain minimally functional hematopoiesis.

It is unlikely that cell-cycle kinetics alone can account for the type of broadly based regenerative hematopoietic response noted here in the continuously irradiated, leukemia-prone dogs. A second, more fundamental process, e.g., "acquired radioresistance," seems indicated: first, to limit and second, to reverse the rate of stem cell loss, which is due directly to radiation sterilization and indirectly to increased differentiative demands.

In our canine model, recovery and subsequent accommodation of the hematopoietic system under continuous irradiation is associated with renewed proliferative activity and expansion of the committed stem cell compartment (Fig. 9). As this renewed activity occurs in the face of continuous irradiation, a change in radiosensitivity of these vital marrow cells is implied.

This concept of "acquired radioresistance" by hematopoietic stem cells was tested directly in vitro. Results clearly showed that stem cells isolated from marrow of leukemia-prone animals during the postrecovery periods exhibited markedly increased resistance, in comparison to the more sensitive cells isolated from either the aplasia-prone animals or from nonirradiated controls (Fig. 10).

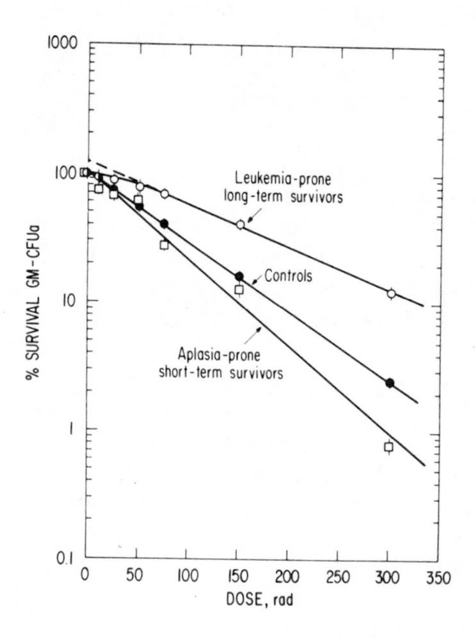

FIG. 10. Postirradiation (in vitro) survival curves of hematopoietic progenitors isolated from bone marrow of short-lived, aplasia-prone dogs; long-lived, leukemia-prone dogs; or nonirradiated controls. From Seed et al. (17).

Sequential testing of the radiosensitivity of stem cells during preleukemic phases, specifically during the transition from the prerecovery to the postrecovery period, revealed that radioresistance was acquired at about the time of the hematopoietic recovery (19). Radioresistance appeared not to be gradually acquired, but appeared with progression from phase I to phase II. This observation tended to suggest that the mode of acquired radioresistance was not through "simple selection" of preexisting (preirradiation) subpopulations, but rather through an alternative route (e.g., gene modification-selection). Simple selection could not be ruled out, however. It is possible that selective outgrowth of a small, preexisting radioresistant population under continuous irradiation required an initial depletion and clearing of the marrow's normally dominant radiosensitive population.

Differential growth responses of radioresistant progenitor clonotypes. Retrospective analysis of the dose-response curves for committed stem cells isolated sequentially from irradiated dogs revealed multiple survival patterns, some of which were associated with specific preclinical phases. Figure 11 illustrates the various patterns observed and lists the limits of radiation sensitivity for each of these patterns. It is reasonable to assume that these survival patterns reflect specific stem cell clonal types. The "frequency of expression" of these clonal types changed during progression of NLL (Fig. 12). The radiosensitive clonal-type "X" was dominant during phase I but rapidly declined with the onset of hematopoietic recovery (phase II) and was replaced initially by radioresistant clonal-types A and B, characterized by elevated D_0 values (suppressed dose-dependent lethality rates) and by elevated D_q values (increased

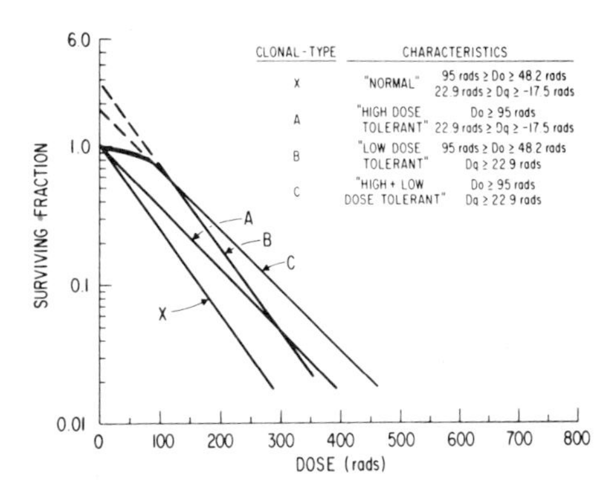

FIG. 11. Schematic of the major postirradiation survival patterns for the radiosensitive and radioresistant stem cell clonal types and a listing of the limits of radiation sensitivity for each clonal type (pattern).

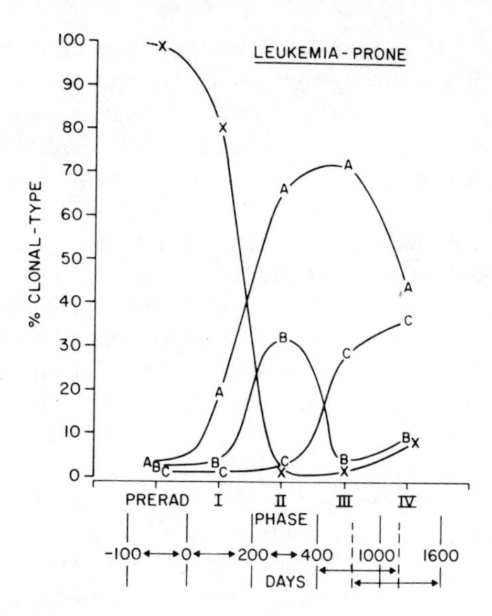

FIG. 12. The sequential change in frequency of expression of radiosensitive (X) and radioresistant (A,B,C) progenitor clonal types during preclinical phase progression of NLL.

sublethal damage capacity), respectively, and later by clonal-type C, with the combined characteristics of elevated D_o and D_q values (Figs. 10 and 11).

This change in frequency of expression of these clonal types suggests that during pathological progression of chronic radiation-induced NLL there is a selective evolution of distinct, radioresistant hematopoietic progenitors.

Survival and Pathological Predisposition as a Function of Stem Cell Radiosensitivity

Considering the above observations concerning changing patterns of radiosensitivity and function of vital hematopoietic stem cells under chronic irradiation, we have formulated the following ideas as part of a working hypothesis on the cellular basis for chronic radiation-induced leukemogenesis. First, hematopoietic stem cells have the potential to maintain two levels of radioresistance: a low constitutive level and an acquired elevated level. Each level dictates a specific stem cell response pattern under continuous irradiation, and, in turn, the divergent hemopathological sequels. The low constitutive radioresistance yields high rates of stem cell lethality and aplastic anemia; the acquired elevated radioresistance yields low rates of stem cell lethality and augments aberrant proliferative processes and NLL. Second, the acquisition of the elevated state promotes leukemia by imparting selective growth advantages to progressively resistant and genetically unstable stem cells. Third, continuous, duration-of-life irradiation, in contrast to the terminated exposure regimens, effectively elicits leukemogenic responses by providing both a constant source of genetic damage and the selective pressure required to clonally amplify preleukemic cells.

In the context of the canine model as presented here, it seems clear that the processes of acquired radioresistance and recovery (repair) of hematopoietic progenitors under continuous irradiation are intimately related to the induction and progression of NLL. Although the basis of these relationships remains to be determined, we speculate here that the intermediary of acquired resistance and leukemogenic transformation is a modified repair process.

It is currently recognized that acquired resistance by mammalian cells to a variety of physicochemical toxicants is mediated by gene amplification and its effect of overproducing target proteins essential for repair, inactivation, etc. (13). The degree of amplification and resistance is directly related to the extent of genetic damage and to the intensity and duration of the selective pressure. Significant genetic instability results from overreplication, the principal mechanism of gene amplification. By analogy, therefore, acquired radioresistance, as we have noted here, might arise through similar genetic and clonal amplifications of repair processes, creating increasing genetic instability, manifested by the progressive loss of proliferative and differentiative control and culminating in overt leukemia.

CONCLUSIONS

Chronic radiation leukemogenesis is both dose-rate and cumulative-dose dependent. These radiation parameters seemingly exert an effect by altering the degree of hematopoietic suppression and the extent of recovery during the period of exposure. The mechanism of recovery appears to be mediated, in part, by the acquisition of radioresistance by vital hematopoietic progenitors. Further, there appears to be a time-dependent, preclinical, phase-specific selection and amplification of resistant clonal types.

Finally, our observations support the concept that chronic radiation leukemogenesis is a "multistaged" process, with radiation protraction serving to promote pathological progression via the acquisition of radioresistance by hematopoietic progenitors.

ACKNOWLEDGMENTS

The authors acknowledge the efforts of Dr. William P. Norris who directed and guided the early work on the survival responses under chronic irradiation. The authors also acknowledge the excellent assistance of Mr. Donald Doyle and Mrs. Carol Fox for computer services and data management; Dr. Calvin Poole, Mr. William Keenan, and Ms. Susan Cullen for clinical and hematological services; and Mr. Gordon Holmblad for assisting in irradiating both animals and cells. We also thank the Argonne Animal Care Specialists for their daily care of the research animals and maintenance of facilities accredited by the American Association of Laboratory Animal Care. This work was supported by the U. S. Department of Energy under contract No. W-31-109-ENG-38 and by the National Cancer Institute, U. S. Department of Health and Human Services under Interagency Agreement Y01-CO-00320.

REFERENCES

1. B.E.I.R. (1980): In: Report of the Committee on Biological Effects of Ionizing Radiation, pp. 398-431. National Research Council, National Academy Press, Washington, DC.
2. Fritz, T.E., Norris, W.P., Tolle, D.V., Seed, T.M., Poole, C.M., Lombard, L.S., and Doyle, D.E. (1978): In: Late Biological Effects of Ionizing Radiation, Vol. 2, pp. 71-82. IAEA-SM-224/206, International Atomic Energy Agency, Vienna.
3. Fritz, T.E., Tolle, D.V., and Seed, T.M. (1984): In: Preleukemic Syndrome, edited by G. Bagby, CRC Press, Inc., Boca Raton, Florida, in press.
4. Fritz, T.E., Seed, T.M., Tolle, D.V., and Lombard, L.S. (1984): In: Life-span Radiation Effects Studies in Animals: What Can They Tell Us?, 22nd Hanford Life Sciences Symposium, U.S. Department of Energy, in press.
5. Gidali, J., Bojtor, I., and Feher, I. (1979): Radiat. Res., 77:285-291.
6. Kaplan, H. (1977): In: Radiation-induced Leukemogenesis and Related Viruses, INSERM Symposium No. 4, edited by J. Duplan, pp. 1-18. Elsevier/North-Holland, Amsterdam.
7. Lamerton, L.F. (1966): Radiat. Res., 27:119-138.
8. Lord, B.I. (1964): Brit. J. Haematol., 10:496-507.
9. Major, I.R., and Mole, R.H. (1978): Nature, 272:455-456.
10. Mole, R.H. (1977): In: Radiation-induced Leukemogenesis and Related Viruses, INSERM Symposium No. 4, edited by J. Duplan, pp. 19-36. Elsevier/ North-Holland, Amsterdam.
11. Mole, R.H., and Major, I.R. (1983): Leukemia Res., 7:295-300.

12. Norris, W.P., and Fritz, T.E. (1972): In: Radiobiology of Plutonium, edited by B.J. Stover, and W.S.S. Jee, pp. 243-260. J.W. Press, University of Utah, Salt Lake City.
13. Schimke, R.T. (1984): Cancer Res., 44:1735-1742.
14. Seed, T.M., Chubb, G.T., and Tolle, D.V. (1981): Scanning Electron Microsc., 1981:61-71.
15. Seed, T.M., Cullen, S.M., Kaspar, L.V., Tolle, D.V., and Fritz, T.E. (1980): Blood, 56:42-51.
16. Seed, T.M., Fritz, T.E., Tolle, D.V., Poole, C.M., Lombard, L.S., Doyle, D.E., Kaspar, L.V., Cullen, S.M., and Carnes, B.A. (1984): In: Responses of Different Species to High Dose Total Body Irradiation, edited by J.J. Broerse and T. MacVittie. Martinus Nijhoff, Netherlands, in press.
17. Seed, T.M., Kaspar, L.V., Tolle, D.V., and Fritz, T.E. (1982): Exp. Hematol., 10:232-248.
18. Seed, T.M., Kaspar, L.V., Tolle, D.V., Poole, C.M., and Fritz, T.E. (1981): Exp. Hematol., 9:68.
19. Seed, T.M., Kaspar, L.V., Tolle, D.V., and Fritz, T.E. (1984): Exp. Hematol., 12:451.
20. Seed, T.M., Tolle, D.V., Fritz, T.E., Devine, R.L., Poole, C.M., and Norris, W.P. (1977): Blood, 50:1061-1079.
21. Seed, T.M., Tolle, D.V., Fritz, T.E., Cullen, S.M., Kaspar, L.V., and Poole, C.M. (1978): In: Late Biological Effects of Ionizing Radiation, Vol. 1, IAEA-SM-224/308, pp. 531-545. International Atomic Energy Agency, Vienna.
22. Sinclair, W.K. (1963): Radiat. Res., 20:288-297.
23. Smith, P.G., and Doll, R. (1982): Br. Med. J., 284:449-460.
24. Tolle, D.V., Fritz, T.E., Seed, T.M., Cullen, S.M., Lombard, L.S., and Poole, C.M. (1982): In: Experimental Hematology Today 1982, edited by S.J. Baum, G.D. Ledney, and S. Thierfelder, pp. 241-249. S. Karger, Basel.
25. Tolle, D.V., Seed, T.M., Fritz, T.E., and Norris, W.P. (1979): In: Experimental Hematology Today 1979, edited by S.J. Baum and G.D. Ledney, pp. 247-256. Springer-Verlag, New York.
26. Upton, A.C. (1977): In: Radiation-induced Leukemogenesis and Related Viruses, INSERM Symposium No. 4, edited by J. Duplan, pp. 37-50. Elsevier/North-Holland, Amsterdam.
27. Upton, A.C., Randolph, M.L., Conklin, J.W., Kastenbaum, M.A., Slatter, M., Melville, G.S., Jr., Conte, F.P., Sproul, J.A., Jr. (1970): Radiat. Res., 41:467-491.
28. Williamson, F.S., Hubbard, L.B., and Jordan, D.L. (1968): In: Annual Report 1968, Biological and Medical Research Division, ANL-7535, pp. 153-156. Argonne National Laboratory, Argonne, Illinois.
29. Wu, C.T., and Lajtha, L.G. (1979): Int. J. Radiat. Biol., 27:41-50.

Carcinogenesis, Vol. 10, edited by E. Huberman and S. H. Barr. Raven Press, New York © 1985.

Biological Basis for Assessing Carcinogenic Risks of Low-Level Radiation

A. C. Upton

Department of Environmental Medicine, New York University Medical Center, New York, New York 10016

It is a privilege to join in paying tribute to Charles Heidelberger. Few have contributed more importantly to cancer research. Although I never had the pleasure of working with him directly, our paths crossed repeatedly at meetings, where I was often stimulated by the fertility of his imagination and the power of his intellect. On one of the first such occasions, a Ciba Foundation symposium on carcinogenesis, Charlie made an observation I shall never forget. After listening to participants of different disciplines discuss their diverse approaches to the study of carcinogenesis, he remarked that "the mechanism of carcinogenesis is a mirror into which we look and see ourselves" (24).

As his statement aptly implies, the problem of carcinogenesis is so broad and complex that it attracts investigators of virtually every discipline. In view of the numbers of scientists who have studied the problem and the magnitude of their efforts, it is not astonishing that our knowledge of the subject has advanced dramatically.

At the same time, one cannot help but be acutely conscious of the gaps in our understanding of carcinogenesis that remain to be filled, especially when one attempts to estimate the carcinogenic risks of low-level ionizing radiation. The uncertainties in risk assessment have prompted growing public concern and litigation.

Since risks at the exposure levels of interest are too small to be measured epidemiologically (36), they can be estimated only by extrapolation, based on assumptions about the dose-incidence relationships and mechanisms of carcinogenesis. Ultimately, therefore, resolution of the problem of assessing the risks associated with low-level radiation will require better understanding of the biology of carcinogenesis. To this end, cogent aspects of the subject are summarized in the following.

FACTORS AFFECTING SUSCEPTIBILITY TO RADIATION CARCINOGENESIS

Genetic Background

Although susceptibility to radiation carcinogenesis is shared by all species and strains of mammals studied to date, susceptibility to the induction of any particular neoplasm varies widely among different species and strains (62). A striking example is the unusually high sensitivity of the mouse ovary to tumor induction. For example, a dose of 100 rad suffices to induce ovarian tumors in nearly half of the females of a susceptible strain (62,63), while such tumors are rarely induced in other species.

In human beings, inherited differences in susceptibility are evident in the dramatic sensitivity of individuals with retinoblastoma, nevoid basal cell carcinoma syndrome, or certain other hereditary diseases (58). It is possible that susceptibility may exist to a lesser degree in heterozygous carriers of these traits and in heterozygous carriers of genes for DNA repair defects (34,51).

The carcinogenic effects of radiation on human populations in different parts of the world, compared on the basis of the numbers of cases per 10^6 person years at risk per unit dose, imply that susceptibility differs little among races or ethnic groups, although this lack of difference is not necessarily true if the relative risks in different populations are compared (49). The absence of an excess of leukemia in A-bomb survivors who were irradiated prenatally, in contrast to prenatally irradiated Caucasian children (46,49), may possibly be related to the lower baseline incidence of leukemia in nonirradiated Japanese children during the postwar period (45).

Differences Among Organs, Tissues, and Cells

Within a given individual, susceptibility to the carcinogenic effects of radiation varies markedly among different organs, tissues, and cells (Table 1). In contrast to chemical carcinogenesis, where such differences are a logical result of pharmacokinetic variables, differences in the radiation dose to target cells and molecules cannot explain the observed variations. Possible explanations meriting further study include differences in the numbers of stem cells at risk in different tissues, differences in cell turnover rates, and differences in repair and renewal capabilities. Whatever the explanation may be, it is noteworthy that susceptibility bears no constant relationship to the baseline incidence of cancer in different organs (Table 1).

Within a given tissue or organ, not all types of cancer are induced with equal frequency. For example, the incidence of chronic lymphocytic leukemia, in contrast to all other types of leukemia, is not detectably increased by irradiation in any population studied to date (49, 62).

TABLE 1. Sensitivity of different tissues to radiation carcinogenesis.[a]

Site or Type of Cancer	Spontaneous Incidence[b] (per 10^6/y) male	female	Excess Cases (per 10^6/y/rem)	Remarks
Breast (female)	----	900		--
Lung, bronchus	690	230		c
Colon	310	340	(>1)	--
Stomach	120	70		--
Leukemia	80[d]	60[d]		d
Thyroid gland	20	60		e
Urinary tract	310	130		--
Pancreas	105	90		--
Lymphoma, multiple myeloma	170	150		f
Liver, biliary tract	50	50		--
Brain, central nervous system	60	40	(0.1-1)	--
Esophagus	50	20		--
Pharynx	15	5		--
Salivary glands	10	10		e
Skin	>1000	<1000		--
Uterus & cervix	--	440		--
Ovary	--	140		--
Parathyroid gland	<5	<5		--
Bone	10	5	(0.01-0.1)	--
Cranial simuses	<5	<5		--
Larynx	80	15		--
Mesothelium	<5	<5		--
Connective tissue, including heart	20	15	uncertain	--
Testis	35	--		--
Prostate	560	--		--
Other	120	120		--

[a]From 7, 30, 49, 61, and 62.
[b]Values represent rounded averages for all ages and races (from 76).
[c]Effect of smoking uncertain.
[d]Excluding chronic lymphatic leukemia.
[e]Low mortality.
[f]Excluding Hodgkin's disease.

Similarly, the thyroid cancers induced by irradiation consist predominantly of well-differentiated papillary and follicular adenocarcinomas, with few tumors of anaplastic types (49,62).

Sex

The susceptibility of endocrine glands and their target organs to radiation carcinogenesis differs between males and females, in keeping with differences in the corresponding baseline tumor rates. A wealth of experimental data implies that these differences are attributable primarily to the promoting effects of hormones (21); however, unexplained sex differences also exist in the susceptibility of nonendocrine organs to neoplasms of other types, e.g., leukemia (11,64).

Age at Exposure

Susceptibility to the carcinogenic effects of radiation varies markedly with age at the time of irradiation, depending on the neoplasm in question (62). In humans, the relative risk of leukemia is increased more markedly by irradiation before birth than by irradiation later in life (48,49,62). The types of leukemia that are induced, however, are also age dependent; those induced by prenatal irradiation are predominantly acute agranulocytic and stem cell leukemias, whereas those induced by exposure later in life include comparable numbers of acute and chronic granulocytic leukemias (49,64). With increasing age at irradiation during adult life, the relative risk of all types of leukemia combined remains essentially constant, while the absolute risk increases with age at irradiation (49) (Fig. 1).

For cancer of the female breast, susceptibility is higher during adolescence and early adult life than at later ages (49). Further, women irradiated during infancy do not develop radiation-induced breast tumors until they reach the fourth decade, with the result that the susceptibility of this age group has been underestimated heretofore (25,60). The data

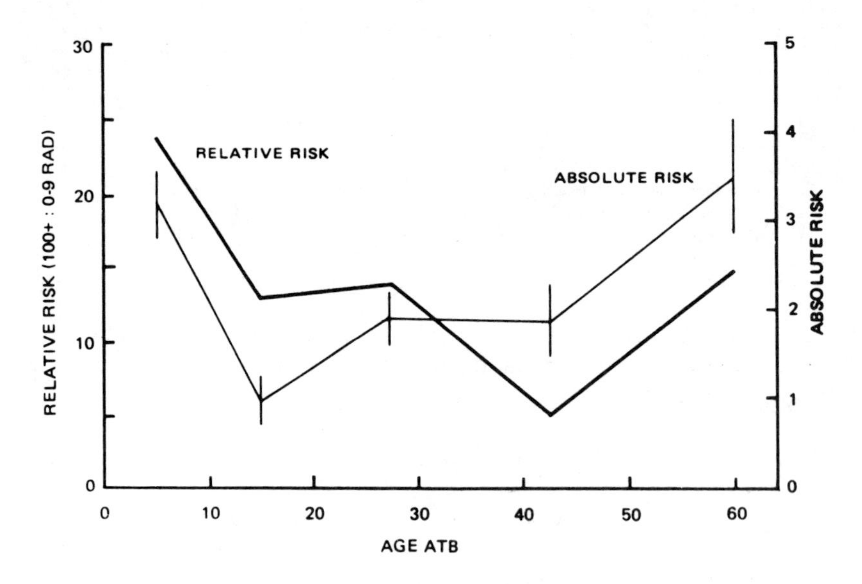

FIG. 1. Age-specific relative risk estimates and absolute risk (excess deaths per million person-year-rad) with 90% confidence intervals. From Beebe et al. (8).

imply that carcinogenesis in the breast, although initiated by irradiation, depends on promoting effects that are associated with age-related hormonal stimulation.

RELATION BETWEEN INCIDENCE AND DOSE

Induction Period

The cancers induced by radiation are typically preceded by a long induction period, the length of which depends on the type of neoplasm in question and age at the time of irradiation. In general, the cancers tend to occur at the ages when cancers of corresponding types characteristically occur in nonirradiated individuals. In populations exposed prenatally, for example, the increased incidence of leukemia peaks within the first five years of life, in parallel with the incidence of juvenile leukemia in the general population (40,46). Similarly, as noted above, the breast cancers induced in women by irradiation during infancy do not appear until more than 30 years later, at an age when breast cancers characteristically appear in the general population.

The induction period for leukemia is generally shorter than that for other forms of cancer. Most cases of chronic granulocytic leukemia appear within 5-15 years, and few later than 25 years, after irradiation. As a result, the temporal distribution of the radiation-induced cases assumes a wave-like pattern (Fig. 2). The patterns for other types of radiation-induced leukemia, although not identical, are also wave-like (Fig. 2).

For cancers other than leukemia, and possibly osteosarcoma (49), the induced cases do not seem to distribute themselves in wave-like patterns with time after irradiation, and their induction periods average more than 15-20 years. From the evidence that is available, it is not clear whether the annual excess of such cases remains constant after a certain length of time (i.e., the "absolute risk" model) or whether it continues indefinitely to increase as a constant fraction of the underlying baseline incidence (i.e., the "relative risk" model) (Fig. 3). The evolving data from studies of atomic bomb survivors (Fig. 4) are more compatible with the latter model than with the former (7,67), as are the results of experiments in laboratory animals (Fig. 5).

With chronic irradiation, the median time to tumor appearance, at least for some forms of cancer, has been shown to vary as a function of the dose rate according to the Blum-Druckrey model, i.e.,

$$t^n d = C \qquad (1)$$

where t is the median time to tumor appearance, d is the daily dose of radiation, and C and n are constants, n being larger than 1 (1).

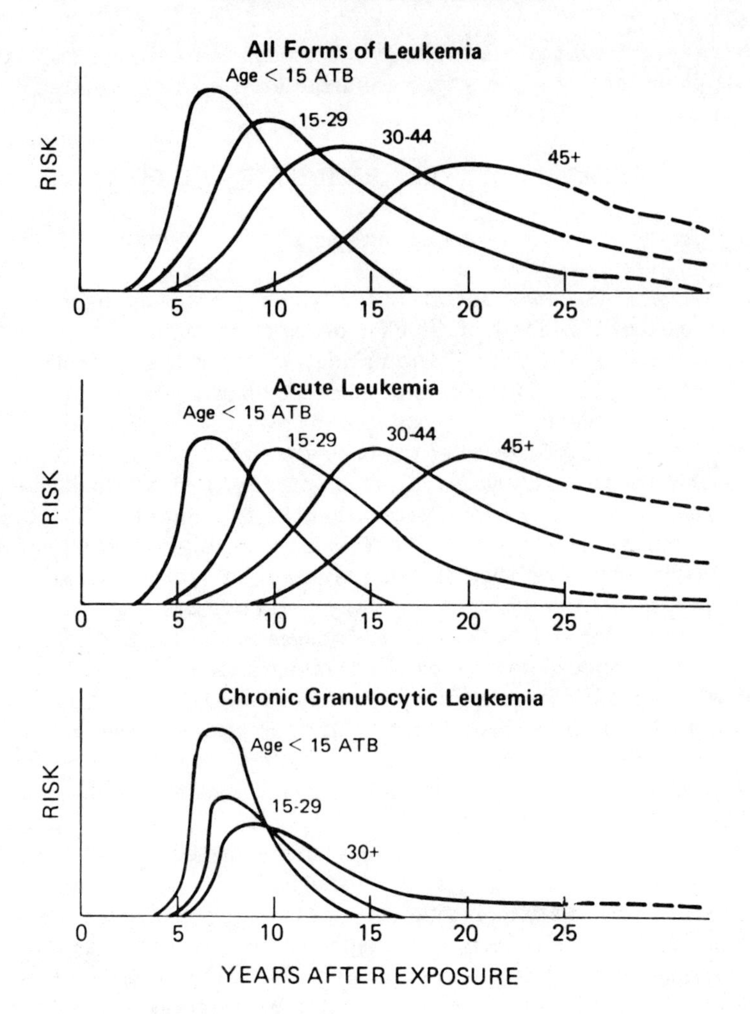

FIG. 2. Temporal distribution of radiation-induced leukemias in A-bomb survivors, in relation to type of leukemia and age at the time of irradiation. From Ichumaru and Ishimaru (29).

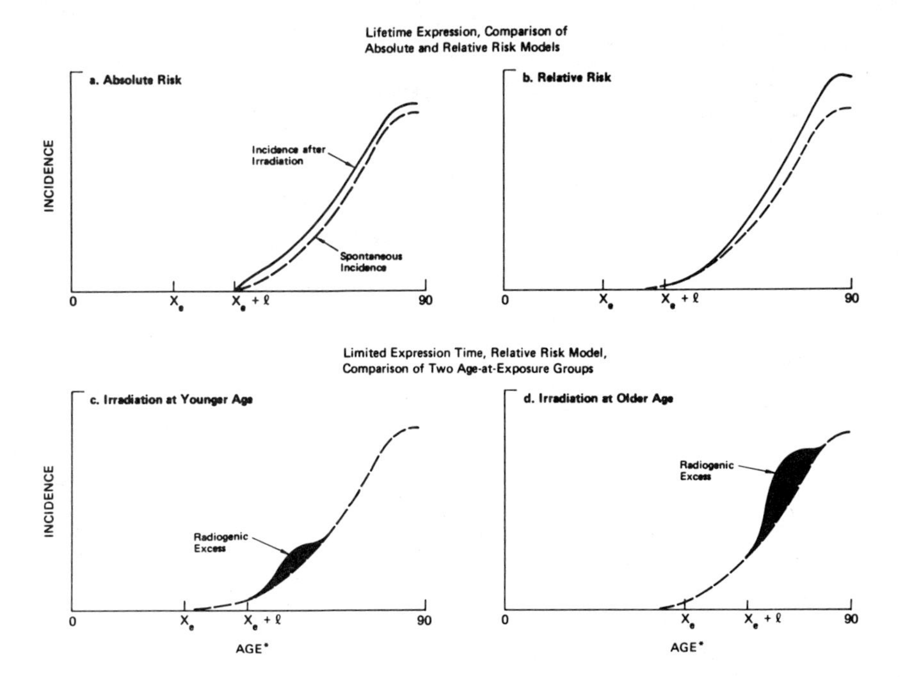

FIG. 3. Distribution of the incidence of radiation-induced cancer with time after irradiation, in relation to baseline age-dependent cancer incidence, as projected for a lifetime by the absolute risk model (a) or the relative risk model (b); or as projected for a shorter period by the absolute risk model (c) or the relative risk model (d). The symbol x denotes the minimal latent period. From National Academy of Sciences (49).

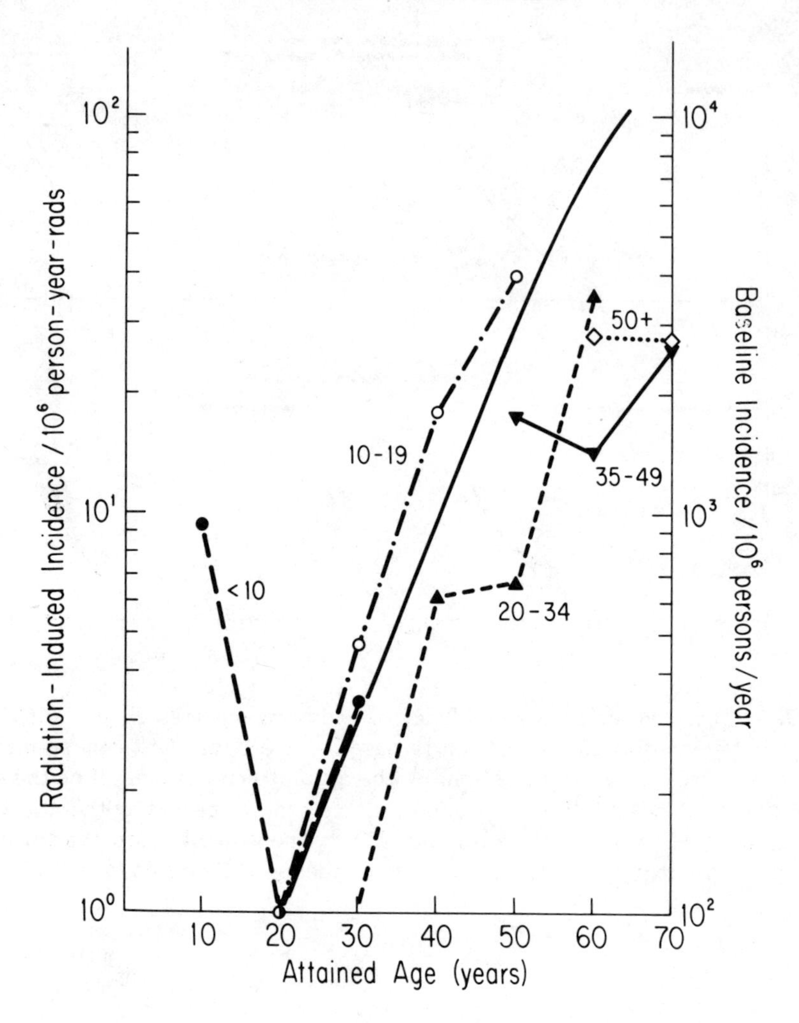

FIG. 4. Incidence of radiation-induced cancers (excluding leukemia) in relation to attained age in different birth cohorts of heavily irradiated Nagasaki A-bomb survivors, as compared with the baseline incidence in the general population of Japan. Data for A-bomb survivors (73) is denoted by dashed curves, with age at time of irradiation of each cohort as indicated. Data for the general population (74) is denoted by the solid curve.

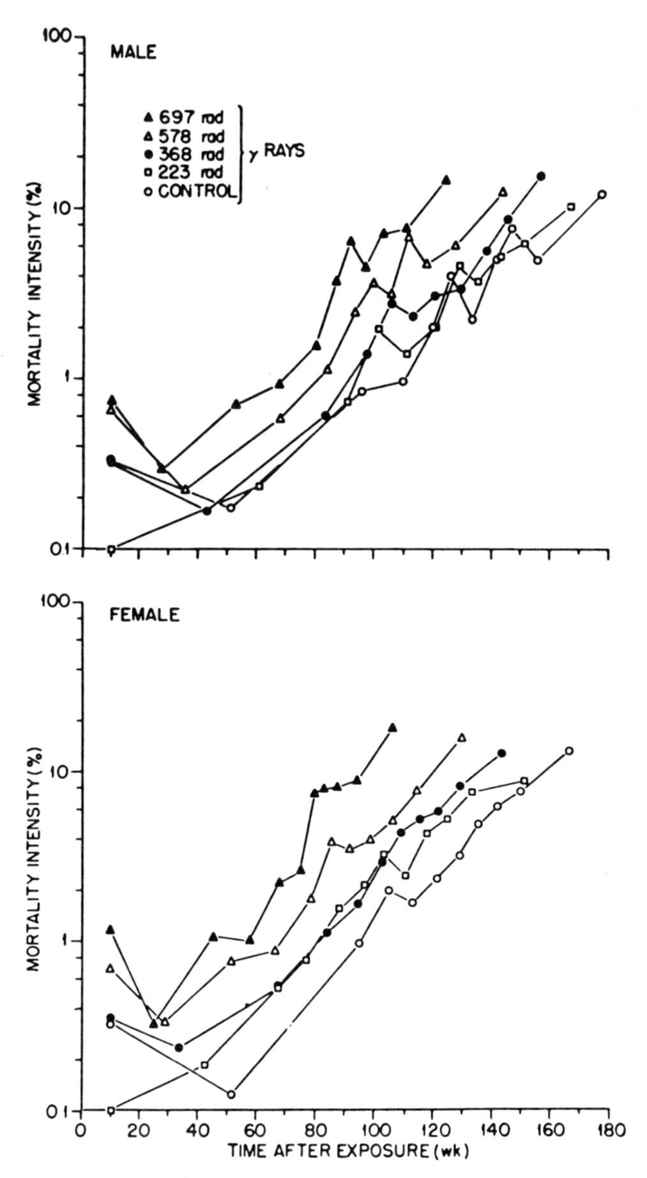

FIG. 5. Death rate in mice in relation to dose of whole-body gamma radiation and time after irradiation. From Upton et al. (69).

Dose

Although irradiation has been observed to induce many types of neoplasms in humans and laboratory animals, all neoplasms are not increased in frequency within a given species or strain (62). Certain neoplasms on the contrary, are decreased in frequency by irradiation (Fig. 6). From the diversity of observed dose-incidence relationships, it is clear that no one mathematical model for relating incidence to dose is universally applicable.

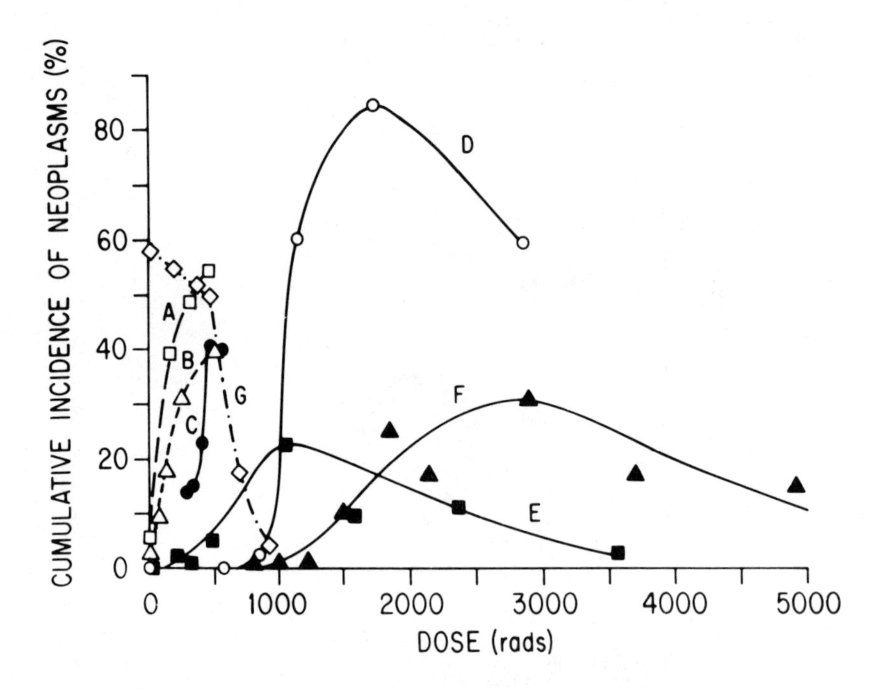

FIG. 6. Dose-incidence curves for different neoplasms in animals exposed to external radiation. A: myeloid leukemia in x-irradiated mice (71); B: mammary gland tumors at 12 months in gamma-irradiated rats (53); C: thymic lymphomas in x-irradiated mice (31); D: kidney tumors in x-irradiated rats (41); E: skin tumors in alpha-irradiated rats (percentage incidence times 10) (17); F: skin tumors in electron-irradiated rats (percentage incidence times 10) (17); and G: **reticulum** cell sarcomas in x-irradiated mice (43). From Upton (68).

For those types of cancer that are known to be induced in humans by high radiation doses, the dose-incidence relationships at lower doses are not well defined. Hence the risks of low-level irradiation can be estimated only by extrapolation. From the existing data, however, it is not possible to distinguish unambiguously among alternative extrapolation models. For this reason, the National Academy of Sciences Advisory Committee on the Biological Effects of Ionizing Radiation (BEIR Committee) used four different models (Fig. 7), and thus derived a range of risk estimates, in its latest report (49). The models in question, all of which assume a nonthreshold dose-incidence relationship, are (i) the linear model, which can be represented by the expression:

$$F(D) = \alpha_0 + \alpha_1 D \tag{2}$$

where F(D), the incidence of cancer at dose D, varies as the sum of the baseline incidence α_0 plus the product of the dose D times a constant risk coefficient (α_1); (ii) the linear-quadratic model, which can be represented by the expression:

$$F(D) = \alpha_0 + \alpha_1 D + \alpha_2 D \tag{3}$$

where F(D), the incidence of cancer at dose D, varies as the sum of the natural incidence (α_0) plus a linear dose term ($\alpha_1 D$) and a squared dose term ($\alpha_2 D^2$); (iii) the quadratic risk model, which can be represented by the expression:

$$F(D) = \alpha_0 + \alpha_2 D^2 \tag{4}$$

where F(D), the excess of cancer at dose D, varies as the sum of the natural incidence (α_0) plus a squared dose term ($\alpha_2 D^2$); and (iv) the linear-quadratic model incorporating a negative exponential to account for saturation of the carcinogenic process at high doses, which can be represented by the expression:

$$F(D) = (\alpha_0 + \alpha_1 + \alpha_2 D^2)\exp(-\beta_1 D - \beta_2 D^2) \tag{5}$$

where the terms are as described above, with the addition of the coefficients β_1 and β_2.

If the induction of a mutation or chromosome aberration in a single somatic cell sufficed to initiate the process of carcinogenesis, the last of the above models would seem more appropriate than the others on purely radiobiological grounds (16,49,65). For leukemia, the revised data for A-bomb survivors (57) and the data for the mouse (4) are compatible with this model. For most cancers other than osteosarcoma, the quadratic model is incompatible with the data.

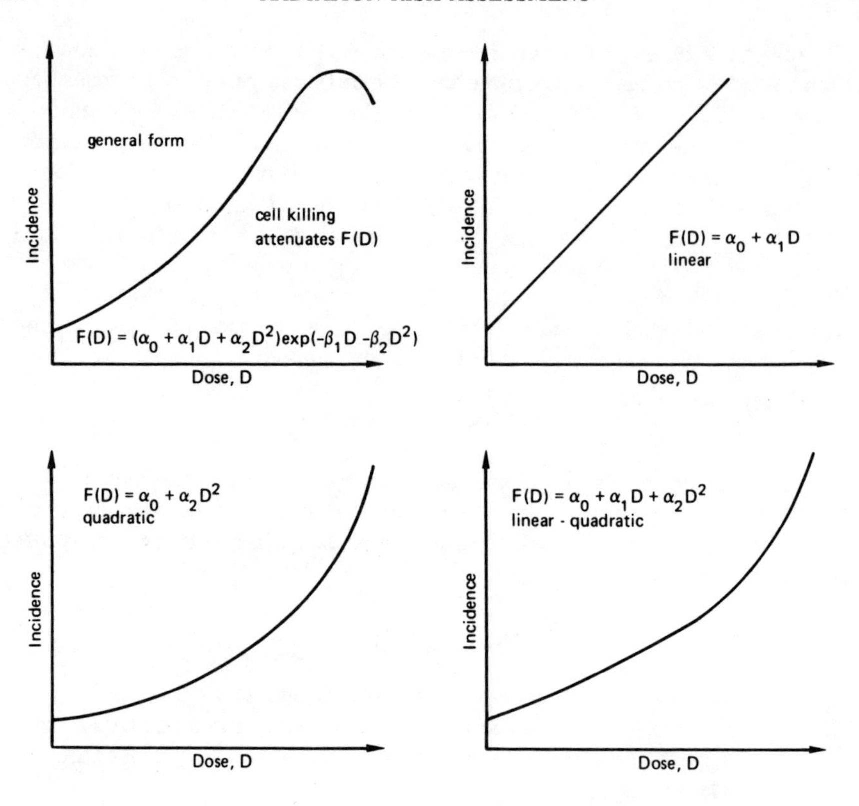

FIG. 7. Alternative mathematical dose-incidence models. From National Academy of Sciences (49).

For cancer of the female breast, the dose-incidence relationship appears to be essentially the same in women surviving A-bomb irradiation (37,38), women who received radiotherapy to the breast for acute postpartum mastitis (55), women whose breasts were irradiated through repeated fluoroscopic examinations of the chest in the course of treatment for pulmonary tuberculosis (14), and luminous dial painters (6). The constancy of the dose-incidence relationship in the four groups, in spite of the marked differences in the duration of their exposures, strongly implies that the linear nonthreshold model is appropriate for carcinogenesis in the female breast (13,36,49).

Evidence favoring a linear nonthreshold model has also been presented for induction of thyroid cancer by irradiation in infancy (56) and for the leukemogenic effects of prenatal irradiation (26).

Dose Rate

In experimental animals, it has been observed repeatedly that a given dose of x-rays or gamma rays is generally several times less tumorigenic if delivered in many small exposures over a period of days or weeks than if delivered in a single exposure lasting only a few minutes (50,62). The effectiveness of high-linear-energy-transfer (LET) radiation (e.g., neutrons or alpha particles), on the other hand, shows little or no decrease (70), but may increase with fractionation or protraction of exposure (59).

The influence of dose rate on carcinogenesis in human populations is not apparent from available data (50).

LET

Studies in laboratory animals have shown high-LET radiation to be more effective than low-LET radiation for induction of essentially all types of neoplasms (62). The difference is especially marked at low doses and low dose rates, since the dose-incidence curve for high-LET radiation is typically linear or convex upward and is relatively independent of the dose rate (in fact, as noted above, the curve for high-LET radiation may actually increase in slope with decreasing rate), whereas the curve for low-LET radiation is typically concave upward and decreases in slope with decreasing dose rate (Fig. 8).

The influence of LET on carcinogenesis in humans appears to be consistent, in general, with the patterns shown in Fig. 8, but the data are fragmentary.

Anatomical Distribution of Dose

Most neoplasms induced by radiation appear to arise from cells that have been altered directly by irradiation. The probability of neoplasia thus generally increases with the number of cells, or volume of tissue, exposed.

In some instances, indirect effects mediated through damage to remote cells or tissues play a role. In the induction of ovarian tumors in the mouse, for example, both ovaries must be exposed to a sterilizing dose of radiation to elicit tumorigenesis in either one (21). Also, in the induction of pituitary tumors in the mouse, radiation-induced damage to target organs can act by disturbing hormonal feedback regulation (21). For induction of thymic lymphomas in the mouse, irradiation of the entire hemopoietic system is necessary to elicit a maximal tumorigenic response (32). Indirect effects have also been implicated in the pathogenesis of osteosarcomas, where the killing of osteoblasts by alpha radiation from locally deposited radium is postulated to promote carcinogenesis through stimulation of regenerative hyperplasia (42).

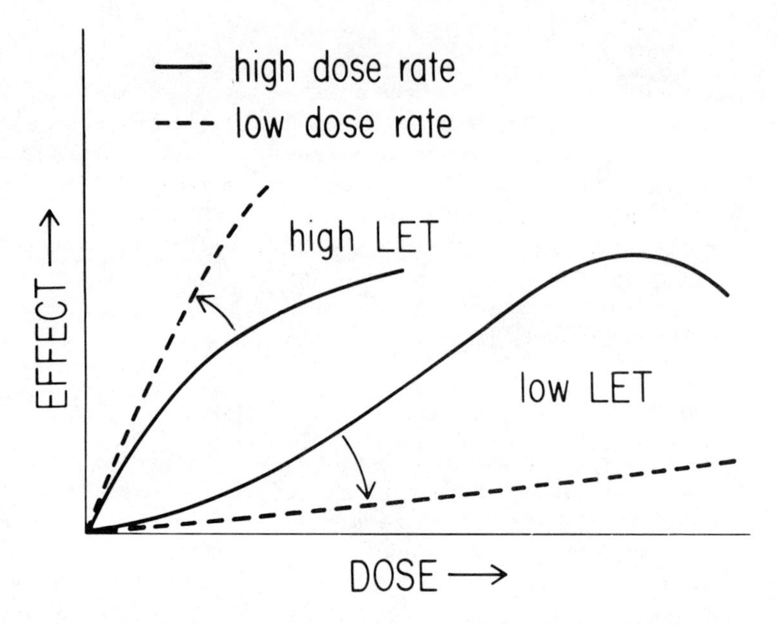

FIG. 8. Schematic dose-incidence curves for carcinogenesis in relation to dose and dose rate of high-LET and low-LET radiation. From Upton (68).

OTHER CHEMICAL AND PHYSICAL AGENTS

Cigarettes

In uranium miners, the magnitude of the increase in the incidence of lung cancer implies that the carcinogenic effects of radiation and cigarette smoking are multiplicative (75). In A-bomb survivors, by contrast, the combined effects of radiation and cigarette smoking appear to be additive rather than multiplicative (12). Since uranium miners are exposed chronically to alpha radiation, whereas the A-bomb survivors were exposed instantaneously to gamma radiation, the two sets of data are not necessarily contradictory. Apart from the promoting effect that cigarette smoking exerts on late stages of carcinogenesis in the lung, it is likely also to affect the dosimetry of inhaled radioactivity through effects on the morphology and function of the respiratory mucosa.

The findings in uranium miners are paralleled by data showing promoting effects of cigarette smoke on radon-induced pulmonary carcinogenesis in rats (18,19).

Ultraviolet Radiation

X-ray therapy to the scalp for tinea capitis in childhood has been observed to cause more basal cell carcinomas of the skin in areas that are exposed to ultraviolet radiation (e.g., ears and forehead) than in areas that are shielded from sunlight by hair (Fig. 9). The dose-incidence data imply that the two types of radiation interact synergistically (23).

Hormones

Experiments in laboratory animals have demonstrated repeatedly that appropriate hormonal stimulation can promote radiation carcinogenesis in endocrine glands and their target organs (21). Depending on the circumstances, synergistic interactions between radiation and hormones are elicited (27).

Chemical Carcinogens

The combined effects of radiation and carcinogenic chemicals have been observed to be additive, synergistic, or antagonistic, depending on the chemicals, doses, sequence of agents, and types of neoplasms in question. Noteworthy examples of additive or synergistic interactions include promoting effects of urethane on the induction of thymic lymphomas in x-irradiated mice (9,10,72), synergistic effects of prenatal exposure to x-rays and ethylnitrosourea on leukemogenesis in mice (52), synergistic effects of beta radiation and 4-nitroquinoline-1-oxide on skin carcinogenesis in mice (28), synergistic effects of x-rays and procarbazine on pulmonary tumorigenesis in mice (3), synergistic effects of benzo[a]pyrene and inhaled plutonium oxide on pulmonary carcinogenesis in rats (44), and additive effects of 3-methylcholanthrene and fission neutrons on the induction of mammary adenocarcinomas in rats (54).

BIOLOGICAL MODELS OF CARCINOGENESIS

Common to present theories of carcinogenesis is the concept that cancer arises as a monoclonal growth through a succession of cellular alterations, or stages (66). Evidence linking these alterations to changes in specific genes is rapidly emerging. Thus, studies of the molecular genetics of cancer cells imply that cancer can result from (i) the homozygous inactivation or deletion of certain genes (47); (ii) the aberrant activation of otherwise normal genes, as may occur through trisomy or chromosomal rearrangements (77); or (iii) the activation of point mutant alleles of otherwise normal genes (35). These types of genetic changes are induced with linear nonthreshold kinetics by irradiation in the low-to-intermediate dose range (62).

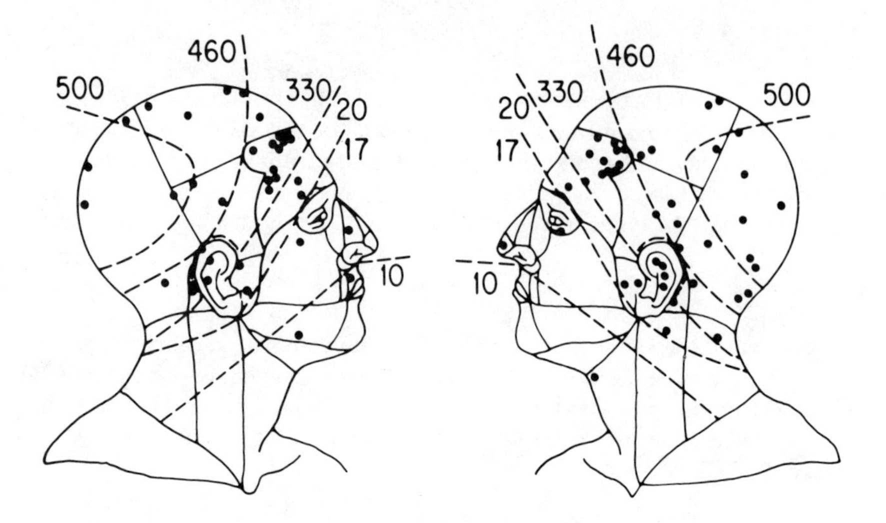

FIG. 9. Distribution of basal cell carcinomas of the skin in relation to x-ray dose (in rad) delivered to the scalp in childhood for treatment of tinea capitis. From Harley et al. (23).

The age-distribution of cancers in vivo (2) and the biology of malignant transformation in vitro (33, 39) imply that more than one alteration, genetic or otherwise, is necessary to cause cancer. From the relation between cancer incidence and age, which can be represented by the expression:

$$I = \delta t^{(n-1)} \tag{6}$$

where the cancer incidence I increases with age (or time after irradiation) t is a power function, and n denotes the number of stages involved (four to six successive steps have generally been inferred to be necessary for carcinogenesis) (2). This range exceeds somewhat the number of activated oncogenes that have been estimated to be necessary for the in vitro transformation of immortalized cells (39); however, it may not be inconsistent with the number of steps that appear to be involved in the in vitro transformation of primary hamster embryo fibroblasts (5).

The powerful methods that are now available for analyzing carcinogenesis at the molecular level should make it possible soon to determine the number and identity of the genetic alterations (including activated oncogenes) that may be involved in radiation-induced neoplasia. Furthermore, if the cancers induced by radiation prove to be distinguishable at the molecular level from those induced by other agents, the problem of risk assessment will be greatly simplified. In one study reported to date, the thymic lymphomas induced in mice by whole-body x-irradiation expressed the K-*ras* oncogene, whereas those induced by 4-nitrosourea expressed the N-*ras* oncogene (22).

With carcinogenesis *in vivo*, the role of extracellular regulatory and homeostatic mechanisms must also be considered. Apart from factors affecting the rate of cell proliferation, which are important *in vitro* (20) as well as *in vivo*, there are additional mechanisms *in vivo* that remain to be fully characterized. For example, the rapid outgrowth of "preleukemic" cells when they are transplanted from an irradiated donor into a newly irradiated recipient (15) remains to be explained.

SUMMARY

Ionizing radiation is carcinogenic to many, if not most, tissues. Its carcinogenicity varies, however, depending on the tissue exposed, conditions of exposure, genetic background, sex, age of the exposed individual, and other factors. The neoplasms induced by radiation also vary in their types and in their times of onset, depending on the age and sex of the exposed individual. The long induction period for radiation carcinogenesis and the enhancing or inhibiting effects of other agents acting after irradiation imply that the induction of cancer is a multistage process, in keeping with experiments on radiation-induced cell transformation *in vitro*. The molecular nature of the steps involved in radiation carcinogenesis remains to be fully elucidated, but it is being rapidly explored through advances in somatic cell genetics and molecular biology. The resulting insights will significantly extend epidemiological data in future attempts to estimate the carcinogenic risks of low-level radiation.

ACKNOWLEDGMENT

The author is grateful to Ms. Maureen Freitag and Ms. Kammy Griffin for assistance in the preparation of the manuscript. Preparation of this report was supported in part by Public Health Service Grant ES00260 from the National Institute of Environmental Health Sciences and Grant CA13343 from the National Cancer Institute, U.S. Department of Health and Human Services, and and Grant SIG0009 from the American Cancer Society.

REFERENCES

1. Albert, R.E., and Altshuler, B. (1973): In: *Radionuclide Carcinogenesis*, edited by C.L. Sanders, R.H. Bresch, J.E. Ballon, and D.D. Mahlum, pp. 233-253. U.S. Atomic Energy Commission, Washington, DC.
2. Armitage, P., and Doll, R. (1954): *Br. J. Cancer*, 8:1-12.

3. Arseneau, J.C., Fowler, E., and Bakemeier, R.F. (1977): J. Nat. Cancer Inst., 59:423-425.
4. Barendsen, G.W. (1978): In: Late Biological Effects of Ionizing Radiation, pp. 263-275. International Atomic Energy Agency, Vienna.
5. Barrett, J.C., Crawford, B.D., and T'so, P.O.P. (1980): In: Mammalian Cell Transformation by Chemical Carcinogens, edited by N. Mishra, V.C. Dunkel, and M. Mehlman, pp. 467-500. Senate Press, New Jersey.
6. Baverstock, K.F., Papworth, D., and Venmaro, J. (1981): Lancet, 1:430-433.
7. Beebe, G.W. (1984): Health Phys., 46:745-762.
8. Beebe, G.W., Kato, H., and Land, C.E. (1978): Radiat. Res., 75:138-201.
9. Berenblum, I., and Trainin, N. (1960): Science, 132:40-41.
10. Berenblum, I., Chen, L., and Trainin, N. (1968): Israel J. Med. Sci., 4:1159-1163.
11. Bizzozero, O.J., Jr., Johnson, K.G., Cicocco, A., Kawasaki, S., and Toyoda, S. (1967): Ann. Intern. Med., 66:522-530.
12. Blot, W.J., Akiba, S., and Kato, H. (1984): In: Atomic Bomb Survivor Data: Utilization and Analysis, edited by R.L. Prentice and D.J. Thompson, pp. 235-248. SIAM Institute for Mathematics and Society, Philadelphia.
13. Boice, J.D., Jr., Land, C.E., Shore, R.E., Norman, J.E., and Tokunaga, M. (1979): Radiology, 131:589-597.
14. Boice, J.D., Jr., Rosenstein, M., and Trout, E.D. (1978): Radiat. Res., 73:373-390.
15. Boniver, J., Decieve, A., Lieberman, M., Honsik, C., Travis, M., and Kaplan, H.S. (1981): Cancer Res., 41:390-392.
16. Brown, J.M. (1977): Radiat. Res., 71:51-74.
17. Burns, F., Albert, R.E., and Heimbach, R.D. (1968): Radiat. Res., 36:225.
18. Chemeaud, J., Perraud, R., Chretien, J., Masse, R., and Lafuma, J. (1980): In: Pulmonary Toxicology of Respirable Particles, edited by C. Sanders, F.T. Cross, G.E. Dagle, and J.A. Mahaffey, pp. 551-558. Proceedings of the 19th Hanford Life Sciences Symposium, Richland, Washington.
19. Cross, F.T., Palmer, R.F., Filipy, R.E., Bush, R.H., and Stuart, B.O. (1978): Pacific Northwest Laboratory Report, PNL-2744.
20. Fernandez, A., Mondal, S., and Heidelberger, C. (1980): Proc. Nat. Acad. Sci., USA, 77:7272-7276
21. Furth, J. (1982): In: Cancer, A Comprehensive Treatise, Vol. 1 (2nd edition), edited by F.F. Becker, pp. 89-134. Plenum Press, New York.
22. Guerrero, I., Calzado, P., Mayer, A., and Pellicer, A. (1984): Proc. Nat. Acad. Sci., USA, 81:202-205.

23. Harley, N., Kolber, A., Shore, R.E., Albert, R.E., Altman, S., and Pasternack, B. (1983): In: Epidemiology Applied to Health Physics, Proceedings of the Health Physics Society Midyear Symposium, Albuquerque, New Mexico, pp. 125-142.
24. Heidelberger, C. (1959): In: Carcinogenesis, Mechanisms of Action, edited by G.E.W. Wolstenhome and M. O'Connor, p. 164. J. & A. Churchill, Ltd., London.
25. Hildreth, N., Shore, R., and Hempelmann, L. (1983): Lancet, 1:273.
26. Holford, R.M. (1975): Health Phys., 28:153-156.
27. Holtzman, S., Stone, J.P., and Shellabarger, C.J. (1981): J. Nat. Cancer Inst., 67:455-459.
28. Hoshino, H., and Tanooka, H. (1975): Cancer Res., 35:3663-3666.
29. Ichumaru, M., and Ishimaru, T. (1975): J. Radiat. Res., 16:89-96.
30. Jablon, S., and Bailar, J. (1980): Prev. Med., 8:219-226.
31. Kaplan, H.S., and Brown, M.B. (1952): J. Nat. Cancer Inst., 13:185-208.
32. Kaplan, H.S., Hirsch, B.B., and Brown, M.B. (1956): Cancer Res., 16:434-436.
33. Kennedy, A.R., Cairns, J., and Little, J.B. (1984): Nature, 307:85-86.
34. Knudson, A.G. (1981): In: Cancer: Achievements, Challenges, and Prospects for the 1980's, edited by J.H. Burchenal and H.F. Oettgen, pp. 381-396. Grune & Stratton, New York.
35. Kukumar, S., Notario, V., Martin-Zanca, D., and Barbacid, M. (1983): Nature 306:658-661.
36. Land, C. (1980): Science, 209:1197-1210.
37. Land, C.E., Boice, J.D., Jr., Shore, R.E., Norman, J.E., and Tokunaga, M. (1980): J. Nat. Cancer Inst., 65:353-376.
38. Land, C.E., and McGregor, D.H. (1979): J. Nat. Cancer Inst., 62:17-21.
39. Land, H., Parada, L.F., and Weinberg, R.A. (1983): Science, 222:771-778.
40. MacMahon, B. (1962): J. Nat. Cancer Inst., 28:1178-1191.
41. Maldague, P. (1969): In: Radiation-Induced Cancer, pp. 439-458. International Atomic Energy Agency, Vienna.
42. Marshall, J.M., and Groer, P.G. (1977): Radiat. Res., 71:149-192.
43. Metalli, P., Covelli, V., DiPaola, M., and Silini, G. (1974): Radiat. Res., 59:21.
44. Metivier, H., Wahrendorf, J., and Masse, R. (1984): Br. J. Cancer (in press).
45. Miller, R.W. (1977): In: Genetics of Human Cancer, edited by J.J. Mulvihill, R.W. Miller, and J.F. Fraumeni, Jr., pp. 1-14. Raven Press, New York.

46. Monson, R.R., and MacMahon, B. (1984): In: Radiation Carcinogenesis: Epidemiology and Biological Significance, edited by J.D. Boice, Jr. and J.F. Fraumeni, Jr., pp. 97-105. Raven Press, New York.
47. Murphree, A.L., and Benedict, W.F. (1984): Science, 223:1028-1033.
48. National Academy of Sciences, (1972): The Effects on Populations of Exposure to Low Levels of Ionizing Radiation. Advisory Committee on the Biological Effects of Ionizing Radiation (BEIR), Washington, DC.
49. National Academy of Sciences, (1980): The Effects on Populations of Exposure to Low Levels of Ionizing Radiation. Advisory Committee on the Biological Effects of Ionizing Radiation (BEIR), Washington, DC.
50. National Council on Radiation Protection and Measurements (1980): Washington, DC.
51. Paterson, M.C. (1978): In: Carcinogenesis: Initiation and Mechanisms of Action, edited by A.C. Griffin and C.R. Shaw, pp. 251-276. Raven Press, New York.
52. Schmal, W., and Kriegel, H. (1978): Zeitschrift fur Krebsforschung und Klinische Onkologie, 91:69-79.
53. Shellabarger, C.J., Bond, V.P., Cronkite, E.P., and Aponte, G.E. (1969): In: Radiation-Induced Cancer, pp. 161-172. International Atomic Energy Agency, Vienna.
54. Shellabarger, C.J., and Straub, R.F. (1972): J. Nat. Cancer Inst., 48:185-187.
55. Shore, R.E., Hemplemann, L., Kowaluk, E., Mansur, P., Pasternack, B., Albert, R., and Haughie, G. (1977): J. Nat. Cancer Inst., 59:813-822.
56. Shore, R.E., Woodward, E.D., and Hemplemann, L.H. (1984): In: Radiation Carcinogenesis: Epidemiology and Biological Significance, edited by J.D. Boice, Jr., and J.F. Fraumeni, Jr., pp. 131-138. Raven Press, New York.
57. Straume, T., and Dobson, R.L. (1981): Health Phys., 41:666-671.
58. Strong, L.C. (1977): In: Genetics of Human Cancer, edited by J.J. Mulvihill, R.W. Miller, and J.F. Fraumeni, Jr., pp. 404-414. Raven Press, New York
59. Thomson, J.F., Williamson, F.S., Grahn, D., and Ainsworth, E.J. (1981): Radiat. Res., 86:573-579.
60. Tokunaga, M., Land, C.E., Yamamoto, T., Asano, M., Tokuoka, S., Ezaki, H., Nishimori, I., and Fujikura, T. (1984): In: Radiation Carcinogenesis: Epidemiology and Biological Significance, edited by J.D. Boice, Jr., and J.F. Fraumeni, Jr., pp. 45-56. Raven Press, New York.
61. United Nations Scientific Committee on the Effects of Atomic Radiation (1972): Report to the General Assembly Official Records: 27th Session, Suppl. No. 25 (A/8725), United Nations, New York.

62. United Nations Scientific Committee on the Effects of Atomic Radiation (1977): Report to the General Assembly, United Nations, New York.
63. Upton, A.C. (1961): Cancer Res., 21:717-729.
64. Upton, A.C. (1968): In: Proceedings of the International Conference on Leukemia-Lymphoma, edited by C.J.D. Zarafonetis, pp. 55-70. Lea & Febiger, Philadelphia.
65. Upton, A.C. (1977): Radiat. Res., 71:51-74.
66. Upton, A.C. (1982): In: Cancer: Principles and Practice of Oncology, edited by V.T. DeVita, Jr., S. Hillman, and S.A. Rosenberg, pp. 33-58. J.B. Lippincott Company, Philadelphia.
67. Upton, A.C. (1984): In: Atomic Bomb Survivor Data: Utilization and Analysis, edited by R.L. Prentice and D.J. Thompson, pp. 280-289. SIAM Institute for Mathematics and Society, Philadelphia.
68. Upton, A.C. (1984): In: Radiation Carcinogenesis: Epidemiology and Biological Significance, edited by J.D. Boice, Jr., and J.F. Fraumeni, Jr., pp. 9-19. Raven Press, New York.
69. Upton, A.C., Kimball, A.W., Furth, J., Christenberry, K.W., and Benedict, W.H. (1960): Cancer Res., 20:1-62.
70. Upton, A.C., Randolph, M.L., and Conklin, J.W. (1970): Radiat. Res., 41:467-491.
71. Upton, A.C., Wolff, F.F., Furth, J., and Kimball, A.W. (1958): Cancer Res., 18:842-848.
72. Vesselinovitch, S.C., Simmons, E.L., Mihailovich, N., Lombard, L.S., and Rao, K.V.N. (1972): Cancer Res., 32:222-225.
73. Wakabayashi, T., Kato, H., Ideda, T., Schull, W.J. (1983): Radiat. Res., 93:112-146.
74. Waterhouse, J., Muir, C., Correa, P., and Powell, J., editors (1976): IARC Publications No. 15, Lyon.
75. Whittemore, A.S., and McMillan, A. (1983): J. Nat. Cancer Inst., 71:489-499.
76. Young, J.L., Percy, C.L., and Asire, A., editors (1981): National Cancer Institute Monograph 57, Washington, DC.
77. Yunis, J.J. (1983): Science, 221:227-236.

Carcinogenesis, Vol. 10, edited by E. Huberman
and S. H. Barr. Raven Press, New York © 1985.

Retinoblastoma Gene: A Human Cancer Recessive (Regulatory?) Susceptibility Gene

W. F. Benedict

Clayton Molecular Biology Program, Childrens Hospital of Los Angeles, Los Angeles, California 90027

Our understanding about the nature of the retinoblastoma (Rb) gene has increased significantly over the last few years. Evidence has continued to accumulate that the retinoblastoma gene is recessive (although susceptibility to the development of retinoblastoma is dominantly inherited), with the loss or inactivation of both Rb alleles being the primary mechanism responsible for the development of this tumor (10). Such a mechanism contrasts with those of gene alteration, amplification, or insertion--mechanisms that are responsible for the expression of known oncogenes.

LOCALIZATION OF THE RB GENE

In 1978 Yunis and Ramsay (15) showed that 13q14 was the usual chromosomal region deleted in the hereditary deletion form of retinoblastoma. This loss of genetic material including the Rb gene in such cases is therefore the first of the two events postulated by Knudson (6) to be necessary for the development of retinoblastoma. The enzyme, esterase D (EsD), whose activity is gene dose dependent, is also located in chromosomal region 13q14 (14). Thus, patients with the deletion form of retinoblastoma have 50% EsD activity in all of their nontumor cells (0% EsD activity from the deleted 13 chromosome and 100% EsD activity from the normal 13 chromosome).

The fact that EsD is also polymorphic made it possible to determine by studying EsD patterns of inheritance in familial retinoblastoma patients that the Rb gene in the nondeletion hereditary form of this tumor was likewise located in chromosomal region 13q14 (13). On the basis of these findings, we believe it is likely that there is one Rb susceptibility locus located in chromosomal region 13q14.

THE RECESSIVE NATURE OF THE RB GENE

While the locus of the Rb gene was being ascertained, evidence for the recessive nature of the Rb gene was being documented. A patient was identified with 50% EsD activity in her nontumor cells and an apparent submicroscopic deletion in one chromosome 13 including both the EsD and Rb loci (3).

In each of two distinct tumor clones from this same patient, a 13 chromosome was shown to be missing. Since no EsD activity was present in the tumor, the normal 13 chromosome must have been lost in the tumor cells. Thus, because neither retinoblastoma clone contained any genetic information at the Rb locus, we suggested that the loss of function from each of the two homologous alleles at the Rb locus could be the two genetic events that were sufficient for tumor development (3).

In parallel studies, we also determined that the loss of a chromosome 13 or a deletion including 13q14 was a nonrandom occurrence in both hereditary and nonhereditary forms of retinoblastoma (2). This finding strengthened our belief that the loss of the second Rb allele was most commonly the second event leading to tumor development. It also added support to our contention that the Rb gene normally had a "suppressor" or "regulatory" function (10).

The other major evidence for the recessive nature of the Rb gene came from the study by Cavenee et al. (4) who used restriction fragment-length polymorphisms present at various locations on chromosome 13 to determine if specific changes within chromosome 13 occurred in the tumor. Cavenee and his associates (4) discovered that a nondisjunctional loss of one 13 chromosome with duplication of the other 13 chromosome or a mitotic recombination resulting in homozygosity at 13q14 was a common feature in the majority of tumors. These studies have been expanded to include more than 20 tumors, and similar results were obtained (Cavenee, personal communication). Consequently, it is now becoming evident that the Rb gene is actually recessive at the tumor level although the susceptibility to retinoblastoma is dominantly inherited. This apparent contradiction can be resolved by realizing that in a given individual who inherits one Rb gene in all nontumor cells, there is an almost certainty of losing function in the normal Rb allele for at least one retinoblast, resulting in tumor development.

MECHANISMS LEADING TO HEMIZYGOSITY OR HOMOZYGOSITY OF THE RB GENE

Various mechanisms that could lead to hemizygosity or homozygosity at the Rb locus are shown in Fig. 1.

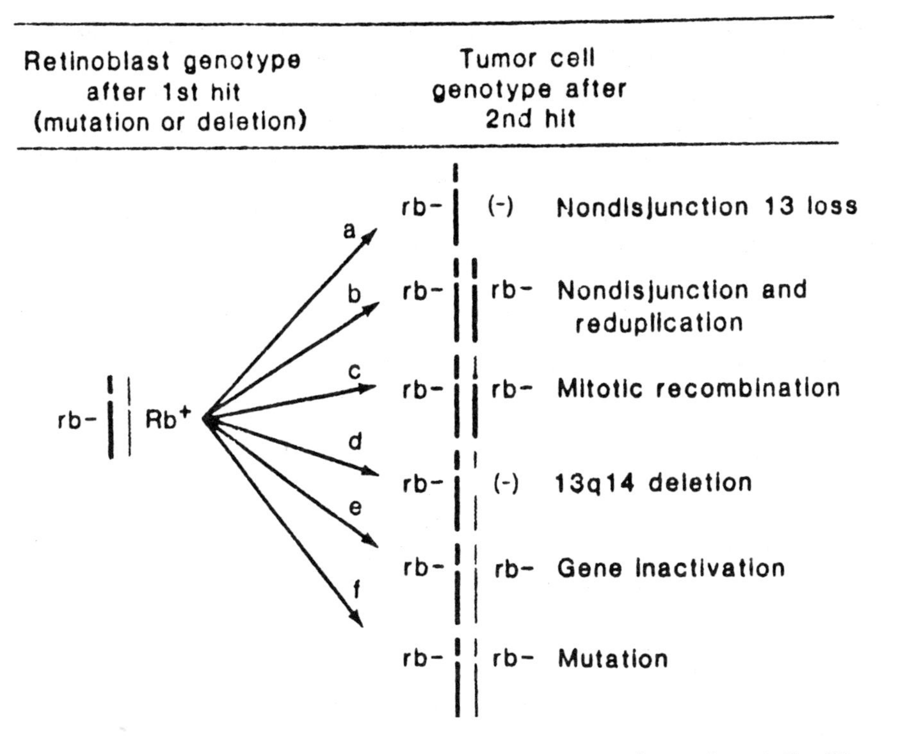

FIG. 1. Mechanisms producing homozygosity or hemizygosity at the Rb locus (13q14) in tumor cells. From Murphree and Benedict (10).

Depicted on the left in Fig. 1 is a retinoblast in which the first "hit" (either a mutation or deletion) has occurred in 13q14 resulting in the loss or inactivation of one Rb allele (rb-). Possible second "hits" or events yielding hemizygosity or homozygosity at the Rb locus are shown on the right. The first possibility (a) is loss of the unaffected 13 chromosome, similar to what we have seen as a nonrandom event in retinoblastoma (2). A loss of the normal 13 chromosome could also be followed by duplication of the affected 13 chromosome (b) resulting in homozygosity at the Rb locus (4).

Mitotic recombination (c) could likewise yield homozygosity at the Rb locus (4). Although this event has not been easily distinguished in mammalian cells, such a mechanism appears to be common in retinoblastomas (Cavenee, personal communication). A 13 chromosome deletion including 13q14 (d) has been seen in retinoblastomas from patients with both the hereditary and nonhereditary forms of the tumor (1,2). Gene inactivation of the Rb locus (d) resulting from the translocation of a 13 chromosome onto an inactive X chromosome also has occurred infrequently in tumors (Benedict et al., unpublished data). Finally, the second event could be a point or frameshift mutation of the remaining Rb allele (f). Point or frameshift mutations are currently not detectable since no probe to measure these mutations is available. However, because the chromosomal events shown as a to d in Fig. 1 appear to occur with high frequency in the tumor (1,2,4), a mutation may not be a common second event in tumor development.

ADDITIONAL ASPECTS OF THE RB AND OTHER RELATED GENES

Other tumors such as Wilms' tumor also appear to result from changes similar to those outlined for retinoblastoma (10). For example, in Wilms' tumor, homozygosity of 11p13 (the locus of the Wilms' tumor susceptibility gene) frequently occurs (5,8,11). Therefore, we feel that the Rb gene is only one of a class of cancer susceptibility genes that are recessive and normally function as "suppressor" or "regulatory" genes.

Individuals who have the hereditary forms of retinoblastoma also are at a very high risk of developing other malignant tumors, particularly osteosarcomas. More than 50% of these patients may have second primary tumors within 30 years after the initial diagnosis of retinoblastoma. Therefore, the Rb gene is an extremely potent cancer susceptibility gene. Studies are now being undertaken to determine if similar mechanisms outlined as second events in the induction of retinoblastomas (Fig. 1) also occur in the induction of second primary tumors found in patients with hereditary retinoblastoma.

ONCOGENES IN RETINOBLASTOMA

In our initial studies we found two retinoblastomas that contained double minute (DM) extrachromosomal fragments (2). Gene amplification of a putative oncogene, N-*myc* (named because of limited nucleotide sequence homology to the c-*myc* gene) has been found in the majority of another neural-crest-derived tumor, neuroblastoma, and the site of amplification was localized to DMs (7,12). Therefore, we looked for N-*myc* amplification in 10 retinoblastomas, including the two tumors containing the DMs. Indeed between 10- to 200-fold amplification of the N-*myc* gene was found in the two tumors with DMs, but not in the eight retinoblastomas without DMs (9).

Perhaps of even greater interest were the results of studies to determine if increased expression of N-*myc* could be found in those tumors in which no amplification of N-*myc* was noted. In all the eight retinoblastomas examined without evidence of N-*myc* amplification, high levels of N-*myc* expression at the RNA level were found (9). The importance of this increased expression in retinoblastoma formation or cell proliferation is currently unknown. However, it is possible that there has been a loss of regulatory control of N-myc expression in these tumors.

SUMMARY

The Rb susceptibility gene located in chromosomal region 13q14 apparently represents one of a class of recessive human cancer genes that can be contrasted to known oncogenes that function following amplification, aberrant insertion, or base substitutions at key sites within their gene. Loss of function of both Rb alleles appears to characterize the mechanism by which the Rb gene produces tumors, which may also include the second primary malignancies seen with high frequency in patients with the hereditary forms of retinoblastoma. Since the Rb gene may have a "suppressor" or "regulatory" function (which could include the regulation of an oncogene), cloning of such a human cancer susceptibility gene would seem highly worthwhile. Our laboratory, as well as several others, is therefore currently attempting to clone this gene.

REFERENCES

1. Balaban, G., Gilbert, F., Nichols, W., Meadows, A.T., and Shields, J. (1982): Cancer Genet. Cytogenet., 6:213-221.
2. Benedict, W.F., Banerjee, A., Mark, C., and Murphree, A.L. (1983): Cancer Genet. Cytogenet., 10:311-333.
3. Benedict, W.F., Murphree, A.L., Banerjee, A., Spina, C.A., Sparkes, M.C., and Sparkes, R.S. (1983): Science, 973-975.
4. Cavenee, W.K., Dryja, T.P., Phillips, R.A., Benedict, W.F., Godbout, R., Gallie, B.L., Murphree, A.L., Strong, L.C., and White, R.L. (1983): Nature, 305:779-784.
5. Fearon, E.R., Vogelstein, B., and Feinberg, A.P. (1984): Nature, 309:176-178.
6. Knudson, A.G. (1971): Proc. Nat. Acad. Sci., USA., 68:820-823.
7. Kohl, N.E., Kanda, N., Schreck, R.R., Bruns, G., Latt, S.A., Gilbert, F., and Alt, F.W. (1983): Cell, 35:359-367.
8. Koufos, A., Hansen, M.F., Lampkin, B.C., Workman, M.L., Copeland, N.G., Jenkins, N.A., and Cavenee, W.K. (1984): Nature, 309:170-172.
9. Lee, W.H., Murphree, A.L., and Benedict, W.F. (1984): Nature, 309:458-460.
10. Murphree, A.L., and Benedict, W.F. (1984): Science, 223:1028-1033.
11. Orkin, S.H., Goldman, D.S., and Sallan, S.E. (1984): Nature, 309:172-174.
12. Schwab, M., Alitalo, K., Klempnauer, K.H., Varmus, H.E., Bishop, M.J., Gilbert, F., Brodeur, G., Goldstein, M., and Trent, J. (1983): Nature, 305:245-248.

13. Sparkes, R.S., Murphree, A.L., Lingua, R.W., Sparkes, M.C., Field, L.L., Funderburk, S.J., and Benedict, W.F. (1983): Science, 219:971-973.
14. Sparkes, R.S., Sparkes, M.C., Wilson, M.G., Towner, J.W., Benedict, W.F., Murphree, A.L., and Yunis, J.J. (1980): Science, 208:1042-1044.
15. Yunis, J.J., and Ramsay, N. (1978): Am. J. Dis. Child., 132:161-163.

Carcinogenesis, Vol. 10, edited by E. Huberman and S. H. Barr. Raven Press, New York © 1985.

Chromosome Abnormalities in Human Leukemia as Indicators of Mutagenic Exposure

J. D. Rowley

The Franklin McLean Research Institute, The University of Chicago, Chicago, Illinois 60637

Nonrandom chromosome changes have been detected in a variety of human cancers. These changes have been studied in most detail in human acute nonlymphocytic leukemia (ANLL). Data from these studies can help to answer the question of whether there are certain clonal chromosomes changes in human acute leukemia cells that are related to particular etiologic agents. A preliminary analysis of data from three groups of patients suggests that the answer may be yes. These groups are (i) patients who have acute leukemia that has developed after therapy with cytotoxic drugs or radiation; (ii) leukemia patients who have been occupationally exposed to potentially mutagenic agents; and (iii) children who have ANLL. The data available for each of these groups will be reviewed in some detail and conclusions that appear tenable will be presented.

ACUTE LEUKEMIA SECONDARY TO CYTOTOXIC THERAPY

The occurrence of ANLL in patients who have previously received cytotoxic therapy (either radiation and/or chemotherapy) for a prior disease is being recognized with increasing frequency (3). It would appear reasonable to consider most, if not all, of these leukemias as therapy induced. Radiation is well recognized for its leukemogenic potential, and many of the drugs used to treat patients are alkylating agents, which are also known for their mutagenic activity.

We have recently completed an analysis of the clinical, morphologic, and cytogenetic data on 63 patients with 2° ANLL whom we studied during the last decade (6). Many of these patients have shown a preleukemic or myelodysplastic phase; the clonal chromosome abnormalities noted in this phase persist when the disease evolves into ANLL although new changes may be superimposed on the original karyotype.

Clinical Correlations

The type of primary disease, mode of primary therapy, and time to initial bone marrow dysfunction are shown in Table 1. Our earlier reports (10,11) describe primarily patients with some type of malignant lymphoma. As the treatment of cancer patients with chemotherapy has increased, the proportion of our patients with some type of carcinoma has also increased; this group now constitutes one-third of the total group of 63 patients. Our data indicate that our chemotherapy patients received prolonged chemotherapy; all had at least seven cycles of therapy with an alkylating agent and many received much more. Eleven of the patients had received only radiotherapy, many of them to the pelvis. Almost one-half of the patients received both types of therapy. Our patients who had radiotherapy only have consistently shown the shortest time from diagnosis to first bone marrow dysfunction: an average of 33 months compared with 56 months for the total group of 63 patients. Given the relatively small numbers in the various groups, this finding is not statistically significant.

TABLE 1. Primary disease, type of therapy, and time to bone marrow dysfunction.

Primary Disease	Number of Patients	Primary Therapy: Chemo-therapy	Primary Therapy: Radio-therapy	Both	Time to First Bone Marrow Dysfunction (months)
Hodgkin's disease	23	4	2	17	63.0
Non-Hodgkin's lymphoma	10	3	1	6	52.5
Other hematologic malignant diseases[a]	6	4	0	2	63.7
Solid tumors[b]	21	8	8	5	65.3
Renal transplants	3	2	0	1	101.0
TOTAL	63	21	11	31	---
Time to bone marrow dysfunction (mos)		50.2 (22-192)	33.0 (10-183)	61.5 (20-132)	56.0 overall

[a]Hairy cell leukemia 1; multiple myeloma 5

[b]Lung carcinoma 2, breast 4, colorectal 4, cervical/endometrial 4, ovarian 2, prostate 2, anus 1, glioblastoma 1, larynx 1

The cytogenetic data in this large series confirm our earlier analysis. First, 61 of the 63 patients (97%) had abnormal karyotypes. These data can be compared with those from our 148 patients (140 of whom could be studied successfully) with ANLL de novo whose chromosomes were analyzed during the same period of time; 53% of these patients had abnormal karyotypes (5). Thus, virtually every patient with secondary ANLL had an abnormal clone, whereas only 53% of patients with ANLL de novo exhibited abnormal karyotypes. We also noted that abnormalities of chromosome Nos. 5 and 7 were unusually frequent in our earlier series. One or both of these chromosomes were abnormal in 55 of the 61 patients with a clonal abnormality; in other words, 90% of those who had a clonal abnormality had a chromosome abnormality, or 87% of all patients with secondary ANLL showed a chromosome abnormality. In contrast, 21 of the 140 patients with ANLL de novo (or 15% of all patients) had an aberration of one or both of these chromosomes.

The chromosome abnormalities that we observed were a loss of the entire chromosome or of part of the long arm of either or both Nos. 5 and 7. We have correlated the type of chromosome aberration with both the primary disease and the primary therapy (Table 2). The data were examined for statistically significant correlations and the following correlations were significant at $p = .05$ or better. Of the 17 patients with abnormalities of both Nos. 5 and 7, 15 had hematology malignant diseases

TABLE 2. Primary disease/therapy and chromosome abnormality.

Primary Disease or Therapy	Chromosome Abnormality (n-61)			
	Abnormality of No. 5[a]	Abnormality of No. 7	Abnormality of No. 5 and 7[a]	Other Abnormality, Not 5 or 7
Disease				
Lymphomas	4	11	14 (7)	3
Other hematologic malignant diseases	2 (1)	3	1	0
Solid tumors	8 (7)	8	2 (2)	3
Renal transplant	0	2	0	0
Therapy				
Chemotherapy only	6 (2)	8	4 (3)	2
Radiation only	5 (5)	2	2	2
Both modalities	3 (1)	14	11 (6)	2

[a]Number in parentheses designates patients with a deletion of No. 5.

as their primary neoplasm ($p = .05$). When only patients with deletions of chromosome No. 5 were considered, the rearrangement was observed more frequently in patients whose initial disease was a solid tumor than in patients with a hematologic primary disease. That is, of 10 patients with a carcinoma and an abnormality of No. 5, nine had a deletion of this chromosome and one patient had a -5, whereas only eight of 21 patients with a hematologic malignant disease had a deletion ($p = .02$).

In the lower section of Table 2, primary therapy subgroups are correlated with the nature of chromosomal abnormalities. Overall, there were no significant associations, although aberrations of chromosome No. 7 were seen somewhat more frequently in patients who had received chemotherapy (37 of 52 patients with an abnormality of No. 7 had chemotherapy with or without radiotherapy, $p = .06$).

Cytogenetic Results

When the specific abnormalities of Nos. 5 and 7 were examined, six patients had a loss of one chromosome No. 5, and eight others had a deletion of the long arm of this chromosome [del(5q)]. Chromosome No. 7 was missing from the cells of 20 patients, and three patients had a deletion of the long arm. An additional 17 patients had abnormalities of both of these chromosomes. In nine of these 17 patients, the aberration of No. 5 was a del(5q), and in four patients, it was loss of a 7q. Thus, 14 patients had a -5, 17 patients had a del(5q), 34 patients had a -7, and seven patients had a del(7q). Although abnormalities of Nos. 5 and 7 were frequently observed together, the occurrence of these anomalies may, in fact, be independent. That is, the number of patients in whom aberrations of both Nos. 5 and 7 were observed was similar to that expected to occur by chance, given the high frequency of patients with involvement of only one or the other of these chromosomes. The most common single abnormality was loss of one No. 7; this karyotype was observed in eight patients. On the other hand, of the 31 patients with abnormalities of No. 5, only one patient had an alteration of this chromosome [del(5q)] as the sole karyotypic change. In eight patients, it was possible to determine whether an abnormality of No. 5 or of No. 7 occurred first. In six of them, the abnormality of chromosome No. 7 occurred first. Clonal abnormalities not involving chromosomes Nos. 5 or 7 were noted in six patients. Two of these patients had the t(15;17)(q22;q21) and secondary acute promyelocytic leukemia, two had only a gain of No. 8, and two others had various other aberrations.

The frequencies of whole chromosome gain or loss and structural rearrangements were analyzed. A gain of chromosomes was infrequent (with the exception of chromosome No. 8, which was noted as a gain in 10 patients); a gain of chromosome Nos. 5 or 7 was never found; however, these two chromosomes were observed as the most frequent losses.

To determine whether the frequency and type of abnormalities observed in patients with therapy-related dysmyelopoietic syndrome (DMPS) and/or ANLL differed from those observed in our own series of patients with ANLL de novo, we used the Fisher's exact test to compare the number of anomalies of each chromosome in the two groups of patients. The frequencies of alterations of seven chromosomes, namely Nos. 1, 4, 5, 7, 12, 14, and 18, were found to differ significantly ($p = < .05$) among these groups of patients (Fig. 1). In each case, these chromosomes were involved more frequently in patients with therapy-related disease. As expected, the greatest difference was found for chromosome Nos. 5 and 7. Fifteen of 140 (11%) patients with de novo ANLL had abnormalities of No. 5 as compared to 31 of 63 (49%) patients with secondary ANLL. Similarly, an abnormality of No. 7 was found in only 12 (9%) de novo ANLL patients and in 41 (65%) previously treated patients. The nature of the abnormalities of Nos. 1, 4, 12, 14, and 18 were similar between the two groups of patients, with chromosome loss and translocations accounting for most of the aberrations.

Deletion of Chromosome No. 5 or No. 7 and Definition of the Critical Region

Seventeen patients were found to have a deletion of the long arm of chromosome No. 5. In each case, the deletion was interstitial rather than terminal. In this type of deletion, two chromosomal breaks occur in the long arm, and the intermediate segment is lost. The breakpoints and segments deleted are illustrated in Fig. 2. In 12 of 17 patients, the proximal breakpoint was in bands q11 to q12, and in most cases (12 patients), the distal breakpoint occurred in bands q33 or q34. The smallest region that was consistently deleted, the critical region, was q23 to q32.

We have had more problems defining the critical region for No. 7 because it is difficult to be certain of the precise breakpoints in three of the seven patients with a 7q-chromosome. Thus, all patients (with one possible exception) are missing 7q32; the possible exception may have an interstitial deletion consisting of 7q32 to 7q35 or a terminal deletion involving 7q34 to 7qter (Fig. 2). If the former interpretation is correct, then all seven patients are lacking band 7q32; if the latter is correct, six of seven are lacking 7q34-q35.

I believe that these observations of therapy- related ANLL provide very provocative insights into the relationship between the chromosome abnormalities in the leukemic cells of an individual patient and the mechanism of leukemogenesis. At present, these relationships are unproven hypotheses; fortunately, they can now be tested.

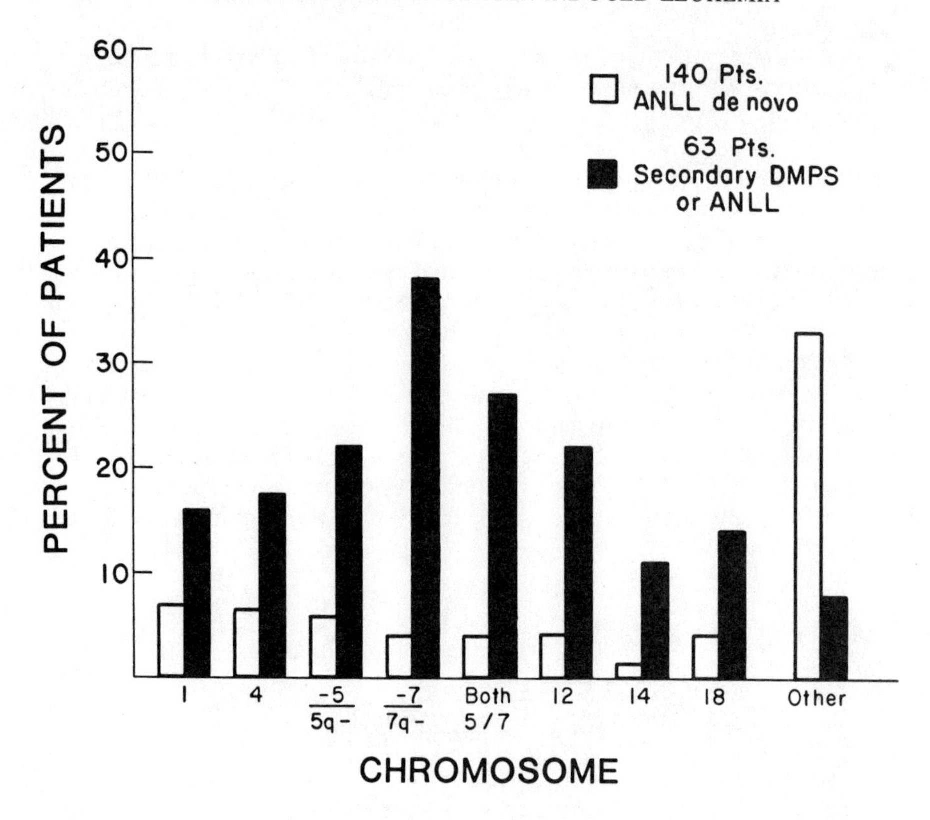

FIG. 1. Comparison of the frequency of involvement of specific chromosomes in patients with ANLL de novo and secondary DMPS or ANLL.

CORRELATION OF KARYOTYPE AND OCCUPATION IN ADULT PATIENTS WITH ACUTE LEUKEMIA

Two other groups of patients who have acute leukemia can also provide information that can be compared with our results with therapy-related ANLL. One study involves a correlation of the occupation of adult patients with ANLL with the karyotype of the leukemic cells. This correlation provides further evidence relating -5 or 5q- and-7 or 7q- to potentially mutagenic exposure. These studies were first reported by Mitelman and his colleagues (7) in Sweden. They divided occupations into those that would be considered as potentially exposing workers to mutagens and those that would not. We conducted a similar study at the University of Chicago (4); patients studied at the Fourth Workshop of Chromosomes and Leukemia were also analyzed by occupation (12). In each instance Dr. Mitelman classified the patient's occupation. Students,

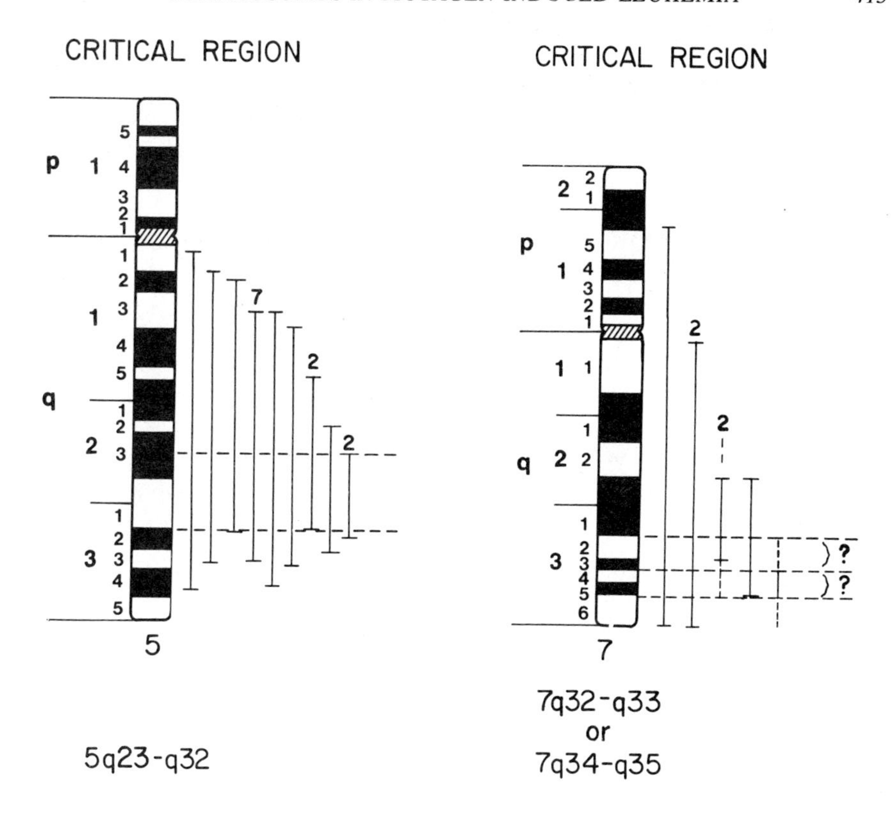

FIG. 2. Diagrammatic illustration of the portion of chromosome 5 (left) or chromosome 7 (right) that is deleted. Each vertical line represents the deletion in a single patient, except those that have a number at the top, which indicates the number of patients with that particular deletion. In chromosome 7, some uncertainty exists about the precise breakpoints, and these segments are indicated by dashed vertical lines. The region that is consistently missing in every patient is indicated by the dashed horizontal lines. This region is called the critical region and encompasses band q23 to q32 in chromosome 5. For chromosome 7, due to the uncertain breakpoints in several patients, the critical region could be 7q32-q33 or q34-q35.

white-collar workers, and housewives were classified as nonexposed, whereas individuals who worked with insecticides, chemicals and solvents, metals or minerals, petroleum products, and ionizing radiation were classified as exposed.

The findings from these three studies, summarized by Golomb et al. (4), showed that 68 of the 236 patients (29%) were classified as exposed.

The frequency of abnormal karyotypes was higher in the exposed than in the nonexposed group, 51 of 68 (75%) compared with 60 of 168 (36%). Perhaps of equal importance was the fact that aberrations of chromosomes 5 and 7 were observed in 37% of the exposed and in only 12% of the nonexposed patients. Relatively similar results were obtained at the Fourth Workshop (12); of 299 patients 30 years or older, 49% of non-exposed compared with 67% of those who were exposed had abnormal karyotypes. In this series, no difference was noted in the frequency of 5q- and/or 7q- between the exposed and nonexposed population. However, -5 and/or -7 was more common in the exposed group (9.7% vs. 6.1%); all of this difference was contributed by the very high frequency of these two aberrations in workers in the chemical industry because these abnormalities were not seen in petroleum or mineral industry workers. In the Workshop analysis, there was no difference in the frequency of clonal abnormalities in exposed and nonexposed patients between the ages of 16 and 29 years.

COMPARISON OF CHROMOSOME PATTERNS IN CHILDREN AND ADULTS WITH ANLL

One way to approach the question of the influence of occupational exposure to toxic agents on the chromosome pattern in leukemic cells is to examine the karyotypes seen in children. Although ANLL is less common in children than in adults, a sufficient number have been examined to provide data.

We and others have noted that the karyotypic pattern of children with ANLL differs from that of adults with the disease (1,9). The Fourth Workshop provides data on 660 patients with ANLL de novo that are classified by age and karyotype (12). Thus, of 102 patients less than 20 years of age, -5 occurred in none, 5q- in two, -7 in one, and 7q- in two. Among 214 patients 60 years old or older, -5 occurred in six, 5q- in 11, -7 in 10, and 7q- in one. Looked at another way, only 5% of patients under 20 years old compared with 13% of those over 60 have some aberration of chromosomes 5 or 7.

SUMMARY

It is from an analysis of these data that I proposed that losses of all or part of the long arm of chromosomes 5 and/or 7 may be indicators of exposure to mutagenic agents (8). Moreover, our analysis of the regions of chromosomes 5 and 7 that are consistently missing can define the critical region of each chromosome with some precision. Looked at from another perspective, the frequency of losses of chromosomes 5 and/or 7 may be an indicator of the proportion of ANLL in each particular population that is related to some mutagenic exposure. These aberrations are most frequent in cells of patients who were previously exposed to various cytotoxic regimens for treatment of a primary (usually malignant) disease; these patients are considered to have 2° ANLL. Among patients with ANLL de novo, abnormalities of chromosomes 5 and 7 are much more frequent in older than younger patients. Finally, among adult patients with ANLL de novo, -5 and -7 are more common in those whose occupations could potentially expose them to mutagenic agents such as chemicals, pesticides, or petroleum products. With regard to chromosome 5, most of these deletions are interstitial and always include 5q23 through q31, which I have called the critical region. Although I am less certain with regard to chromosome 7, it appears that the critical region that is consistently deleted may be 7q32 or 7q34-35.

FUTURE PROSPECTIVES

New models have been developed to explain deletions or somatic crossing-over in Wilms' tumor or retinoblastoma (2). If one applies these models to ANLL, the loss of part or all of Nos. 5 and/or 7 would result in homozygosity or hemizygosity for some, as yet unknown, essential genes located in the critical region of chromosome Nos. 5 or 7. The new techniques for recombinant DNA now make it possible to isolate the critical regions of 5 and 7 and then to define the nature of the gene or genes that are involved in the transformation to leukemia. This is clearly not an easy task, but the techniques are available that make it possible.

If I am correct that these chromosome regions contain genes that are especially vulnerable to mutagenic exposure, for whatever reason, then this finding has very important implications for public health policy, as well as for medical management of patients with malignant disease. With regard to the apparent specificity of the two regions on chromosomes 5 and 7, there are no obvious genetic markers that have been detected that would provide insight into the increased propensity of this region to be lost in ANLL associated with mutagenic exposure. There are, however, a number of possibilities. It may be that a particular recessive allelic variant of the critical gene occurs in a certain proportion of the population. Loss of the homologous normal allele due to a deletion of the chromosome carrying this normal allele or somatic crossing-over leading to duplication of the recessive variant allele could lead to its expression. In either case, one step in the multistage process leading to leukemic transformation would be accomplished. Alternatively, because many of the patients with 2° ANLL have had a series of exposures to mutagens often over a number of years, they may have normal alleles at this locus initially, one of which becomes a mutant "leukemogenic" allele as the result of therapy. As before, subsequent loss of the normal allele would result in leukemia. If the former possibility is correct, at least for a fair proportion of the population at risk, and if the variant allele can be distinguished from the normal one on the basis of some restriction enzyme polymorphism, then it would be possible to identify some individuals who are at increased risk of leukemia by screening for this DNA polymorphism. In addition to patients who would be treated for a malignant disease, individuals whose occupation exposes them to potential mutagens could be screened. Thus, pursuit of the biological significance of these recurring chromosome changes can potentially lead to development of major genetic markers for a preventable type of leukemia.

REFERENCES

1. Benedict, W.F., Lange, M., Greene, J., Derensceny A., Alfi, O.S. (1979): Blood, 54:818-823.
2. Cavenee, W.K., Dryja, T., Phillips, R.S., Benedict, W.F., Godbout, R., Gallie, B.L., Murphree, A.L., Strong, L.C., White, R.L. (1983): Nature, 305:779-784.
3. Coleman, C.N., William, G.J., Flint, A., Glatstein, E.J., Rosenberg, S.A., Kaplan, H.S. (1977): N. Engl. J. Med., 297:1249-1252.
4. Golomb, H.M., Alimena, G., Rowley, J.D., Vardiman, J.W., Testa, J.R., Sovik, C. (1982): Blood, 60:404-411.
5. Larson, R.A., Le Beau, M.M., Vardiman, J.W., Testa, J.R., Golomb, H.M., and Rowley, J.D. (1983): Cancer Genet. Cytogenet., 10:219-236.
6. Le Beau, M.M., Albain, K.S., Vardiman, J.W., Blough, R., Golomb, H.M., Rowley, J.D. (in preparation)
7. Mitelman, F., Nilsson, P.G., Brandt, L., Alimena, G., Gastaldi R., Dallapicola, B. (1981): Cancer Genet. Cytogenet., 4:187-214.
8. Rowley, J.D. (1983) In: Chromosomes and Cancer, edited by J.D. Rowley and J.E. Ultmann, pp. 140-159. Bristol-Myers Symposia Series, Vol. 5, Academic Press, New York.
9. Rowley, J.D., Alimena, G., Garson, O.M., Hagemeijer, A., Mitelman, F., Prigogina, E.L. (1982): Blood, 59:1013-1022.
10. Rowley, J.D., Golomb, H.M., Vardiman, J.W. (1977): Blood, 50:759-770.
11. Rowley, J.D., Golomb, H.M., Vardiman, J.W. (1981): Blood, 58:759-767.
12. The Fourth International Workshop on Chromosomes in Leukemia. (1984): Cancer Genet. Cytogenet., 11:251-360.

Carcinogenesis, Vol. 10, edited by E. Huberman and S. H. Barr. Raven Press, New York © 1985.

Relationship of Chromosomal Alterations to Gene Expression in Carcinogenesis

N. C. Popescu and J. A. DiPaolo

Laboratory of Biology, Division of Cancer Etiology, National Cancer Institute, Bethesda, Maryland 20205

TYPES OF CHROMOSOME CHANGES IN HUMAN CANCER

In the last two decades, various human cancers have been extensively characterized for their chromosome constitution. Even with the limitations of conventional staining, specific and consistent chromosome alterations have been identified in certain human leukemias and solid tumors (57). Cancer cytogenetics has benefited more than any other chromosome-related research area from the introduction of banding techniques. This technical development led to the identification of specific chromosome changes in a number of human cancers. Furthermore, it has become possible to determine the precise origin and formation of chromosome rearrangements. For example, the Philadelphia chromosome (Ph^1) in chronic myeloid leukemia (CML) involves not only chromosome 22, as was thought for a long time, but also other chromosomes, most commonly chromosome 9. Interchanges between these autosomes occur without loss of chromosome material. Similarly, a reciprocal translocation involving chromosomes 8 and 14 consistently occurs in Burkitt lymphoma (BL).

Other types of alterations can lead to substantial or minute losses of chromosomes or chromosome segments. In meningioma, the specific change consists of a loss of an entire chromosome 22. Deletions of large segments of chromosomes 1 and 3 are characteristic in neuroblastoma and small-cell carcinoma of the lung, respectively. In Wilm's tumor or retinoblastoma, deletions of the size of a single chromosome band occur in the short arm of chromosomes 11 and 13, respectively. Chromosome duplication or nonrandom nondisjunction are also frequent alterations in human cancer; however, these features appear to be associated with evolving clonal subpopulations in advanced stages of neoplastic development (45). An illustrative example is the appearance of additional chromosomes 8, 17, 19, and Ph^1 in CML entering the acute phase. In other malignancies, duplications of chromosome segments may be important in early stages of tumor development. For example the isochromosome 6p, i(6p) occurs with a high frequency in retinoblastoma (7); recently this abnormality was also identified in several cases of malignant melanoma (MM) (5). Homogeneously staining regions (HSR) and their cytological derivative, double minutes (DM), are separate alterations from structural or numerical abnormalities. Their uniformly stained appearance on banded chromosomes makes them distinguishable from the rest of the complement. The relevance of these alterations to acquisition of resistance to specific chemotherapeutic agents and neoplastic development is currently being defined.

HUMAN CELLULAR TRANSFORMING GENES AND CHROMOSOME CHANGES

Although consistent and specific chromosome alterations have been demonstrated in several forms of human cancer, the role of these changes in the induction and maintenance of malignancy remained elusive until the discovery of cellular transforming genes capable of inducing neoplastic development after transfection into nonmalignant cell recipients (9,15,70).

The chromosome localization of human proto-oncogenes is essential for studying the interaction of these proto-oncogenes with other genetic elements. More than 20 human proto-oncogenes have been localized on human chromosomes (55,56,72); four proto-oncogenes have been mapped (51) with a trypsin banding procedure through the emulsion in conjunction with *in situ* chromosome hybridization (52) (Fig. 1).

The precise localization led to the realization that, in certain human cancers, cellular proto-oncogenes coincide with the breakpoints involved in translocations or deletions (55); frequently these breakpoints occur at fragile chromosome sites (39). As a result of chromosome translocations, proto-oncogenes can be transposed to different chromosomes near active promoter or enhancer sequences and can cause alterations in gene expression. Conclusive evidence linking proto-oncogene expression with specific chromosome translocations has been obtained in BL and CML. The most common translocation in BL (8;14) involves the c-*myc* locus on chromosome 8 and the immunoglobulin heavy chain locus on chromosome 14 (16, 18,30,42). Two other translocations present in fewer numbers are the 8;22 and 2;8, which also involve the c-*myc* gene at either of the immunoglobulin light chain loci (29,31,68). In all cases, the translocated *myc* gene appears activated due to its insertion into an immunoglobulin locus. Transcription of the *myc* gene takes place from the translocated allele to the exclusion of the other allele on the normal, nonrearranged chromosome 8. The breakpoint on chromosome 8, while always at band 8q24, is found to be quite variable at the nucleotide level (1,68). This variability is also true for the break in the immunoglobulin locus. Activation of the *myc*

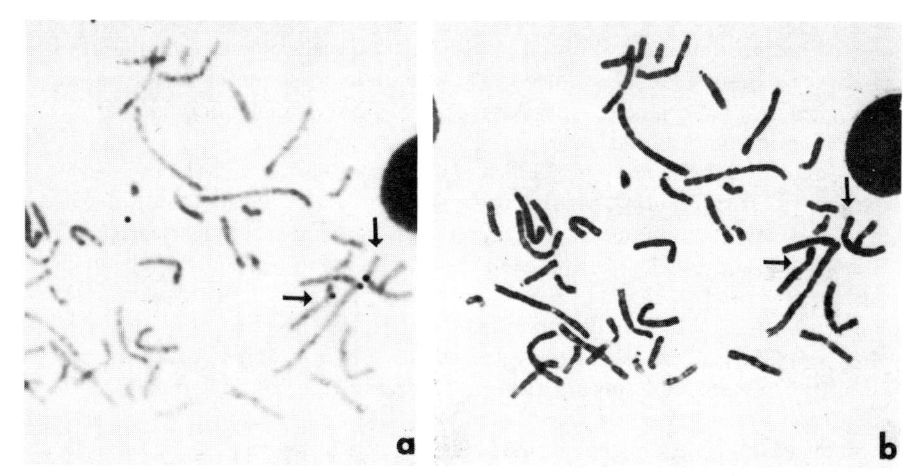

FIG. 1. (a) Silver grains after in situ hybridization of human chromosomes with a c-Ki-ras-2 probe, which also has homology to c-Ki-ras-1. (b) G-banding produced with trypsin permits the localization of grains to bands 6p11-12 and 12p11.1-12.1 (arrows) where c-Ki-ras-1 and c-Ki-ras-2, respectively, were assigned.

proto-oncogene is thought to involve either increased transcription or somatic mutation of the translocated myc allele (54). Reciprocal 9;22 translocation in CML is another case of a chromosome rearrangement responsible for changes in proto-oncogene expression. The human cellular homolog of the transforming sequence of Abelson murine leukemia virus (c-abl) located on the terminal segment of the long arm of chromosome 9 is translocated to chromosome 22 adjacent to the site of the immnunoglobulin light gene cluster (20). This c-abl transposition, which places a proto-oncogene in a transcriptionally active site, occurs only in Ph[1] positive and not in Ph[1] negative cases. The latter suggests that this alteration may be the event responsible for the initiation of the neoplastic state and that Ph[1] negative cases are a distinctive subclass of CML (4).

Recently a cellular homolog of the erbA gene of avian erythroblastosis has been mapped to human chromosome band 17q21-22 (19). The close proximity of the c-erbA proto-oncogene and the breakpoints of translocations involving chromosome 17 in acute promyelocytic leukemia and poorly differentiated acute leukemia indicates that activation of this proto-oncogene may also be the result of chromosome translocation (19).

Another mechanism considered important for proto-oncogene activation is gene amplification associated with HSR and DM. C-myc gene was found highly amplified in a human promyelocytic cell line in HSR of chromosome 8 (14,17) and in a human colon carcinoma cell line in HSR and DM of an altered X chromosome (2). Similarly, in a mouse adrenocortical tumor the

part of the infected cell DNA and be responsible for the induction and maintenance of the transformed phenotype. Recently, a relatively rare form of leukemia was found to be caused by a retrovirus, the human T-cell leukemia/lymphoma virus (HTLV) (48). Although oncogenic viruses are suspected to cause some neoplasia, epidemiological data indicates that a number of environmental chemical or physical agents may be responsible for the majority of human cancers (26). Chemical or physical carcinogens also interact with the genetic material of the target cell and certain tumors induced by chemical carcinogens have activated oncogenes (3,34,64).

It is generally accepted that neoplastic cell transformation is a multistage process with at least two distinguishable stages. Initiation, an irreversible cell alteration invariably precedes promotion, a reversible event that occurs as a result of exposure to noncarcinogenic agents. Morphological cell transformation is an obligatory step in the process leading to neoplasia, and our laboratory has demonstrated transformation of embryo- or fetus-derived cells from rat, guinea pig, Syrian hamster, human, and established cell lines. Carcinogen-treated cultures of animal cells develop colonies characterized by a random crisscross orientation which, when isolated and grown, produce tumors upon injection into appropriate hosts (24). In contrast, cells from normal colonies with a regularly oriented cell arrangement fail to result in tumors. The development of tumors in vivo after injection of in vitro transformed cells is not the result of the host tissue proliferation since the in vivo tumors and the in vitro transformed cells exhibited identical chromosome constitutions (25).

A major objective in our laboratory is to identify chromosome and DNA alterations that regulate gene expression responsible for neoplastic transformation. This process involves studies with carcinogen-induced chromosome alterations in the initial stages of transformation and the identification of critical chromosome rearrangements of transformed cells. Sister chromatid exchanges (SCE) and chromosome aberrations are cellular responses that occur shortly after exposure of the cell to carcinogens. Their relevance to the initiation of neoplastic transformation has recently been reviewed (53). Experiments with hamster embryo cells (HEC) in which a variety of physical or chemical carcinogens (49) or split doses of carcinogens are used have demonstrated that DNA lesions involved in SCE formation may be essential to the induction of transformation (50). DNA changes involved in SCE or chromosome aberrations might also cause proto-oncogene alterations by a mutational event or by transposition of promoting or enhancing sequences near transforming genes.

As in human leukemias or solid tumors, rodent and human cells transformed in vitro or tumors induced in vivo with known carcinogens have an abnormal chromosome constitution with structural and numerical alterations (25,58). A fundamental question is whether specific changes are associated with the process of neoplastic cell transformation. At present, one must also consider the influence of chromosome alterations on proto-oncogene structure and function. A major obstacle is apparent for consideration of this issue. Most cellular transforming genes have been characterized and localized in human cells. The inoculation of human cells initiated in vitro by carcinogens into appropriate hosts usually results in the formation of nodules, which increase in mass by a limited amount and then frequently regress (22). In contrast, rodent cells can be readily transformed in vitro, evolving into highly tumorigenic cell lines (24). Yet, in rodents, with the exception of mouse and rat, few proto-oncogenes have been isolated and localized. Despite these limitations, a study of the cytogenetics of experimentally induced cancer is generating relevant information for human cancer. HEC transformed in vitro have been a source of tumor material for chromosome studies in several laboratories studying the mechanisms of neoplastic transformation. In the early 1970s, our laboratory analyzed a number of clones derived from chemically or virally transformed cells and concluded that chromosome alterations are not specific for the etiological agent (25). This lack of specificity has been confirmed in several systems, including mouse leukemia in which trisomy 15 is induced by a variety of chemical and physical carcinogens and oncogenic viruses (12). The results of other investigations with HEC have suggested that the mechanism involved in transformation has a chromosomal basis because the balance between chromosomes with genes for expression and supression of malignancy determines the behavior of the cell (8,71). Other investigators consider the absence of one chromosome 15 essential for transformation of HEC (47). Although monosomy 15 was observed in some of our cell lines, transformation occurred without this abnormality.

Recently neoplastic transformation of HEC was demonstrated with sodium bisulfite, a nonmutagen at neutral pH (23). Inoculation of 5×10^6 cells isolated from morphologically altered colonies produced colonies in agarose within 10 days and progressively growing fibrosarcomas in nude mice. Five neoplastic cell lines are chromosomally abnormal with structural and numerical alterations and an increased frequency of polyploid cells. Some cell lines had extensive chromosome alterations with as many as six marker chromosomes while others were less heterogeneous (Fig. 2). None of these lines has monosomy 15. The abnormal chromosome constitution of these transformed cell lines has a special significance because sodium bisulfite at concentrations effective in producing transformation did not cause detectable DNA damage [as indicated by excision or postreplication repair analysis (27)], induced a minimal increase in SCE, and caused a low frequency of chromosome aberrations at lethal concentrations. Therefore, stable chromosome alterations in transformed clones may be considered independent of the initial insult with bisulfite.

Our extensive data on HEC conclusively demonstrates a lack of specificity of both structural and numerical chromosome alterations in transformation with chemical carcinogens or oncogenic viruses. Other investi-

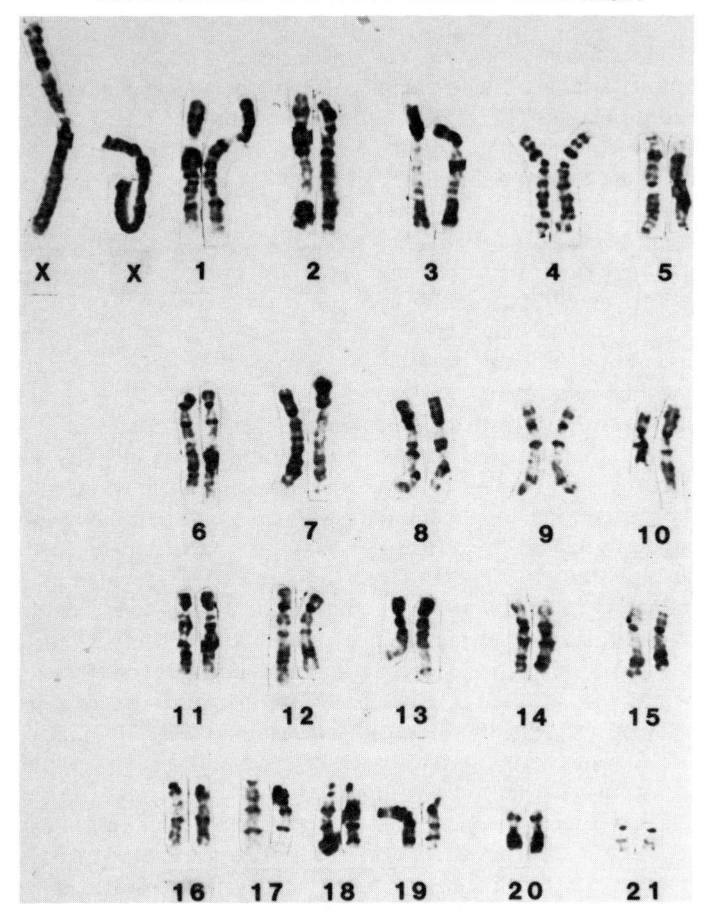

FIG. 2. Pro-metaphase G-band karyotype from a HEC line transformed with sodium bisulfite. An interstial deletion of one chromosome 1 and a deletion on the long arm of chromosome 10 are the only abnormalities. observed.

gators studying the same species also concluded that "cytogenetic changes in transformed Syrian hamster cells, especially those induced by chemical carcinogens exhibit considerable heterogeneity" (40). In contrast, chromosome changes in certain mouse and rat tumors are specific and, as in human leukemia or lymphomas, involve similar proto-oncogene transposition. Chromosome changes associated with murine plasmacytomas (MP) and leukemia (ML) have been investigated (11). Both BL and MP are B-cell-derived malignancies with similar types of chromosome alterations. In MP, chromosome 15, which carries c-myc, is involved in a reciprocal translocation with chromosomes 6 and 12 at the loci of K and the heavy chain immunoglobulin genes respectively (35,46,66). The translocated c-myc appears to be activated by its insertion into an immunoglobulin locus and the regulatory control of the immunoglobulin gene. Trisomy 15 is a specific, and frequently the only, abnormality in spontaneous leukemia or leukemia induced by a variety of chemical or physical carcinogens and oncogenic viruses (12,37,58). Studies with the induction of leukemia in mouse with Robertsonian fusions between 15 and other autosomes (61) have shown that only the distal end of 15 at the site of the myc gene is critical for the development of leukemia (11).

Myc gene was also implicated in a third mammalian species. Rat ileocoecal immunocytomas exhibit specific translocations that contain the myc gene between chromosomes 6 and 7; translocations may cause rearrangement and amplification of the myc gene (11). The most notable alteration in rat neoplasia, however, is trisomy 2 or structural alterations of this chromosome. Once considered specific for 7, 12-dimethylbenz[a]-anthracene (DMBA), alterations involving chromosome 2 occur in sarcomas, carcinomas, and leukemias induced by a variety of carcinogens (58). Recent chromosome studies have shown that other alterations are potentially important in rat neoplasia. Rat mammary epithelial cells were transformed in vitro with DMBA or in combination with promoter 12-O-tetradecanoyl-phorbol-13-acetate (33). Cells from transformed cell lines, which produce carcinomas after inoculation into nude mice, had specific structural alterations of chromosomes 1 and 3 and trisomy of chromosomes 14 and X. Trisomy 2 occurred in one line at advanced passages. The abnormal chromosome originating from chromosome 3 was present in the majority of the cells in all cell lines examined (Fig. 3). Chromosome 3 carries rRNA genes as indicated by specific silver staining for nucleolar organizer regions (NOR). As a result of marker formation, NOR regions of chromosome 3 were eliminated and those from normal chromosome 3 were inactivated. Interestingly, other groups have reported activation of c-abl and c-H-ras proto-oncogenes associated with translocation to NOR of chromosomes 3, 11, and 12 in rat leukemia induced by DMBA (63). This fact coupled with the specific involvement of the chromosome segment carrying NOR on chromosome 3 in rat mammary carcinogenesis suggests the importance of this particular site.

Only a limited number of human fibroblast cell lines derived from normal cells transformed in vitro by chemical and physical carcinogens or oncogenic viruses have been obtained. Thus, conclusions concerning the specificity of chromosome changes are premature. Until recently, very few cases of sarcoma had been studied (43). Currently data exists for 29 cases (6). The majority of the tumors have a diploid or near diploid range with a distinctive monoclonal karyotype and a frequent involvement of chromosomes 1 and 2 (6). Chromosome 1 is also nonrandomly involved in carcinoma of the breast, bladder, cervix, and ovary (43,57,72). Changes of chromosome 1 are important because three proto-oncogenes, N-ras, B-lym, and c-fgr are located on its short arm. The possiblity that this chromosome also contains genes for supression of malignancy has been suggested (62).

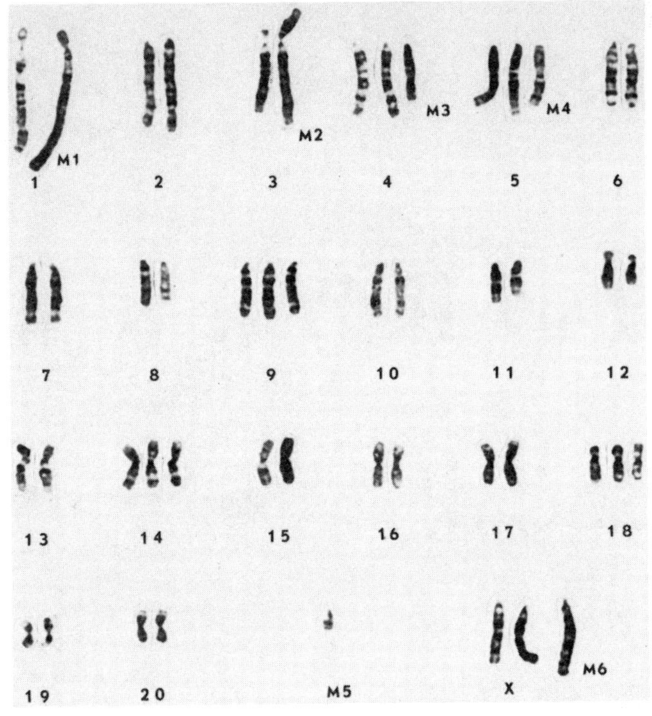

FIG. 3. Pro-metaphase G-band karyotype from a DMBA-transformed rat mammary epithelial cell line with several abnormal chromosomes and trisomies. M_2 (derived from chromosome 3) was present in all lines examined.

Although human cells transformed in vitro are considerably less heterogeneous in terms of chromosome alterations than human tumors derived from the same type of tissue, aneuploidy and other characteristic alterations occur as a result of carcinogen insult. These alterations include a changed cell morphology, an increased life-span, an acquired ability to grow in semisolid medium, and an ability to produce nodules in an appropriate host. A karyotype (unpublished results) from a cell line derived from foreskin fibroblasts exposed to ultraviolet (UV) radiation and isolated from agarose is shown (Fig. 4). Eighty percent of the cells have an abnormal chromosome 22 resulting from a duplication of a segment of the long arm at the locus of the c-sis gene. A foreskin-derived cell line initiated with aflatoxin B_1 has three markers, derived from chromosomes 1, 11, and X. Alterations of chromosome 11 and 22 are at the site of H-ras-1 and c-sis proto-oncogenes, respectively. These proto-oncogenes are related to or are in close proximity to sequences involved in the regulation of cell growth and development. C-sis has a close structural and functional relationship to platelet growth factor (28,36). The insulin-like growth factors IGF-I and IGF-II have recently been localized on chromosomes 11 and 12, respectively (10,69); IGF-II is located on the short arm

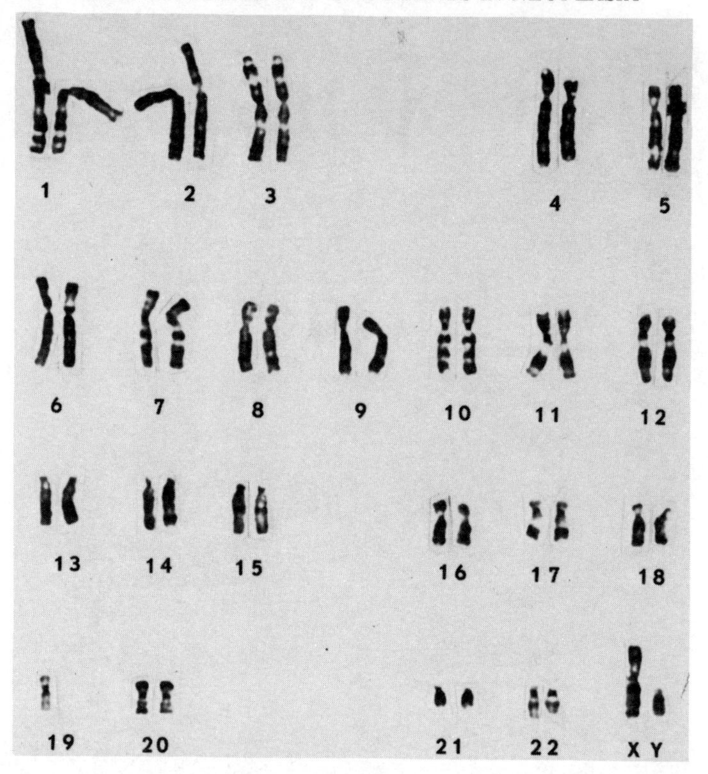

FIG. 4. G-band karyotype from a foreskin cell line initiated with UV. An abnormal 22 chromosome is the only structural abnormality; one chromosome 19 is missing in this particular cell.

of chromosome 11 with H-ras-1 and the insulin structural gene (69). Therefore, as in human cancer, chromosome alterations of experimentally induced neoplasia involve chromosomes carrying proto-oncogenes.

In the final analysis, mutation, chromosome translocation, and formation of HSR or DM are associated with structural alterations and expression of cellular transforming genes. The elucidation of their function is essential in understanding their role in carcinogenesis.

REFERENCES

1. Adams, J., Gerondakis, S., Webb, E., Corcoran, L.M., and Cory, S. (1983): Proc. Nat. Acad. Sci., USA, 80:1982-1986.
2. Alitalo, K., Schwab, M., Lin, C.C., Varmus, H., and Bishop, J.M. (1983): Proc. Nat. Acad. Sci., USA, 80:1707-1711.
3. Balmain, A., and Prangell, L.B. (1983): Nature, 303:72-74.

4. Bartram, C.R., de Klein, A., Hagemeijer, A., van Agthoven, T., van Kessel, A.G., Bootsma, D., Grosveld, G., Ferguson-Smith, M.A., Davies, T., Stone, M., Heisterkamp, N.M., Stephenson, J.R., and Groffen, J. (1983): Nature, 306:277-280.
5. Becher, R., Gibas, Z., Karakovsis, C., and Sandberg, A.A. (1983): Cancer Res., 43:5010-5016.
6. Becher, R., Wake, N., Gibas, Z., Ochi, H., and Sandberg, A.A. (1984): J. Nat. Cancer Inst., 72:823-831.
7. Benedict, W.F., Banerjee, A., Mark, C., and Murphree, A.L. (1983): Cancer Genet. Cytogenet., 10:311-333.
8. Benedict, W.F., Rucker, N., Mark, C., Kouri, R.E. (1975): J. Nat. Cancer Inst., 54:157-162.
9. Bishop, J.M. (1983): Annu. Rev. Biochem., 52:301-354.
10. Brissenden, J.E., Ullrich, A., and Francke, U. (1984): Nature, 310:781-784.
11. Caccia, N.C., Mak, T.W., and Klein, G. (1984): J. Cell Physiol., 3:199-208.
12. Chan, F.P., Ball, J.K., and Sergovich, F.R. (1979): J. Nat. Cancer Inst., 62:605-610.
13. Chang, E.H., Gonda, M.A., Ellis, R.W., Scolnick, E.M., and Lowy, D.R. (1982): Proc. Nat. Acad. Sci., USA, 79:4848-4852.
14. Collins, S., and Groudine, M. (1982): Nature, 298:679-681.
15. Cooper, G.M. (1982): Science, 217:801-806.
16. Dalla-Favera, R., Bregni, M., Erikson, J., Patterson, D., Gallo, R.C., and Croce, C.M. (1982): Proc. Nat. Acad. Sci., USA, 79:7824-7827.
17. Dalla-Favera, R., Wong-Stall, F., and Gallo, R.C. (1982): Nature, 299:61-63.
18. Davis, M., Malcolm, S., and Rabbits, T.H. (1984): Nature, 308:286-288.
19. Dayton, A.I., Selden, J.R., Laws, G., Dorney, D.J., Finan, J., Tripputi, P., Emanuel, B.S., Rovera, G., Nowell, P.C., and Croce, C.M. (1984): Proc. Nat. Acad. Sci., USA, 81:4495-4499.
20. DeKlein, A., Van-Kessel, A.G., Grosveld, G., Bartram, C.R., Hagemeijer, A., Bootsma, D., Spurr, N.K., Heisterkam, N., Groffen, J., and Stephenson, J. (1982): Nature, 300:765-767.
21. Der, C.J., Krontiris, T.G., and Cooper, G.M. (1982): Proc. Nat. Acad. Sci., USA, 79:3637-3640.
22. DiPaolo, J.A. (1983): J. Nat. Cancer Inst., 70:3-8.
23. DiPaolo, J.A., DeMarinis, A.J., and Doniger, J. (1981): Cancer Lett., 12:203-208.
24. DiPaolo, J.A., Nelson, R.L., and Donovan, P.J. (1969): Science, 165:917-918.
25. DiPaolo, J.A., and Popescu, N.C. (1976): Am. J. Pathol., 85:709-726.
26. Doll, R. (1978): Cancer Res., 38:3573-3583.
27. Doniger, J., O'Neill, R., and DiPaolo, J.A. (1982): Carcinogenesis, 3:27-32.

28. Doolittle, R.F., Hunkapiller, M.W., Hood, L.E., Devare, S.G., Robbins, K.C., Aaronson, S.T., and Antoniades, H.A. (1983): Science, 221:275-277.
29. Erikson, J., Ar-Rushdi, A., Drwinga, H.L., Nowell, P.C., and Croce, C.M. (1983): Proc. Nat. Acad. Sci., USA, 80:820-824.
30. Erikson, J., Finan, J., Nowell, P.C., and Croce, C.M. (1982): Proc. Nat. Acad. Sci., USA, 79:5611-5615.
31. Erikson, J., Martinis, J., and Croce, C.M. (1981): Nature, 294:173-175.
32. Gilbert, F. (1983): Nature, 303:475.
33. Greiner, J.W., Evans, C.H., and DiPaolo, J.A. (1983): Cancer Res., 43:273-278.
34. Guerrero, I., Calzada, P., Mayer, A., and Pellicer, A. (1984): Proc. Nat. Acad. Sci., USA, 81:202-205.
35. Hengartner, H., Meo, T., and Muller, E. (1978): Proc. Nat. Acad. Sci., USA, 75:4494-4498.
36. Josephs, S.F., Ratner, L., Clarke, M.F., Westin, E.H., Reitz, M.S., and Wong-Staal, F. (1984): Science, 225:636-639.
37. Klein, G. (1981): Nature, 294:313-318.
38. Kraus, M.H., Yuasa, Y., and Aaronson, S.A. (1984): Proc. Nat. Acad. Sci., USA, 17:5384-5389.
39. Lebeau, M.M., and Rowley, J.D. (1984): Nature, 308:607-608.
40. Li, S., and Pathak, S. (1983): Anticancer Res., 3:117-120.
41. Marshall, C.J., Vousden, K.H., and Phillips, D.H. (1984): Nature, 310:586-589.
42. Mitchell, K.F., Battey, J., Hollis, G.F., Moulding, C., Taub, R., and Leder, P. (1984): J. Cell Physiol. Supp., 3:171-177.
43. Mitelman, F., and Levan, G. (1981): Hereditas, 95:79-139.
44. Müller, R., and Verma, I.M. (1984): Curr. Top. Microbiol. Immunol., 112:73-115.
45. Nowell, P.C. (1976): Science, 194:23-28.
46. Ohno, S., Babonitis, M., Wiener, F., Soira, J., and Klein, G. (1979): Cell, 1001-1007.
47. Pathak, S., Hsu, T.C., Trentin, J.J., Butel, J.S., and Panigrahy, B. (1981): In: Genes, Chromosomes, and Neoplasia, edited by F.E. Arrighi, P.N. Rao, and E. Stubblefield, pp. 405-418. Raven Press, New York.
48. Poiesz, B.J., Ruscetti, R.W., Gazdar, A.F., Bunn P.A., Minna, J.D., and Gallo, R.C. (1980): Proc. Nat. Acad. Sci., USA, 77:7415-7419.
49. Popescu, N.C., Amsbaugh, S.C., and DiPaolo, J.A. (1981): Int. J. Cancer, 28:71-77.
50. Popescu, N.C., Amsbaugh, S.C., and DiPaolo, J.A. (1984): Cancer Res., 44:1933-1938.
51. Popescu, N.C., Amsbaugh, S.C., DiPaolo, J.A., Tronick, S.R., Aaronson, S.A., and Swan, D.C. (1984): Somat. Cell Mol. Genet., in press.

52. Popescu, N.C., Amsbaugh, S.C., Swan, D.C., and DiPaolo, J.A. (1984): Cytogent. Cell Genet., in press.
53. Popescu, N.C., and DiPaolo, J.A. (1982): In: Progress and Topics in Cytogenetics, Vol. 2, edited by A.A. Sandberg, pp. 425-460. Alan R. Liss, Inc., New York.
54. Rabbits, T.H., Forster, A., Hamlyn, P., and Baer, R. (1984): Nature, 309:592-597.
55. Rowley, J.D. (1979): Nature, 301:290-291.
56. Rowley, J.D. (1984): Cancer Res., 3159-3169,
57. Sandberg, A.A. (1980): The Chromosomes in Human Cancer and Leukemia. Elsevier Press, Amsterdam.
58. Sasaki, M. (1982): Cancer Genet. Cytogenet., 5:153-172.
59. Schwab, M., Alitalo, K., Varmus, H.E., and Bishop, J.M. (1983): Nature, 303:497-501.
60. Shimizu, K., Goldfarb, M., Perucho, M., and Wigler, M. (1983): Proc. Nat. Acad. Sci., USA, 80:383-387.
61. Spira, J., Wiener, F., Ohno, S., and Klein, G. (1979): Proc. Nat. Acad. Sci., USA, 76:6619-6621.
62. Stoler, A., Beaird, J., and Bouck, N. (1984): Proc. Am. Assoc. Cancer Res., 25:70.
63. Sugiyama, T., Takhaski, R., Mihara, H., Yamaguchi, T., Chen, H.L., and Maeda, S. (1984): Current Topics in Cancer Cell Biology (Third Int. Congress Cell Biol.), edited by S. Seno and Y. Okada, pg. 40. Academic Press, Japan.
64. Sukumar, S., Notario, V., Martin-Zanca, D., and Barbacid, M. (1983): Nature, 306:658-661.
65. Sukumar, S., Pulciani, S., Doniger, J., DiPaolo, J.A., Evans, C.H., Zbar, B., and Barbacid, M. (1984): Science, 223:1197-1199.
66. Swan, D., D'Eustachio, F., Leinwand, J., Seidman, D., Keithley, D., and Ruddle, F.H. (1979): Proc. Nat. Acad. Sci., USA, 76:2735-2739.
67. Tainsky, M.A., Cooper, C.S., Giovanella, B.C., and Vande Woude, G.F. (1984): Science, 225:643-645.
68. Taub, R., Kirsch, L., Morton, C., Lenoir, C., Swan, D., Tronick, S., Aaronson, S., and Leder, P. (1982): Proc. Nat. Acad. Sci., USA, 79:7837-7841.
69. Tricoli, J.V., Rall, L.B., Scott, J., Bell, G.I., and Shows,T.B. (1984): Nature, 310:784-786.
70. Weinberg, R.A. (1982): Cell, 30:3-4.
71. Yamamoto, T., Rabinowitz, Z., and Sachs, L. (1973): Nature, 243:247-250.
72. Yunis, J.J. (1983): Science, 221:227-236.

Carcinogenesis, Vol. 10, edited by E. Huberman and S. H. Barr. Raven Press, New York © 1985

Mammalian Cell Mutation and Polycyclic Hydrocarbon Carcinogenesis

P. Brookes, H. W. S. King, and M. R. Osborne

Chemical Carcinogenesis Section, Institute of Cancer Research, London SW3 6JB, United Kingdom

The present publication, dedicated to the memory of the life and work of Charles Heidelberger, seems an appropriate place in which to review some current ideas on the role in carcinogenesis of polycyclic hydrocarbon-DNA reactions and mutation. It was a mutual interest in this area of research that led one of the present authors (P.B.) to first meet and subsequently to have the privilege of working with and becoming a close friend of Charles Heidelberger. In retrospect it is interesting to reflect that this happy situation would not have developed had our views on the significance of hydrocarbon-DNA reaction been fully shared by the McArdle group in the early 1960s. It was the presentation of contrary views on this subject at the 1964 Gatlinburg Symposium (8,19) that led to the invitation of one of us (P.B.) to repeat some experiments in Madison and subsequently to spend a year (1965-66) working in McArdle. It was during this very happy and productive period that full appreciation of Charles Heidelberger as both scientist and human being developed and continued until his untimely death in January 1983.

BACKGROUND

The study by Lawley and Brookes (28,29) of the mechanism of action of antitumor alkylating agents focused our attention on the dramatic biological consequences resulting from very low levels of DNA damage. Therefore, when contemplating the spectrum of cellular changes involved in malignant transformation induced by the wide range of chemical carcinogens having no common structural features, we hypothesized that the unifying factor might be their ability to react with DNA.

Polycyclic hydrocarbons were selected to test this hypothesis because they were classical carcinogens lacking chemical reactivity, for which the mutagenicity data was either negative or controversial. To our delight and surprise, evidence of covalent binding of a series of carcinogenic hydrocarbons to the DNA of mouse skin was found after the application of carcinogenic doses (7).

During the subsequent decade, many laboratories reported on a number of exotic ways by which many carcinogens could be metabolized to yield electrophilic derivatives having the ability to bind covalently to DNA (30,31). Rather surprisingly in the case of the hydrocarbons, the search for the ultimate carcinogen proved to be long and tedious and was probably delayed by enthusiasm for the K-region as the site for hydrocarbon activation. The concept of the K-region epoxide as the carcinogenic metabolite was certainly attractive since such derivatives were formed as microsomal metabolites (17) and were capable of reacting with DNA (2). However, we were opposed to their primary importance in carcinogenesis, at least by benzo[a]pyrene (BAP), since chromatographic separation of hydrocarbon-nucleoside products, following enzymic degradation of DNA with BAP bound *in vivo*, gave a profile (Fig. 1) quite distinct from that obtained by reaction with DNA of the K-region epoxide (2,26). The fractionation of the products, essential for this analysis, was achieved with methanol-water gradient elution on LH-20 Sephadex (1). This simple technical innovation proved to be extremely valuable and overcame a problem that had frustrated Brookes and Heidelberger (5).

The breakthrough in identifying the ultimate carcinogen from BAP came when Peter Sims and his colleagues (41) isolated the 7,8-diol-9,10-epoxide of BAP (BPDE), the *anti*-isomer of which was shown to be the

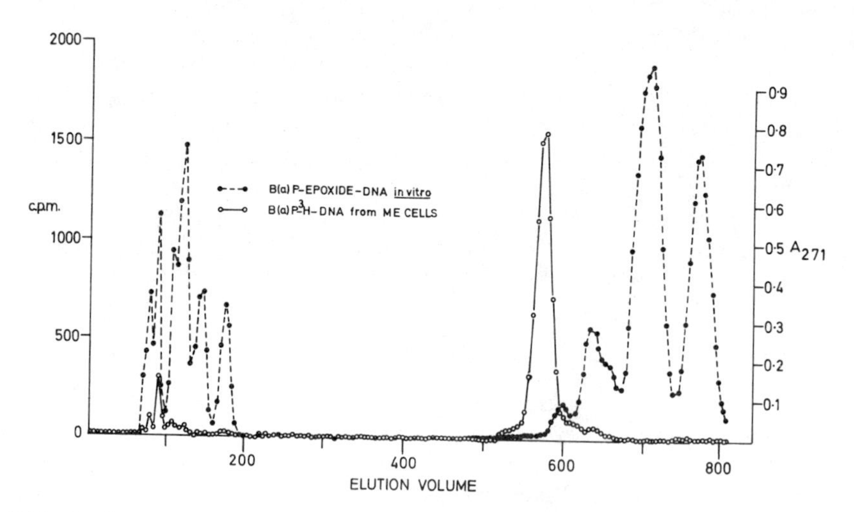

FIG. 1. Sephadex LH-20 column fractionation of an enzyme digest of DNA reacted with BAP-K-epoxide (UV_{271}, ---●---) and DNA from mouse embryo cells exposed to ^{3}H-BAP (^{3}H activity, ---O---. [B(a)P = BAP] See Baird et al. (2) for full details.

derivative responsible for the major product of in vivo DNA binding of BAP (35).

The importance of diol epoxides and the complexities resulting from the possibility of both geometric (anti- and syn-) and stereo-isomerism (+ and -) inspired organic chemists to devise ingenious schemes for the synthesis of these derivatives (Fig. 2) of BAP (18,48) and later other hydrocarbons (12,22,40).

The availability of this range of diol epoxides provided a wonderful opportunity for structure-activity studies in relation to DNA products (36,37), mutagenesis in bacterial (46,47) and mammalian cells (4,9,20,32), and, of course, tumorigenesis (10,12,23,42). Virtually all the early studies involved only BPDE, but information on other hydrocarbon diol epoxides is becoming available (12).

To summarize the wealth of data on BPDE would prove difficult, but in relation to the present topic the following points seem worth considering:

1. The major site of reaction with DNA of all four isomers of BPDE is the 2-amino group of guanine, and for the (+) anti-isomer, this reaction produces essentially the only product. Minor products of reaction at the N-7 and O-6 positions of guanine are found mainly with the (-) anti-isomer.

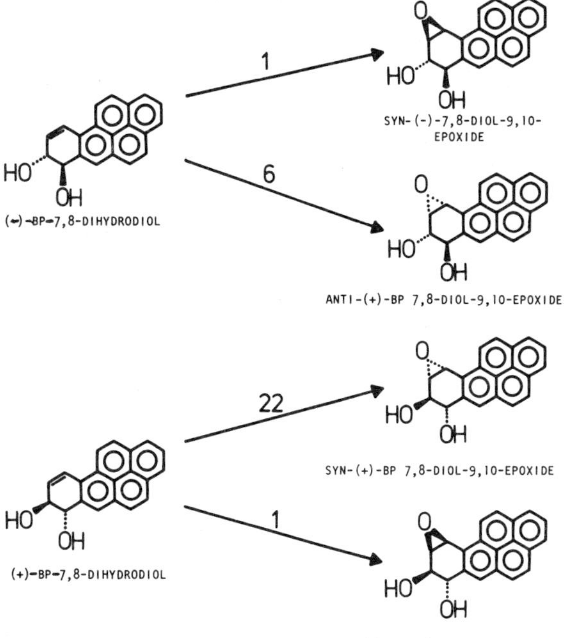

FIG. 2. Formulas of BPDE isomers indicating their derivation, in the proportion indicated, from the (+) and (-) BAP-7,8-dihydrodiols.

2. All four BPDE isomers are potent mutagens for various strains of Escherichia coli and Salmonella typhimurium. In general the syn-isomers are more active than the anti-derivatives.

3. Anti-BPDE is a much more potent carcinogen than syn-BPDE for both the skin and lung of mice. Furthermore, the activity is solely due to the (+) anti-isomer.

4. In V79 Chinese hamster cells, anti-BPDE is a much more efficient mutagen than syn-BPDE at both the hgprt- and ouabain loci. Equal extents of DNA reaction of (+) anti and (-) anti-BPDE result in equal cytotoxicity, but the (+) anti-isomer is considerably more mutagenic (Fig. 3).

Supporters of the somatic mutation theory of carcinogenesis must be encouraged by the data outlined above. The selective metabolism of BAP to the (+) anti-diol epoxide and the ability of this compound to penetrate to the nucleus and to react selectively with DNA to produce a lesion that is powerfully mutagenic but nontoxic could not have been expected. Equally supportive is the correlation of carcinogenicity and mammalian cell mutation induction with (+) and (-) anti-BPDE.

However, those less enthusiastic about these ideas might very reasonably raise a number of questions. For example:

1. Are the results with BAP applicable to other hydrocarbons and how relevant are they to other classes of carcinogen?

2. What is the nature of mutations induced in mammalian cells by (+) anti-BPDE?

3. What can chemical carcinogenesis studies tell us about the relevant genetic loci involved in malignant transformation?

4. Is there any connection between chemical carcinogenesis and the oncogene concept?

5. What is the relevance of chemical carcinogenesis to human cancer?

An attempt to provide answers to at least some of these questions resulted in the experiments reported in the next section.

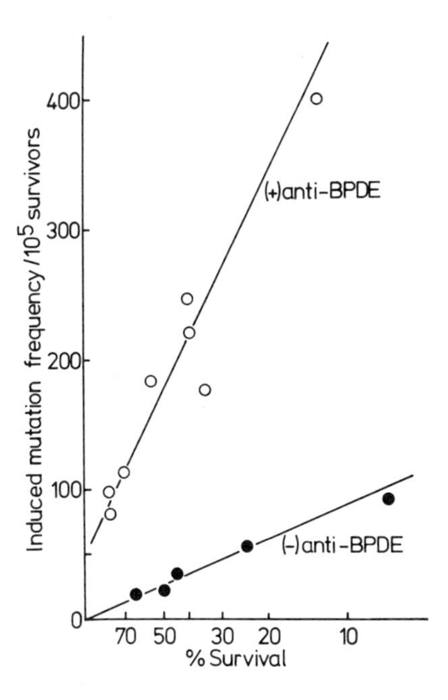

FIG. 3. Induction of 8-azaguanine resistant (AG^r) mutants versus survival for V79 cells treated with either (+) or (-) anti-BPDE. From Brookes and Osborne (9), where full details can be found.

MUTAGENICITY OF 3-METHYLCHOLANTHRENE-DIOL EPOXIDE

Our initial studies (25) of 3-methylcholanthrene (3MC) binding to DNA in vivo implicated the 9,10-diol-7,8-epoxide of both 3MC and 1-hydroxy-3MC. The synthesis of trans-9,10-dihydroxy-anti-7,8-epoxy-7,8,9,10-tetrahydro-3-methylcholanthrene (anti-3MCDE) was reported by Jacobs et al. (22) and generously supplied to us by Professor R.G. Harvey. This compound, like BPDE, was found to react with DNA mainly at the N-2-guanine position although other minor products were observed (34). It was therefore of interest to assess the mutagenicity of anti-3MCDE in comparison with that of anti-BPDE.

Chinese hamster V79 cell in the logarithmic phase of growth were treated with various doses of anti-3MCDE in dimethyl sulfoxide solution for 1 h before harvesting and plating for cell survival and mutation to 8-azaguanine resistance, as previously described (32).

A linear dose response for mutation induction was found even at low nontoxic doses. The plot of mutation versus survival was very similar to that previously reported for anti-BPDE and suggests that exceptional mutagenic efficiency is not unique to the BAP ring structure.

In a separate study of the effect of substituents at the 1-position of BPDE on mutagenicity, 1-isopropyl-BPDE (kindly provided by Professor R. G. Harvey) was tested essentially as described above. Significantly higher doses were required, but the overall dose response for mutation was again of the type shown in Fig. 4.

NATURE OF DIOL EPOXIDE-INDUCED MUTATIONS

The unexpected correlation between mutation in V79 cells and carcinogenic activity for two compounds so closely related chemically as the (+) and (-) isomers of BPDE encouraged us to speculate that the DNA change responsible for the mutations might also be of relevance for the mechanism of tumor induction.

To gain information on the nature of the mutational event, we isolated a series of AG^r mutants induced by anti-BPDE and a similar series induced by treatment of V79 cells with methylnitrosourea (MNU). This latter compound would be predicted to induce single base-pair transition mutation, predominantly GC —> AT resulting by mispairing of O^6-methylguanine with thymine. Previous studies (33) had established that for methylating agents, including MNU, the yield of AG^r mutants in V79 cells was directly related to the level of O^6-alkylation in the cellular DNA.

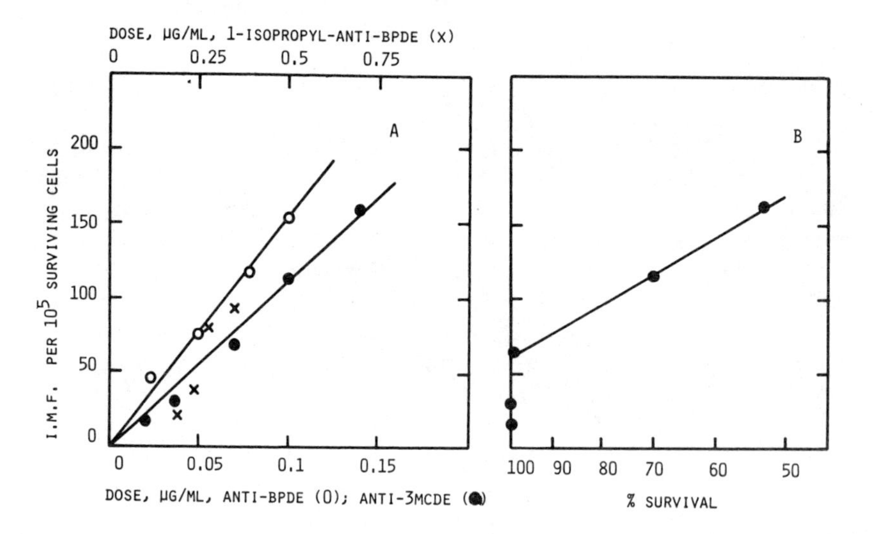

FIG. 4. Mutation versus dose (A) and survival (B) of V79 cells treated with diol epoxides of BAP and 3MC.

The two series of mutants were compared for the following properties (6):

1. The level of hypoxanthine guanine phosphoribosyl transferase (HGPRT) enzyme activity in cell lysates as a proportion of that in wild-type V79 cells.

2. The ability of the mutant HGPRT enzyme (when present) to use AG as substrate.

3. The presence or absence of protein precipitable by antisera to the purified HGPRT-enzyme, i.e., cross-reacting material (CRM), and the molecular weight of such protein.

4. The isoelectric point of the HGPRT in CRM-plus mutants.

As judged by the above criteria, no significant difference was found between the series of AG^r mutants induced by BPDE or MNU. For the CRM-plus mutants, the precipitated protein had the same molecular weight as the normal HGPRT (M_r = 25K), as judged by the position of the band in a polyacrylamide gel. However, for some mutants the protein was either more acidic or more basic than the wild-type enzyme (pI = 6.3) by approximately one charge unit, consistent with the substitution of a neutral amino acid for one having a negative or positive charge (Fig. 5).

Consideration of all the above data led to the conclusion that the HGPRT-plus and/or CRM-plus mutants could have resulted from single base-pair changes in the coding region of the hgprt-gene.

The serious limitation of our earlier report (6) was that only AG^r mutants retaining some enzyme activity or CRM could be examined, and these mutants represented less than 50% of all mutants. Study of the HGPRT-minus: CRM-minus mutants became possible following the isolation of c-DNA clones of mouse and hamster HGPRT by Caskey and coworkers in Houston (3,27). We are particularly grateful to Professor C. T. Caskey for providing us with the appropriate plasmids that enabled us to examine the genomic status and m-RNA product of the hgprt-gene in these anti-BPDE-induced mutants (24).

Restriction enzyme digestion of the DNA isolated from a series of 11 HGPRT-minus: CRM-minus mutants, subsequent gel electrophoresis, and Southern blotting revealed the same set of restriction fragments as found with the DNA of wild-type V79 cells, e.g. Fig. 6. Therefore, within this limited series, no gross deletions or chromosomal rearrangements involving the hgprt-gene were apparent.

The m-RNA from the same series of mutants was examined by the Northern blotting technique, and for seven of the 11 was found to be indistinguishable in molecular weight or quantity from that found in V79 cells. Since these apparently normal m-RNAs were isolated from HGPRT-

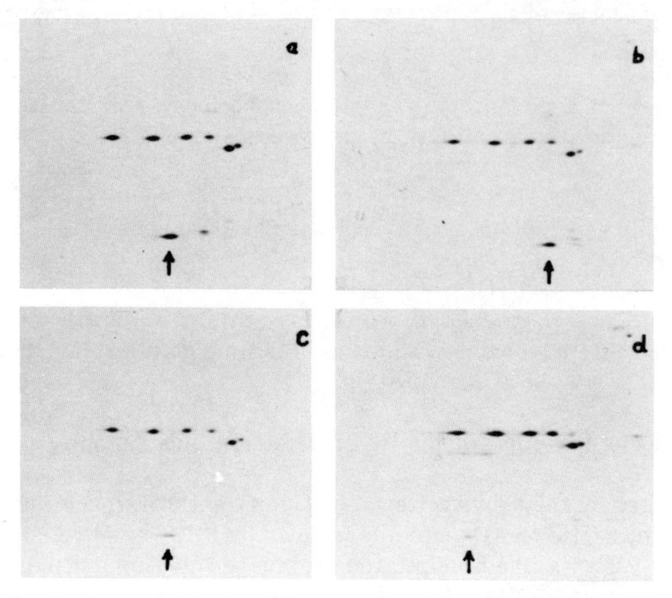

FIG. 5. Two-dimensional electrophoresis of ^{35}S-labeled HGPRT isolated by adsorption onto Sephadex 4B-bound HGPRT-antibody. The basic side of the gel is on the left. The major HGPRT spot is arrowed. The six prominent spots of greater M_r are coprecipitants unrelated to HGPRT: (a) V79; (b) mutant with more acidic HGPRT; (c) mutant with wild-type HGPRT; (d) mutant with more basic HGPRT. Full details in Ref. 6.

minus: CRM-minus mutants, we conclude that they must be incapable of translation, perhaps as a result of a base-pair change. In the remaining four mutants studied, the amount of m-RNA was significantly reduced, and in two mutants the size of the molecule was reduced by about 300 base pairs (Fig. 7). No similar reduction was seen in the Southern analysis of these mutants (e.g., Fig. 6) suggesting the possibility of splice site mutation and perhaps an increased rate of degradation.

A further property of the HGPRT-minus: CRM-minus mutants consistent with their generation by a point mutation was the fact that all were revertible to hypoxanthine:aminopterin:thymidine (HAT) resistance by ethylnitrosourea (ENU). The frequency of reversion, both spontaneous and induced, varied considerably. One mutant, designated B19, showed a very low frequency of spontaneous reversion (less than 10^{-7}) but this increased to 10^{-4} on treatment with ENU, making the isolation of revertants relatively easy. A series of such revertants was examined further as discussed below.

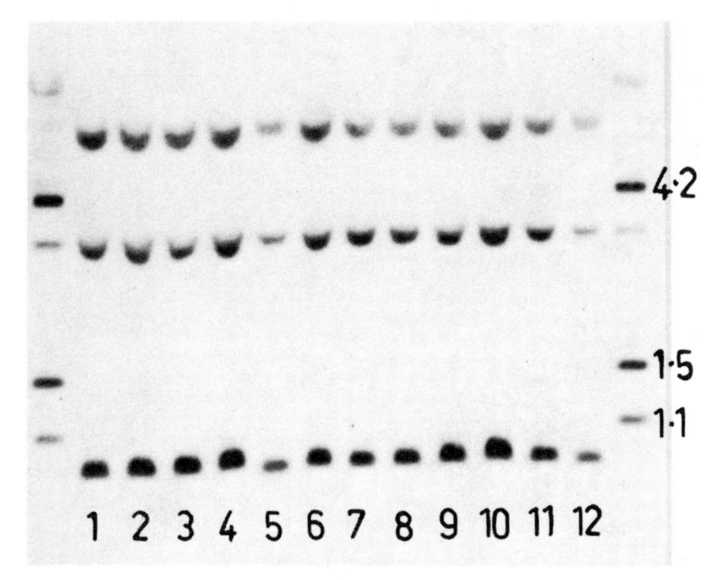

FIG. 6. HGPRT-gene restriction fragments (Pst I) obtained from V79 (Lane 1) and 11 AG^r, CRM-negative mutants (Lanes 2-12). Size markers are shown on each side of the gel.

HAT-RESISTANT REVERTANTS OF MUTANT B19

Eight revertants of mutant B19 were selected at random for detailed study (24). Assay of the HGPRT activity in cell lysates of the revertants showed a variation from 2-34% of the wild-type enzyme level. It has been previously reported (3,16) that reversion of HGPRT-minus mutants resulted from an increase in copy number of the mutant hgprt-gene. Southern analysis of the DNA from mutant B19 in comparison with that of the revertants showed no evidence of increased gene copy number.

Immunoprecipitation of ^{35}S-labeled CRM with HGPRT antisera and subsequent gel electrophoresis indicated that all the revertants possessed a 25K protein as found in V79. However, isoelectric focusing on non-denaturing gels and localization of the HGPRT by its enzyme activity (Fig. 8) showed that all eight revertants contained enzyme with identical pI = 7.0, which was more basic than the wild-type enzyme (pI = 6.3).

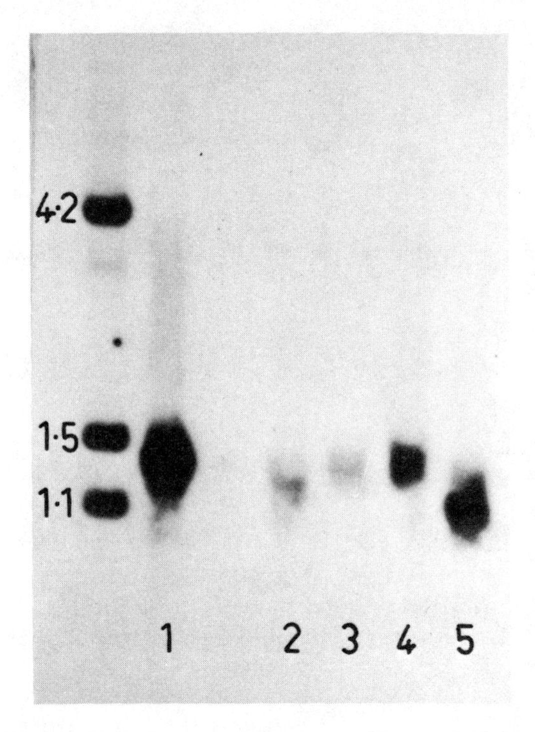

FIG. 7. HGPRT-mRNA from four mutants (Lanes 2-5) having reduced amount and/or size of m-RNA compared to that found normally (Lane 1). Size markers are shown on the left of the gel. Full details in Ref. 24.

A consideration of the above data on mutant B19 and its revertants led us to speculate on a sequence of base-pair changes that could account for the findings as shown below.

Cell type	V79 ——>		B19 ——>		Revertants
HGPRT	M_r = 25K; pI = 6.3		$HGPRT^-$:CRM^-		M_r = 25K; pI = 7.0
Gene	- GAA -	BPDE>	- TAA -	ENU>	- AAA -
Amino Acid	Glutamic		Nonsense		Lysine

The very high frequency of ENU-induced reversion of B19 (approximately 10^{-4}) would imply a mutational hot spot in the hgprt-gene. Alternatively one might speculate that the revertants represent a spectrum of ocre or amber suppressors resulting from ENU-induced mutation in the gene(s) coding for lysine t-RNA. Experiments to investigate these speculations are in progress.

We consider that all the evidence on the range of diol epoxide-induced AG^r mutants is consistent with the conclusion that single base changes are a major factor in their causation.

CHEMICAL CARCINOGENESIS AND ONCOGENES

The recognition that an oncogenic retrovirus contains within its genome a DNA sequence derived from the host, and furthermore that some of these same sequences are implicated in human tumor development, has had a major impact on carcinogenesis research over the past five years (13,43). What impact do these findings have for the somatic mutation hypothesis? An obvious way to merge these different ideas would be to suggest that DNA reaction to chemical carcinogens leads to oncogene activation, either by mutation within the oncogene itself or in a regulatory sequence or by causing gene rearrangement.

In the case of the *ras*-oncogene, direct evidence of activation by point mutation in a number of human tumors has been reported as summarized below.

Oncogene	Tumor	Amino Acid	Codon	Ref.
H-*ras*	Bladder	Gly ——> Val[12]	GGC ——> GTC	38,44
	Lung	Glun ——> Leu[61]	CAG ——> CTG	49
K-*ras*	Lung	Gly ——> Cyst[12]	GGT ——> TGT	11,39
	Colon	Gly ——> Val[12]	GGT ——> GTT	11
N-*ras*	Neural	Glun ——> Lys[61]	CAA ——> AAA	45

Each of these mutations involves either an A —> T or G —> T transversion, which was the type of change reported by Eisenstadt et al. (15) to be induced by BPDE in the *lac* I gene of a *uvr* B^- strain of *E. coli*, and incidentally, the change postulated to explain the reversion studies discussed above.

CHEMICAL CARCINOGENS AND HUMAN CANCER

Even the most enthusiastic research worker in the field of chemical carcinogenesis must have some concern about the slogan that "80% of human cancer is caused by chemicals." While this may be true within a very wide definition of "chemicals," it is worth noting how few cancers can be attributed to the classical carcinogens as used by research workers (21).

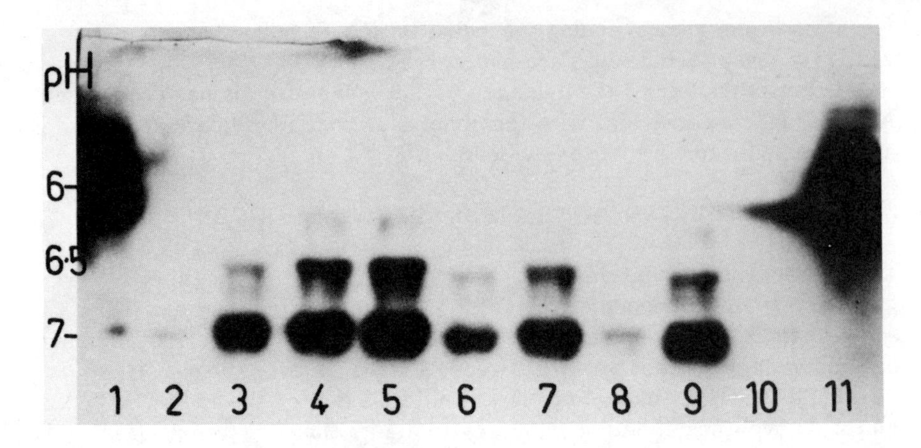

FIG. 8. Isoelectrofocusing of HGPRT enzyme of V79 (Lanes 1 and 11), mutant B19 (Lane 2), and eight revertants (Lanes 3-10). The HGPRT activity was detected by the conversion of ^{14}C-labeled hypoxanthine to inosine monophosphate. Full details in Ref. 24.

The ubiquitous contamination of the environment with carcinogenic hydrocarbons generated by combustion of fossil fuels has not been linked with any increase in human disease. Even the relevance of hydrocarbons in tobacco smoke to the undoubted carcinogenicity of this material has been questioned, while Doll and co-workers (14) found only a small increase in lung cancer incidence in coke-oven workers exposed to very high levels of BAP. Similarly, the role of the aromatic amines and the alkyl-nitrosamines as causes of human cancer remains uncertain. Undoubtedly some human tumors can be attributed to these carcinogens but it seems unlikely that these carcinogens are responsible for the major cancers, namely those of bronchial tract, colon, breast, stomach, and liver. In the case of liver tumors, it seems probable that several factors are involved, in particular aflatoxin B_1 and hepatitis B virus. This multifactorial causation of human cancer may tend to obscure the role of low doses of carcinogens whose activity is often demonstrable in experimental animals only when high doses are used.

Despite the current emphasis on the importance of "life-style," genetic makeup, and oncogenes in human cancer, the advances of the past 20 years in experimental chemical carcinogenesis should not be underestimated. The knowledge gained from these studies must ultimately help towards an understanding of the mechanism of carcinogenesis, however caused.

REFERENCES

1. Baird, W.M., and Brookes, P. (1973): Cancer Res., 33:2378-2385.
2. Baird, W.M., Harvey, R.G., and Brookes, P. (1975): Cancer Res., 35:54-57.
3. Brennand, J., Chinault, A.C., Konecki, D.S., Melton, D.M., and Caskey, C.T. (1982): Proc. Nat. Acad. Sci., USA, 79:1950-1954.
4. Brookes, P. (1977): Mutat. Res., 39:257-284.
5. Brookes, P., and Heidelberger, C. (1969): Cancer Res., 29:157-165.
6. Brookes, P., King, H.W.S., Mundy, C.R., and Newbold, R.F. (1982): Carcinogenesis, 3:687-692.
7. Brookes, P., and Lawley, P.D. (1964): Nature, 202:781-784.
8. Brookes, P., and Lawley, P.D. (1964): J. Cell. Comp. Physiol. Suppl. 1, 64:111-127.
9. Brookes, P., and Osborne, M.R. (1982): Carcinogenesis, 3:1223-1226.
10. Buening, M.K., Wislocki, P.G., Levin, W., Yagi, H., Thakker, D.R., Akagi, H., Koreeda, M., Jerina, D.M., and Conney, A.H. (1978): Proc. Nat. Acad. Sci., USA, 75:5358-5361.
11. Capon, D.J., Seeburg, P.H., McGrath, J.P., Hayflick, J.S., Edman, U., Levinson, A.D., and Goeddel, D.V. (1983): Nature, 303:507-513.
12. Conney, A.H. (1982): Cancer Res., 42:4875-4917.
13. Cooper, G.M. (1982): Science, 218:801-806.
14. Doll, R., Vessey, M.P., Beasley, P.W.R., Buckley, A.P., Fear, E.C., Fisher, P.E.M., Gammon, E.J., Gunn, W., Hughes, G.O., Lee, K., and Norman-Smith, B. (1972): Br. J. Ind. Med., 29: 394-420.
15. Eisenstadt, E., Warren, A.J., Porter, J., Atkins, D., and Miller, J.H. (1982): Proc. Nat. Acad. Sci., USA, 79:1945-1949.
16. Fuscoe, J.C., Fennick, R.G., Kruh, G., and Konecki, D. (1983): Mol. Cell. Biol., 3:1086-1095.
17. Grover, P.L., Hewer, A., and Sims, P. (1972): Biochem. Pharmacol., 21:2713-2726.
18. Harvey, R.G., and Fu, P.P. (1978) In: Polycyclic Hydrocarbons and Cancer, edited by H.V. Gelboin and P.O.P. Ts'o, pp. 133-166. Academic Press, New York.
19. Heidelberger, C. (1964): J. Cell. Comp. Physiol. Suppl. 1, 64:129-148.
20. Huberman, E., Sachs, L., Yang, S.K., and Gelboin, H.V. (1976): Proc. Nat. Acad. Sci., USA, 73:607-611.
21. I.A.R.C. Monograph Supplement 4. (1982): Chemicals, Industrial Processes and Industries Associated with Cancer in Humans. World Health Organization, Lyon, France.
22. Jacobs, S.A., Cortez, C., and Harvey, R.G. (1983): Carcinogenesis, 4:519-522.
23. Kapitulnik, J., Wislocki, P.G., Levin, W., Yagi, H., Jerina, D.M., and Conney, A.H. (1978): Cancer Res., 38:354-358.
24. King, H.W.S., and Brookes, P. (1984): Carcinogenesis, 5:965-970.

25. King, H.W.S., Osborne, M.R., and Brookes, P. (1977): Int. J. Cancer, 20:564-571.
26. King, H.W.S., Thompson, M.H., and Brookes, P. (1975): Cancer Res., 35:1263-1269.
27. Konecki, D.S., Brennand, J., Fuscoe, J.C., Caskey, C.T., and Chinault, A.C. (1982): Nucleic Acids Res., 10:6763-6775.
28. Lawley, P.D., and Brookes, P. (1963): Exp. Cell Res. Suppl., 9:512-520.
29. Lawley, P.D., and Brookes, P. (1965): Nature, 206:480-483.
30. Miller, J.A. (1970): Cancer Res., 30:559-576.
31. Miller, E.C., and Miller, J.A. (1976): In: Chemical Carcinogens, Monograph 173, edited by C.E. Searle. American Chemical Society, Washington, DC.
32. Newbold, R.F., Brookes, P., and Harvey, R.G. (1979): Int. J. Cancer, 24:203-209.
33. Newbold, R.F., Warren, W., Medcalf, A.S.C., and Amos. J. (1980): Nature, 283:596-599.
34. Osborne, M.R. (1984): In: Polycyclic Aromatic Hydrocarbons, Vol. 8, edited by A.J. Dennis and M. Cooke. Battelle Press, Colombus, Ohio.
35. Osborne, M.R., Beland, F.A., Harvey, R.G., and Brookes, P. (1976): Int. J. Cancer, 18:362-368.
36. Osborne, M.R., Harvey, R.G., and Brookes, P. (1978): Chem. Biol. Interact., 20:123-130.
37. Osborne, M.R., Jacobs, S., Harvey, R.G., and Brookes, P. (1981): Carcinogenesis, 2:553-558.
38. Reddy, E.P., Reynolds, R.K., Santos, E., and Barbacid, M. (1982): Nature, 300:149-152.
39. Shimizu, K., Birnbaum, D., Ruley, M.A., Fasano, O., Suard, Y., Edlund, L., Taporowsky, E., Goldfarb, M., and Wigler, M. (1983): Nature, 304:497-506.
40. Sims, P., and Grover, P.L. (1981): In: Polycyclic Hydrocarbons and Cancer, Vol. 3, edited by H.V. Gelboin and P.O.P. Ts'o, pp. 117-182. Academic Press, New York.
41. Sims, P., Grover, P.L., Swaisland, A., Pal, K., and Hewer, A. (1979): Nature, 252:326-327.
42. Slaga, T.J., Bracken, W.J., Gleason, G., Levin, W., Yagi, H., Herina, D.M., and Conney, A.H. (1979): Cancer Res., 39:67-71.
43. Slamon, D.J., deKerion, J.B., Verma, I.M., and Cline, M.J. (1984): Science, 224:256-262.
44. Tabin, C.J., Bradley, S.M., Bargmann, C.I., Weinberg, R.A., Papageorge, A.G., Scolnick, E.M., Dhar, R., Lowry, D.R., and Chang, E.H. (1982): Nature, 300:143-149.
45. Taporowsky, E., Shimizu, K., Goldfarb, M., and Wigler, M. (1983): Cell, 34:581-586.

46. Wislocki, P.G., Wood, A.W., Chang, R.L., Levin, W., Yagi, H., Hermandez, O., Dansette, P.M., Jerina, D.M., and Conney, A.H. (1976): Cancer Res., 36:3350-3357.
47. Wood, A.W., Chang, R.L., Levin, W., Yagi, H., Thakker, D.R., Jerina, D.M., and Conney, A.H. (1977): Biochem. Biophys. Res. Commun., 77:1389-1396.
48. Yagi, H., Hernandez, O., and Jerina, D.M. (1975): J. Am. Chem. Soc., 97:6881-6883.
49. Yuasa, Y., Srivastava, S.K., Dunn, C.Y., Rhim, J.S., Reddy, E.P., and Aaronson, S.A. (1983): Nature, 303:775-779.

Carcinogenesis, Vol. 10, edited by E. Huberman
and S. H. Barr. Raven Press, New York © 1985.

Role of Intercalation in Polycyclic Aromatic Hydrocarbon Carcinogenesis

R. G. Harvey, M. R. Osborne, *J. R. Connell, *S. Venitt,
*C. Crofton-Sleigh, *P. Brookes, J. Pataki, and †J. DiGiovanni

*The Ben May Laboratory for Cancer Research, The University of Chicago, Chicago, Illinois 60637; *Institute of Cancer Research, Pollards Wood Research Station, Chalfont St. Giles, Bucks HP8 4SP, United Kingdom; †The University of Texas System Cancer Center, Science Park, Smithville, Texas 78957*

It is fitting to discuss the mechanism of carcinogenesis of polycyclic aromatic hydrocarbons (PAH) in a symposium in honor of Charles Heidelberger. Hydrocarbons were one of his major research interests from the earliest days.

It is now generally accepted that PAH require activation by the P-450 microsomal enzymes to exert their mutagenic and carcinogenic effects (3, 11,27). Considerable evidence has implicated DNA as the critical cellular target. Benzo[a]pyrene (BAP) has been most intensively investigated, and its principal carcinogenic metabolite has been identified as the highly reactive diol epoxide derivative known as (+)-anti-BPDE. The anti-isomer has the epoxide ring on the opposite molecular face to the benzylic hydroxyl group; the syn isomer has these groups on the same face. The (+)-enantiomer of anti-BPDE has the absolute stereochemistry shown in Fig. 1. Covalent binding of (+)-anti-BPDE to DNA and RNA in mammalian cells has been shown to take place principally on the 2-NH_2 group of guanosine (14-16,21,22,29,33). The available evidence suggests that other PAH undergo similar metabolic activation to diol epoxide derivatives, which also bind covalently to nucleic acids. However, the specific details of the process whereby PAH-DNA adduct formation leads to tumor induction still remain largely unknown.

This chapter will review some recent findings on the mechanism of PAH-DNA interaction and present new evidence concerning the possible role of intercalation in the mechanism of covalent binding of active PAH metabolites to nucleic acids.

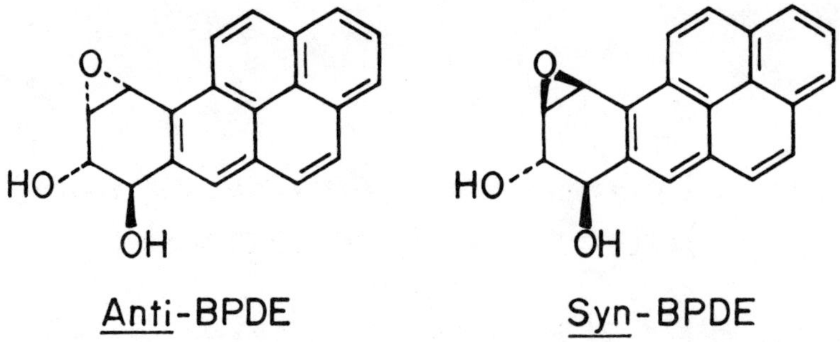

Anti-BPDE   *Syn*-BPDE

FIG. 1. Structures of the active carcinogenic metabolite of benzo[a]-pyrene, (+)-*trans*-7,8-dihydroxy-*anti*-9,10-epoxy-7,8,9,10-tetrahydrobenzo-[a]pyrene (*anti*-BPDE), also known as (+)-7β, 8α-dihydroxy-9α, 10α-epoxy-7,8,9,10-tetrahydrobenzo[a]pyrene, and the *syn* diastereomer.

MECHANISM OF COVALENT BINDING

Kinetic studies of the reaction between *anti*-BPDE and native DNA were conducted in collaboration with Geacintov (6,7) with ultraviolet (UV) fluorescence and electric linear dichroism techniques. The findings (Fig. 2) indicate that the hydrocarbon undergoes relatively rapid intercalation, which is complete in < 5 msec, between the base pairs of the nucleic acid. The initial intercalation complex undergoes rate-determining protonation to yield an intercalated triol carbonium ion intermediate, which decomposes to products via two pathways (8,9,20,31). The major path, which accounts for 90-95% of the *anti*-BPDE, is hydrolysis to the tetraols. DNA catalyzes this hydrolysis (7), and the tetraols remain associated with DNA in a noncovalent intercalation complex (13). The alternative, more biologically important, pathway entails covalent binding of the carbonium ion intermediate to DNA. Although the overall rate of reaction of *anti*-BPDE is dependent upon pH, temperature, ionic strength, and other factors, the ratio of the rates of the two paths is independent of these variables (8,9).

UV fluorescence, linear dichroism, and fluorescence quenching experiments indicate that the aromatic ring system of the major covalent *anti*-BPDE adduct is located external to and approximately parallel to the DNA helix (4,5,19). This adduct arises from covalent linkage to the 2-NH_2 of dG and the hydrocarbon moiety is believed to lie in the minor groove of the DNA helix (2,25). Although the weight of current evidence supports this model, a different conclusion was reached by Hogan et al. (12) who studied the covalent adducts derived from reaction of *anti*-BPDE with fragments of sonicated DNA of defined length. They observed that

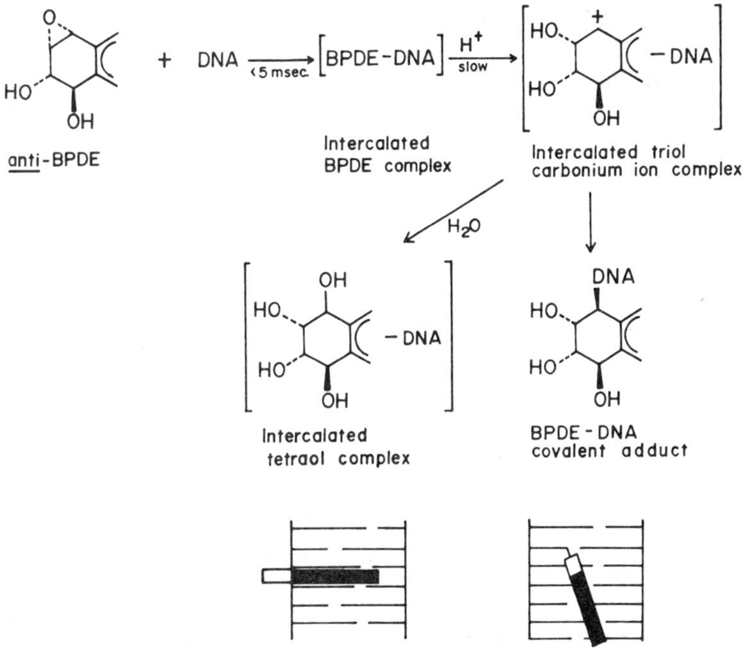

FIG. 2. Proposed mechanism of hydrolysis and covalent binding of anti-BPDE to native DNA based on kinetic studies (6-9).

the fluorescence of the covalently bound pyrenyl moiety was not strongly quenched by acrylamide, a known quencher of aromatic fluorescence, and they concluded that the hydrocarbon molecule was intercalated within the DNA helix. However, recent studies by Geacintov et al. (10) have shown that reduced accessibility of covalently bound BPDE to fluorescence-quenching agents in sonicated DNA is due to concentration-dependent self-association of the DNA fragments. Self-association of high-molecular-weight DNA was found to be markedly less, and significant quenching of the fluorescence by acrylamide and molecular oxygen was observed. These findings are consistent with the external-binding model for the principal BPDE-DNA adduct structure.

An alternative mechanism of interaction of PAH diol epoxides with DNA involving formation of tetraols via an intercalated intermediate (Fig. 3) and formation of PAH-DNA adducts by direct covalent binding at an external site, without prior intercalation, cannot be entirely excluded. Indeed, theoretical molecular modeling studies indicate that reorientation of a BPDE adduct from an intercalated to an external site would entail a

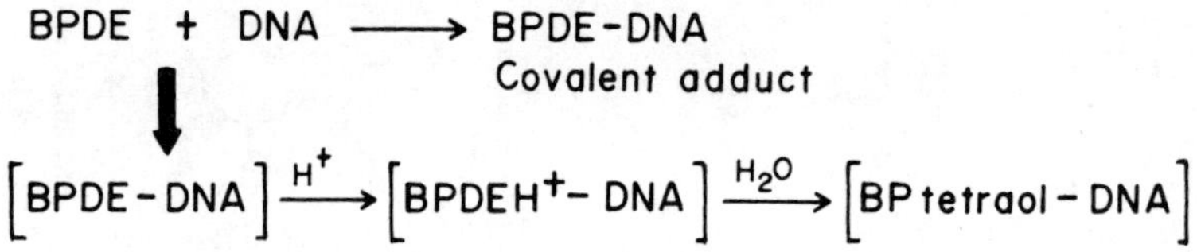

$$\text{BPDE} + \text{DNA} \longrightarrow \text{BPDE-DNA (Covalent adduct)}$$

$$\downarrow$$

$$[\text{BPDE-DNA}] \xrightarrow{H^+} [\text{BPDEH}^+\text{-DNA}] \xrightarrow{H_2O} [\text{BP tetraol-DNA}]$$

FIG. 3. Alternative dual pathway mechanism of hydrolysis and covalent binding of anti-BPDE to DNA.

complex reaction pathway involving considerable readjustment of DNA conformation (18,30).

BIOLOGICAL PROPERTIES OF 1-ALKYLBENZO[a]PYRENES

To gain insight into the question of whether intercalation is essential to the mechanism of covalent binding of diol epoxide metabolites to DNA, we examined the effects of alkyl substitution on the biological activity and DNA binding of BAP. Our experimental approach was based on the assumption that in the associated complex the highly polar diol epoxide ring remains on the outside of the helix, while the hydrophobic aromatic ring system enters and associates with the purine and pyrimidine bases aided by π-π interactions. Groups attached to the aromatic ring system that substantially increase molecular thickness may be expected to sterically interfere with or entirely block intercalation of the fused aromatic ring system, thereby inhibiting biological activity. Accordingly, we synthesized a series of 1-alkylbenzo[a]pyrene derivatives bearing alkyl groups with increasing steric requirements, methyl, ethyl, isopropyl, tert-butyl (Table 1). It was anticipated that alkyl groups in the 1-position remote from the benzo ring would interfere minimally with metabolic activation in the latter molecular region.

Syntheses of 1-Alkylbenzo[a]pyrenes

Syntheses of the 1-alkyl derivatives of benzo[a]pyrene was accomplished from BAP (Fig. 4), making use of the previously reported Friedel-Crafts acylation of BAP in the 1-position with acetyl chloride and $AlCl_3$ (35). Oxidation of 1-acetyl-BAP with sodium hypochlorite (35) gave BAP 1-carboxylic acid, which in turn underwent reduction with trichlorosilane and tri-n-propylamine (17) to yield 1-methyl-BAP (52%), m.p. 178-179°C, lit. 178-178.5°C (1). Reduction of 1-acetyl-BAP (1 mmol) by the Wolff-Kishner method with hydrazine (0.2 mL) and KOH (20 mg) in refluxing diethylene glycol (10 mL) for 4.5 h followed by chromatography on Florisil gave 1-ethyl-BAP (76%), m.p. 116-118°C lit. 112°C (35). Reaction of

TABLE 1. Steric requirements of 1-alkyl groups in the 1-alkylbenzo[a]pyrenes.

1-Alkyl Group	Thickness (A°)
Methyl	4.0
Ethyl	4.0 - 5.5[a]
Isopropyl	4.0 - 5.9[a]
tert-Butyl	5.9

Ring Thickness = 3.4 A°

[a]Figures represent minimum and maximum values for freely rotating alkyl groups.

1-acetyl-BAP with methylmagnesium bromide and acidic dehydration furnished 1-isopropenyl-BAP (35). Hydrogenation of the latter in ethyl acetate in the presence of a palladium catalyst at 20 psig for 24 h provided 1-isopropyl-BAP plus its 4,5-dihydro derivative [20% by nuclear magnetic resonance (NMR)]. This product was heated at reflux with 2,3-dichloro-5,6-dicyanoquinone (DDQ) for 5 min and purified by chromatography and crystallization to yield pure 1-isopropyl-BAP (68%), m.p. 135-136°C, lit. 128-130°C (26). 1-tert-butyl-BAP was synthesized from 1-isopropenyl-BAP via epoxidation and acid-catalyzed rearrangement to the related aldehyde followed by methylation α to the carbonyl with KH and methylation on this site with methyl iodide. Reduction of the resulting aldehyde by the Wolff-Kishner method furnished 1-tert-butyl-BAP (75%), m.p. 97.5-99°C (23). The 500 MHz proton NMR spectra of the 1-alkyl-BAP derivatives were consistent in all cases with their structural assignments.

Mutation Assays

If intercalation is involved in the mechanism of covalent binding of PAH diol epoxide metabolites to DNA, an inverse relation between alkyl group size in the 1-alkylbenzo[a]pyrenes and their carcinogenicities is anticipated. Since mutagenicity assays are conveniently used as rapid screening tests for potential carcinogens, mutation experiments were conducted in both bacterial and mammalian cells. Benzo[a]pyrene and its 1-alkyl derivatives were assayed for mutagenicity in plate incorporation tests in three bacterial strains, Salmonella typhimurium TA 100 and TA 98 and Escherichia coli WP2uvrA(pKM101) with methods previously described (32). Tests were performed in the absence of S9 liver microsomes and with 4, 10, and 30% (by volume) of S9. Neither BAP nor its 1-alkyl

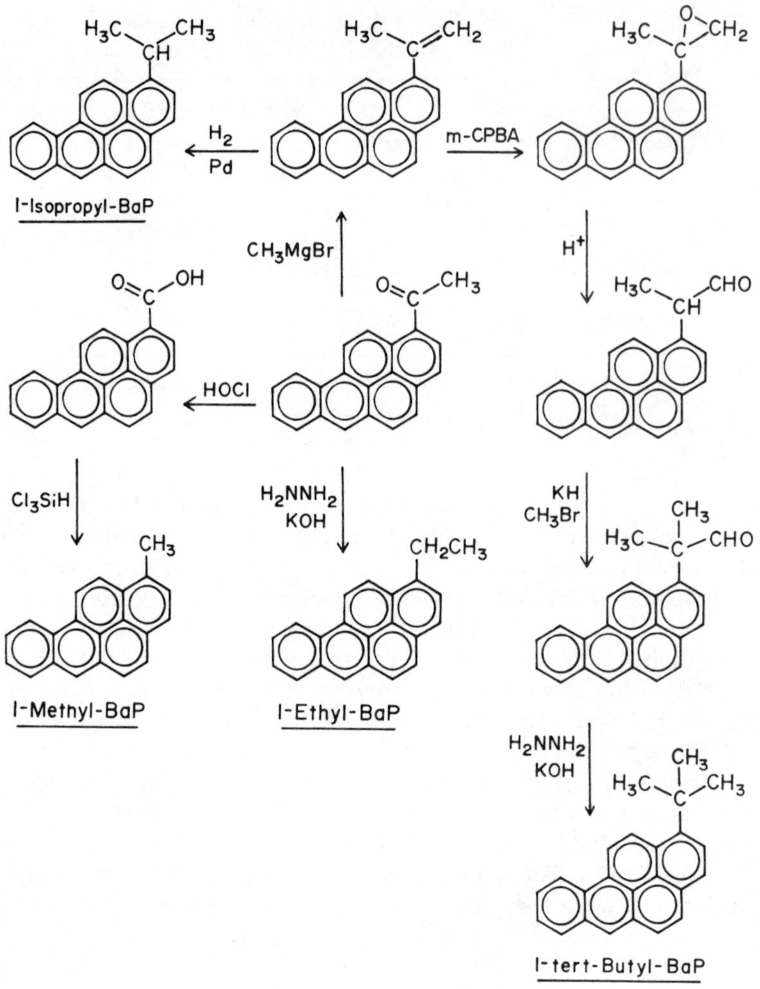

FIG. 4. Syntheses of the 1-alkylbenzo[a]pyrenes.

derivatives were mutagenic in the absence of S9-mix in the three strains of bacteria tested. For those hydrocarbons that were mutagenic in the presence of S9-mix, optimum results were obtained when 10% S9 was used in the S9-mix. Doses above 2.5 μg per plate were toxic, as indicated by reductions in the numbers of induced mutants and by thinning of the bacterial lawn. Figure 5 shows the mutagenic response of the three bacterial strains when this level of S9 was used.

BAP and 1-methyl-BAP were mutagenic in all three strains, the methyl derivative being less mutagenic than BAP to *S. typhimurium* TA 98 and *E. coli* WP2uvrA(pKM101); both compounds were equally mutagenic to *S. typhimurium* TA 100, there being no significant difference in slope for the two compounds. 1-Ethyl-BAP was a weaker mutagen, producing statistically significant dose-related, but smaller, increases in mutation in *S. typhimurium* TA 98 and *E. coli* WP2uvrA(pKM101); the slight increase in revertants per plate obtained with TA 100 was not significant. 1-isopropyl-BAP and 1-*tert*-butyl-BAP were not mutagenic to any of the three bacterial strains under the conditions of the assay. In general, substitution of alkyl groups larger than methyl in the 1-position of BAP was associated with a loss of mutagenic activity, and the larger the alkyl substituent, the greater the reduction in mutagenicity.

The results of cell-mediated mutation assays, with irradiated BHK21 Syrian hamster cells used as the feeder layers and V79 Chinese hamster cells as the target cells are shown in Table 2. The findings with BAP were similar to those reported earlier (34), while 1-methyl-BAP was about twice as effective as BAP at both killing cells and inducing mutation. 1-ethyl-BAP was clearly less mutagenic than either BAP or 1-methyl-BAP. Although there appeared to be some effect of dose on both survival and mutation, the relative mutagenicities were independent of the dosage. 1-isopropyl-BAP and 1-*tert*-butyl-BAP were not mutagenic under the conditions used in the assay.

Tumorigenicity Experiments

Tumor initiation studies with the 1-alkyl-BAPs were carried out on the skins of female SENCAR mice. One week after initiation with a 200 nmol dose of each hydrocarbon, the mice began to receive twice weekly topical applications of 3.4 nmol of the promoter 12-0-tetradecanoylphorbol-13-acetate (TPA). The results of these studies (Table 3) correlate well with the mutagenicity data. 1-methyl-BAP was slightly more active than the parent hydrocarbon in inducing papillomas on the skins of female SENCAR mice at the dosage tested. This finding agrees with data reported earlier by Slaga et al. (28). 1-ethyl-BAP was less effective as a tumor initiator than either BAP or 1-methyl-BAP, and 1-isopropyl-BAP and 1-*tert*-butyl-BAP exhibited only minimal tumorigenic activity. Thus, the activities of this series of 1-alkyl-BAP derivatives as tumor initiators and mutagens show a parallel relationship, both decreasing markedly as the size of the alkyl group in the 1-position increases from methyl to the bulky *tert*-butyl group.

These findings are consistent with, but do not prove, the hypothesis that intercalation is an essential step in the covalent binding of the carcinogenic diol-epoxide metabolites of PAH to DNA.

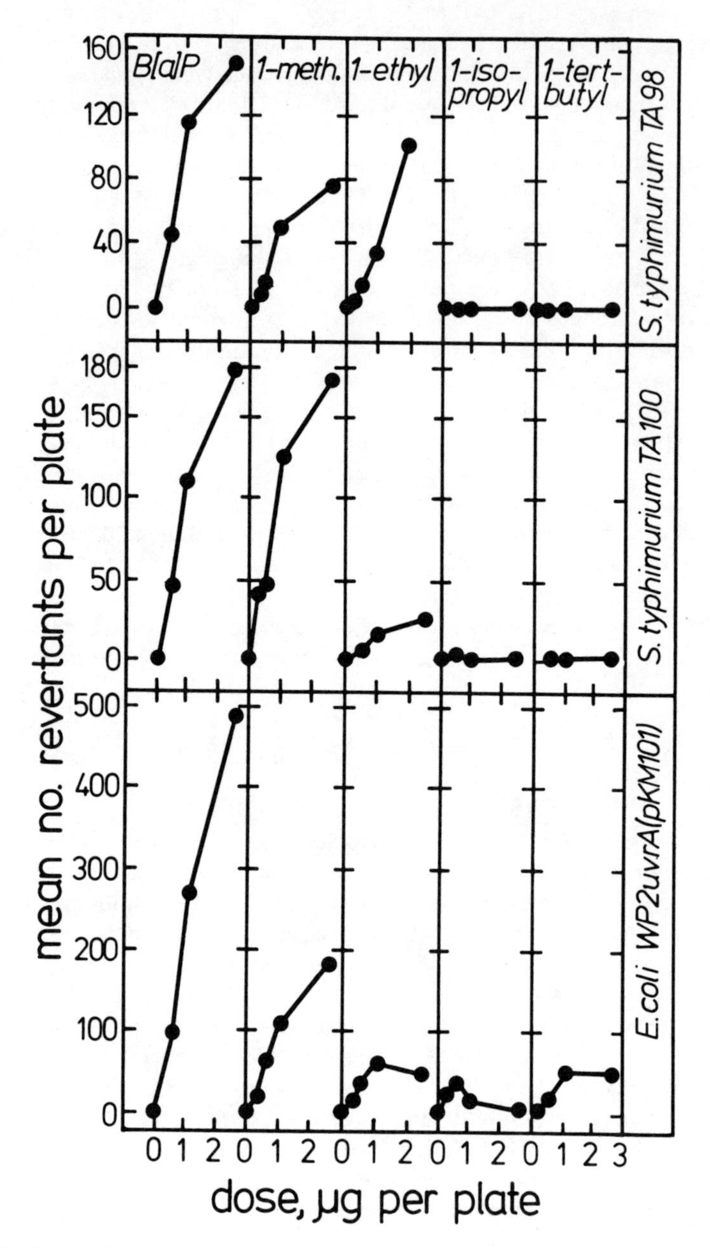

FIG. 5. Plate-incorporation assays of 1-alkylbenzo[a]pyrene derivatives in S. typhimurium TA 100 and TA 98 and E. coli WP2uvrA(pKM101). Arclor 1254-induced rat-liver S9 (10% in S9 mix) was used in all assays. Each point represents the mean number of revertants per plate (triplicate plates) after subtraction of the zero-dose values.

TABLE 2. Cell-mediated mutagenicity and cytotoxicity of 1-alkyl-benzo[a]pyrenes in V79 Chinese hamster cells.[a]

BAP Derivative[b]	Dose (μg per mL)	Survival (%)	AG^r mutants per 10^5 survivors
BAP	-	100	2
	0.01	99	6
	0.10	98	52
	1.00	62	139
1-Me-BAP	0.01	91	14
	0.05	73	80
	0.10	51	151
	0.50	21	258
	1.00	14	288
1-Et-BAP	0.10	82	33
	1.00	77	30
	10.00	61	24
1-i-Pr-BAP	1.00	93	5
	10.00	76	5
1-t-Bu-BAP	1.00	96	4
	10.00	79	8

[a]A feeder-layer of x-irradiated BHK21 Syrian hamster cells provided the source of metabolic activation.
[b]Me - methyl, Et = ethyl, i-Pr = isopropyl, t-Bu = tert-butyl.

Metabolism Experiments

As the decrease in the biological activities of the 1-alkyl-BAPs with increasing size of the alkyl group might be a consequence of steric inhibition of metabolism or alteration of the enzymatic pathways to favor detoxification over activation, it was necessary to investigate the metabolism of these hydrocarbons.

Metabolism of the 1-alkyl-BAPs with rat liver microsomes was carried out by established procedures, and the ethyl acetate soluble metabolites were analyzed by high-performance liquid chromatography (HPLC) on a reverse phase C18 column. Metabolism of BAP, as a reference standard, furnished the 4,5-, 7,8-, and 9,10-dihydrodiols, the 3- and 9-phenols, and

TABLE 3. Tumor-initiating activities of 1-alkyl-benzo[a]pyrenes.[a]

Initiator[b]	Dose (nmol)	Mice with papillomas (%)	Papillomas per mouse (no.)[c]
BAP	200	90	5.37
1-Me-BAP	200	100	6.97
1-Et-BAP	200	73	1.43
1-i-Pr-BAP	200	52	0.97
1-t-Bu-BAP	200	57	0.97

[a]Thirty female SENCAR mice were used for each experimental group. One week after initiation, mice began receiving twice weekly applications of 3.4 nmol TPA. Mice receiving the acetone vehicle at the time of initiation, followed by TPA promotion, had 0.1 papillomas per mouse with 10% of the mice bearing papillomas.

[b]Me = methyl, Et = ethyl, i-Pr = isopropyl, t-Bu = tert-butyl.

[c]Data are summarized after 20 weeks of promotion.

the 1,6- and 3,6-quinones as described previously by numerous investigators (3,11,27). Under similar conditions, the 1-alkyl-BAPs were metabolized to similar extents except that the metabolism of 1-tert-butyl-BAP was considerably slower. The HPLC elution profile of the metabolites of 1-methyl-BAP is typical (Fig. 6). The peaks were identified by their relative retention times in comparison with the corresponding metabolites of BAP (Fig. 7) and by their UV fluorescence spectra, and in some cases by comparison with authentic synthetic compounds. The patterns of metabolism were quite similar to that of BAP except that oxidative metabolism also takes place on the alkyl group. In the case of 1-methyl-BAP, the 1-hydroxy derivative is a major metabolite. Also, metabolism in the K-region to the 4,5-dihydrodiol is significantly reduced. More importantly, the 7,8-dihydrodiols are detected as major metabolites in all cases, independent of alkyl group size.

Syntheses of the 7,8-dihydrodiol of 1-isopropyl-BAP and its anti diol epoxide were undertaken to obtain a sample of the former to confirm its identify as a microsomal metabolite and to obtain sufficient quantities of both compounds for biological experiments. Full details of these syntheses will be reported elsewhere (24). The HPLC retention time and the UV fluorescence spectrum (Fig. 8A) of the authentic 7,8-dihydrodiol matched those of the metabolite of 1-isopropyl-BAP identified as the 7,8-dihydrodiol, confirming its structural assignment.

Metabolism of the 7,8-dihydrodiol of 1-isopropyl-BAP with liver microsomes gave as the principal product the tetraols, which must have

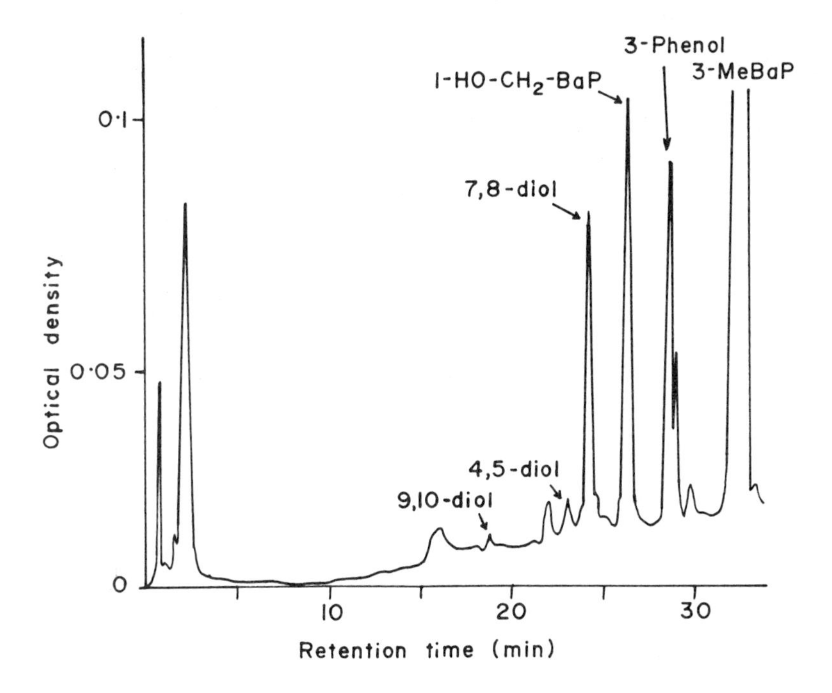

FIG. 6. HPLC chromatogram of the 1-methylbenzo[a]pyrene metabolites. The ethyl acetate soluble fraction of the total metabolites was dissolved in 50% methanol-water and applied to a Waters Microbondapak C18 column. Elution with a linear gradient of 30-100% methanol in water at 2 mL/min for 30 min gave the elution profile shown. The main UV-absorbing peaks were collected and UV absorption and fluorescence spectra were recorded.

arisen via hydrolysis of the corresponding diol epoxide metabolite. Further confirmation was provided by hydrolysis of the synthetic anti diol epoxide of 1-isopropyl-BAP. The UV fluorescence spectrum of the tetraols obtained from hydrolysis of the synthetic diol epoxide (Fig. 8B) was closely similar to that of the tetraols obtained from microsomal oxidation of the dihydrodiol. These findings confirm that the 7,8-dihydrodiol of 1-isopropyl-BAP is fully capable of undergoing microsomal metabolism to the corresponding bay region diol epoxide, which is believed to be its probable ultimate carcinogenic metabolite. Therefore, the lack of activity of 1-isopropyl-BAP as a mutagen in bacterial and hamster V79 cells and its weak activity as a tumor-initiator on mouse skin cannot be due to steric inhibition of metabolic activation by the isopropyl group.

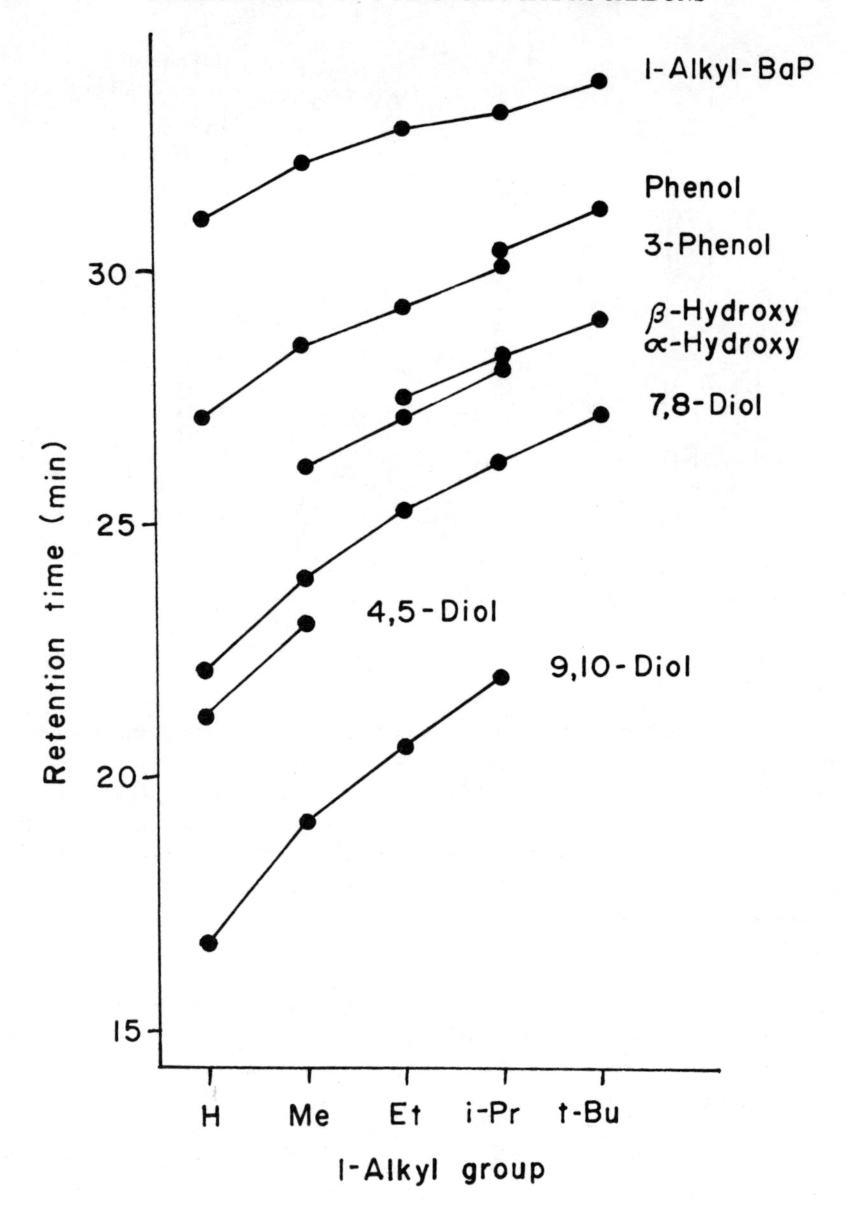

FIG. 7. The relative retention times of the 1-alkyl-BAPs in comparison with those of BAP.

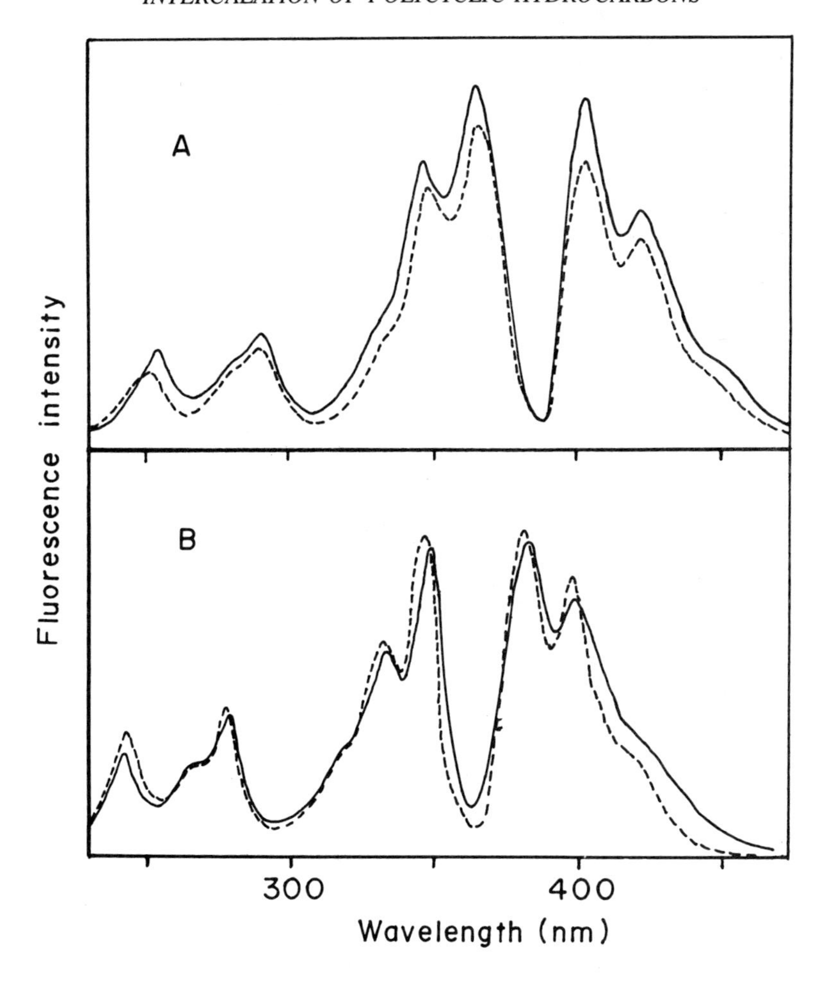

FIG. 8. Fluorescence spectra: (A) The synthetic 7,8-dihydrodiol of 1-isopropyl-BAP (——) and the metabolite of 1-isopropyl-BAP assigned this structure (-----). (B) The mixture of isomeric tetraols formed by hydrolysis of the synthetic racemic anti diol epoxide of 1-isopropyl-BAP (——) and the tetraols isolated as metabolites of the 7,8-dihydrodiol of 1-isopropyl-BAP (-----).

DISCUSSION

All the foregoing evidence is consistent with the hypothesis that intercalation of the active diol epoxide metabolites of the 1-alkyl-BAPs into DNA is involved in the covalent binding of these reactive intermediates to DNA, and this intercalation is a critical event in the mechanism of PAH carcinogenesis. The mutagenicity data for both bacterial (Fig. 5) and mammalian cells (Table 2) correlates well with the tumor-initiating activities of the 1-alkyl-BAPs on mouse skin (Table 3). And both these biological properties show an inverse relationship with the steric requirements of the alkyl group in the 1-position of BAP (Table 1). The differences in the biological activities of the 1-alkyl-BAPs cannot be due to steric interference with metabolic activation of the larger members of the series, since all the 1-alkyl-BAPs undergo metabolism to approximately similar extents and the 7,8-dihydrodiol is a major metabolite in all cases. Moreover, the 7,8-dihydrodiol of 1-isopropyl-BAP was shown to undergo further metabolism to the corresponding bay region diol epoxide. It is likely that similar enzymatic activation takes place with other members of the series. The simplest interpretation of these findings is that substitution of an alkyl group larger than ethyl into the 1-position of BAP leads to formation of diol epoxide metabolites, which are biologically inactive due to steric interference by the alkyl group with intercalation between the base pairs of the DNA helix.

It is conceivable, though less likely, that intercalation is not critically involved, and the presence of large alkyl groups in the 1-position of BAP somehow alters the site of external attack of the active metabolites on DNA to a molecular region that yields adducts that do not result in mutation and cancer. However, it is difficult to see the molecular basis for such an effect, since the 1-alkyl group is quite remote from the reactive center of the PAH molecule. In any case, additional experimental studies are in progress to determine whether the diol epoxide derivative of 1-isopropyl-BAP binds covalently to native DNA and to determine its mutagenic and tumorigenic properties. If the intercalation hypothesis is correct, minimal covalent binding and biological activity may be expected.

ACKNOWLEDGMENTS

Investigations at the University of Chicago were supported by Grants BC-132 and IN-41-W3 from the American Cancer Society and by Public Health Service Grant CA 36097 from the National Cancer Institute, Department of Health and Human Services. Research at the Institute for Cancer Research was supported by the U.K. Medical Research Council, the Cancer Research Campaign, and Grant CA 25807 from the National Cancer Institute, U.S. Department of Health and Human Services.

REFERENCES

1. Adelfgang, J.L., and Daub, G.H. (1955): J. Am. Chem. Soc., 77:3297-3300.
2. Beland, F. (1978): Chem. Biol. Interact., 22:329-339.
3. Conney, A.H. (1982): Cancer Res., 42:4875-4917.

4. Geacintov, N.E., Gagliano, A.G., Ivanovic, V., and Weinstein, I.B. (1978): Biochemistry, 17:5256-5262.
5. Geacintov, N.E., Ibanez, V., Gagliano, A.G., Yoshida, H., and Harvey, R.G. (1980): Biochem. Biophys. Res. Commun., 92:1335-1342.
6. Geacintov, N.E., Yoshida, H., Ibanez, V., and Harvey, R.G. (1981): Biochem. Biophys. Res. Commun., 100:1569-1577.
7. Geacintov, N.E., Yoshida, H., Ibanez, V., and Harvey, R.G. (1982): Biochemistry, 21:1864-1869.
8. Geacintov, N.E., Ibanez, V., Benjamin, M.J., Hibshoosh, H., and Harvey, R.G. (1983): In: Polynuclear Hydrocarbons: Physical and Biological Chemistry, 7th International Symposium, pp. 559-570. Battelle Press, Columbus, Ohio.
9. Geacintov, N.E., Hibshoosh, H., Ibanez, V., Benjamin, M.J., and Harvey, R.G. (1984): Biophys. Chem., 20:121-133.
10. Geacintov, N.E., Ibanez, V., Zinger, D., Yoshida, H., Santella, R., Grunberger, D., and Harvey, R.G. (1984): Carcinogenesis, in press.
11. Harvey, R.G. (1981): Acc. Chem. Res., 14:218-227.
12. Hogan, M.E., Dattagupta, N., and Whitlock, J.P., Jr. (1981): J. Biol. Chem., 256:4504-4513.
13. Ibanez, V., Geacintov, N.E., Gagliano, A.G., Brandimarte, S., and Harvey, R.G. (1980): J. Am. Chem. Soc., 102:5661-5666.
14. Jeffrey, A.M., Jennette, K.W., Blobstein, S.H., Weinstein, I.B., Beland, F.A., Harvey, R.G., Kasai, H., Muira, I., and Nakanishi, K. (1976): J. Am. Chem Soc., 98:5714-5716.
15. Jeffrey, A.M., Weinstein, I.B., Jennette, K.W., Grzeskowiak, K., Nakanishi, K., Harvey, R.G., Autrup, H., and Harris, C. (1977): Nature, 269:348-350.
16. Jennette, K.W., Jeffrey, A.M., Blobstein, S.H., Beland, F.A., Harvey, R.G., and Weinstein, I.B. (1977): Biochemistry, 16:932-938.
17. Li, G.S., Ehler, D.F., and Benkeser, R.A. (1977): In: Organic Syntheses, Vol. 56, edited by G.H. Büchi, pp. 83-87. John Wiley and Sons, New York.
18. Lin, J-H., LeBreton, P.R., and Shipman, L.L. (1980): J. Phys. Chem., 84:642-649.
19. MacLeod, M.C., Mansfield, B.K., and Selkirk, J.K. (1982): Carcinogenesis, 3:1031-1037.
20. Meehan, T., Gamper, H., and Becker, J.F. (1982): J. Biol. Chem., 257:10479-10485.
21. Nakanishi, K., Kasai, H., Cho, H., Harvey, R.G., Jeffrey, A.M., Jennette, K.W., and Weinstein, I.B. (1977): J. Am. Chem. Soc., 99:258-260.
22. Osborne, M.R., Beland, F.A., Harvey, R.G., and Brookes, P. (1976): Int. J. Cancer, 18:362-368.

23. Pataki, J., Konieczny, M., and Harvey, R.G. (1982): J. Org. Chem., 47:1133-1136.
24. Pataki, J., and Harvey, R.G., manuscript in preparation.
25. Pulkrabek, P., Leffler, S., Weinstein, I.B., and Grunberger, D. (1977): Biochemistry, 16:3127-3132.
26. Salerni, L., Engel, J.F., and Downs, J.J. (1966): J. Pharm. Sci., 55:115-116.
27. Singer, B., and Grunberger, D. (1983): Molecular Biology of Mutagens and Carcinogens. Plenum Press, New York.
28. Slaga, T.J., Iyer, R.P., Lyga, W., Secrist, A., III, Daub, G.H., and Harvey, R.G. (1980): In: Polynuclear Aromatic Hydrocarbons: Chemistry and Biological Effects, edited by A. Bjorseth and A.J. Dennis, pp. 753-769. Battelle Press, Columbus, Ohio.
29. Straub, K.M., Meehan, T., Burlingame, A.L., and Calvin, M. (1977): Proc. Nat. Acad. Sci., USA, 74:5285-5288.
30. Subbiah, A., Islam, S.A., and Neidle, S. (1983): Carcinogenesis, 4:211-215.
31. Undeman, O., Lycksell, P., Graslund, A., Astlind, T., Ehrenberg, A., Jernstrom, B., Tjerneld, F., and Norden, B. (1983): Cancer Res., 43:1851-1860.
32. Venitt, S., and Crofton-Sleigh, C. (1981): In: Progress in Mutation Research, Evaluation of Short-Term Tests for Carcinogens, Vol. 1, edited by F.J. deSerres and J. Ashby, pp. 351-366. Elsevier/North Holland, New York.
33. Weinstein, I.B., Jeffrey, A.M., Jennette, K., Blobstein, S., Harvey, R.G., Harris, C., Autrup, H., Kasai, H., and Nakanishi, K. (1976): Science, 193:592-595.
34. Wigley, C.B., Newbold, R.F., Amos, J., and Brookes, P. (1979): Int. J. Cancer, 23:691-696.
35. Windaus, A., and Raichle, K. (1939): Liebigs Ann. Chem., 537:157-170.

Carcinogenesis, Vol. 10, edited by E. Huberman and S. H. Barr. Raven Press, New York © 1985.

Stabilization of Z-DNA Conformation by Chemical Carcinogens

*†D. Grunberger, †R. M. Santella, †L. H. Hanau, and ‡B. F. Erlanger

**Department of Biochemistry, and †Division of Environmental Sciences, School of Public Health, and ‡Department of Microbiology, Columbia University Cancer Center/Institute of Cancer Research, New York, New York 10032*

It has become evident that induction of cancer is a multistep process consisting of an initiation event followed by several stages of progression involving promoters and possibly viruses (43).

It appears that the critical event in the initiation stage of carcinogenesis is modification of DNA by chemical or physical agents, with most chemical carcinogens requiring metabolic activation (19). Since DNA is the carrier of genetic information and mutation occurs as a consequence of its modification, studies on the mechanism of action of carcinogens have generally focused on the ability of the carcinogen to covalently modify DNA and on the effects of chemical modification on DNA structure and function (20,37).

The potent liver carcinogen, N-2-acetylaminofluorene (AAF) has been used extensively to study the effects of carcinogen binding on DNA structure (14). The major DNA adduct obtained after *in vitro* modification of DNA by the N-acetoxy derivative of AAF is *N*-(deoxyguanosin-8-yl)-AAF (Fig. 1) (14), although *in vivo* part of the C8-dG adduct is deacetylated (2,15). A minor adduct, N-(deoxyguanosin-N^2yl)-AAF (45) is also formed *in vitro*.

Our previous studies have indicated that binding of the AAF residue to the C8 position of deoxyguanosine is associated with major conformational changes in DNA structure, resulting in a conformation termed "base displacement" (8,44). In this model, the glycosidic N9-C1' bond is altered from the *anti* conformation, characteristic of B-DNA, to the *syn* conformation. In addition, the AAF residue is inserted into the helix and is stacked parallel with adjacent bases; the modified guanine residue is dis-

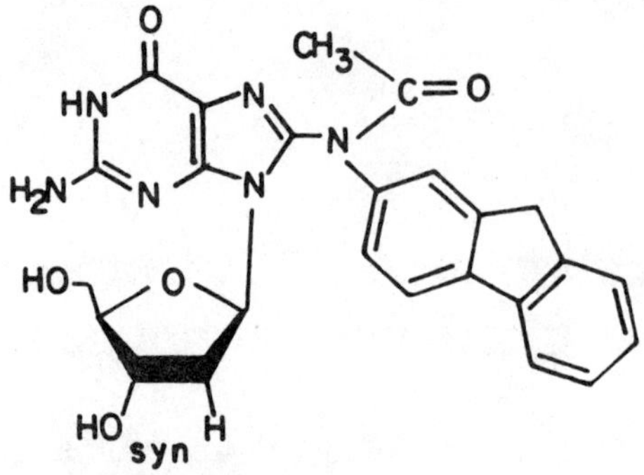

FIG. 1. Structure of N-(deoxyguanosine-8-yl)N-acetyl-2-aminofluorene in syn conformation.

placed to the outside of the helix. A similar model referred to as the "insertion-denaturation" model has been presented (7).

Recently, a new double-stranded conformation of left-handed DNA, designated Z-DNA, has been described (41). In left-handed poly-(dG-dC).poly(dG-dC) the guanosine residues are in syn conformation, whereas the deoxycytidine residues are in anti conformation. There is evidence that Z-DNA can occur both in crystal form and in solution.

There are two distinct forms of poly(dG-dC).poly(dG-dC) in solution, and a cooperative transition between them occurs at high salt or ethanol concentrations (29,30). These conformational changes can be demonstrated by circular dichroism (CD) spectra. The CD spectrum of the high-salt form is virtually an inversion of that of the low-salt (B-DNA) form. Thus, varying the salt or ethanol concentration of a solution of poly(dG-dC).poly(dG-dC) makes it possible to convert the molecules from a right-handed to a left-handed form. It is also possible to influence the equilibrium by chemical modification of DNA. Bromination and methylation of poly(dG-dC).poly(dG-dC) in the C8 and N7 position of dG, respectively, stabilize Z-DNA (16,23). Likewise, it is stabilized by methylation (1) or bromination of cytosine in the C5 position (22). Moreover, in vivo Z-DNA can be stabilized by specific binding proteins (28) or by an appropriate degree of negative supercoiling of DNA (38). Stabilization of Z-DNA structure by specific anti Z-DNA antibodies has also been reported (18,26).

The wide distribution of purine-pyrimidine sequences in eukaryotic genomes has prompted the suggestion that a corresponding B-Z equilibrium may be involved in gene expression (13). It appears to be of considerable interest therefore to study conditions for the stabilization of Z-DNA by chemical carcinogens.

RESULTS

CD Spectra of AAF-modified poly(dG-dC).poly(dG-dC) and poly(dG-m^{5}dC).poly(dG-m^{5}dC)

As a direct consequence of the syn conformation, the C8 position of the dG residues in Z-conformers is exposed on the outer surface of the molecule (42). Therefore, modification of this position in the Z-conformers by AAF should be less hindered and could lead to the stabilization of the Z-form.

We have explored the conformational changes induced in poly-(dG-dC).poly(dG-dC) by AAF modification (35). Reaction of acetoxy-AAF with poly(dG-dC).poly(dG-dC) produces only the C8-dG adduct. Figure 2 presents the CD spectra of poly(dG-dC).poly(dG-dC) in aqueous buffer and in 60% ethanol. In aqueous solution, a B-DNA spectrum is seen, whereas in 60% ethanol, the CD spectrum is inverted and is characteristic of Z-DNA. Also shown is the spectrum of poly(dG-dC).poly(dG-dC) modified by AAF to the extent of 28%. With this high level of modification, even in aqueous solution the CD spectrum is characteristic of Z-DNA. With a lower level of modification (3%) the polymer still shows a CD characteristic of B-DNA but inversion to that of Z-DNA occurs at a lower ethanol concentration (45%) than required for unmodified polymer. On the other hand, poly(dG).poly(dC), a homopolymer that cannot undergo the B to Z transition (30), showed no changes in CD spectra even after high levels of AAF modification.

It has been shown that the methylated polynucleotide poly-(dG-m^{5}dC).poly(dG-m^{5}dC) undergoes a transition from B to Z at much lower salt concentrations than required to invert the nonmethylated form (1). Because the dinucleotide sequence m^{5}dC-dG occurs frequently in eukaryotic DNA and the presence of methylated C within a structural gene has been implicated in the inhibition of transcription of certain eukaryotic genes (31), it was of interest to investigate the possible conformational changes of poly(dG-m^{5}dC).poly(dG-m^{5}dC) resulting from AAF modification (34). Figure 3 shows the CD spectra of poly-(dG-m^{5}dC).poly(dG-m^{5}dC) with various levels of AAF modification. With increasing levels of AAF binding, there is a decrease in the positive band at 280 nm and an increase in the negative band at 255 nm. At modification levels above 6%, a negative band appears at 295 nm, which is characteristic of Z-DNA. With 10.4% modification, the polymer was completely in the Z-form. Thus, both methylation at the C5 position of dC and AAF modification at the C8 position of dG favor stabilization of the Z-DNA conformation.

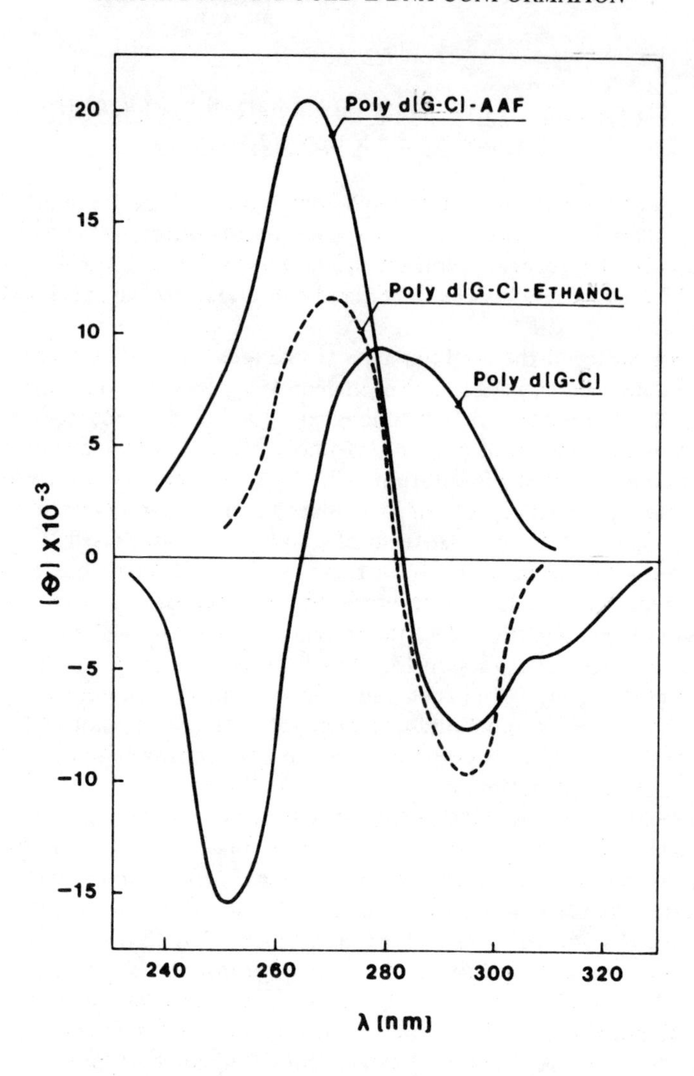

FIG. 2. Circular dichroism (CD) spectra of poly(dG-dC).poly(dG-dC) in 1 mM phosphate buffer, in 60% ethanol, and modified by AAF to an extent of 28% in 1 mM phosphate buffer. From Santella et al. (35).

Recognition of AAF-poly(dG-dC).poly(dG-dC) by Nuclease S_1 and Anti-cytidine Antibodies

Previous studies on the susceptibility of AAF-modified B-DNA to S_1 nuclease digestion have provided evidence for localized regions of denaturation at the sites of AAF modification (6,46). To obtain additional

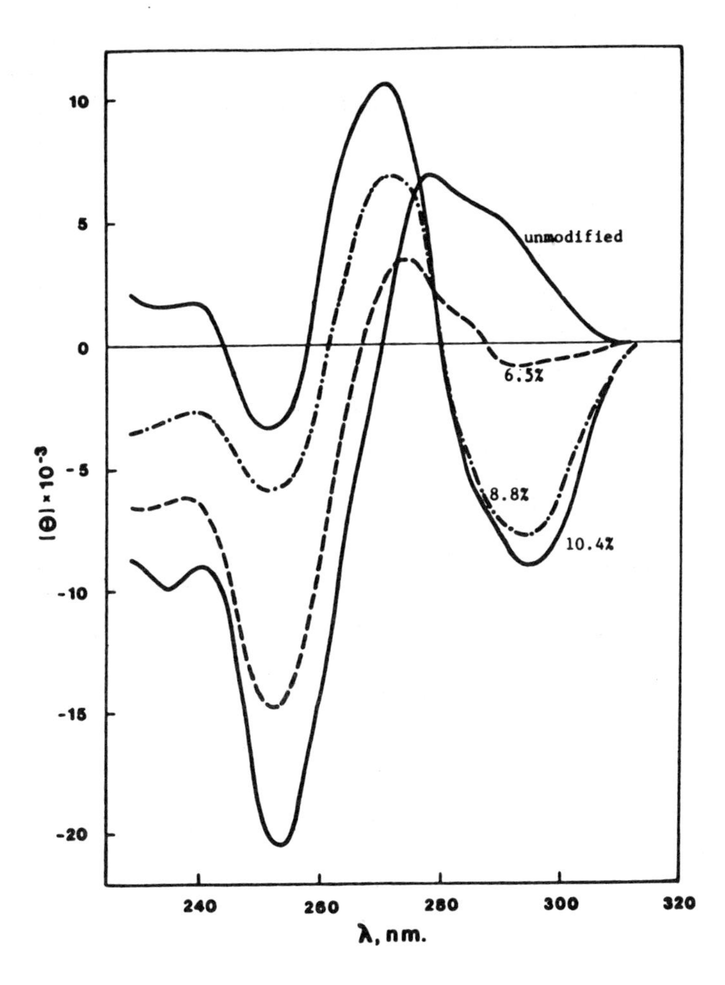

FIG. 3. CD spectra of poly(dG-m^{5}dC).poly(dG-m^{5}dC) in 50 mM NaCl, 5 mM Tris (pH 8.0) with various levels of AAF modification as indicated. From Santella et al. (34).

information on the conformations of AAF-modified poly(dG-dC).poly-(dG-dC) and poly(dG-m^{5}dC).poly(dG-m^{5}dC), we investigated their susceptibilities to S_1 nuclease digestion (33). Figure 4 shows that denatured DNA and AAF-modified DNA were readily digested, indicating that they contain significant amounts of single-stranded regions. In contrast, AAF-modified poly(dG-dC).poly(dG-dC) and poly(dG-m^{5}dC).poly(dG-m^{5}dC) are essentially resistant to S_1 nuclease digestion in accord with a Z-DNA type conformation.

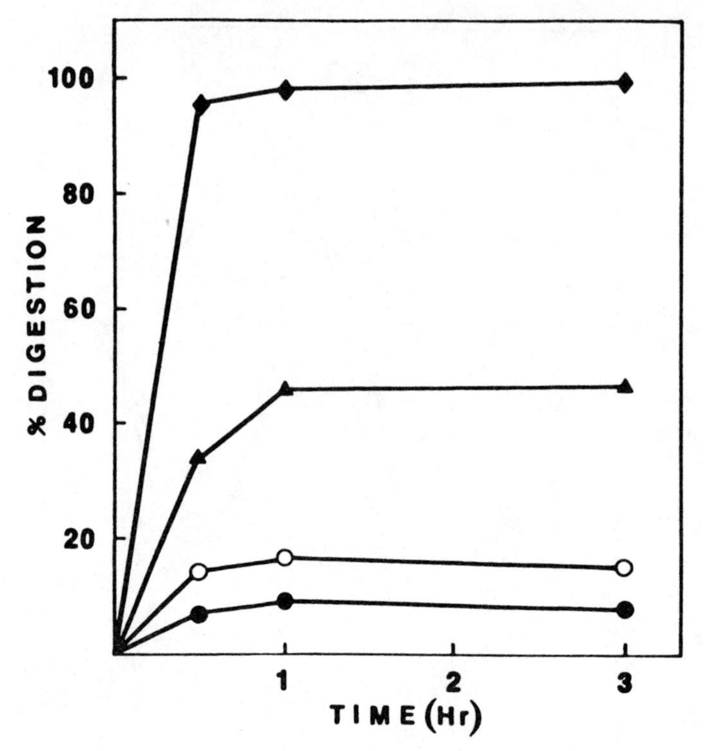

FIG. 4. Nuclease-S_1 digestion of nucleic acids modified with AAF. (◆) denatured calf thymus DNA; (▲) AAF-DNA, 8.5% modified; (O) poly(dG-m dC).poly(dG-m^{5}dC); (●) AAF-poly(dG-dC).poly(dG-dC), 28% modified. From Santella et al. (34).

Similar results were obtained with anticytidine antibodies that react only with bases in single-stranded regions (4). These antibodies precipitate denatured [^{3}H] DNA but not native DNA, and the addition of unlabeled denatured DNA inhibits the precipitation of the radioactive tracer. Shown in Fig. 5 is the inhibition of precipitation of [^{3}H] denatured DNA by anti-C antibodies by AAF-modified DNA and AAF-modified poly-(dG).poly(dC) but not by AAF-modified poly(dG-dC).poly(dG-dC).

These results are consistent with those of S_1 nuclease digestion experiments and with the conclusion that AAF-modified poly(dG-dC).poly-(dG-dC) does not contain any significant single-stranded regions. In turn, the CD spectrum indicated that the AAF-modified poly(dG-dC).poly(dG-dC) adopts a Z-DNA type conformation.

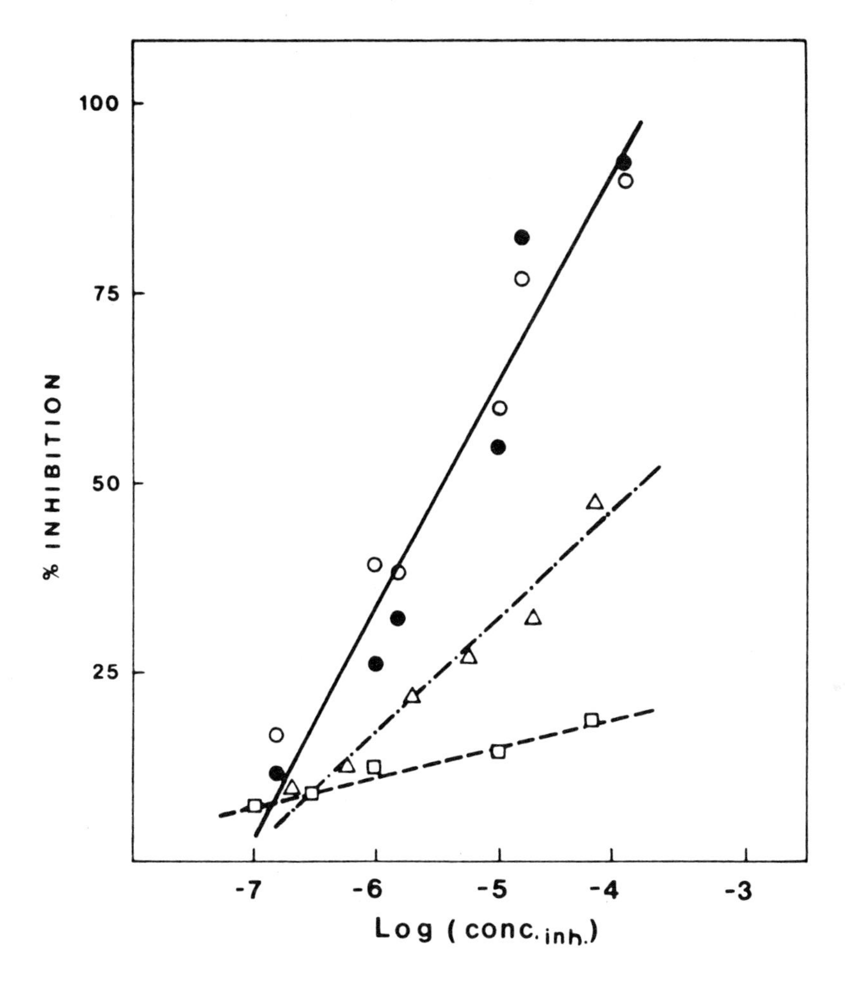

FIG. 5. Radioimmunoassay at nonequilibrium conditions in which the binding of anti-C (antibody to [^{3}H] denatured DNA) was measured in the presence of various concentrations of (●) dDNA; (○) AAF-DNA, 11% modified; (□) AAF-poly(dG-dC).poly(dG-dC), 21% modified; (△) poly(dG).poly(dC), 5% modified. From Santella et al. (33).

Antibodies to Z-DNA Raised by AAF-poly(dG-dC).poly(dG-dC)

A major advance in the effort to determine the biological role of the left-handed Z-DNA helix came about with the finding that Z-DNA, unlike B-DNA, is immunogenic (16). The first antibodies were elicited by immunization with poly(dG-dC).poly(dG-dC) in which the 8-position of dG and the 5-position of dC were brominated (16). Subsequently, Z-specific antibodies were also raised to (dien) PtCl -poly(dG-dC) (17).

Since the inverted CD spectrum of AAF-poly(dG-dC).poly(dG-dC) indicated the ability of this polymer to assume a Z-conformation, we have raised antibodies to AAF-poly(dG-dC).poly(dG-dC) to determine whether it has unique Z-DNA determinants of its own (10).

The immunization of rabbits was carried out by a complex of methylated bovine serum albumin and AAF-poly(dG-dC).poly(dG-dC) emulsified in an equal volume of Freund's adjuvant. Antibodies from the serum 14 days after the fifth boost were purified by affinity chromatography on an Affi-Gel-102-AAF-poly(dG-dC).poly(dG-dC) or Br-poly(dG-dC).poly(dG-dC) column. Exposure of the antisera to Br-poly(dG-dC).poly(dG-dC) immunoadsorbent yielded two fractions: a supernatant containing antibodies that did not bind to the immunoadsorbent and a fraction that bound and was subsequently eluted with 1.8 M KSCN.

Figure 6 shows that the fraction that bound to a Br-poly(dG-dC).poly(dG-dC) column and was then eluted with 1.8 M KSCN bound [^{3}H]poly(dG-dC).poly(dG-dC) in 3.5 M NaCl and to [^{3}H]AAF-poly(dG-dC).poly(dG-dC). The supernatant, on the other hand, bound only to AAF-poly(dG-dC).poly(dG-dC).

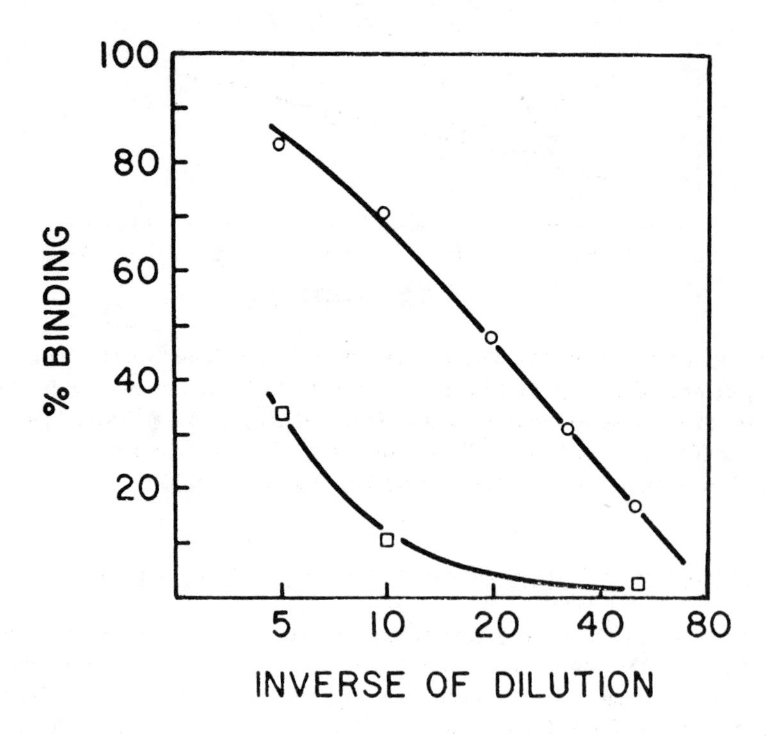

FIG. 6. Binding of Br-poly(dG-dC).poly(dG-dC) purified fraction to [^{3}H]AAF-poly(dG-dC).poly(dG-dC) (○) and to [^{3}H]poly(dG-dC).poly(dG-dC) in 3.5 M NaCl (□). From Hanau et al. (10).

TABLE 1. Maximal inhibition of binding of [^{3}H]AAF-poly(dG-dC).poly-(dG-dC) with the Br-poly(dG-dC).poly(dG-dC)-purified fraction.

Inhibitor	Maximal inhibition (%)
AAF-poly(dG-dC).poly(dG-dC)	78
BR-poly(dG-dC).poly(dG-dC)	73
AAF-poly(dG).poly(dC)	0

In radioimmunoassay inhibition studies (Table 1) binding of [^{3}H]AAF-poly(dG-dC).poly(dG-dC) with the Br-poly(dG-dC).poly(dG-dC)-purified fraction was inhibited about equally (approximately 75%) by AAF-poly-(dG-dC).poly(dG-dC) and by Br-poly(dG-dC).poly(dG-dC). On the other hand, binding of [^{3}H]AAF-poly(dG-dC).poly(dG-dC) was not inhibited by AAF-poly(dG).poly(dC).

Thus, the purified preparation includes a population of antibodies that recognizes to an equal extent determinants that are shared by AAF-poly-(dG-dC).poly(dG-dC), Br-poly(dG-dC).poly(dG-dC), and poly(dG-dC).poly-(dG-dC) in high salt. The supernatant (Table 2, Fraction A), on the other hand, as shown by inhibition studies, recognized determinants only on AAF-poly(dG-dC).poly(dG-dC) and did not recognize Br-poly(dG-dC).poly-(dG-dC). Similarly, the supernatant fraction did not bind to poly(dG-dC).poly(dG-dC) in high salt. Only a small percentage of the binding of the supernatant could be identified as being specific for a non-Z-DNA epitope, as shown by the fact that a 10-fold excess of AAF-poly(dG).poly-(dC) inhibited only about 20% of the binding (Table 2, Fraction A). We

TABLE 2. Inhibition studies on a fraction not bound by Br-poly-(dG-dC).poly(dG-dC) (Fraction A) and on a subfraction subsequently purified on an AAF-poly(dG-dC).poly(dG-dC) immunoadsorbent (Fraction B).

Inhibitor	Maximal inhibition (%)	
	Fraction A	Fraction B
AAF-poly(dG-dC).poly(dG-dC)	89.5	95.5
Br-poly(dG-dC).poly(dG-dC)	0	4.0
AAF-poly(dG).poly(dC)	20.0	21.0

conclude that unique Z-conformational determinants on AAF-poly-(dG-dC).poly(dG-dC) are being detected in the supernatant fraction.

An AAF-poly(dG-dC).poly(dG-dC) immunoadsorbent was further used to purify the populations of antibodies that did not bind to the Br-poly-(dG-dC).poly(dG-dC) immunoadsorbent. Binding properties of this subfraction (Table 2, Fraction B) reflected those of its source. Binding to [^{3}H]AAF-poly(dG-dC).poly(dG-dC) was inhibited almost completely by AAF-poly(dG-dC).poly(dG-dC) but not by Br-poly(dG-dC).poly(dG-dC); a maximal inhibition of 21% was obtained by AAF-poly(dG).poly(dC). Thus, there is a population of antibodies that recognizes AAF-poly(dG-dC).poly-(dG-dC) and yet is not bound by poly(dG-dC).poly(dG-dC) in high salt, by Br-poly(dG-dC).poly(dG-dC), or AAF-poly(dG).poly(dC).

The antiserum, when fractionated on an AAF-poly(dG-dC).poly(dG-dC) immunoadsorbent, yielded a specifically purified fraction that reflected the binding characteristics of the serum. Figure 7 shows that this fraction bound [^{3}H]poly(dG-dC).poly(dG-dC) in 3.5 M NaCl and [^{3}H]AAF-poly-(dG-dC).poly(dG-dC). Furthermore, binding to [^{3}H]poly(dG-dC).poly-(dG-dC) in 3.5 M NaCl could be completely inhibited by Br-poly-(dG-dC).poly(dG-dC) (data not shown).

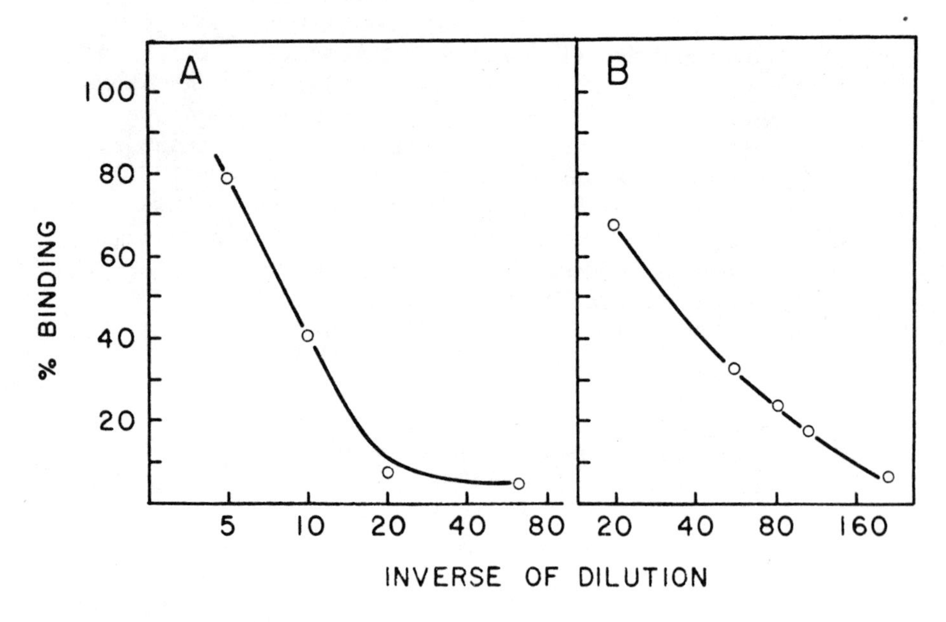

FIG. 7. Binding of AAF-poly(dG-dC).poly(dG-dC) purified fraction to [^{3}H]poly(dG-dC).poly(dG-dC) in 3.5 M NaCl (A) and to [^{3}H]AAF-poly-(dG-dC).poly(dG-dC) (B). From Hanau et al. (10).

Taken together, immunization experiments with AAF-poly-(dG-dC).poly(dG-dC) yielded at least two different populations of Z-specific antibodies. First, there were populations that were stimulated by conformational characteristics of Z-DNA shared in common by poly-(dG-dC).poly(dG-dC) in 3.5 M NaCl, Br-poly(dG-dC).poly(dG-dC), and AAF-poly(dG-dC).poly(dG-dC). Also elicited was another population of antibodies that was stimulated by structural characteristics unique to AAF-poly(dG-dC).poly(dG-dC). This fraction, which was not bound by the Br-poly(dG-dC).poly(dG-dC) immunoadsorbent, did not react with Br-poly-(dG-dC).poly(dG-dC) or poly(dG-dC).poly(dG-dC) in 3.5 M NaCl. After purification by binding to, and elution from, an AAF-poly(dG-dC).poly(dG-dC) immunoad-sorbent, its binding properties remained unchanged. Only a small fraction recognized a non-Z, AAF-bearing polynucleotide, AAF-poly-(dG).poly(dC). The rest, therefore, must have been produced in response to immunogenic conformational characteristics unique to AAF-poly(dG-dC).poly(dG-dC). Since B-DNA is not immunogenic, a portion of the population reacting with AAF-poly(dG-dC).poly(dG-dC) appears to be directed toward unique Z-DNA determinants of this polynucleotide. Our finding, therefore, is added to the reports of others (18,21,47) concerning Z-DNA polymorphism as detected immunochemically.

Perfect and Distorted Z-DNA Conformations of AAF-poly(dG-dC).poly(dG-dC)

The two different populations of anti-Z-DNA antibodies elicited by AAF-poly(dG-dC).poly(dG-dC) raised the possibility of existence of two Z-type conformer populations of AAF-modified poly(dG-dC).poly(dG-dC). Minimized potential energy calculations for dCpdG-AAF were used to predict two conformations for the AAF-modified poly(dG-dC).poly(dG-dC) (3).

In the first model, the DNA backbone conformation matches that found in the dCpdG residue of Z forms in every respect (Fig. 8a). In this model, the carcinogen is situated at the Z helix exterior in a flexible position, and it causes no denaturation or other deformation to the Z form (33). Since this conformation is shared by other Z-DNA conformers, the first antibody population will recognize all Z-DNA polymers sharing these characteristics.

The other population of antibodies that reacts only with AAF-poly-(dG-dC).poly(dG-dC) has to recognize special antigenic determinants unique to this modified polymer. Broyde and Hingerty (3) suggest a new model in which the carcinogen is stacked with the adjacent bases. The model (11) shown in Fig. 8b incorporates the Z-like base displaced conformer into the Z-II form of DNA (42). To incorporate the modified dCpdG into this Z form, some alterations of the perfect Z conformation are required. The modified dG residue is at the exterior of the helix and the bulky AAF is parallel to and partly stacked with the adjacent cytidine on the same strand. In addition, in the case of double-stranded DNA, the cytidine

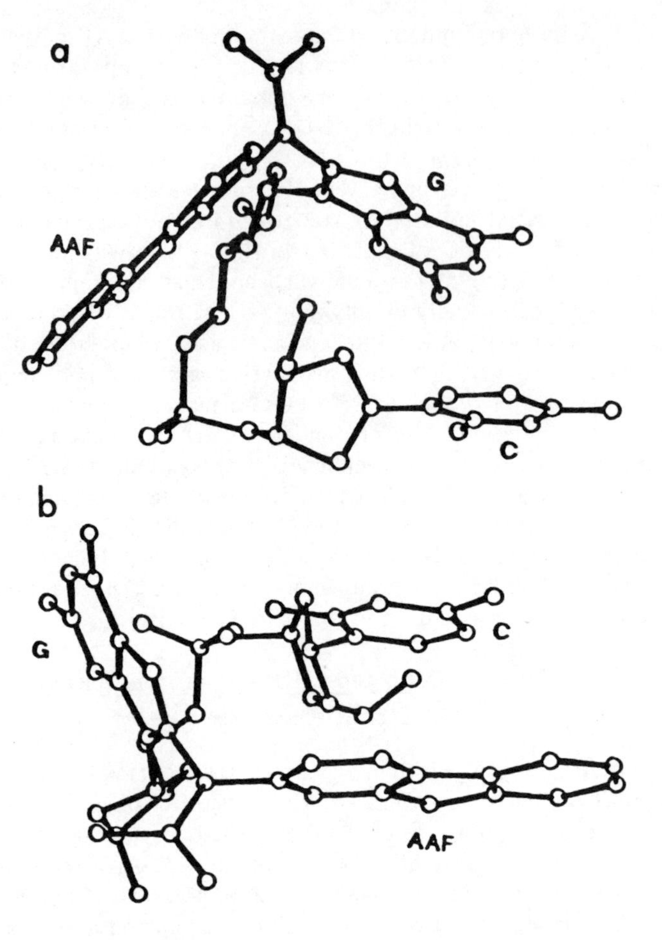

FIG. 8. Minimum energy conformation of dCpdG-AAF. (a) Perfect Z-type conformation with AAF at exterior of DNA backbone. (b) Distorted Z-like conformer with AAF stacked with the adjacent base. From Santella et al. (33) and Hingerty and Broyde (11).

opposite the AAF was rotated from anti to syn causing a rupture of that base pair. It is possible that the low but measurable single strandedness detected by S_1 nuclease digestion and anti-C antibodies (Figs. 4 and 5) for the Z form polymers is a reflection of the base-displaced Z-DNA model in which one base pair is ruptured. Since this conformation represents a distorted Z-DNA, this form may be the antigenic determinant for that population of antibodies that react uniquely with AAF-poly(dG-dC).poly(dG-dC).

Although the above model is consistent with immunological data, we do not exclude alternative explanations for the apparent polymorphism of Z-DNA.

DISCUSSION

The inverted CD spectra, the S_1 nuclease digestion experiments, and the findings with anti-cytidine antibody indicate that in AAF, modified poly(dG-dC).poly(dG-dC) and poly(dG-m^{5}dC).poly(dG-m^{5}dC) adopt a Z-DNA type conformation (32,33,34,35). Since populations of antibodies raised by immunization with AAF-poly(dG-dC).poly(dG-dC) recognized conformational characteristics of Z-DNA shared in common by poly(dG-m^{5}dC).poly-(dG-m.dC) and poly(dG-dC).poly(dG-dC) in high salt, and by Br-poly(dG-dC).poly(dG-dC), there is no doubt that AAF modification of polymers with alternating purine-pyrimidines stabilizes a left-handed Z-type conformation.

Of critical importance in evaluating the possible roles of Z-DNA *in vivo* is the demonstration of alternating purine-pyrimidine sequences in the eukaryotic genome. Such sequences have been found in introns of the human actin (9) and fetal gamma globin genes (39), in the 3' flanking region of the mouse Vk immunoglobulin gene (25), and in the 5' flanking region of the human parathyroid hormone gene (40).

In spite of the wealth of knowledge about the structure and distribution of these sequences, very little is yet known about their function. It has recently been proposed that Z-DNA may play a role in gene expression (13). Furthermore, Nordheim and Rich (27) have identified several stretches of alternating purine-pyrimidines in the enhancer sequence from SV-40 that are able to assume the Z-DNA conformation and have suggested that the enhancer function might be dependent on the presence of Z-DNA. On the other hand, it has been shown that removal of a 12 base-pair-long stretch of alternating purine-pyrimidine residues from the 5' flanking sequence of a transcriptionally inactive tRNA^{Met} gene resulted in activation of this gene (12). Similarly, a strong inhibitory effect occurred when potential Z-DNA sequences were placed in the flanking regions of a tRNA^{Pro} gene (36). In other systems it has been reported that Z-DNA impairs the formation of normal chromatin structure (24) and that it has lower affinity for regulatory DNA binding proteins (5). Despite all these findings, the biological function of Z-DNA is still not clear. However, if the effects of AAF modification on the conformation of alternative purine-pyrimidine sequences that we have seen in our *in vitro* studies also occur *in vivo*, then AAF modification could markedly influence the biological role of those sequences. Further studies are required to determine how these alternative conformations, if they affect biological function, relate to the process of carcinogenesis.

ACKNOWLEDGMENTS

This work was supported by Public Health Service Grants CA 21111, CA 13696, 5T-32-G17-07367, and AI-06860 from the National Institutes of Health, U.S. Department of Health and Human Services. We thank Dr. S. Broyde for valuable discussions and Ms. S. Allen for typing this manuscript.

REFERENCES

1. Behe, M., and Felsenfeld, G. (1981): Proc. Nat. Acad. Sci., USA, 78:1619-1623.
2. Beland, F.A., Dooley, K.L., and Jackson, C.D. (1982): Cancer Res., 42:1348-1354.
3. Broyde, S., and Hingerty, B. (1984): In: Molecular Basis of Cancer, edited by R. Rein. Alan R. Liss, Inc., New York (in press).
4. Erlanger, B.F., and Beiser, S.M. (1964): Proc. Nat. Acad. Sci., USA, 52:433-437.
5. Fried, M.G., Wu, H.M., and Crothers, D.M. (1983): Nucleic Acids Res., 11:2479-2494.
6. Fuchs, R.P.P. (1975): Nature, 257:151-152.
7. Fuchs, R.P.P., and Daune, M. (1972): Biochemistry, 11:2659-2666.
8. Grunberger, D., and Weinstein, I.B. (1976): In: Biology of Radiation and Carcinogenesis, edited J.M. Yuhas, R.W. Tennant, and J.D. Regan, pp. 175-187. Raven Press, New York.
9. Hamada, H., and Kakunaga, T. (1982): Nature, 298:396-398.
10. Hanau, L.H., Santella, R.M., Grunberger, D., and Erlanger, B.F. (1984): J. Biol. Chem., 259:173-178.
11. Hingerty, B., and Broyde, S. (1982): Biochemistry, 21:3243-3252.
12. Hipskind, R.A., and Clarkson, S.G. (1983): Cell, 34:881-890.
13. Klysik, J., Stirdivant, S.M., Larson, J.E., Hart, P.A., and Wells, R.D. (1981): Nature, 290:672-677.
14. Kriek, E., Miller, J.A., Juhl, U., and Miller, E.C. (1967): Biochemistry, 6:177-182.
15. Kriek, E., and Westra, J.G. (1980): Carcinogenesis, 1:459-468.
16. Lafer, E.M., Möller, A., Nordheim, A., Stollar, B.D., and Rich, A. (1981): Proc. Nat. Acad. Sci., USA, 78: 3546-3550.
17. Malfoy, B., and Leng, M. (1981): FEBS Lett., 132:45-48.
18. Malfoy, B., Rousseau, N., and Leng, M. (1982): Biochemistry, 21:5463-5467.
19. Miller, E.C. (1978): Cancer Res., 38:1479-1496.
20. Miller, E.C., and Miller, J.A. (1981): Cancer, 47:2327-2345.
21. Möller, A., Gabriels, J.E., Lafer, E.M., Nordheim, A., Rich, A., and Stollar, B.D. (1982): J. Biol. Chem., 257:12081-12085.

22. Möller, A., Nordheim, A., Kozlowski, S.A., Patel, D.J., and Rich, A. (1984): Biochemistry, 23:54-62.
23. Möller, A., Nordheim, A., Nichols, S.R., and Rich, A. (1981): Proc. Nat. Acad. Sci., USA, 78:4774-4781.
24. Nickol, J., Behe, M., and Felsenfeld, G. (1982): Proc. Nat. Acad. Sci., USA, 79:1771-1775.
25. Nishioka, Y., and Leder, P. (1980): J. Biol. Chem., 255:3691-3694.
26. Nordheim, A., Pardue, M.L., Lafer, E.M., Möller, A., Stollar, B.D., and Rich, A. (1981): Nature, 294:417-422.
27. Nordheim, A., and Rich, A. (1983): Nature, 303:674-679.
28. Nordheim, A., Tesser, P., Azorin, F., Kwon, Y.H., Miller, A., and Rich, A. (1982): Proc. Nat. Acad. Sci., USA, 79:7729-7733.
29. Pohl, F.M. (1976): Nature, 260:743-745.
30. Pohl, F.M., and Jovin, T.M. (1972): J. Molec. Biol., 67:375-396.
31. Razin, A., and Riggs, A.D. (1980): Science, 210:604-610.
32. Sage, E., and Leng, M. (1980): Proc. Nat. Acad. Sci., USA, 77:4597-4601.
33. Santella, R.M., Grunberger, D., Broyde, S., and Hingerty, B.E. (1981): Nucleic Acids Res., 9:5459-5467.
34. Santella, R.M., Grunberger, D., Nordheim, A., and Rich, A. (1982): Biochem. Biophys. Res. Comm., 106:1226-1232.
35. Santella, R.M., Grunberger, D., Weinstein, I.B., and Rich, A. (1981): Proc. Nat. Acad. Sci., USA, 78:1451-1455.
36. Santoro, C., Costanzo, F., and Ciliberto, G. (1984): EMBO J., 3:1553-1559.
37. Singer, B., and Kusmierek, J.T. (1982): Annu. Rev. Biochem., 52:655-693.
38. Singleton, C.K., Klysik, J., Stirdivant, S.M., and Wells, R.D. (1982): Nature, 99:312-316.
39. Slightom, J.L., Blechl, A.E., and Smithies, O. (1980): Cell, 21:627-638.
40. Vasicek, T.J., McDevitt, B.E., Freeman, M.W., Fennick, B.J., Hendy, G.N., Potts, J.T., Rich, A., and Kronenberg, H.M. (1983): Proc. Nat. Acad. Sci., USA, 80:2127-2131.
41. Wang, A.H., Quigley, G.J., Kolpak, F.J., Crawford, J.L., Van Boom, J.A., Van der Marel, G., and Rich, A. (1979): Nature, 282:680-686.
42. Wang, A.H.J., Quigley, G.J., Kolpak, F.J., Van der Marel, J., van Boom, J.H., and Rich, A. (1981): Science, 211:171-176.
43. Weinstein, I.B. (1981): J. Supramol. Struc. Cell. Biochem., 17:99-120.
44. Weinstein, I.B., and Grunberger, D. (1974): In: Chem. Carcinogenesis, Part A, edited by P.O.P. Tso and J.A. DiPaolo, pp. 217-235. Marcel Dekker, New York.
45. Westra, J.G., Kriek, E., and Hittenhausen, H. (1976): Chem. Biol. Interact., 15:149-164.

46. Yamasaki, H., Pulkrabek, P., Grunberger, D., and Weinstein, I.B. (1977): Cancer Res., 37:3756-3760.
47. Zarling, D.A., Arndt-Jovin, D.J., Robert-Nicoud, M., McIntosh, L.P., Tomae, R., and Jovin, T.M. (1984): J. Mol. Biol., 176:369-415.

Carcinogenesis, Vol. 10, edited by E. Huberman and S. H. Barr. Raven Press, New York © 1985.

In Vitro Models of Mutagenesis

B. S. Strauss, K. Larson, D. Sagher, S. Rabkin, R. Shenkar, and J. Sahm

Department of Molecular Genetics and Cell Biology, The University of Chicago, Chicago, Illinois 60637

One of the thumb-marked chapters in our copy of the 1975 Annual Reviews of Biochemistry is the essay on chemical carcinogenesis written by Dr. Charles Heidelberger (8). This chapter appeared at the start of what has been an exciting period in cancer research, and it is interesting to see the topics that Heidelberger thought important: "Chemicals as a major cause of human cancer," "Chemical oncogenesis in cell cultures," "Do chemical carcinogens 'switch on' oncogenic viruses?" Dr. Heidelberger saw the importance of the phenomenon of mutation for carcinogenesis, and he was particularly excited by the development of experimentally controlled cell-transformation systems. He concluded in 1975 by saying, "...there are exciting recent developments in chemical carcinogenesis, and that while very few final and definitive answers to questions about mechanisms have yet emerged, there are now model systems available in cell cultures that promise to provide such answers." As Heidelberger predicted, the cellular transformation systems have provided experimental tools for the exciting oncogene analysis of recent years, and although there are still few final and definitive answers, we may be closer to them.

It is easy enough to demonstrate that treatment of cells with carcinogens blocks DNA synthesis (15). It is also possible, although not so easy, to demonstrate that most carcinogens are mutagens for mammalian cells (1). It is a hypothesis that these phenomena are related, that is, that the lesion that blocks DNA synthesis is also the lesion responsible for mutation. This last phrase has become more complex than was thought some years ago because "responsible" has two meanings in bacteria where, for ultraviolet (UV) irradiation at least, the lesion that blocks DNA synthesis, the cyclobutane dimer, is also responsible for inducing a metabolic state in which mutation is possible (26), but another lesion, the 6-4 dimer, is the site at which most mutation occurs (2). Since the second lesion also appears to block DNA synthesis and to be recognized by the excision repair system (6), this sophistication may be unnecessary since one can suppose that both carcinogenesis and mutagenesis occur as a result of the initial block to synthesis that results, directly or indirectly, in the insertion of errors into DNA.

For the past several years we have been engaged in the study of a model system that would have the characteristics required of a mutation system. It would block the progression of DNA synthesis at growing points and it would permit bypass by translesion synthesis with the insertion of "targeted" errors opposite the lesion. In this paper we will discuss this model in terms of both new and previously published data from this laboratory. We would also like to demonstrate the applicability of the hypothesis to data in the literature and to point out an opportunity for an *in vivo* test that arises from our experiments.

MODEL SYSTEM

Our model is dependent on the observation (12) that DNA synthesis stops exactly one nucleotide before the site of the lesion when a single-stranded circular DNA used as a primer is reacted with carcinogen. The basic protocol that permits this conclusion is shown in Fig. 1. A multi-reacted template is annealed with an appropriate restriction fragment (RF) prepared from viral RF and used in a DNA synthesis reaction with [^{32}P]-labeled deoxynucleotide triphosphates (dNTPs). After a period of time the reaction is stopped; the product is then restricted, denatured, and sequenced. Termination bands are seen before the expected reaction sites for DNA treated with ultraviolet radiation (UV), benzo[a]pyrene (BAP), and N-acetoxy acetyl-aminofluorene (AAF) (13). Since it is a question whether synthesis past lesions occurs more easily on double-stranded templates, we have also used a protocol (16) in which a nicked (-) strand is annealed to a reacted (+) viral strand and the 5'OH end is either nick translated or strand displaced (Fig. 2). We purify the annealed double-stranded template away from other molecules in the mixture by gel electrophoresis (Fig. 3) and then electroelute from the gel. When such templates are reacted with dimethyl sulfate (DMS) and used for synthesis with either *Escherichia coli* pol I (Klenow fragment) [polI Kf] or with avian myleoblastosis virus (AMV) reverse transcriptase, a pattern of bands is seen that we interpret as stops before the sites blocking DNA synthesis--in this particular template almost exclusively adenines (Fig. 4). We understand that these results are not universal but refer only to the exact reaction conditions of our experiments and that there are suggestions of easier bypass on double-stranded templates (16). We will return to this point later. It is possible to specify some of the factors that are involved in termination. We suppose that termination does not depend on the 3'>5' exonuclease activity of prokaryotic polymerases because we observe ter-

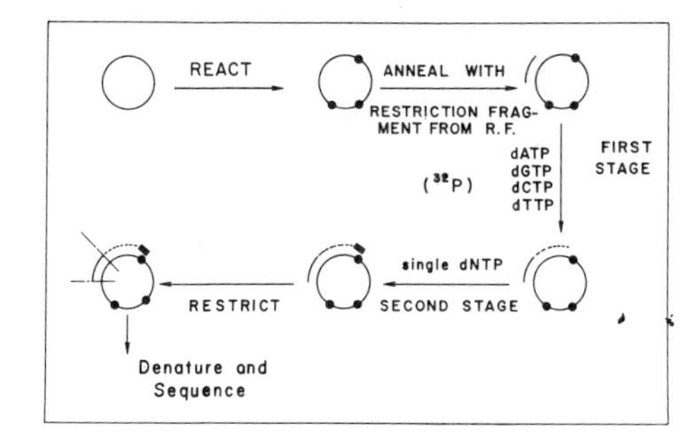

FIG. 1. Protocol for first and second stage reactions in which reacted, circular, single-stranded virus DNA is used as a template (12,13,18,19, 20,23).

mination at all lesions with enzymes such as AMV reverse transcriptase (14) and DNA polymerase alpha (12), which have no detectable nuclease activity. (Second-stage gels provide a useful method of looking for traces of contaminating nuclease as seen by the degradation of the first-state product that occurs on incubation with enzyme in the absence of dNTPs.) The exact site at which the major portion of termination is observed in a reaction mixture containing dNTPs does depend on nucleolytic activity because addition of high concentrations of dNMPs, known to inhibit nuclease activity, makes termination opposite rather than one nucleotide before lesions a more likely event (19). The exact

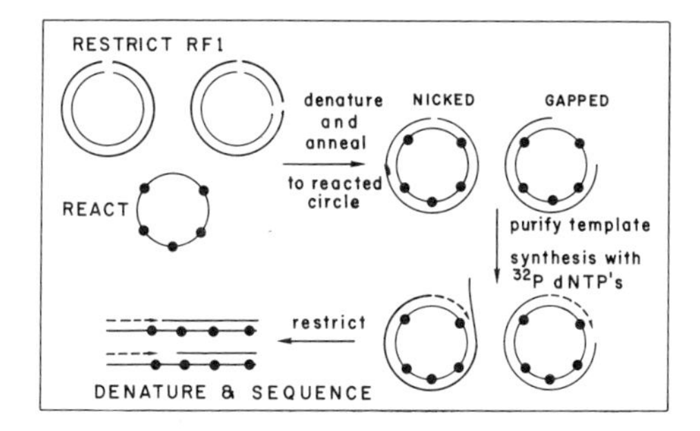

FIG. 2. Protocol for synthesis in which double-stranded DNA is used, with synthesis by nick translation or strand displacement. The figure shows the steps required for preparation of nicked or gapped templates from single-strand virus and RF. Either viral (+) or (-) strands can be reacted with carcinogen.

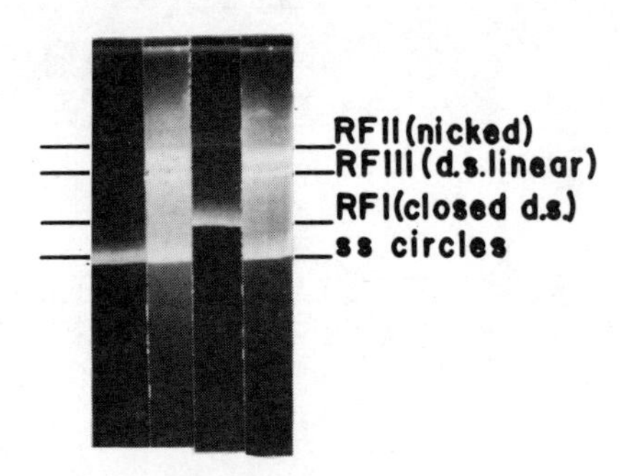

FIG. 3. Purification of nicked double-stranded molecules prepared according to the protocol shown in Fig. 2. The gel shows the electrophoretic separation of single-stranded (s.s.) circles, supercoils (RFI), linear full-length double-stranded (d.s.) molecules (RF III), and nicked circles (RF II). Electrophoresis of M13mp2 molecules was for 5 h at 80 V on a 1% agarose slab gel. Sections of the gel containing RF II are cut and electroeluted to provide purified substrate.

site of termination does depend on the polymerase (Fig. 5) since with the same lesion and template, different polymerases may have different patterns of termination.

It is clear that the nature of the lesion plays a major role in termination. We have previously demonstrated that both T4 DNA polymerase and AMV reverse transcriptase terminate DNA synthesis at different positions with acetyl aminofluorene and with aminofluorene adducts (14). The importance of the nature and position of the same adduct can be seen in an experiment in which single-stranded DNA templates (Fig. 1) were treated with DMS after priming and then heated for 60 minutes at 65°C (Fig. 6). As with the double-stranded substrate, stops occur one nucleotide before the adenines in the template. Heating of both double-stranded (Fig. 4) and single-stranded (Fig. 6) methylated templates, but not control templates, results in additional termination sites, one nucleotide before the position of guanines in the template. These results should come as no surprise since we have known for many years (9) that 7-methylguanine is the major alkylation product of reaction with DMS. We therefore interpret these results as follows: 3-methyladenine is a block to DNA synthesis but the major alkylation product, 7-methylguanine, is ignored by the polymerase [and by the replication apparatus in vivo (17)]. Heating produces apurinic sites at the sites of both adenines and guanines, and these are blocks to replication (e.g., 20). The new bands therefore represent apurinic/apyrimidinic (AP) sites produced at guanines. The results illustrate the importance of location of the lesion; even a group as small

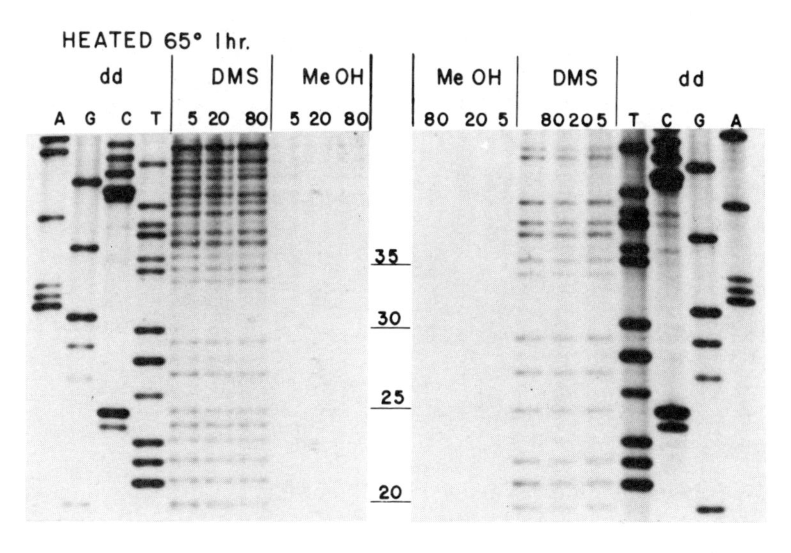

FIG. 4. Termination of synthesis on alkylated double-stranded templates. Double-stranded M13mp2 nicked at the EcoR1 site and prepared as shown in Figs. 2 and 3 was reacted with 10 mM DMS for 30 min at 25C. The DNA (0.1 μg) was incubated with 0.3 units E. coli polI (holoenzyme) and 0.6 μM [^{32}P] dNTPs for 10 min at 30°C followed by an additional period with 100 μM dNTPs as shown. Heating was 65°C for 60 min before polymerase treatment. The template sequence starts from the EcoR1 digestion site as follows: 3'A(O)GCATT(5)AGTAC(10)CAGTA(15)-TCGAC(20)AAAGG(25)ACACA(30)CTTTA(35)ACAAT(40)AGGCG(45)-AGTGT(50)TAAGG(55)TGTGT(60)TGTAT(65)GCTCG(70)GCCTT(75)-GGTAT(80)TTCAC(85)ATTTC(90)GG5'.

as a methyl can be an efficient block to DNA synthesis when it is placed in the appropriate position.

Addition of nucleotides opposite lesions depends on the concentration and the identity of the nucleotide. Although the Kms for addition of nucleotides on damaged templates range from about 0.1 to 4 μM, addition of an incorrect base to a position opposite a lesion may have a Km that is 100 times greater [(19); Fig. 7]. In addition, even though lesions may be apparently noninformative, as in the case of AP sites, there is nontheless selectivity for particular deoxynucleotides. We have previously reported on the preference of polymerases for adenine (18, 20), a preference that seems to extend to the in vivo situation (3,4,5,11). One can actually see the gradation of response with structural change in the nucleotide by comparing the incorporation of dATP, dGTP, and 2-aminopurine deoxynucleoside triphosphate (Fig. 8). We thank Dr. Myron Goodman for his

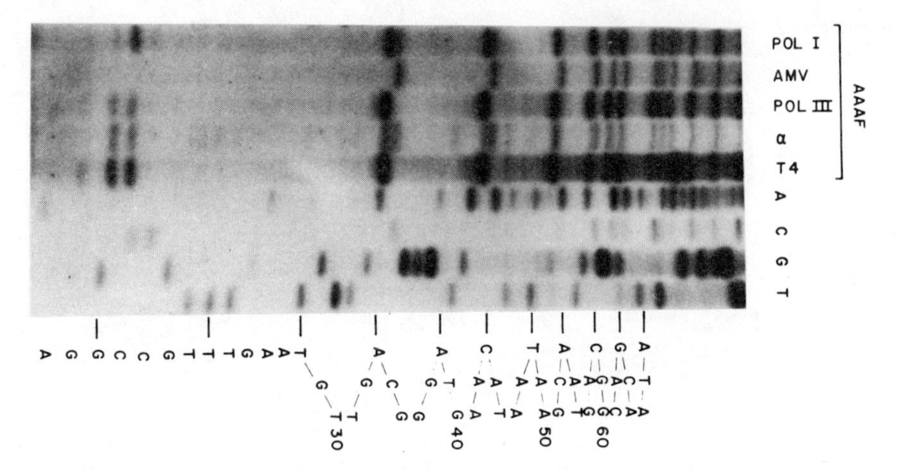

FIG. 5. Polyacrylamide gel analysis of the products synthesized on an N-acetoxy AAF-treated ϕ X 174 template by E. coli pol I, pol III, T4 DNA polymerase, AMV reverse transcriptase and a human polymerase α. From Moore et al. (13).

generous gift of this compound. Incorporation opposite the putative AP site can be seen to be determined by both enzyme and nucleotide (Fig. 8).

BYPASS

We have pointed out previously (24) that a model for mutation requires both an error-generating mechanism plus a mechanism for the bypass of lesions. Different mechanisms have been suggested for bypass, which in bacteria can occur by recombination (7). In vivo it is difficult to demonstrate conclusively that translesion synthesis does occur, although recent data make it very likely (27). Different lesions are bypassed in our in vitro system with widely different efficiencies. Bypass of AP sites by some enzymes occurs readily (22). Bypass of UV-induced lesions at pyrimidine:pyrimidine sites in DNA (most likely cyclobutane dimers) occurs at particular sites only (18). Bypass of N-guanin-8-yl-acetyl-2-aminofluorene sites is critically dependent on the sequence of bases following the adduct as studied by our two-stage protocol (Fig. 1). As one of several studied, we have determined bypass in the template sequence: 3'ACTGCCCTACTTGTA5'. At other sites (19), we find that DNA with a nucleotide added opposite a reacted site is not a good substrate for further elongation. In addition, dG is added very poorly by either polI (Kf) or by DNA polymerase alpha opposite an acetyl-aminofluorene-reacted site. However, in the presence of Mn^{++} dG is not only added opposite the lesion but also very quickly bypasses to at least the third cytosine in the sequence (Fig. 9; 23). We assume that the presence of a sequence of three

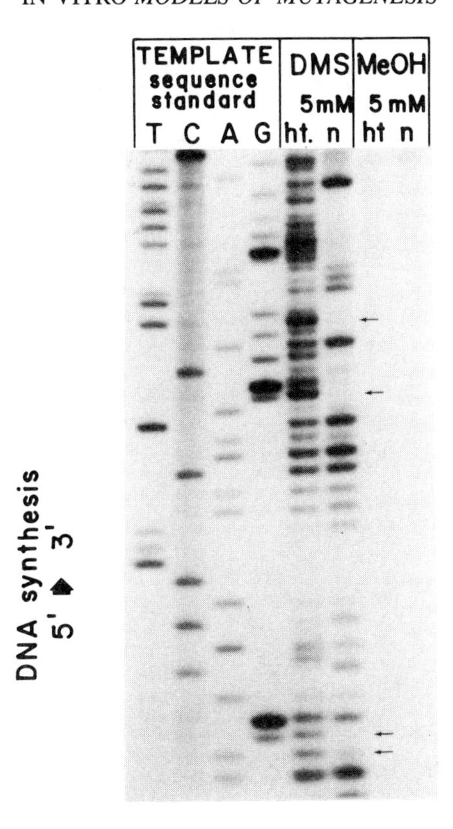

FIG. 6. Sites of methylation-induced blocks to DNA synthesis on a single stranded DNA template. M13mp2 DNA hybridized with a priming fragment was treated with 5 mM DMS for 30 min at 25°C. The DNA was precipitated, and a portion was heated for 60 min at 65°C. The sample was reprecipitated and used for a first-stage reaction with E. coli polI (Kf). Arrows indicate termination sites developed as a result of the heating.

bases complementary to the single dNTP added in our second-stage reaction promotes the fast bypass. Our guess is that similar structural features will be found to account for the observations that some sites are preferred for bypass.

It has been supposed that bypass may be more easily accomplished in vitro on double-stranded templates (16). We have not yet observed a slow bypass similar to that observed in second-stage reactions on AAF-dG templates at high dNTP concentration (19), with either DMS-treated templates (Fig. 4) or UV-irradiated templates. Using either polI (Kf) (Fig. 10) or AMV reverse transcriptase, we observe termination bands when the (+) template, but not the (-) primer strand is irradiated. In these experiments, either double-stranded nicked RF was irradiated and then hybridized to unreacted (+) strand, or the irradiated (+) strand was hybridized

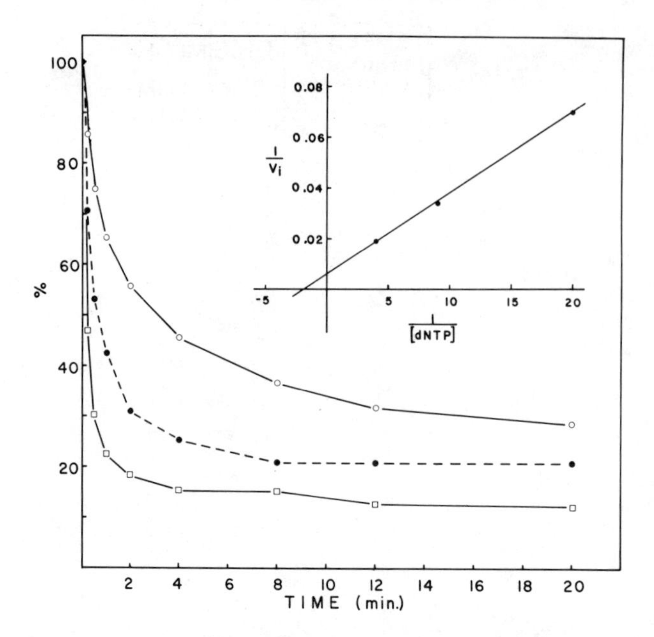

FIG. 7. Time course of dATP insertion opposite AAF-dG reaction sites by polI (Kf)/Mn^{2+} (0.035 unit) at 50, 105, or 250 μM. The decrease in initial stage I (Fig. 1) product at one particular site is plotted as a function of time. Insert: plot of the reciprocal of initial velocity as a function of the reciprocal of dATP concentration. The apparent Km for dATP is 555 μM. From Rabkin and Strauss (19).

with unreacted nicked RF and the products were separated by electrophoresis. We observe no significant change in the relative intensity of the termination bands with time that might lead to a conclusion of bypass. This result is even true for DMS and its heated template (Fig. 4). Heating does lead to the incorporation of some nucleotide opposite the lesion (e.g., positions 36 and perhaps 28, Fig. 4), but there is no obvious increase in the proportion of radioisotope opposite the lesion with time. It is true that our methodology is designed to detect termination and is relatively insensitive for the detection of low frequencies of bypass. It is also correct that the results of such experiments are critically dependent on the concentration of nucleotides, the amount of enzyme and substrate, and the ratios of these factors. There is also an effect of the lesion: 3-methyladenine is not bypassed but 7-methylguanine is not even a termination signal. Nonetheless, on purified templates, which we know to be double-stranded by their electrophoretic mobility (Fig. 3), at least two lesions are not more readily bypassed than in single-stranded DNA. Trans-

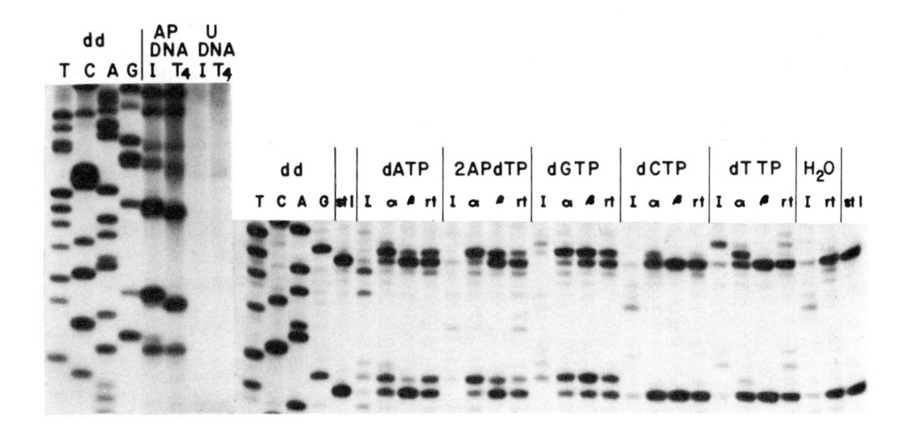

FIG. 8. Insertion of dNTPs opposite AP sites produced by the removal of cytosine from DNA. Left: Formation of first-stage product from bisulfite-treated M13 DNA with or without uracil-N-glycosylase treatment. Right: Elongation of first-stage product synthesized with T4 DNA polymerase. Isolated first-stage product was incubated with 200 μM nucleotide and the following enzymes: E. coli polI (Kf) (I) 0.2 units (10 min at room temperature); polymerase alpha (15 min at 35°C); polymerase beta (Novikoff hepatoma) 0.03 units (30 min at 35°C); and AMV reverse transcriptase 8 units (20 min at 35°C). 2APdTP (2 aminopurine deoxynucleoside triphosphate) provided by Dr. Myron Goodman.

lesion synthesis may be possible but under special conditions analogous to those found effective with single-stranded templates.

MODEL FOR MUTAGENESIS

The series of experiments from this laboratory has led us to the following model (24): Translesion synthesis is possible when biologically relevant components are used at attainable concentrations. Certain lesions alter nucleotides so that they are functionally noninstructional for Watson-Crick base pairing. Polymerases tend to add purines, particularly adenine, opposite such noninstructional sites. As a result, treatments such as UV-irradiation, which produce lesions at the site of pyrimidines, tend to produce transitions, whereas agents such as benzo[a]pyrene diol epoxide, which damage purines, tend to produce transversions. The model is compatible with one proposed by Schaaper et al. (21) who suggested AP sites as common intermediates in the mutagenic action of a variety of carcinogens. Both models predict that a variety of agents should produce

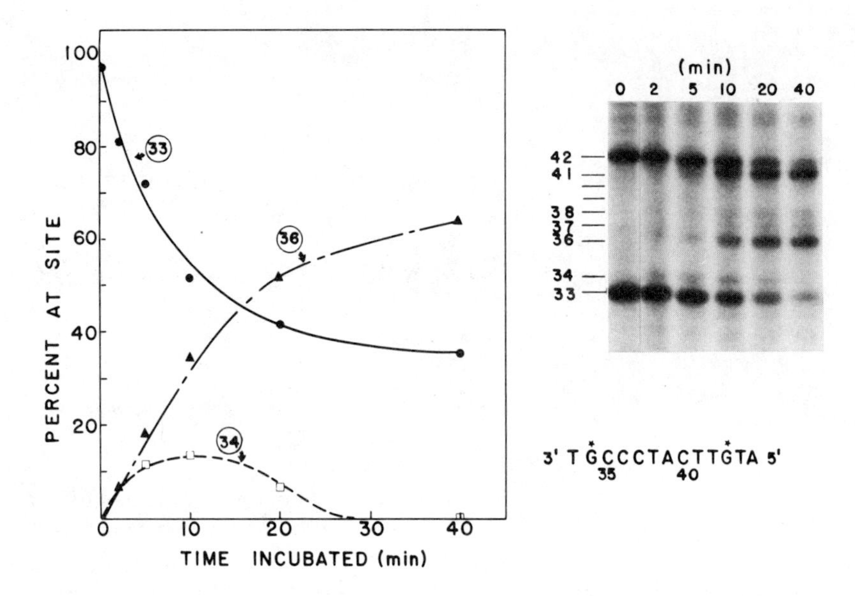

FIG. 9. Time course of dGTP insertion opposite an AAF-dG adduct by polI (Kf)/Mn^{2+}. Stage I product was synthesized to before the lesion (Fig. 1) and the isolated stage I product was incubated with dGTP (180 μm). The percentage of the initial stage I product is plotted for positions 33, 34, and 36 as determined by scanning the autoradiograph of the gel (insert). From Strauss et al. (23).

mainly G:C>T:A transversions in DNA and studies in bacteria seem to confirm this hypothesis for both induced (3,4,5,10) and spontaneous (non-targeted) mutations (11).

What about the situation in mammalian cells? One of the interesting genes for which an analysis is available is the ras oncogene in its various forms. Ras can be activated by a single base change at one of a number of sites. The result of analyses of six spontaneous ras activations has been tabulated by Yuasa et al. (28). One is an A:T>G:C transition. The other five are transversions (three G:C>T:A; one A:T>T:A; and one C:G>A:T), all interpretable as due to the insertion at replication of an adenine opposite an apurinic site. This situation is to be compared with activation of the gene as a result of nitrosomethylurea (NMU) treatment of rats (25) in which a transition of the type G:C>A:T is observed in several cases. Assuming the major mutagenic product of NMU treatment to be the informational O6-methylguanine, the C:G>A:T change is expected. Although the data are limited, they indicate the kind of conclusions that additional results on sequence changes in oncogenes might give. If, as additional data are accumulated, the majority of changes in the activated ras gene continue to appear as transversions, particularly

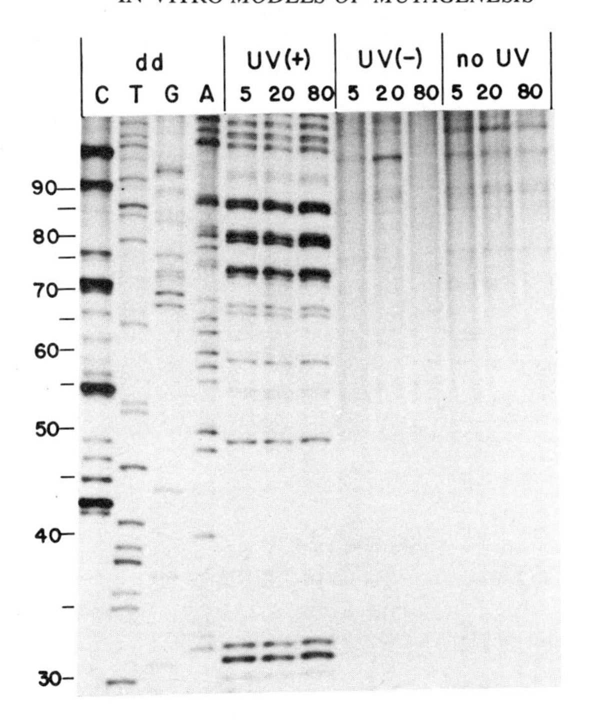

FIG. 10. Synthesis on double-stranded nicked DNA in which the protocol of Fig. 2 was used with either the (+) or the (-) strand irradiated with UV at 1 kJ/m^2. E. coli polymerase I. Reaction conditions are as described in the legend to Fig. 4. The times indicated are the period of chase in 100 μM dNTPs following an initial 10 min labeling at 0.6 μM dNTP. The sequence numbering is from the EcoR1 site, and the sequence is given in the legend to Fig. 4.

those in which adenine is added opposite an adenine or guanine, the explanation will be more likely that mutation is due either to spontaneous or induced AP sites or to the polymerase selected addition of bases opposite noninstructive lesions. Transitions of G:C>A:T, particularly when it is likely that the guanine has been altered in an informational way, signal mutation due to mispairing. However, even transitions can indicate the active role of polymerase, as when a purine is added opposite an altered base (as in UV-induced damage) since the change may indicate the preference of the polymerase rather than the residual information content of the altered base. Analysis of the changes observed in the activation of oncogenes is therefore not only important for an understanding of the mechanism of their activation but also may provide evidence for a less-simplistic model of the mechanism of mutation in human cells, one in which the protein plays an active role in base selection.

SUMMARY

The bypass of lesions in DNA with insertion of nucleotides opposite damaged bases has been studied as a model for mutagenesis in an in vitro system. Lesions introduced by dimethyl sulfate at adenines and by ultraviolet light at pyrimidine dimers act as termination sites on both double- and single-stranded DNA templates. Base selection opposite noninformational lesions is, in part, a property of the polymerases: different polymerases have different selectivities although all polymerases tested seem to prefer purines. The ability to insert "incorrect" bases is determined in part by the sequence 5' to the lesion on the template strand. The hypothesis that damaged purines tend to result in transversions can be applied to published data on activation of the c-ras oncogene.

ACKNOWLEDGMENT

The work reported from our laboratory was supported by Public Health Service Grants GMO7816 and CA32436, National Institutes of Health, U.S. Department of Health and Human Services, and by a grant from the U.S. Department of Energy (DE-ACO2-76).

REFERENCES

1. Aust, A., Drinkwater, N., Debien, K., Maher, V., and McCormick, J. (1984): Mutat. Res., 125:95-104.
2. Brash, D., and Haseltine, W. (1982): Nature, 298:189-192.
3. Brouwer, J., van de Putte, P., Fichtinger-Schepman, A., and Reedilk, J. (1981): Proc. Nat. Acad. Sci., USA, 78:7010-7014.
4. Eisenstadt, E., Warren, A., Porter, J., Atkins, D., and Miller, J. (1982): Proc. Nat. Acad. Sci., USA, 79:1945-1949.
5. Foster, P., Eisenstadt, E., and Miller, J. (1983): Proc. Nat. Acad. Sci., USA, 80:2695-2698.
6. Franklin, W., and Haseltine, W. (1984): Proc. Nat. Acad. Sci., USA, 81:3821-3824.
7. Hanawalt, P., Cooper, P., Ganesan, A., and Smith, C. (1979): Annu. Rev. Biochem., 48:783-836.
8. Heidelberger, C. (1975): Annu. Rev. Biochem., 44:79-121.
9. Lawley, P., and Brooks, P. (1963): Biochem. J., 89:127-138.
10. Miller, J. (1983): Annu. Rev. Genetics, 17:215-238.
11. Miller, J., and Low, K. (1984): Cell, 37:675-682.
12. Moore, P., Bose, K., Rabkin, S., and Strauss, B. (1981): Proc. Nat. Acad. Sci., USA, 78:110-114.
13. Moore, P., and Strauss, B. (1979): Nature, 278:674-666.
14. Moore, P., Rabkin, S., Osborn, A., King, C., and Strauss, B. (1982): Proc. Nat. Acad. Sci., USA, 79:7166-7170.

15. Painter, R.B. (1981): In: DNA Repair. A Laboratory Manual of Research Procedures, Vol. 1B, edited by E. Friedberg and P. Hanawalt, pp. 569-574. Marcel Dekker, New York.
16. Piette, J., and Hearst, J. (1983): Proc. Nat. Acad. Sci., USA, 80:5540-5544.
17. Prakash, L., and Strauss, B. (1970): J. Bacteriol., 102:760-766.
18. Rabkin, S., Moore, P., and Strauss, B. (1983): Proc. Nat. Acad. Sci., USA, 80:1541-1545.
19. Rabkin, S., and Strauss, B. (1984): J. Mol. Biol., 178:569-594.
20. Sagher, D., and Strauss, B. (1983): Biochemistry, 11:3610-3617.
21. Schaaper, R., Glickman, B., and Loeb, L. (1982): Cancer Res., 42:3480-3485.
22. Schaaper, R., Kunkel, T., and Loeb, L. (1983): Proc. Nat. Acad. Sci., USA, 80:487-491.
23. Strauss, B., Rabkin, S., and Sagher, D. (1984): In: Genes and Cancer. UCLA Symposia on Molecular and Cellular Biology, New Series, volume 17, edited by J. Bishop, M. Greaves, and J. Rowley. Alan R. Liss, Inc., New York (in press).
24. Strauss, B., Rabkin, S., Sagher, D., and Moore, P. (1982): Biochimie, 64:829-838.
25. Sukumar, S., Notario, V., Martin-Zanca, D., and Barbacid, M. (1983): Nature, 306:658-661.
26. Walker, G. (1984): Microbiol. Rev., 48:60-93.
27. White, J., and Dixon, K. (1984): Mol. Cell. Biol., 4:1286-1292.
28. Yuasa, Y., Gol, R., Chang, A., Chui, I., Reddy, E., Tronick, S., and Aaronson, S. (1984): Proc. Nat. Acad. Sci., USA, 81:3670-3674.

Carcinogenesis, Vol. 10, edited by E. Huberman and S. H. Barr. Raven Press, New York © 1985.

Estimation of the Potencies of Chemicals that Produce Genetic Damage

A. R. Peterson and H. Peterson

University of Southern California Comprehensive Cancer Center, and The Institute for Toxicology, The School of Pharmacy, Cancer Research Laboratory, Los Angeles, California 90033

In 1975, Dr. Heidelberger and the authors of this paper undertook to develop procedures for the early detection of carcinogens. Our desiderata were sensitive, versatile, and reproducible procedures that will produce data more rapidly than will long-term carcinogenesis studies in animals and will be more relevant to carcinogenesis than are mutagen assays in bacteria (14). Several lines of evidence indicated that carcinogenesis is initiated by DNA damage, and a cell line from C3H 10T1/2 clone 8 of mouse embryo fibroblasts had been developed in Dr. Heidelberger's laboratory for studies of oncogenic transformation, which was shown to be highly relevant to carcinogenesis (8,30,31). Therefore, we considered that procedures for measuring DNA damage, mutagenesis, and oncogenic transformation in the C3H 10T1/2 cells could be adapted for screening carcinogens. The methods that we chose to adapt, their end points, and the criteria of genetic damage that these represent are shown in Table 1. Our approach was to apply the latter methods to obtain dose-response data from a set of compounds selected for the diversity of their carcinogenic potentials. The mutagenesis assay in which 8-azaguanine (AG) resistance is used as an end point is particularly versatile and sensitive (3), but AG^r mutants could not be detected in the tetraploid C3H 10T1/2 cells (28). However, we considered that mutations to AG resistance detected in V79 Chinese hamster cells could be related through associations with DNA damage, DNA-synthesis inhibition, and cytotoxicity to oncogenic transformation in C3H 10T1/2 cells. Moreover, we considered that the strength and form of these associations could provide inform on mechanisms of carcinogenesis (19).

TABLE 1. Short-term tests for carcinogens.

End Point	Criterion
Inhibition of [^{3}H]-dThd-incorporation into DNA	Inhibition of DNA synthesis
Decreased molecular weight of DNA	DNA damage and repair
Decreased cloning efficiency	Cytotoxicity
8-Azaguanine-resistant colonies, ouabain-resistant colonies	Mutations
Foci of morphologically transformed cells	Oncogenic transformation

In the present article we describe the progress of this work from its beginnings with Dr. Heidelberger to the present stage where assays for DNA-synthesis inhibition, cytotoxicity, and mutagenesis can be combined in a three-tier screen of the potency of chemicals that produce genetic damage. While this work was in progress, Painter and Howard (18) and Carver et al. (5) published associations between mutagenesis and DNA-synthesis inhibition and between mutagenesis and cytotoxicity, respectively. These publications greatly influenced the development of our ideas. However, the present article is not a review of the literature on methods for determining genotoxic potency and assessing genotoxic risk in human populations. These areas are reviewed in the series of Banbury Reports (e.g., 15), the monographs issued by the International Agency for Research on Cancer (e.g., 10), the volumes of Advances in Modern Toxicology (e.g., 6), and reports of the Gene-Tox Program of the United States Environmental Protection Agency (EPA) (e.g., 3).

SHORT-TERM TESTS FOR CARCINOGENS: CONCEPTS AND PROCEDURES

The concept that short-term tests can be used to detect carcinogens presupposes that laws can be established relating the end points of these tests to the end point of carcinogenesis, which is cancer. Now a law is a proposition asserting a relation of uniform association, which can be established by experiment. This assertion, uniform association, cannot be ed, but it can be paraphrased: if end points A, B, C, D, and E are observed, then end point Z can be observed (4). However, if A, B, C, D, and E are observed in cultured cells and Z is cancer in humans, these end points may not be uniformly associated. For example, DNA damage will not always be associated with the induction of AG^r mutants, because the hypoxanthine-guanine phosphoribosyl transferase gene may not be damaged. Or if this gene is damaged, its replication may not be aberrant, the damage may be repaired, or damage elsewhere in the genome may produce cytotoxicity. Nevertheless, if the tests for DNA damage and mutagenesis are applied to enough cells, some will be mutated, and DNA damage may thus be shown to be associated with mutagenesis. The test for DNA damage can then be used to detect possible mutagens, and the results of this test can be confirmed by the assays for AG^r mutants. Thus, although associations among the end points in Table 1 may not be uniform, they may be strong enough to allow use of the rapid tests for quick identification and the slower assays for confirmation of potential carcinogens. Moreover, establishing the strength and forms of the associations will also provide information on mechanisms of DNA damage, inhibition of DNA synthesis, cytotoxicity, mutagenesis, and oncogenic transformation.

A prerequisite for this approach of establishing associations is that the assay procedures must be validated. Much of our work was directed to establishing procedures whose end points provided observations of the chosen criteria that were as accurate, reproducible, and unambiguous as possible. An example of a small part of this work is our study of the use of 8-AG-resistance as a genetic marker. We found that the serum used in the culture medium contained dialyzable components, presumably hypoxanthine, that inhibited the cytotoxicity of AG. Thus, colonies in medium containing low concentrations of AG were not AG^r mutants (26,27). We have used dialyzed serum in our procedures for AG^r mutants ever since (22). A further prerequisite is that all end points of the short-term tests should be comparable. For this reason we avoided procedures whose end points could be observed only after treatment of cells with high concentrations of carcinogens or with inhibitors. For example, we avoided using the inhibitor hydroxyurea to measure DNA repair (33); instead, we used the alkaline-sucrose-sedimentation procedure, which is sensitive enough to allow detection of DNA damage and DNA repair in cells treated with low (LD_5-LD_{90}) concentrations of alkylating agents (28).

ASSOCIATIONS SUGGESTED BY DOSE-RESPONSE DATA

Our procedures and concepts were developed and validated through a study of the effects of four monofunctional alkylating agents: N-methyl-N'-nitro-N-nitrosoguanidine (MNNG), methyl methane sulfonate (MMS), and their ethyl homologs, ENNG and EMS. With these compounds we generated dose-response curves for each of the criteria listed in Table 1. The V79 cells and C3H 10T1/2 cells were treated with alkylating agents for two hours in these experiments.

The AG^r-mutant frequencies in V79 cells and single-strand breaks in the DNA of V79 cells and C3H 10T1/2 cells increased as linear functions of the dose of alkylating agent (28). The linearity was established by regression analyses, which also yielded numbers for the slopes of the curves. The numbers represent the mutant frequency per unit of concentration (AG^r mutants x 10^{-5} survivors x μM^{-1}) and the single-strand break frequency per unit of concentration (single-strand breaks x 10^{-8} daltons x μM^{-1}).

The colony-forming frequency and the incorporation of [^{3}H]-dThd into DNA were expressed as fractions of the respective controls. These end points are termed the surviving fraction (S) and the fractional incorporation (I), respectively. These fractions decreased exponentially with increasing concentrations of alkylating agents, and straight lines were fitted to plots of ln S against dose and ln I against dose (20). The slopes of these lines represent the doses of alkylating agent that reduce the surviving fraction (D_0) and the incorporation of [^{3}H]-dThd into DNA (C_0) by 37% in the linear portions of the plots. The reciprocals of these numbers (D_0^{-1} and C_0^{-1}) represent the cytotoxicity and inhibition of DNA synthesis, respectively, produced by 1 μM of the alkylating agent.

The slopes of the dose-response curves for mutations and single-strand breaks, and the reciprocals of the slopes of the curves for DNA-synthesis inhibition and cytotoxicity, are estimates of the activities of the alkylating agents in producing those end points. Figure 1 shows that the activities in inhibiting DNA synthesis correlate well with the activities in producing cytotoxicity, mutations, and alkali-labile lesions, which are responsible for most of the single-strand breaks. However, these alkylating agents did not induce oncogenic transformation of asynchronous C3H 10T1/2 cells (23) when we used the procedures developed by Reznikoff et al. (30,31). Moreover, MMS did not induce Oua^r mutants in C3H 10T1/2 cells (23), and polycyclic hydrocarbons that induced transformation, mutations, and cytotoxicity in C3H 10T1/2 cells did not induce detectable single-strand breaks in those cells (Peterson, unpublished observations).

From these results we concluded that (i) the existing procedures for assays of oncogenic transformation were unsuitable for use as short-term tests for monofunctional alkylating agents; (ii) ouabain resistance was not a suitable marker for mutagenesis associated with the types of genetic damage produced by MMS; and (iii) alkaline-sucrose-sedimentation analysis could not be used to detect DNA damage induced by biologically relevant doses of carcinogenic polycyclic hydrocarbons. The latter procedure is most useful for detecting compounds that produce high frequencies of single-strand breaks or alkali-labile lesions (28).

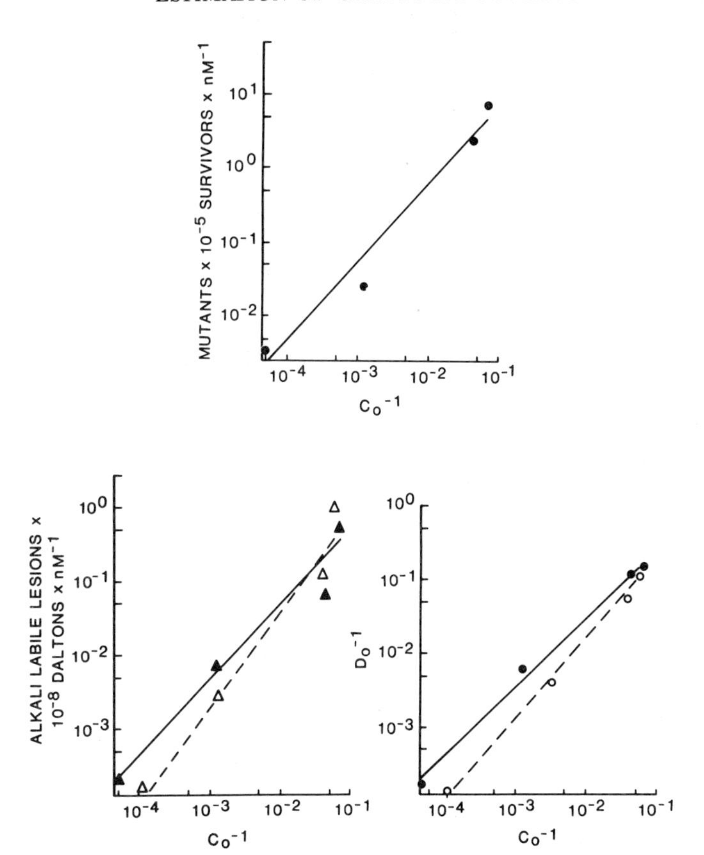

FIG. 1. Relationships between mutagenic, cytotoxic (D_0^{-1}), and DNA-damaging activities and DNA-synthesis inhibition (C_0^{-1}) of monofunctional alkylating agents. Solid symbols: results obtained with V79 cells. Open symbols: results obtained with C3H 10T1/2 cells. The relative activities of the alkylating agents are MNNG>ENNG>MMS>EMS. From Peterson (20).

From the dose-response data in Fig. 1 we concluded that inhibition of DNA synthesis, cytotoxicity, and mutagenesis may be uniformly associated.

THE STRENGTH AND FORM OF THE ASSOCIATION

The linear dose responses for mutagenesis are consistent with target theory, which predicts the following relationship:

$$\mu = aD \tag{1}$$

where μ = mutant frequency, a is a function of the target size, and D is dose (13). The semilogarithmic, dose-response relationships for cytotoxicity and inhibition of DNA synthesis are also consistent with target theory:

$$\ln S = D/D_0 + \ln b \tag{2}$$
$$\ln I = D/C_0 + \ln c \tag{3}$$

where D_0 and C_0 are functions of target size, and b and c are constants (7,20). Thus, plots of μ against ln S and of ln I against ln S would constitute a more rigorous test of the strength of the associations than is the comparison of activities in Fig. 1. According to target theory, if μ, S, and I are uniformly associated, then the ratios μ/ln S and ln I/ln S should be invariant. Invariance of these ratios would indicate that the types of lesions that produce mutations may also produce cytotoxicity and inhibition of DNA synthesis.

The latter ratios are shown in Table 2. It can be seen that the mutant frequencies induced by MNNG were at least 13-fold greater than those induced by equitoxic concentrations of MMS. Therefore, we conclude that μ and S are not uniformly associated. These data suggest that (i) the types of lesions that induce mutagenesis are different from the types that induce cytotoxicity or (ii) the mechanisms of cytotoxicity are different from those of mutagenesis. We have demonstrated that MNNG produces ~10-fold more O^6 methyl guanine in the DNA of V79 cells than does an equitoxic concentration of MMS. We have also shown that one round of DNA replication following treatment with MNNG is necessary for maximum mutagenesis, but not even one round of DNA replication is necessary for maximum cytotoxicity (24). These findings and other evidence, which has recently been reviewed (21), strongly indicate that (i) alkylation of oxygens in purines and pyrimidines in DNA produces mutagenic lesions and (ii) cytotoxicity and mutagenesis occur by quite different mechanisms.

Table 2 shows that equitoxic doses of MMS and MNNG produce equal inhibition of DNA synthesis. This finding suggests a strong association between cytotoxicity and inhibition of DNA synthesis. This association remained essentially uniform in experiments with V79 cells and C3H 10T1/2 cells treated with MNNG, MMS, ENNG, and EMS (20).

The strong association between inhibition of DNA synthesis and cytotoxicity affords an opportunity to discuss a major pitfall in this approach: it is tempting to conclude that the inhibition of DNA synthesis is the cause of the cytotoxicity. We were delivered from this temptation by further experiments, the comments of our referees, and collaboration with Dr. Luminita Ibric. Dr. Ibric showed that the DNA-synthesis inhibition may be uniformly associated with poly(ADP-ribose) synthesis. Our

TABLE 2. Mutagenesis and inhibition of DNA synthesis by equitoxic concentrations of MNNG and MMS in V79 cells.

Mutagen	ln I/ln S	μ/ln S
MMS	0.19[a] ± 0.05	3.97 ± 0.60
MNNG	0.41[a] ± 0.08	53.0 ± 6.75

[a]Students' t test shows a probability of $0.1>p>0.05$ that the difference between these means is due to chance. Data from Peterson (20).

interpretation of this result is that single-strand breaks resulting from N-alkylation of DNA purines stimulate poly(ADP-ribose) synthesis, which inhibits DNA synthesis (Ibric and Peterson, manuscript in preparation). This inhibition is reversible (Peterson and Mulkins, manuscript in preparation). Far from causing cytotoxicity, this inhibition of DNA synthesis may allow time for strand breaks to be repaired and may thereby reduce the probability of cytotoxicity.

We have shown that inhibition of DNA synthesis, cytotoxicity, and mutagenesis are not uniformly associated. However, the forms of the dose responses produced by monofunctional alkylating agents are consistent with target theory. Moreover, the associations among these dose responses are strong enough to suggest a common target, namely, DNA. This interpretation is illustrated by the Venn diagrams in Fig. 2. The set of DNA lesions that increase the frequency of AG^r mutants may not induce the inhibition of DNA synthesis or cytotoxicity. The set of cytotoxic lesions may include DNA lesions that induce the inhibition of DNA synthesis, but all lesions that induce this inhibition may not be DNA lesions. However, observations of end points of mutagenesis, cytotoxicity, and inhibition of DNA synthesis can be considered to indicate that DNA damage has occurred. In view of the strong evidence indicating a role for genetic damage in carcinogenesis (reviewed in ref. 9), the latter end points should be useful indicators of carcinogenic potential. Oncogenic transformation would undoubtedly be a more relevant indicator and more useful in studies of the mechanism of carcinogenesis. However, we are still in the process of developing methods for unambiguous quantitation of oncogenic transformation of C3H 10T1/2 cells.

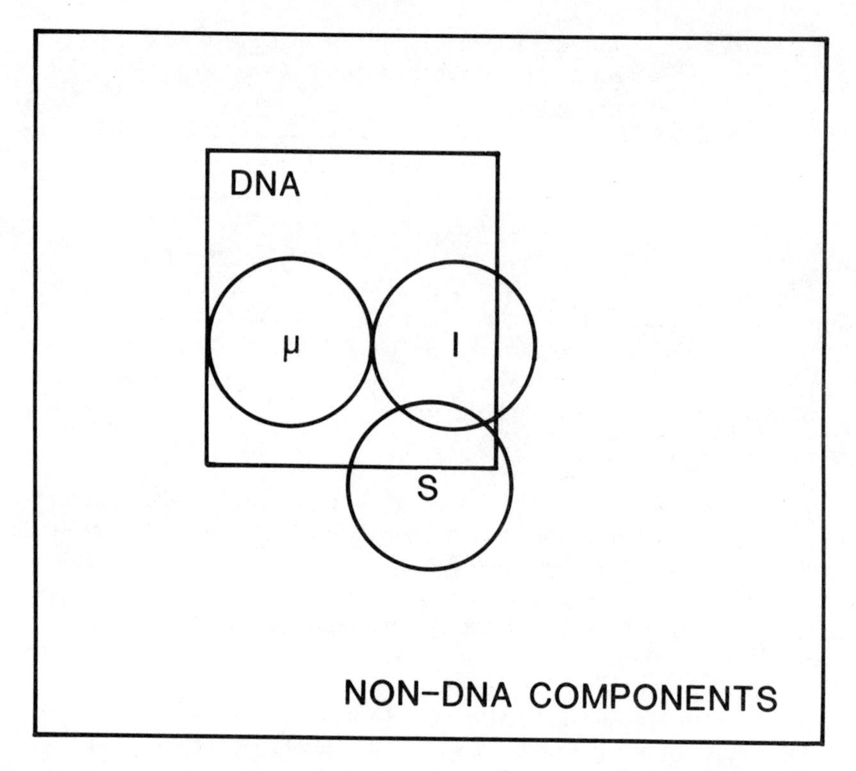

FIG. 2. Interpretation of associations between DNA-synthesis inhibition (I), cytotoxicity (S), and mutagenesis (μ). The large square represents the set of all cellular components. The smaller square represents the set of DNA molecules. The circles represent sets of molecules that are critical targets for I, S, and μ.

A THREE-TIER SCREEN FOR GENOTOXIC POTENCY

Although our specific aim was to develop short-term tests for carcinogens, Fig. 3 shows that our procedures detect chemicals that produce genotoxicity, which presumably includes carcinogenesis. Genotoxicity is here defined as modification of DNA structure that results in genome malfunction. Genotoxic chemicals, then, are chemicals that directly modify DNA, or, by perturbing the replication, repair, recombination, or methylation of DNA, induce DNA modification. We reasoned that our procedures for detecting the inhibition of DNA synthesis, cytotoxicity, and mutations might provide a three-tier screen for grading chemicals according to their genotoxic potency.

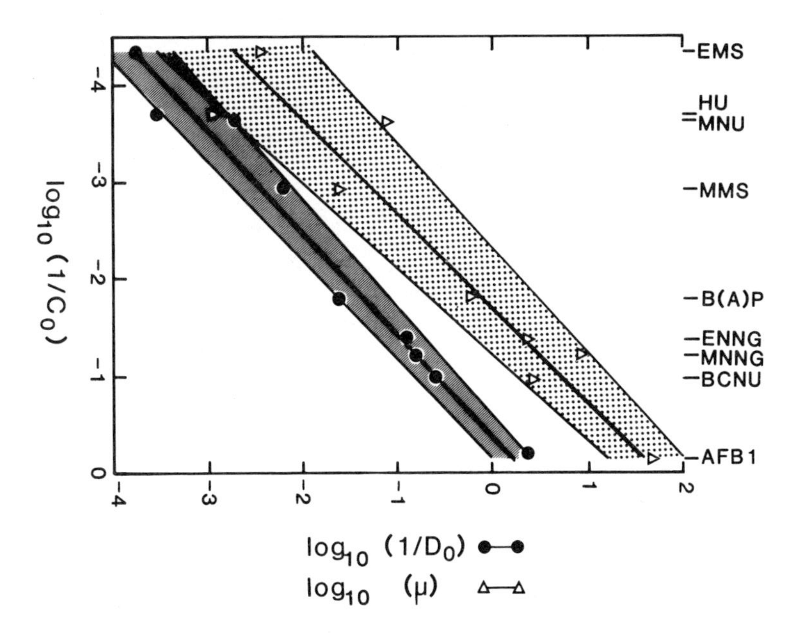

FIG. 3. Relationships between mutagenic (μ = μ/D), cytotoxic (D_0), and DNA-synthesis-inhibition (C_0) activities produced in V79 cells by aflatoxin B_1 (AFB_1), MNNG, ENNG, benzo[a]pyrene (BAP), 1,3,-Bis-2-chloroethyl-1-nitrosourea (BCNU), MMS, methyl nitrosourea (MNU), hydroxyurea (HU), and EMS. Units of activity are the same as in Fig. 1.

The procedure for applying the screen is as follows: Exponentially growing cultures of V79 cells are treated for 2 h with a wide range of concentrations of the chemical whose potency is being evaluated. Parallel treatment with 6.8 μM MNNG and 0.5% acetone are included as positive and negative controls, respectively. Following the treatment, the medium is changed, then the cells are washed and incubated for 45 min with fresh medium. The cells are next incubated with [^{3}H]-dThd, and the incorporation of [^{3}H]-dThd into the DNA is determined. Thus, within 6 h a compound's genotoxic activity can be detected, and the dose range most likely to produce cytotoxicity and mutagenesis can be estimated. On the basis of this estimate, cultures are treated with three narrow-range doses of the chemical, and the inhibition of DNA synthesis, cytotoxicity, and the frequency of AG^r mutants are assayed. The usual positive and negative controls are included in these assays. [For compounds requiring metabolic activation, in collaboration with Dr. A. Uwaifo, we used post-mitochondrial supernatants from rat liver (12).] Within 4 days these assays are set up in duplicate. Within 48 h the dose-response for DNA-synthesis inhibition has been established. Within 9 days the cytotoxic activity has been measured, and within 3 weeks the mutagenic activity has been determined.

Figure 3 shows the relationships among the activities of nine selected chemicals in producing DNA-synthesis inhibition, cytotoxicity, and mutations in V79 cells. The association between cytotoxicity and inhibition of DNA synthesis is strongest (coefficient of correlation, r=0.97). But the association with mutagenesis is also strong (r=0.93). Even hydroxyurea induced mutations. We have also shown that 5-fluorodeoxyuridine induces mutations when the cells are incubated with deoxycytidine during the treatment (25). The data obtained with hydroxyurea and 5-fluorodeoxyuridine add to an increasing weight of evidence that compounds that perturb nucleotide metabolism produce genotoxicity (21). Such compounds are genotoxic agents even though they may not directly damage DNA. Thus, Fig. 3 shows that compounds representing a diversity of chemical structures and mechanisms of action can be graded on a scale of genotoxic activity spanning at least four orders of magnitude.

The three-tier screen meets with three important criteria for systems of short-term tests: it provides for rapid identification, confirmation, and quantitation of genotoxic potency. The fourth and most important criterion of the usefulness of short-term tests is that potency measurements obtained with these tests can be extrapolated to risk estimates in humans. An indication that measurements of mutagenic potency in V79 cells might be of use in estimating carcinogenic potency was provided in an exercise that was undertaken by the members of the EPA-commissioned Gene-Tox Committee for evaluating the V79-cell-mutagenesis assay (3). The mutagenic "potencies" of eight compounds were calculated from the pooled measurements produced by two or more independent laboratories. The carcinogenic "potencies" of these chemicals were calculated by the method of Meselson and Russell (16) from published measurements of carcinogenesis in the most sensitive rodents (2,16).

Figure 4 shows that carcinogenic "potencies" of six of the chemicals correlated well with their mutagenic "potencies". This correlation could not be extended to dimethyl nitrosamine (DMN) and diethyl nitrosamine (DEN) until a more suitable system for metabolic activation of these compounds had been developed. Following this development (11), the mutagenic "potencies" of DMN and DEN fell into line with their carcinogenic "potencies". It should be noted that these "potencies" are actually activities.

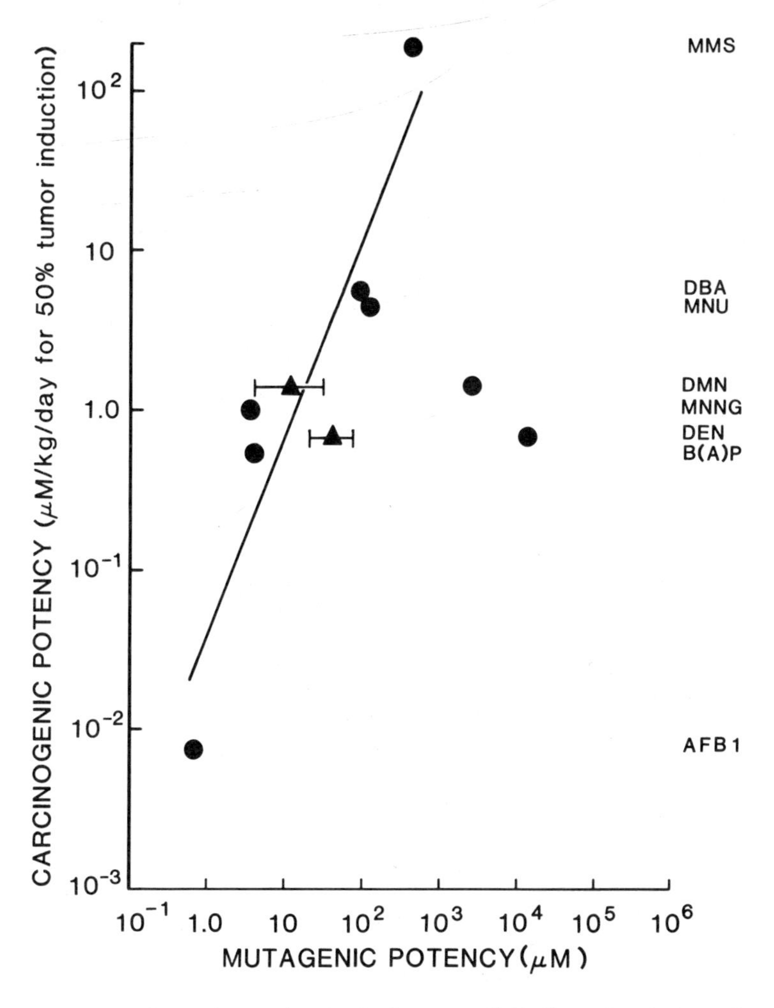

FIG. 4. Comparison of carcinogenic "potency" (daily animal doses in μM/kg/day that give 50% cumulative single risk incidences of induced cancer after 2 y of exposure: means of data from Refs. 2 and 16) with mutagenic "potency" (the dose in micromoles per liter that gives a mutant frequency 10 times the spontaneous mutant frequency in V79 cells). ● represents means of the mutant frequencies measured by at least two independent laboratories with resistance to AG, 6-thioguanine, or ouabain used as markers with cell-mediated or microsome-mediated activation systems (3). ▲ represents means of mutant frequencies obtained by Jones and Huberman (11) with activation by primary rat hepatocytes. The straight line was fitted by linear regression. AFB_1:aflatoxin B_1; BP:benzo[a]pyrene; DBA:7,12-dimethylbenz(a)anthracene; DEN:diethyl nitrosamine; DMN:dimethyl nitrosamine. (From Peterson, Bhuyan, Bradley, Huberman, and Langenbach, unpublished.)

The aforementioned exercise cannot be performed with measurements of carcinogenic potency in human beings, because such measurements are not generally possible. However, there is a great deal of epidemiological data on the genetic effects of ionizing radiation in human populations. Therefore, it was proposed that the REC doses of genotoxic chemicals be determined. The REC dose is the dose that produces genotoxicity equivalent to that associated with exposure to 1 rad of gamma rays. It was postulated that the REC dose would provide a risk estimate relative to the known risks resulting from exposure of humans to ionizing radiation (6).

For the REC dose to be determined, it is necessary to know not only the concentration of the chemical but also the length of time for which the cells are exposed to that concentration. This dose-time integral, D^*, can be calculated from the formula:

$$D^* = D \left(\int_0^t e^{-kt} \cdot dt \right)$$

where D is the initial concentration of chemical, t is the treatment time, and k is the pseudo-first-order rate constant calculated from the half-life of the compound in the medium. The half-lives of alkylating agents in culture medium were determined by Dr. Paul Spears, who measured the rate of disappearance of activity towards 4-(4'-nitrobenzyl)pyridine (23) and nicotinamide (method of Nelis and Sinsheimer, ref. 17). We established detailed dose-response curves for DNA-synthesis inhibition, cytotoxicity, and mutagenesis induced by cobalt-60 gamma rays in V79 cells. In these experiments MNNG was used as a positive control, so the REC dose of MNNG could be calculated. As MNNG was used as a positive control in all the screening assays, the REC doses of all the chemicals screened could also be calculated.

Plotting the reciprocals of the REC doses for cytotoxicity (REC_S) and mutagenesis (REC_μ) against the reciprocals of the REC doses for inhibition of DNA synthesis (REC_I) produced relationships similar to those shown in Fig. 3. Therefore, we use the following definitions:

Mutagenic potency = $1/REC_\mu$

Cytotoxicity potency = $1/REC_S$

Potency for DNA-synthesis inhibition = $1/REC_I$

Genotoxic potency = $(1/REC_\mu + 1/REC_S + 1/REC_I)/3$

Thus, the genotoxic potency of aflatoxin B_1 is:

$$\frac{1}{1.77\times10^{-4}} + \frac{1}{1.47\times10^{-3}} + \frac{1}{1.2\times10^{-3}}/3 = 2.39\times10^{3}\ (\mu M.h)^{-1}$$

By contrast, the genotoxic potency of EMS is 9.41×10^{-2} $(\mu M.h)^{-1}$. Thus, the genotoxic potency of aflatoxin B_1 is four orders of magnitude greater than that of EMS.

ENVOI

The three-tier screen for genotoxic potency that we have developed may not be suitable for grading some compounds that require metabolic activation. Furthermore, the screen may produce erroneous results with complex mixtures that contain compounds that inhibit DNA synthesis or produce cytotoxicity by nongenotoxic mechanisms. Therefore, we use the screen for research with pure compounds that do not require metabolic activation. However, there are many possible mechanisms by which compounds may induce genotoxicity without directly interacting with nucleophilic sites in DNA (1,29). For example, in our first application of the screen, we and Dr. Sevanian showed that the mutagenic activity of cholesterol 5α, 6α epoxide is destroyed by an epoxide hydrolase that converts the epoxide to a nonmutagenic triol. The implication of this finding is that compounds that inhibit the hydrolase may induce genotoxicity by preventing the inactivation of the endogenous mutagen cholesterol epoxide (32). Such inhibitors may not be electrophiles and need not react with DNA; they will, nevertheless, be classified as genotoxic compounds.

We are currently using the screen to further study the genotoxic potency of cholesterol oxidation products and to determine the genotoxic effects of compounds that perturb nucleotide pools. In these areas alone there is much work to be done with our screen, which not only provides estimates of genotoxic potency but also provides information on mechanisms of genotoxicity.

ACKNOWLEDGMENTS

This work was initially supported by Public Health Service Grant CA-21036 from the National Cancer Institute, U.S. Department of Health and Human Services to Professor Charles Heidelberger. It continued with support under Contract 956398 from California Institute of Technology Jet Propulsion Laboratory to Dr. A. R. Peterson. It continues with support from Grant BC-441 from the American Cancer Society to Dr. Peterson and from Public Health Service Grant ES 03466, National Institutes of Health, U.S. Department of Health and Human Services to Drs. Sevanian and Peterson. The continuation of the work that Dr. Heidelberger started is our best dedication to his memory.

REFERENCES

1. Auerbach, C. (1976): Mutation Research. Chapman and Hall, London.
2. Bralow, S.P., Gruenstein, M., Meranze, D.R., Bonakdarpour, A., and Shimkin, M.D. (1970): Cancer Res., 30:1215-1222.
3. Bradley, M.O., Bhuyan, B., Francis, M.C., Langenbach, R., Peterson, A., and Huberman, E. (1981): Mutat. Res., 87:81-142.
4. Campbell, N.R. (1957): Foundations of Science, pp. 71-79. Dover, New York.
5. Carver, J.H., Hatch, F.T., and Branscomb, E.W. (1979): Nature, 279:154-156.
6. Drake, J.W. (1978): In: Advances in Modern Toxicology, Vol. 5, edited by W.G. Flamm and M.A. Mehlman, pp. 9-26. Wiley, New York.
7. Elkind, M.M., and Whitmore, G.F. (1967): Radiobiology of Cultured Mammalian Cells. Gordon and Breach, New York.
8. Heidelberger, C. (1975): Ann. Rev. Biochem., 44:79-121.
9. Heidelberger, C. (1982): In: Mechanisms of Chemical Carcinogenesis, edited by C.C. Harris and P.A. Cerutti, pp. 563-375. Alan R. Liss, Inc., New York.
10. IARC Monographs on The Evaluation of The Carcinogenic Risk of Chemicals to Humans. Supplement 2, (1980): International Agency for Research on Cancer, Lyon.
11. Jones, C.A., and Huberman, E. (1980): Cancer Res., 40:406-411.
12. Krahn, D.F., and Heidelberger, C. (1977): Mutat. Res., 46:27-44.
13. Lawley, P.D. (1974): Mutat. Res., 23:283-295.
14. McCann, J., Choi, E., Yamasaki, E., and Ames, B.N. (1975): Proc. Nat. Acad. Sci., USA, 72:5135-5139.
15. McElheny, V.K., and Abrahamson, S., editors (1979): Banbury Report 1. Assessing Chemical Mutagens: The Risk to Humans, Cold Spring Harbor Laboratory, Cold Spring Harbor, New York.
16. Meselson, M., and Russell, K. (1977): In: Origins of Human Cancer, edited by H.H. Hiatt, J.D. Watson, and J.A. Winsten, pp. 1473-1481. Cold Spring Harbor Laboratory, Cold Spring Harbor, New York.
17. Nelis, H.J.C.F., and Sinsheimer, J.E. (1981): Anal. Biochem., 115:151-157.
18. Painter, R.B., and Howard, R. (1978): Mutat. Res., 54:113-115.
19. Peterson, A.R. (1978): In: In Vitro Carcinogenesis, Guide to the Literature, Recent Advances and Laboratory Procedures, edited by U. Saffiotti and H. Autrup, pp. 205-211. U.S. Dept. of Health, Education and Welfare, Washington, DC.
20. Peterson, A.R. (1980): Cancer Res., 40:682-688.
21. Peterson, A.R., Danenberg, P.V., Ibric, L.L.V., and Peterson, H. (1984): In: Genetic Consequences of Nucleotide Pool Imbalance, edited by F.J. deSerres. Plenum, New York (in press).
22. Peterson, A.R., Krahn, D.F., Peterson, H., Heidelberger, C., Bhuyan, B.K., and Li, L.H. (1976): Mutat. Res., 36:345-356.

23. Peterson, A.R., Landolph, J.R., Peterson, H., Spears, C.P., and Heidelberger, C. (1981): Cancer Res., 41:3095-3099.
24. Peterson, A.R., and Peterson, H. (1982): Proc. Nat. Acad. Sci., USA, 79:1643-1647.
25. Peterson, A.R., Peterson, H., and Danenberg, P.V. (1983): Biochem. Biophys. Res. Commun., 110:573-577.
26. Peterson, A.R., Peterson, H., and Heidelberger, C. (1974): Mutat. Res., 24:25-33.
27. Peterson, A.R., Peterson, H., and Heidelberger, C. (1975): Mutat. Res., 29:127-137.
28. Peterson, A.R., Peterson, H., and Heidelberger, C. (1979): Cancer Res., 39:131-138.
29. Potter, V.R. (1981): Perspectives Biol. Med., 24:525-542.
30. Reznikoff, C.A., Bertram, J.S., Brankow, D.W., and Heidelberger, C. (1973): Cancer Res., 33:3239-3249.
31. Reznikoff, C.A., Brankow, D.W., and Heidelberger, C. (1973): Cancer Res., 33:3231-3238.
32. Sevanian, A., and Peterson, A.R. (1984): Proc. Nat. Acad. Sci., USA, 81:4198-4202.
33. Smith, C.A., and Hanawalt, P.C. (1976): Biochim. Biophys. Acta, 432:336-347.

Carcinogenesis, Vol. 10, edited by E. Huberman
and S. H. Barr. Raven Press, New York © 1985.

Potential Use of Gradient Denaturing Gel Electrophoresis in Obtaining Mutational Spectra from Human Cells

W. G. Thilly

Genetic Toxicology Department, Massachusetts Institute of Technology, Cambridge, Massachusetts 02139

As a postdoctoral student of Charlie Heidelberger, I was frequently exposed to his ethic that we are responsible for our own work and, if we find ourselves in a logical trap, it is also our responsibility to plan an escape. When I first met Charlie in 1971, he was worried because the chemical transformation of Syrian hamster embryo cells and mouse prostate cells was occurring at frequencies higher than were observed for forward mutations in nonessential genes in bacteria and human cells. He distrusted, therefore, the popular model of in vitro transformation proceeding via gene mutation as a rate-limiting step. His view at the time was that human cancers were caused by chemicals, viruses, and/or something else; he wasn't easily led to participating in consensus views.

Somewhat in this spirit, I would like to offer for this memorial volume some serious doubts I have about the present status of the field of genetic toxicology. Basically, I fear that we may have accelerated so fast in terms of the development of model systems and their application in the form of short-term assays that we may have gone off the road that leads to contribution to the public good.

To begin, it seems to me that two facts have been consciously or unconsciously associated to form what I call the central premise of genetic toxicology:

Fact 1: People suffer from diseases caused by genetic changes.

Fact 2: Humans are exposed to chemicals and radiation, which cause genetic changes in human cells in culture.

Central Premise: The environmental mutagens to which humans are exposed cause a significant portion of their genetic changes. The basic idea is that we are exposed to many chemical mutagens and that each does its bit in adding to the rate of genetic change.

The application of this premise at the scientific level has been the development of a series of assays designed to detect the potential of a substance to cause genetic changes. Microbiologists have chipped in with histidine reversion assays, phage induction assays, drug resistance assays, and the like, while cytogeneticists have recorded innumerable chromosomal gaps, translocations, and deletions. Curiously, there are relatively few studies of aneuploidy.

Naturally, using these systems makes one feel that one is pulling in the right direction, but the feeling has a fair chance of being illusory. While one might hope to mimic some of the biochemical processes encountered in humans, one could not realistically expect to mimic them all in any particular assay system. What might be a simple metabolic problem for a liver cell could be an unsolvable muddle for a colonic cell, and vice versa.

Furthermore, all of these kinds of studies have been performed for short durations, usually less than a day, with enough mutagen used to assure a positive response and, inevitably, a publication. That this may not be reasonable is appreciated by most researchers. Despite having given more than a decade's effort to developing some of these protocols, I am forced to conclude that these test systems will not be capable of predicting the genetic effects in humans with a useful degree of certainty. Even when test systems consist of human cells or subcellular components of human tissue, I cannot produce objective evidence of their "relevance." The problem is one of the uncertainty of what I have come to think of as "subjunctive genetic toxicology."

Subjunctive genetic toxicology is the practice of identifying potential genetic hazards. The phenomenon is well known to readers. If raspberry juice or diesel exhaust can induce genetic change in bacterial cells, one must conclude that these substances could potentially cause genetic changes in fruit lovers or garage mechanics. However, no such assay can tell us what, if anything, is actually causing genetic change in these groups.

Sooner or later one has to recognize the difference between faith (intuition) and knowledge (data) in these matters. In some ways the public is ahead of the toxicologists. Most people are skeptical of testing protocols that involve high doses in rodents. I imagine that we as a group would get the old horse laugh from the taxpayers if we testified to our simple faith that observations in *Drosophila* or *Tradescantia* equip us to prescribe what should or should not be in breakfast cereals, etc.

So how does genetic toxicology change its voice from the subjunctive to the indicative? There are several lines of study in which information is sought to reveal the reality of human exposure to chemicals and the genetic result in humans. One answer lies in the application of Seymour

Benzer's finding that individual mutagenic stimuli cause specific patterns of mutation within DNA sequences (4).

MUTATIONAL SPECTRA

In 1961, Seymour Benzer reported that when mutants in the RII sequence of 14 bacteriophage were isolated, the positions of the mutations were not randomly distributed but instead were clustered in positions (hot spots), which were characteristic of the mutagenic stimuli to which the bacteriophages were exposed (4). This remarkable observation was often cited over the subsequent 15 years, but in my reading no significant advance or even an independent confirmation occurred until Jeffrey Miller published his observations on the specificity of mutagens in causing nonsense mutations in the E. coli gene for the lac operon repressor, lac I (5,7,8,19,20,23). Miller's observations for the amber or ochre mutations induced by several different mutagens and for mutations occurring during growth in unadulterated bacterial revertant medium (spontaneous) are reproduced in Fig. 1. As in Benzer's work, it is clear that Miller found this partial mutational spectrum to be sensitive to differences among mutagenic stimuli. I use the word "partial" to underscore the fact that all attempts to map the kinds and positions of mutations are limited by selection steps used to identify mutants.

Such restrictions have not obscured the differences among spectra caused by mutagens and chemicals, however, as Miller's data demonstrate. Skopek and Hutchinson (24) applied the techniques of DNA sequencing to obtain the spectrum of all expressed mutations in the CI region of phage lambda in E. coli, an approach that permits, at the cost of considerable effort, the identification of the kind and position of all noncryptic mutations within that genetic region.

In the context of trying to find out if environmental chemicals or spontaneous changes are responsible for gene mutations in humans, our group has proposed that one might use the mutational spectra found in a human blood cell population as an indicator of the kind(s) of causes of those mutations. In a sense this is genetic spectroscopy: the patterns of genetic changes caused in human cells by specific environmental chemicals are theoretically available through research effort, and these could be compared to the spectra actually found in samples of human blood cells (25).

Recognizing that the work necessary to test the validity of this concept in human cells by direct DNA sequencing would be prohibitive, an indirect means was devised and tested in the Ph.D. dissertation of Dr. Phaik-Mooi Leong (17), now working in Jeffrey Miller's laboratory at UCLA. The general indirect approach is outlined in Fig. 2.

Figure 2 portrays the idea that if a toxin acts by blocking the formation of an essential single polypeptide gene product, an opportunity is

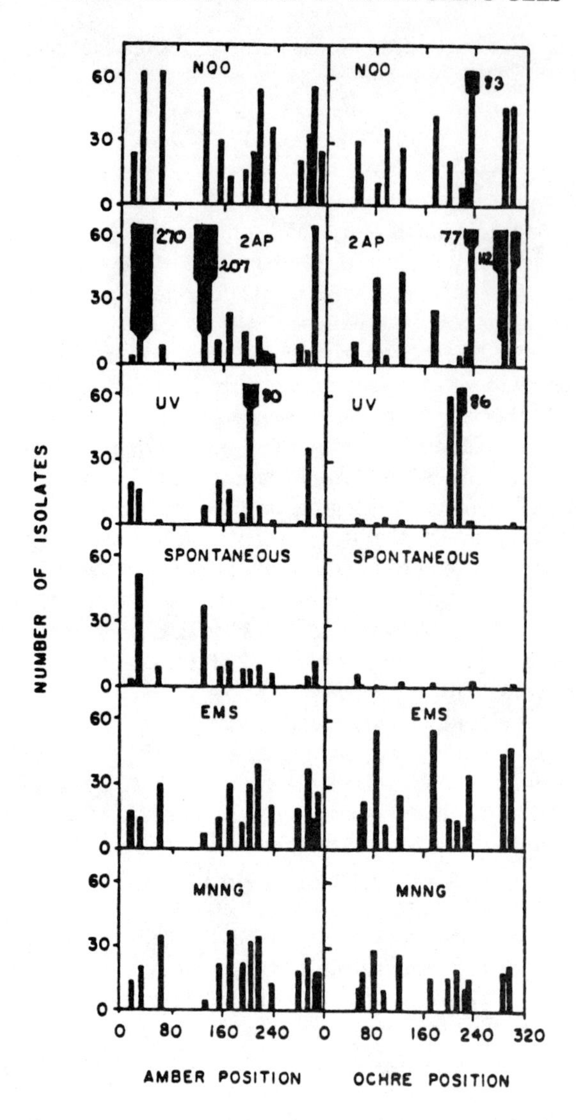

FIG. 1. Distribution of spontaneous and induced mutations in the lac I system of E. coli. Suppressible amber and ochre mutants in the lac I gene were selected and their approximate location in the gene determined by crosses with deletion mutants. The percentage of the amber and ochre mutants found at each site is illustrated. Data from Coulondre and Miller (7,8).

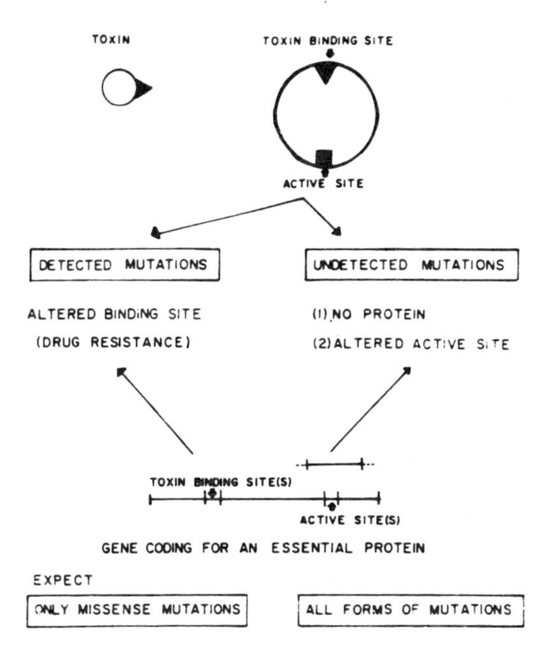

FIG. 2. Schematic representation of forward selection systems that monitor specific kinds of mutations (missense mutations) at relatively few nucleotide pairs. Taken from Leong (17).

afforded to those interested in mutational spectra. In the first place, only restricted kinds of mutation (missense or small additions or deletions of three base pairs) would be expected to confer resistance to such a toxin. Second, because nonsense or frameshift mutations would simultaneously erase the essential function of the enzyme, only those mutations in specific positions constituting the toxin binding site(s) would be expected to markedly reduce the sensitivity of the essential enzyme to the toxin. Thus, the mutations to toxin resistance would be restricted in kind to certain missense mutations and in place to certain positions. The actual kinds and positions would be defined by the structure of each different essential protein and the binding sites within it for an inactivating toxin. On the basis of modeling of the data of Benzer, Miller, and Skopek, Leong predicted that knowledge of the frequency of missense mutations within a gene would be sufficient for mutational spectra purposes if one knew that the mutations lay within any sequence consisting of 30 or fewer base pairs. Since she expected that specific binding sites for toxins were defined by fewer than 10 amino acids, she reasoned that the set of positions and kinds of mutations that produced loss of toxin A binding site in essential protein a would produce a different apparent sensitivity to mutation than a set defined by the binding site of toxin B in essential protein b.

She next set out to devise a set of quantitative mutation assays based on loss of toxin binding sites in essential proteins in cultured human B cells. She used a set of three such assays to test the hypothesis that mutational spectra (clearly demonstrated in E. coli and its phages) could be used in human cells to differentiate among populations in which the primary cause of mutation was spontaneous change or any of a number of chemical mutagens.

Table 1 is a small abstract I have made from her thesis to provide evidence here for my own conviction that her hypothesis has been tested and found to be essentially correct.

The data show that the alkylating agents ethylmethane sulfonate (EMS) and methylnitronitrosoguanidine (MNNG), which are known from Miller's data (Fig. 1) to yield similar but statistically different mutational spectra based on the set of base pair substitution (amber) mutations studied, are differentiable by the "small marker" spectrum approach. This differentiation is possible because EMS induces a significantly greater number of mutations to ouabain resistance (OUA^R) than does MNNG relative to the number of general mutations to 6-thioguanine resistance (TG^R) induced. The responses at the other two loci (PPT^R, DRB^R) are similar, however, as expected for chemicals that yield similar spectra for the lac I gene in E. coli. The spectrum resulting from simply permitting mutants to increase spontaneously in the culture over many generations is obviously different from EMS- and MNNG-induced mutational spectra and indicates that missense mutations affecting these sites are rare events relative to whatever mutations predominate in spontaneous human cell mutagenesis.

The essential lesson I have drawn from this work is that human cell populations can indeed be used to derive mutational spectra of sufficient precision to resolve differences among populations mutated spontaneously and those mutated by many, but presumably not all, chemical mutagens.

It is a simple step of logic to move from a demonstration in cultured B cells to experiments with cell populations that can be obtained by direct human blood sampling. In her work, Dr. Leong was able to treat greater than 10^8 colony-forming human B cells and induce mutant fractions of 10^{-6} for the small markers she used.

Facing up to the absolute need for sufficient numbers of colony-forming cells from human blood samples has been made somewhat easier by the development of protocols for cloning T cells by the immunology community. Among genetic toxicologists, Albertini (1,2) and Morley (6,9,15,26) have reported colony-formation frequencies in the 10-50% range for fresh T cells, and we have reproduced their observations and extended them to the point that it seems reasonable to think of a milliliter of human blood as yielding greater than 2.5×10^5 colony-forming T

TABLE 1. Ratios of mutant fractions at small markers to 6-thioguanine resistance.[a]

Mutagen	OUA^R/TG^R	PPT^R/TG^R ($\times 10^{-2}$)	DRB^R/TG^R
Spontaneous (40 generations)	0.7 ± 0.2	0.1 ± 0.1	0.1 ± 0.1
EMS (170 μM x 4 h)	13 ± 1	0.8 ± 0.3	1.5 ± 0.5
MNNG (70 nM x 1 h)	3 ± 0.8	2.5 ± 0.7	0.4 ± 0.4

[a]Values are means from several experiments ± 99% confidence intervals. The data given are the ratios of the induced mutant fractions for ouabain resistance (OUA^R), podophyllotoxin resistance (PPT^R), and 5,6-dichloro ribofuranosyl benzimidazole resistance (DRB^R) to induced mutant fractions for 6-thioguanine resistance (TG^R). The numerators OUA^R, PPT^R, and DRB^R are used as indicators of specific missense mutations at specific codons in essential genes that are assumed to code for those amino acids constituting the binding site for each toxin. The common denominator TG^R is used as an indicator of a general or non-specific mutation induced spontaneously or by chemical treatment. Resistance to thioguanine is assumed to arise from any of the many kinds of mutations affecting protein structure in the nonessential gene coding for hypoxanthine guanine phosphoribosyl transferase (HGPRT). Thus, each ratio is an indication of the relative amount of specific mutation induced by a particular mutagenic experience. Mutagens that in general cause missense mutations would be expected to yield high ratios in one or more such assays, while chemicals that do not cause such mutations would be expected to yield small ratios.

cells, although some experiments yield twice that number. Thus, 10^8 colony-forming units require $10^8/2.5 \times 10^5$ or 400 mL of blood (nearly a pint!) to permit the search for the expected 1-10 small marker mutants within a population that has experienced nothing excessive in terms of everyday mutagen exposure. Of course, the integral amount of mutation in a human life might yield higher values, but Morley and Albertini have measured fractions of cells mutant for the "large marker" TG and found them to be between 2 and 20×10^{-6}. It seems prudent to us to devise an additional approach that would yield the mutational spectra data desired from blood samples of a size that would not be life threatening. Thus, while it is our present intention to actually measure for OUA^R, PPT^R, and DRB^R in large (pint) samples from volunteers (faculty only!), we are placing our major effort on an approach, the essentials of which I would like to describe here.

There are two reasons to present in detail the following research program so early in its life. First, it seems unfair to castigate the present state of genetic toxicology without at least suggesting certain approaches that might conceivably answer our needs. Second, I would like to solicit suggestions from those skilled in the art of DNA manipulation who might spot some erroneous assumption or unnecessarily difficult path that a laboratory long practiced in cellular manipulation would be expected to make as it begins to practice quantitative molecular manipulations.

There are three technical advances necessary to this approach. First, we need to be able to grow large numbers of cells mutant in or about the structural gene for HGPRT. Second, we need to know the DNA sequences of the portions of the hgprt gene (exons) coding for protein and portions of the sequences of the nonstructural portions (introns) immediately flanking the exon. Third, we need reliable means to recover restriction fragments of the hgprt gene in sufficient yield and purity to permit recognition of mutant fragments by their behavior when annealed with wild-type DNA sequences on gradient denaturing electrophoretic gels.

The first consideration is the ability to grow cells directly from human blood samples under conditions in which one would expect most, if not all, of the growing cells to contain mutations in the DNA coding for the structure of the enzyme HGPRT. On the basis of our experiences using Albertini's and Morley's protocols for growing T cells, we are confident but have not yet demonstrated that cells that are resistant to 6-thioguanine can be grown en masse from cultures derived from a few hundred milliliters of blood. In no $6TG^R$ human cell have we detected HGPRT activity. The work of the Capecchis (27) with CHO cells indicated that about 50% of cells resistant to this drug synthesized a protein that was immunologically similar to normal HGPRT but without its enzyme activity, which is essential for conferring sensitivity to 6-thioguanine.

Thus, we had an expectation based on analogy that $6TG^R$ cells should contain a fair fraction of mutants in the structural coding region of the hgprt gene, since we expect that only a mutant in the structural coding region will give rise to any protein product resembling the normal product in an immune reaction. Furthermore, in these proceedings, Peter Brooke has reported that some 75% of the 8-azaguanine-resistant cells in his laboratory studies are making an mRNA that hybridizes to a cDNA probe derived from mRNA for HGPRT from normal cells and is approximately the same size as normal HGPRT mRNA. Since 8-azaguanine resistance is also mediated by loss of most HGPRT activity, this result is a further important indication that a set of mutants resistant to 6-thioguanine or

8-azaguanine will consist of a fair fraction, now estimated to be 75%, which is mutant in the structural hgprt gene. Mutations outside the structural gene, such as control regions and/or introns, would be expected to yield little or no HGPRT mRNA or an mRNA of markedly different size, respectively.

Thus, we believe the first essential element, the ability to grow large numbers of hgprt structural gene mutants, is within the grasp of existing technology.

The second step requires that we obtain knowledge of the DNA sequences of certain portions of the hgprt gene. The hgprt gene has been studied at many laboratories including those of Caskey at Baylor and Jolly at the University of California at San Diego Medical School, who studied its organization and DNA sequence in mouse, hamster, and human cells (13,14,18,21). This work has shown hgprt to be about 35,000 base pairs long in humans, with nine short sequences that code for the protein (654 base pairs) (C. T. Caskey, personal communication) interspersed with the relatively long sequences of noncoding material called introns. The general structure is indicated in Figs. 3 and 4, which give the size and order of the exons (coding regions) and introns, respectively. Both laboratories have isolated cDNA complementary to normal HGPRT mRNA and have generously shared the cDNA in the form of bacterial plasmids with other interested laboratories, including ours. A cDNA-containing plasmid permits one to return to the human genome and select for complementary DNA sequences to all or part of the cDNA. For instance, for our studies it is necessary to know the DNA sequences on either side of each exon for several hundred base pairs. We will use the cDNA as a probe to isolate a set of lambda phages that carry about 9,000 base pair sequences derived

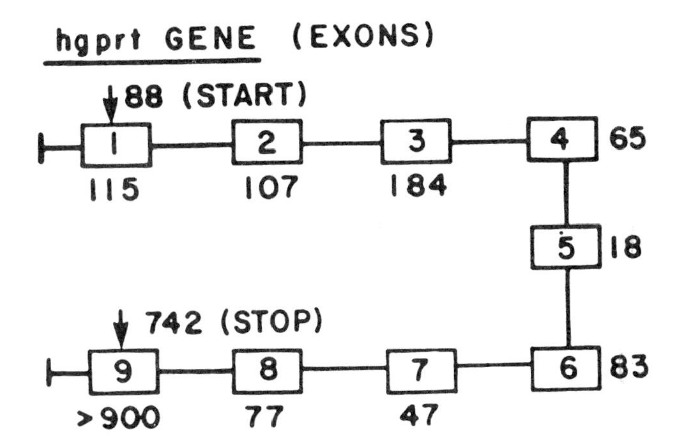

FIG. 3. Organization of mouse hgprt gene's exons.

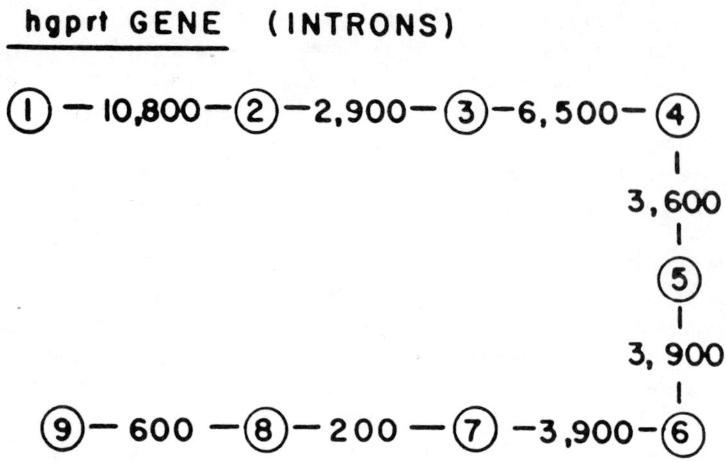

hgprt GENE (INTRONS)

(1) — 10,800 — (2) — 2,900 — (3) — 6,500 — (4) — 3,600 — (5) — 3,900 — (6) — 3,900 — (7) — 200 — (8) — 600 — (9)

FIG. 4. Organization of mouse hgprt gene's introns.

from the human X chromosome complementary to some fraction of the cDNA. This lambda library has been supplied by our colleague Professor Sam Latt of Children's Hospital of Boston and the Harvard Medical School. This set of hgprt fragments will then be further probed with sequences specific for each exon to obtain subsets of phage-carrying sequences containing each particular exon. The appropriate phage DNAs will then be expanded by growth in a fermentor, and restriction fragments originating in a specific exon and terminating in the introns to either side will be isolated and their sequences determined. This process, while iterative, is to be performed in parallel for several exons. The goal is to discover the DNA sequences on both sides of each exon for about 200-250 base pairs.

Given those sequences, we can calculate the melting behavior of any DNA fragment such as those generated by restriction enzyme digests. If we find "appropriate" melting behavior for the majority of exon sequences, we should be able to devise means to separate mutants of the exon sequences based on their melting point behavior when they are used to form heteroduplexes with complementary wild-type DNA sequences. This expectation is based wholly on the work of Leonard Lerman, who has developed the technique of gradient denaturing gel electrophoresis to the point permitting the separation of double-stranded DNAs of 100-150 base pairs even if they differ by a single base pair substitution (10,16). Figures 5, 6, 7, 8, and 9 outline the basic idea.

The first thing to appreciate is that DNA melting, the dissociation of the two strands, is an equilibrium process. The probability of a particular sequence being in the melted or unmelted form as a function of tempera-

GRADIENT DENATURING GEL ELECTROPHORESIS

DOUBLE STRANDED DNA CONTAINING A HIGH AND A LOW MELTING POINT REGION

FIG. 5. Diagram of DNA fragment suitable for study of mutant sequence by gradient denaturing gel electrophoresis (GDGE).

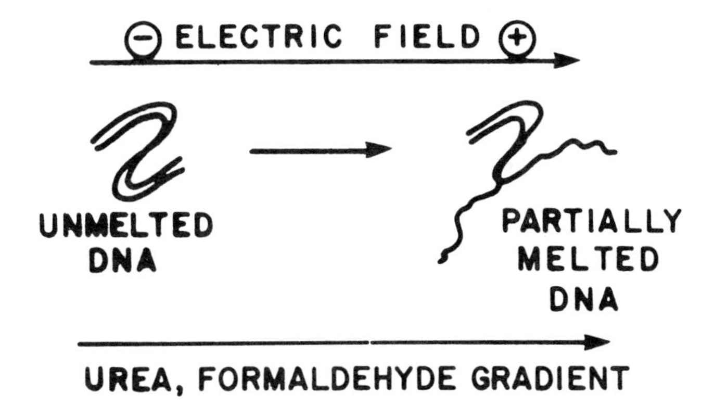

FIG. 6. A low-melting-point region will melt when it reaches a particular point in a gradient denaturing gel.

ture can be approximated by recursive calculations in which each AT or GC base pair is modeled as either melted or not with a fixed probability at a particular temperature (3,11,12,22). Because melting of a DNA sequence requires all of the base pairs to be in the melted form at a given time, any particular sequence of equal length has a fair probability of not having the same melting temperature as any other sequence. Thus, any sequence would be expected to contain "isomelting regions" of different melting points. This idea is shown schematically in Fig. 5.

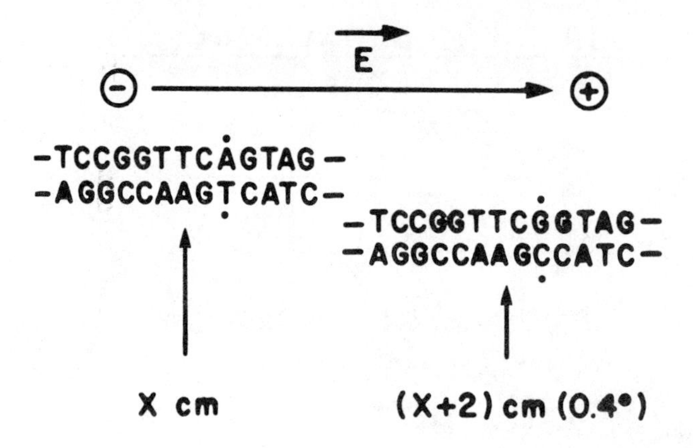

FIG. 7. Two sequences differing by a single base pair are expected to have different melting points and behavior on a gradient denaturing gel.

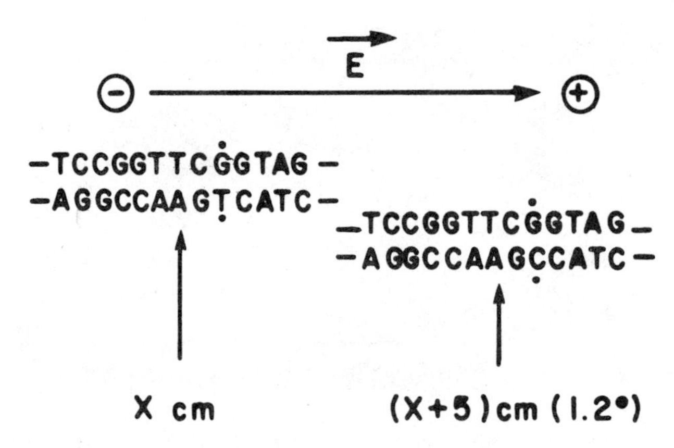

FIG. 8. A single mismatch in the sequence of Fig. 7 is expected to have a larger effect on behavior on a gradient denaturing gel than a base-pair substitution.

When a double-stranded DNA migrates through an electric field on an electrophoretic gel, it is a compact hydrodynamic structure relative to the same material in a melted form. Normally the solution in electrophoretic gels is uniform, but Lerman has devised means to set up a stable gradient of increasing solute (e.g., urea/formaldehyde) concentration in the direction of DNA motion (Fig. 6).

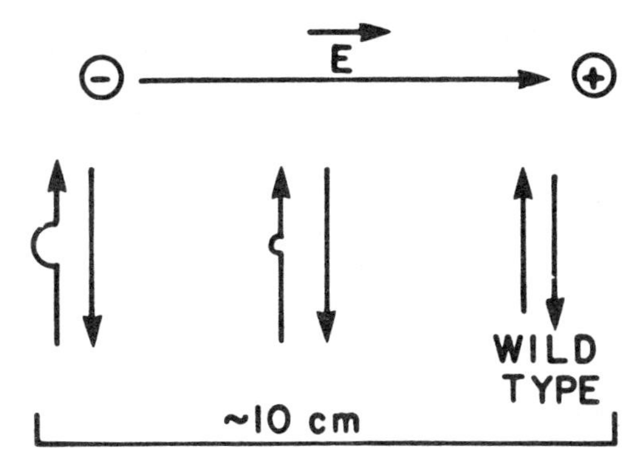

FIG. 9. Different mutant sequences are expected to yield successively greater displacements from wild-type sequences depending on the nature of the mutations.

Thus, as a double-stranded DNA molecule is drawn by the electric field into higher and higher concentrations of urea/formaldehyde, it will eventually reach a point at which the concentration is high enough to melt the lower-melting-point region. The melting results in an essentially immobile structure within the gel so that the position at which the DNA molecule stops on the gradient gel is determined by its melting point, which is uniquely determined by its nucleotide sequence.

Figure 7, drawn from Lerman's work, shows that a single base pair substitution within a low melting point sequence of some 100 base pairs has changed the expected melting point by 0.4°C and resulted in about a 2-cm displacement under appropriate denaturing gel conditions. Figure 8 is my own conception, again drawing on conversations with Professor Lerman, of what would be expected were a mismatch introduced into the same molecule at the same position as the base pair substitution of Fig. 7. Here, because the mismatch was assumed to exist in the melted form for the purpose of estimating its effect on melting within the sequence, a considerable displacement is predicted, greater in any case than would be expected for a similarly placed base-pair substitution.

This expectation leads to the idea that if a mixture of DNA sequences derived from point mutations within the same restriction fragment were permitted to anneal with a complementary wild-type sequence, the melting point of each type of heteroduplex would differ depending on the kind and position of each mutation. As indicated in Fig. 9, mutations in sequences that create disturbances in base pairing affecting several base pairs would be expected to give greater displacements from wild-type sequences than mismatches involving a single base pair.

Figure 10 indicates our expectation of what this might look like on an actual gradient denaturing gel. The basic idea is that if different kinds of mutations were distributed over a sequence in which mutations were detectable and for which a suitable wild-type probe DNA existed, then a pattern would appear on the gel arising from the diversity of mutant DNA sequences. A particular set of mutants would reproducibly be seen as a particular pattern on the gel.

Referring back to the discussion of the human X-chromosomal gene hgprt, we note that its exons are of a size appropriate for probing by gradient denaturing gel electrophoresis, since the shortest exon (#5) has 18 base pairs and the longest (#3) has 184 base pairs in the reading frame for the HGPRT protein.

Leonard Lerman was kind enough to carry out the calculations for the melting behavior of the cDNA sequences determined by Jolly for human hgprt (Fig. 11). An interesting detail is a very high melting point (>80°C) region in the untranscribed leading sequence up to base pair 88 and an abrupt transition to a 69-71°C melting point for the entire cDNA sequence to and somewhat beyond the stop codon at base pair 742. Of particular initial interest is the fact that exon 3 happens to have restriction sites that include 165 of its 184 base pairs and a melting point transition (69.5-71°C) some 40 base pairs into this restriction fragment. Our present work involves isolating and screening $hgprt^-$ human cells for mutants in the roughly 125-base-pair sequence that we should be able to examine with a suitable complementary DNA probe on gradient gels as a first means to test the sensitivity and precision of the overall approach in recognizing patterns generated from mixtures of known human cell mutants. Since the entire coding sequence of hgprt is approximately 654 base pairs, this particular (165 base pair) probe would be expected to be useful for somewhat less than 20% of all mutants, assuming that mutable sites are distributed evenly over the hgprt reading frame. To find mutants in the entire reading frame of each exon, we will need to know the flanking sequences for the exons with the aim of finding higher melting point regions adjacent to exons, permitting recognition of the mutants on gradient gels in melting point exon sequences. The basic problem being attacked is depicted in Fig. 12 for exon 3 of hgprt. The 184 base pair sequence and the available internal restriction fragment are depicted above the melting profile of the exon, the sequence of which is known, and the desired melting profile for useful flanking sequences, which are unknown at present.

We currently see the order of this work in progress as a series of discrete steps.

(a) Isolate from a lambda library of the human X-chromosome a set of vectors containing the hgprt exons and flanking sequences.

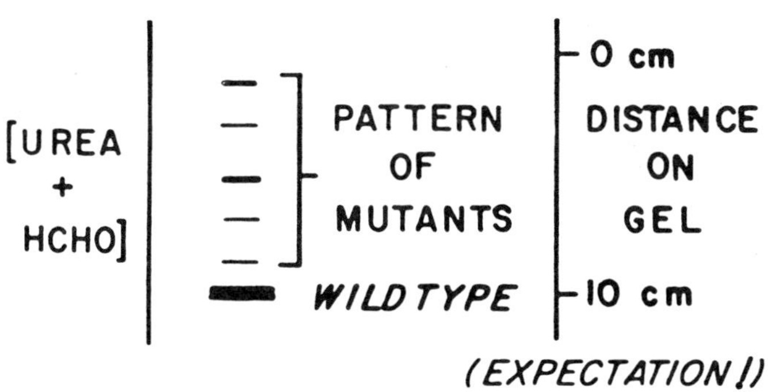

FIG. 10. Expectation of a pattern on a gradient denaturing gel generated by the behavior of a set of mutants in exon 3 of hgprt.

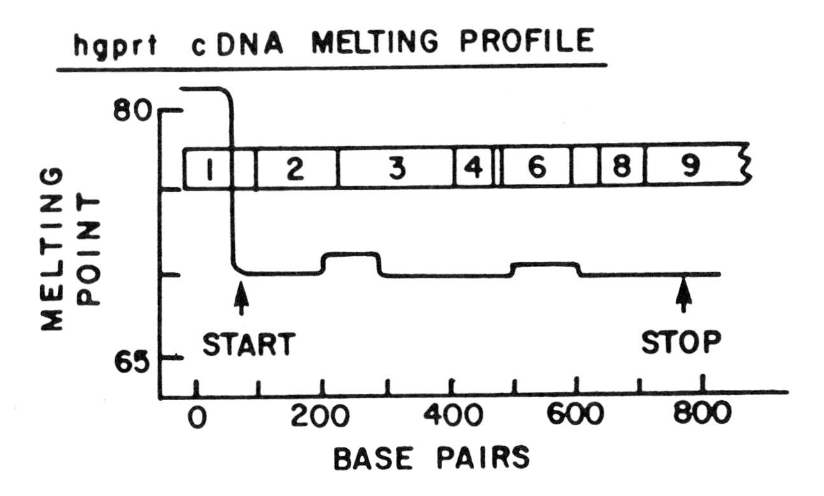

FIG. 11. Diagram of hgprt exon structure in cDNA (after Caskey) and melting point profile calculated by Lerman from the sequence of Jolly.

(b) Determine the sequences of the flanking sequences (up to 250 base pairs).

(c) Generate restriction maps of the fragments containing flanking sequences and exons.

(d) Discover by calculation which, if any, restriction fragments would represent domains of high and low melting points suitable for analysis by gradient denaturing gel electrophoresis.

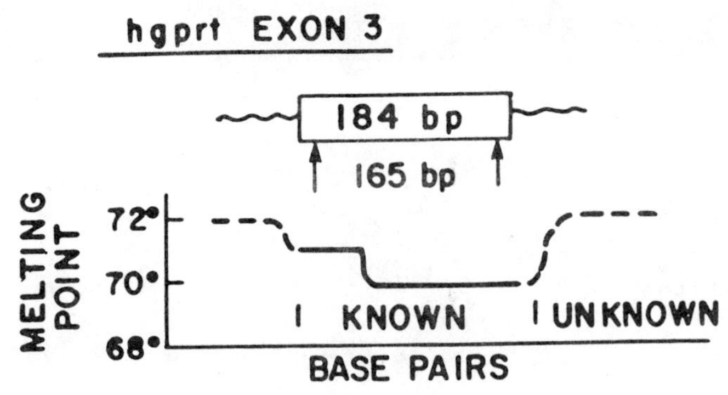

FIG. 12. Diagram illustrating the need to know the sequences flanking exon 3 of hgprt to discover appropriate restriction fragments which might yield higher melting point sequences outside the exon.

(e) Create probes for the suitable restriction fragments for as much of the hgprt structural gene as possible.

(f) Use the probes to characterize mutational spectra in mutants derived from cultured human cell populations (tests for accuracy and precision).

(g) Use the probes to characterize the mutational spectra in hgprt in human T cell populations derived from fresh blood samples.

(h) Use the probes to compare the mutational spectra in T cell populations in persons exposed to high concentrations of a particular known mutagenic substance to the spectra produced by exposing the person's cells to the substance in vitro.

(i) Develop further studies toward application as an epidemiological tool.

Now, the fact that these steps represent a lot of work is not unknown to us who are beginning to follow them. It is because a series of complex steps are required that they have been spelled out in some detail to invite criticism and suggestions. It is also clear to me, as it must be to you, that this paper does not discuss the nuts and bolts of how, if the technical approach suggested were developed, it would be applied in a useful epidemiological program. The discussion is not present because it has not yet been thought out; the American Cancer Society is sponsoring a collaborative effort with Professor Brian McMahon of the Harvard School of Public Health to consider this aspect as thoroughly as possible.

There are, of course, many intelligent people who believe that the central premise stated early in this paper is correct and that present systems for identifying mutagens are in fact leading to measures that protect human health in a real and not illusory way. These scientists may in fact be correct. Just because the central premise--that environmental mutagens cause a fair fraction of human genetic lesions--is untested does not mean that it is untrue.

If, however, one argues that the central premise is so clearly true that it need not be tested, then I call on the proponents of that argument to present the evidence that forms the basis of their convictions; I have looked and found none.

If one argues that the difficulty and expense of devising means to directly test the untested premise is too great, I call for an audit. What is the cost of the present subjunctive methodologies that tell us what chemical might, or then again, might not be a genetic hazard to people? Should we add to that the cost of uncertainty in developing new materials for human use? Should we count the cost of public anxiety when we as scientists cannot provide reliable measurements? How shall we account for the cost of litigation for "toxic torts" in which the courts are asked to weigh the opinions but not the evidence provided by "expert witnesses"?

I have derived a premise of my own about these costs: it will be a lot cheaper to recognize the need for means to diagnose the actual causes of human genetic change and fund a number of efforts to find a practical means than to argue interminably in the legislature and the courts.

ACKNOWLEDGMENTS

The opportunity to develop these ideas has been provided by funding from the Department of Energy Grant AC02-77EV04627, National Institute of Environmental Health Sciences (NIEHS) Program Project Grant 2-PO1-ES01640, NIEHS Program Project Grant 5-PO1-ES00597, NIEHS Center Grant 2-P30-ES02109, American Cancer Society Grant SIG-10-1, and a grant from the Dana Foundation.

REFERENCES

1. Albertini, R.J., Sylvester, D.L., Dannenberg, B.D., and Allen, E.F. (1982): Basic Life Sci., 21:403-424.
2. Albertini, R.J., Castle, K.L., and Borcherding, W.R. (1982): Proc. Nat. Acad. Sci., USA, 79:6617-6621.
3. Amirikyan, B.R., Vologodskii, A.V., and Lyubchenko, Y.L. (1981): Nucleic Acids Res., 9:5469-5482.
4. Benzer, S. (1961): Proc. Nat. Acad. Sci., USA, 47:403-415.
5. Calos, M.P., and Miller, J.H. (1982): J. Mol. Biol., 153:39-66.
6. Chrysostomou, A., Seshadri, R., and Morley, A. (1984): Scand. J. Immunol., 19:293-296.
7. Coulondre, C., and Miller, J.H. (1977a): J. Mol. Biol., 117:525-575.
8. Coulondre, C., and Miller, J.H. (1977b): J. Mol. Biol., 117:577-606.

9. Dempsey, J.L., Morley, A.A., Seshadri, R.S., Emmerson, B.T., Gordon, R., and Bhagat, C.I. (1983): Hum. Genet., 64:288-290.
10. Fischer, S.G., and Lerman, L.S. (1983): Proc. Nat. Acad. Sci., USA, 80:1579-1583.
11. Fixman, M., and Freire, J.J. (1977): Biopolymers, 16:2693-2704.
12. Gotoh, O., and Tagashira, Y. (1981): Biopolymers, 20:1033-1042.
13. Jolly, D.J., Okayama, H., Berg, P., Esty, A.C., Filpula, D., Bohlen, P., Johnson, G.G., Shively, J.E., Hunkapillar, T., and Friedman, T. (1983): Proc. Nat. Acad. Sci., USA, 50:477-481.
14. Koneki, D.S., Brennand, J., Fuscoe, J.C., Caskey, T.C., and Chinault, A.C. (1982): Nucleic Acids Res., 10:6763-6775.
15. Kutlaca, R., Seshadri, R.S., and Morley, A.A. (1984): Exp. Hematol., 12:339-342.
16. Lerman, L.S., Fischer, S.G., Hurley, I., Silverstein, K., and Lumelsky, N. (1984): Annu. Rev. Biophys. Bioeng., 13:399-423.
17. Leong, P.-M. (1984): Means of Recognizing the Causes of Mutation in Human Cell Populations: The Normalized Small Marker Mutational Spectra. Ph.D. Thesis, Massachusetts Institute of Technology, Cambridge.
18. Melton, D.W., Konecki, D.S., Brennand, J., and Caskey, C.T. (1984): Proc. Nat. Acad. Sci., USA, 81:2147-2151.
19. Miller, J.H., Coulondre, C., and Farabaugh, P.J. (1978): Nature, 274:770-775.
20. Miller, J.H., Ganem, D., Lu, P., and Schmitz, A. (1977): J. Mol. Biol., 109:275-301.
21. Patel, P.I., Nussbaum, R.L., Framson, P.E., Ledbetter, D.H., Caskey, C.T., and Chinault, A.C. (1984): Somatic Cell Mol. Genet., 10:483-493.
22. Poland, D. (1974): Biopolymers, 13:1859-1871.
23. Schmeissner, U., Ganem, D., and Miller, J.H. (1977): J. Mol. Biol., 109:303-326.
24. Skopek, T.R., and Hutchinson, F. (1982): J. Mol. Biol., 159:19-33.
25. Thilly, W.G., Leong, P.-M., and Skopek, T.R. (1982): In: Banbury Report 13. Indicators of Genotoxic Exposure in Man and Animals, edited by B.A. Bridges, B.E. Butterworth, and I.B. Weinstein, pp. 453-465. Cold Spring Harbor Laboratory, Cold Spring Harbor, New York.
26. Trainor, J., and Morley, A.A. (1983): J. Immunol. Methods, 65:369-372.
27. Wahl, G.M., Hughes, S.H., and Capecchi, M.P. (1975): J. Cell. Physiol., 85:307-320.

Subject Index

AAF, *see* N-2-Acetylaminofluorene
Aberrations, in gene transcription in tumor cells, 182–184
Abnormal clones, emergence of, in human diploid cells, 346–348
Abnormalities, chromosome, in human leukemia, 409–418
Accommodation, recovery and, acquired radioresistance of hematopoietic elements as primary mechanism of, 373–375
N-2-Acetylaminofluorene-poly (dG-dC).poly(dG-dC)
  antibodies to Z-DNA raised by, 471–475
  perfect and distorted Z-DNA conformations of, 475–477
  and poly(dG-m^{5}dC).poly(dG-m^{2}dC), CD spectra of, 467–469
  recognition of, by nuclease S_1 and anticytidine antibodies, 468–471
Acquired radioresistance of hematopoietic elements, as primary mechanism of recovery and accommodation, 373–375
Activated oncogenes, and tumor promoters, synergy between, 180,182
Activation
  or induction, of C-kinase, absence of, 281–283
  mechanism of, of *ras* oncogenes, 42–45
  metabolic, of chemical carcinogenesis, *in vitro* systems to study organ and species differences in, 135–145
  of oncogenes, 38–40
  of protein kinase C, 178–181
Acute (nonlymphocytic) leukemia
  chromosome patterns in children and adults with, comparison of, 416
  correlation of karyotype and occupation in adult patients with, 414–416
  secondary to cytotoxic therapy, 409–415
Adducts
  nucleic acid, detection of, 170–171
  urinary excretion of, 173–174
Adult patients, with acute leukemia, karyotype and occupation in, 414–416
Adults, children and, with ANLL, 416
Age, at exposure, and susceptibility to radiation carcinogenesis, 384–385
Age-response dependencies, 320–321
Alkenylbenzenes, and aromatic amines, 93–105
1-Alkylbenzo(a)pyrenes, 452–461
Alterations, chromosomal, *see* Chromosomal alterations
Amines, aromatic, alkenylbenzenes and, 93–105
4-Aminoazobenzene, 101–103
Analogous patterns of benzo(a)pyrene metabolism, in human and rodent cells, 123–131
Analogs, endogenous, for phorbol ester tumor promoters
  diacylglycerols as, 255–259
  receptors and, 249–260
Analysis
  of cellular DNA, 171–173
  chemical, direct, quantification of carcinogens by, 167–170
Anatomical distribution of dose, 393
Animal, intact, apparent discrepancies between *in vitro* studies and *in vivo* studies with N-nitrosos proline and N-nitrosocimetidine in, 110–115
ANLL, *see* Acute (nonlymphocytic) leukemia
Antibodies
  anti-cytidine, nuclease S_1 and, recognition of AAF-poly(dG-dC).poly(dG-dC) by, 468–471
  to Z-DNA, raised by AAF-poly(dG-dC).poly(dG-dC), 471–475
Anti-oncogenes, 19–20
Antioxidants, free radical scavengers and, protection by, 308–310
Aplasia-prone subgroups, and leukemia-prone, preclinical responses of, 369–373
Aromatic amines, alkenylbenzenes and, 93–105
Assay(s)
  for chemically induced mutation to ouabain resistance in C3H 10T½ cells, 211–218
  liver cell-mediated, species differences in data from, 145
  mutation, 1-alkylbenzo(a)pyrenes and, 453–457
Assay system, transformation, 318–319
Ataxia-telangiectasia, cancer in, 355–361

BALB 3T3 cells, enhancement of chemically induced transformation of, 279–281
Benzo(a)pyrene metabolism, analogous patterns of, in human and rodent cells, 123–131
Binding, covalent, mechanism of, 450–452
Biochemical dosimetry, chemical and, 167–174
Biological basis, for assessing carcinogenic risks of low-level radiation, 381–397

Biological models, of carcinogenesis, 395–397
Biological properties, of 1-alkylbenzo(a)pyrenes, 452–461
Biologically active metabolite, of vitamin D_3, induction of macrophage cell differentiation in HL-60 cells by, 265–268
Biotransformation, of xenobiotics and steroid hormones
  effects of cigarette smoking on, 153–156
  effects of diet on, 156–161
Blood responses, of aplasia-prone and leukemia-prone subgroups, 369–370
Bone-marrow responses, of aplasia-prone and leukemia-prone subgroups, 370–372
B16 melanoma cells, stimulation of melanin synthesis of, 279
Bypass, of lesions, 486–491

Cancer
  in ataxia-telangiectasia, 355–361
  development of, 20
  evolution and, 51–62
  human, *see* Human cancer *entries*
  nonviral, and cancer induced by replication-competent retroviruses, 18–19
Carcinogen-induced DNA damage, cellular responses to, 177–179
Carcinogen metabolism, drugs for predicting, 149–152
Carcinogenesis
  biological models of, 395–397
  chemical, *see* Chemical carcinogenesis
  evolutionary multi-hit concept of, 81–82
  gene expression in, relationship of chromosomal alterations to, 419–428
  and mutagenesis by N-nitroso compounds, DNA dosimetry in, 115–119
  polycyclic (aromatic) hydrocarbon
    intercalation in, 449–462
    mammalian cell mutation and, 433–444
  radiation, susceptibility to, 382–385
Carcinogenic metabolites, electrophilic and, ultimate, of some alkenylbenzenes and aromatic amines, 93–105
Carcinogenic risks, of low-level radiation, biological basis for assessing, 381–397
Carcinogens
  chemical, *see* Chemical carcinogens
  short-term tests for, 496–497
CD spectra, *see* Circular dichroism spectra
Cell culture studies, 177–185
Cell differentiation, *see* Differentiation
Cell-mediated mutagenesis
  of salmonella, 141–144
  of V79 cells, 136–141
Cell mutation, mammalian, and polycyclic hydrocarbon carcinogenesis, 433–444
Cell transformation, *see* Transformation
Cells
  BALB 3T3, enhancement of chemically induced transformation of, 279–281
  C3H 10T1/2, *see* C3H 1OT1/2 *entries*
  diploid, human, 337–352
  hamster embryo, enhancement of chemically induced cell transformation in, by 1,25-$(OH)_2D_3$, 266–270
  HL-60, *see* HL-60 *entries*
  human, *see* Human cells
  melanoma, B16, stimulation of melanin synthesis of, 279
  mouse skin, 201–209
  organs, tissues and, 382–383
  rodent, and human, radiogenic transformation of, oncogenes and cellular controls in, 303–313
  R1B6, phenotypic surface characteristics of, 294–295, 297–298
  stem, radiosensitivity of, survival and pathological predisposition as function of, 376–377
  transformation of normal into malignant, in culture, 23–25
  tumor, aberrations in gene transcription in, 182–184
  V79, cell-mediated mutagenesis of, 136–141
Cellular changes, and initiation, 202–205
Cellular controls, oncogenes and, in radiogenic transformation of rodent and human cells, 303–313
Cellular DNA, analysis of, 171–173
Cellular mechanisms, of hematopoietic recovery, 373–376
Cellular responses
  to carcinogen-induced DNA damage, 177–179
  in chronic radiation leukemogenesis, 363–378
Cellular transforming genes, *see* Transforming genes
Changes
  cellular, associated with initiation, 202–205
  chromosome, *see* Chromosomal alterations
  molecular, and initiation, 205–209
Chemical agents, and physical, 394–397
Chemical analysis, direct, 167–170
Chemical carcinogenesis, 51–58
  in mouse skin cells, 201–209
  and oncogenes, 443
Chemical carcinogens
  "epigenetic," 72–74
  and human cancer, 443–444
  in vitro systems to study organ and species differences in metabolic activation of, 135–145
  and radiation, 395
    transformation of normal into malignant cells in culture by, 23–25
  stabilization of Z-DNA conformation by, 465–477
  that are not significantly mutagenic, 219,221
  and tumor promoters, mechanism of action of, cell culture studies on, 177–185
Chemical dosimetry, and biochemical, 167–174

Chemically induced multistep neoplastic transformation, in C3H 10T1/2 cells, mechanisms of, 211–221
Chemically induced mutation, to ouabain resistance, in C3H 10T1/2 cells, comparison of chemical transformation and, 218–220
Chemically induced transformation
of BALB 3T3 cells, enhancement of, 279–281
in hamster embryo cells, by 1,25-$(OH)_2D_3$, enhancement of, 266–270
Chemicals
genotoxic, exposure to, chemical and biochemical dosimetry of, 167–174
that produce genetic damage, 495–507
Children, and adults, with ANLL, comparison of chromosome patterns in, 416
Chromosomal alterations
in human cancer, types of, 419–420
human cellular transforming genes and, 420–422
relationship of, to gene expression in carcinogenesis, 419–428
Chromosomal rearrangements, X-ray-induced, in long-term cultures, 341–345
Chromosomal translocation, 38–40
Chromosome abnormalities, in human leukemia, 409–418
Chromosome No. 5 or 7, in acute leukemia secondary to cytotoxic therapy, 413–415
Chromosome patterns, in children and adults with ANLL, comparison of, 416
Chronic radiation leukemogenesis, 363–378
Cigarette smoking, stimulatory effect of, on biotransformation of xenobiotics and steroid hormones, 153–156
Cigarettes, combined with radiation, 394
Circular dichroism spectra, of AAF-modified poly(dG-dC).poly(dG-dC) and poly(dG-m^5dC).-poly(dG-m^5dC), 467–469
Clinical correlations, of acute leukemia secondary to cytotoxic therapy, 410–412
Clones
abnormal, emergence of, in human diploid cells, 346–348
with prolonged life-span, emergence of, in human diploid cells, 348–351
Clonotypes, progenitor, radioresistant, differential growth responses of, 375–376
Conformations, Z-DNA, perfect and distorted, of AAF-poly(dG-dC).poly(dG-dC), 475–477
Control(s)
of cell differentiation, and cell transformation *in vitro*, by PMA and 1,25-$(OH)_2D_3$, 263–270
cellular, oncogenes and, in radiogenic transformation of rodent and human cells, 303–313
normal, uncoupling of, growth factors, differentiation factors and, 25–27
Covalent binding, mechanism of, 450–452
Critical region, deletion of chromosome No. 5 or No. 7 and, in acute leukemia secondary to cytotoxic therapy, 413–415
C3H 10T1/2 cells, mechanisms of chemically induced multistep neoplastic transformation in, 211–221
C3H 10T1/2 CL8 cells, inhibition and enhancement of oncogenic cell transformation in, 225–233
Culture(s)
long-term, X-ray-induced chromosomal rearrangements in, 341–345
transformation of normal into malignant cells in, 23–25
Cultured human cells, control of cell differentiation in, after treatment with tumor-promoting agents, 264–265
Cytogenic results, of acute leukemia secondary to cytotoxic therapy, 412–414
Cytotoxic therapy, acute leukemia secondary to, 409–415

Damage
DNA, carcinogen-induced, cellular responses to, 177–179
genetic estimation of potencies of chemicals that produce, 495–507
transformation, sublethal and subeffective, by γ-rays, repair of, 326–328
Data, dose-response, 497–502
Decreased G_T synthesis, in response to tumor promoters, 239
Deletion, of chromosome No. 5 or 7, in acute leukemia secondary to cytotoxic therapy, 413–415
Dependence, transformation, on seeding density, 322–324
Dependencies, age-response, 320–321
Detection, of nucleic acid adducts, 170–171
Development
of assay for chemically induced mutation to ouabain resistance in C3H 10T1/2 cells, 211–218
of cancer, 20
neoplastic
differentiation as epigenetic characteristic of, 74–76
natural history of, 68–71
Diacylglycerols, as endogenous phorbol ester analogs, 255–259
Diet, effects of, on biotransformation of xenobiotics and steroid hormones, 156–161
Differential growth responses, of radioresistant progenitor clonotypes, 375–376
Differentiation
enzyme markers of, expression of, 293–296
as epigenetic characteristic of neoplastic development, 74–76
induction of, by phorbol esters, HL-60 variant reversibly resistant to, 287–300

Differentiation *(contd.)*
regulation of, by 1α,25(OH)$_2$D$_3$, 277–280
terminal, of epidermal keratinocytes, 278
and transformation, control of, *in vitro*, by PMA and 1,25-(OH)$_2$D$_3$, 263–270
Differentiation factors, and uncoupling of normal controls, 25–27
1α,25-Dihydroxyvitamin D$_3$ (1α,25(OH)$_2$D$_3$
as immunoregulatory hormone, 277–278
PMA and, 263–270
regulation of cell differentiation by, 277–280
regulation of tumor promotion by, 279–284
Diol epoxide-induced mutations, 438–442
Diploid cells, human, mechanisms of malignant transformation of, 337–352
Direct chemical analysis, 167–170
Distorted Z-DNA conformations, perfect and, of AAF-poly(dG-dC).poly(dG-dC), 475–477
Distribution, anatomical, of dose, in radiation carcinogenesis, 393
DNA, cellular, analysis of, 171–173
DNA damage, carcinogen-induced, cellular responses to, 177–179
DNA dosimetry, pharmacokinetics and, in relating *in vitro* to *in vivo* actions of N-nitroso compounds, 109–119
Z-DNA, antibodies to, raised by AAF-poly(dG-dC).-poly(dG-dC), 471–475
Z-DNA conformation(s)
perfect and distorted, of AAF-poly(dG-dC).-poly(dG-dC), 475–477
stabilization of, 465–477
Dose, in radiation carcinogenesis
anatomical distribution of, 393
incidence and, relation between, 385–394
Dose rate
effects of, leukemia subtypes, incidences and, with chronic radiation, 366–368
in radiation carcinogenesis, 393
Dose-response data, 497–502
Dosimetry
chemical and biochemical, of exposure to genotoxic chemicals, 167–174
DNA, pharmacokinetics and, in relating *in vitro* to *in vivo* actions of N-nitroso compounds, 109–119
Drugs, as probes for predicting carcinogen metabolism, 149–152

Electrophilic metabolites, and carcinogenic, ultimate, of some alkenylbenzenes and aromatic amines in mouse liver, sulfuric acid esters as, 93–105
Electrophoresis, gradient denaturing gel, in obtaining mutational spectra from human cells, 511–527
Embryo cells, hamster, enhancement of chemically induced cell transformation in, by 1,25-(OH)$_2$D$_3$, 266–270
Emergence
of abnormal clones, in human diploid cells, 346–348
of clones with prolonged life-span, in human diploid cells, 348–351
Endogenous phorbol ester analogs
diacylglycerols as, 255–259
receptors and, 249–260
Enhancement
of chemically induced transformation
of BALB 3T3 cells, 279–281
in hamster embryo cells, by 1,25-(OH)$_2$D$_3$, 266–270
inhibition and, of oncogenic cell transformation, in C3H 10T1/2 CL8 cells, 225–233
Enzyme markers, of differentiation, 293–296
Epidermal keratinocytes, stimulation of terminal differentiation of, 278
Epigenetic characteristic of neoplastic development, differentiation as, 74–76
"Epigenetic" chemical carcinogens, 72–74
Epigenetics, 65–77
Esters, sulfuric acid, as ultimate electrophilic and carcinogenic metabolites of some alkenylbenzenes and aromatic amines in mouse liver, 93–105
Estimation, of potencies of chemicals that produce genetic damage, 495–507
Evolution
and cancer, 51–62
of oncogenes, 58–62
Evolutionary concept of carcinogenesis, 81–82
Excretion, of adducts, urinary, 173–174
Exposure
age at, and susceptibility to radiation carcinogenesis, 384–385
to genotoxic chemicals, 167–174
mutagenic, and chromosome abnormalities in human leukemia, 409–418
Expression
of enzyme markers of differentiation, 293–296
gene, in carcinogenesis, relationship of chromosomal alterations to, 419–428

Fission-spectrum neutrons, 327–331,333
Free radical scavengers, and antioxidants, protection by, 308–310

Gamma-rays, repair of sublethal and subeffective transformation damage by, 326–328
Gene(s)
human cancer recessive (regulatory?) susceptibility, 403–407
and membrane signals, involved in neoplastic transformation, 235–245
normal, and oncogenes, 26,28
promotion sensitivity, 240–244
ras, detection of, as human transforming genes, 41–42

retinoblastoma, 403–407
transforming, *see* Transforming genes
Gene expression, in carcinogenesis, 419–428
Gene transcription, in tumor cells, 182–184
Genetic background, and susceptibility to radiation carcinogenesis, 382
Genetic damage, estimation of potencies of chemicals that produce, 495–507
Genetic factors, in transformation, 306
Genetic mechanisms of oncogenesis, 15–20
Genetics, and epigenetics, of neoplasia, 65–77
Genotoxic chemicals, exposure to, 167–174
Gradient denaturing gel electrophoresis, in obtaining mutational spectra from human cells, 511–527
Growth factors, differentiation factors, and uncoupling of normal controls, 25–27
Growth responses, differential, of radioresistant progenitor clonotypes, 375–376
G_T synthesis, *see* Trisialoganglioside synthesis

Hamster embryo cells, enhancement of chemically induced cell transformation in, by 1,25-$(OH)_2D_3$, 266–270
HAT, *see* Hypoxanthine:aminopterin:thymidine
Heidelberger, C., 1–13
Hematopoietic recovery, 373–376
Hematopoietic stem cell responses, of aplasia-prone and leukemia-prone subgroups, 372–373
Hemizygosity, or homozygosity, of RB gene, mechanisms leading to, 404–406
Hepatocarcinogenesis, single-hit versus multi-hit concept of, 81–91
Heterogeneity, in phorbol ester recognition and response, 250–253
History, natural, of neoplastic development, 68–71
HL-60 cells, induction of macrophage cell differentiation in, by biologically active metabolite of vitamin D_3, 265–268
HL-60 variant, reversibly resistant to induction of differentiation by phorbol esters, 287–300
Homozygosity, hemizygosity or, of RB gene, mechanisms leading to, 404–406
Hormonal effects, in transformation, 306, 308
Hormone(s)
combined with radiation, 395
immunoregulatory, 1α,25$(OH)_2D_3$ as, 277–278
steroid, xenobiotics and, biotransformation effect of cigarette smoking on, 153–156
effects of diet on, 156–161
Human cancer
chemical carcinogens, and, 443–444
chromosome changes in, types of, 419–420
Human cancer recessive (regulatory?) susceptibility gene, 403–407
Human cells
cultured, control of cell differentiation in, after treatment with tumor-promoting agents, 264–265
mutational spectra from, gradient denaturing gel electrophoresis and, 511–527
ras proto-oncogenes of, 40
and rodent
analogous patterns of benzo(a)pyrene metabolism in, 123–131
radiogenic transformation in, oncogenes and cellular controls in, 303–313
Human diploid cells, mechanisms of malignant transformation of, 337–352
Human leukemia, chromosome abnormalities in, 409–418
Human malignancies
oncogenes in, activation of, by chromosomal translocation, 38–40
transforming genes of, 35–45
Human transforming genes
and chromosome changes, 420–422
detection of *ras* genes as, 41–42
N-Hydroxy-2-acetylaminofluorene, 98–102
1′-Hydroxysafrole, 95–100
Hypoxanthine:aminopterin:thymidine-resistant revertants of mutant B19, 441–444

Immunoregulatory hormone, 1α,25$(OH)_2D_3$ as, 277–278
In vitro actions of N-nitroso compounds, relation of *in vivo* actions to, pharmacokinetics and DNA dosimetry in, 109–119
In vitro cell transformation, radiogenic, 303–305
In vitro control, of cell differentiation and cell transformation, by PMA and 1,25-$(OH)_2D_3$, 263–270
In vitro models of mutagenesis, 481–492
In vitro radiogenic transformation, 310–312
In vitro systems, to study organ and species differences in metabolic activation of chemical carcinogens, 135–145
In vivo actions of N-nitroso compounds, relation of *in vitro* actions to, pharmacokinetics and DNA dosimetry in, 109–119
Incidence(s)
and dose, relation between, in radiation carcinogenesis, 385–394
and dose-rate effects, leukemia subtypes and, with chronic radiation, 366–368
Individuals, different, drugs for predicting carcinogen metabolism in, 149–152
Induction
activation or, of C-kinase, 281–283
of differentiation, by phorbol esters, HL-60 variant reversibly resistant to, 287–300
of macrophage cell differentiation, in HL-60 cells, by biologically active metabolite of vitamin D_3, 265–268

Induction period, in radiation carcinogenesis, 385–389
Inhibition
and enhancement, of oncogenic cell transformation, in C3H 10T1/2 CL8 cells, 225–233
of tumor promotion, in mouse skin, 282–284
Initiation, 69–70
cellular changes associated with, 202–205
molecular changes associated with, 205–209
and promotion, skin tumor, 189–192
Intact animal, apparent discrepancies between *in vitro* studies and *in vivo* studies with N-nitroso proline and N-nitrosocimetidine in, 110–115
Intercalation, in polycyclic aromatic hydrocarbon carcinogenesis, 449–462
Interindividual differences in, metabolism of xenobiotics, 147–161
Isolation, of TPA-resistant variants, 290–292

Karyotype, and occupation, in adult patients with acute leukemia, 414–416
Keratinocytes, epidermal, stimulation of terminal differentiation of, 278
Kinase activity, association of, with phorbol ester receptor, 252–254
C-Kinase, absence of activaton or induction of, 281–283

LET, *see* Linear-energy-transfer
Leukemia
acute, *see* Acute (nonlymphocytic) leukemia
human, chromosome abnormalities in, as indicators of mutagenic exposure, 409–418
Leukemia-prone subgroups, aplasia-prone and, preclinical responses of, 369–373
Leukemia subtypes, incidences, and dose-rate effects, with chronic radiation, 366–368
Leukemogenesis, chronic radiation, 363–378
Life cycle, of retroviruses, 15–18
Life-span, prolonged, emergence of clones with, in human diploid cells, 348–351
Linear-energy-transfer, 393–394
Liver, mouse, some alkenylbenzenes and aromatic amines in, sulfuric acid esters as ultimate electrophilic and carcinogenic metabolites of, 93–105
Liver cell-mediated assays, species differences in data from, 145
Localization, of RB gene, 403
Long-term cultures, X-ray-induced chromosomal rearrangements in, 341–345
Low-level radiation, carcinogenic risks of, biological basis for assessing, 381–397

Macrophage cell differentiation, in HL-60 cells, induction of, by biologically active metabolite of vitamin D_3, 265–268

Malignancy(ies)
human
oncogenes in, 38–40
transforming genes of, 35–45
reversibility of
oncogenes and, 29–31
origin and, 23–31
Malignant cells, transformation of normal into, in culture, 23–25
Malignant transformation of human diploid cells, mechanisms of, 337–352
Mammalian cell mutation, and polycyclic hydrocarbon carcinogenesis, 433–444
Markers, enzyme, of differentiation, 293–296
Mechanism(s)
of action, of chemical carcinogens and tumor promoters, 177–185
of activation, of *ras* oncogenes, 42–45
cellular, of hematopoietic recovery, 373–376
of chemically induced multistep neoplastic transformation in C3H 1OT1/2 cells, 211–221
of covalent binding, 450–452
genetic, of oncogenesis, 15–20
leading to hemizygosity or homozygosity of RB gene, 404–406
of malignant transformation of human diploid cells, 337–352
in multistage skin tumorigenesis, 189–197
Melanin synthesis, of B16 melanoma cells, stimulation of, 279
Membrane effects, and activation of protein kinase C, by tumor promoters, 178–181
Membrane signals, genes and, involved in neoplastic transformation, 235–245
Metabolic activation, of chemical carcinogens, *in vitro* systems to study organ and species differences in, 135–145
Metabolism
benzo(a)pyrene, analogous patterns of, in human and rodent cells, 123–131
carcinogen, in different individuals, drugs as probes for predicting, 149–152
xenobiotic, *see* Xenobiotic metabolism
Metabolism experiments, with 1-alkylbenzo(a)-pyrenes, 457–461
Metabolite(s)
electrophilic and carcinogenic, ultimate, of some alkenylbenzenes and aromatic amines in mouse liver, sulfuric acid esters as, 93–105
of vitamin D_3, biologically active, induction of macrophage cell differentiation in HL-60 cells by, 265–268
3-Methylcholanthrene-diol epoxide, 437–438
Migrant populations, xenobiotic metabolism in, 160
Misrepair, repair and
in radiation-induced neoplastic transformation, 317–334
in transformation due to fission-spectrum neutrons, 327–331,333

Molecular changes, and initiation, 205–209
Morphologic transformation, of human diploid cells, 338–342
Mouse liver, some alkenylbenzenes and aromatic amines in, sulfuric acid esters as ultimate electrophilic and carcinogenic metabolites of, 93–105
Mouse skin, tumor promotion in, inhibition of, 282–284
Mouse skin cells, chemical carcinogenesis in, 201–209
Multi-hit concept of carcinogenesis, 81–82
Multistage skin tumorigenesis, 189–197
Multistage tumor promotion, 192–194
Multistep neoplastic transformation, chemically induced, in C3H 1OT1/2 cells, mechanisms of, 211–221
Mutagenesis
  cell-mediated
    of salmonella, 141–144
    of V79 cells, 136–141
  chemical carcinogens that do not significantly produce, 219,221
  in vitro models of, 481–492
  by N-nitroso compounds, carcinogenesis and, DNA dosimetry in, 115–119
Mutagenic exposure, chromosome abnormalities in human leukemia as indicators of, 409–418
Mutagenicity, of 3-methylcholanthrene-diol epoxide, 437–438
Mutant B19, HAT-resistant revertants of, 441–444
Mutation(s)
  cell, mammalian, and polycyclic hydrocarbon carcinogenesis, 433–444
  diol-epoxide-induced, nature of, 438–442
  to ouabain resistance, chemically induced, in C3H 1OT1/2 cells
    chemical transformation and, 218–220
    development of assay for, 211–218
Mutation assays, 1-alkylbenzo(a)pyrenes and, 453–457
Mutational spectra, from human cells, gradient denaturing gel electrophoresis in obtaining, 511–527

Natural history, of neoplastic development, 68–71
Neoplasia, epigenetics of
  facts and theories on, 66–68
  genetics and, 65–77
Neoplastic development
  differentiation as epigenetic characteristic of, 74–76
  natural history of, 68–71
Neoplastic transformation
  genes and membrane signals in, 235–245
  multistep, chemically induced, in C3H 1OT1/2 cells, mechanisms of, 211–221
  radiation-induced, 317–334
Neutrons, fission-spectrum, repair/misrepair in transformation due to, 327–331,333
N-Nitroso compounds, 109–119
N-Nitrosocimetidine, 112–115
N-Nitrosoproline, 110–112
Nonevolutionary concept of hepatocarcinogenesis, 83–91
Nonviral cancer, and cancer induced by replication-competent retroviruses, oncogenes in, 18–19
Normal cells, transformation of, into malignant cells, 23–25
Normal controls, uncoupling of, growth factors, differentiation factors and, 25–27
Normal genes, and oncogenes, 26,28
Nuclease S_1, and anti-cytidine antibodies, recognition of AAF-poly(dG-dC).poly(dG-dC) by, 468–471
Nucleic acid adducts, detection of, 170–171

Occupation, karyotype and, in adult patients with acute leukemia, 414–416
Oncogene(s)
  activated, and tumor promoters, 180,182
  and cellular controls, in radiogenic transformation of rodent and human cells, 303–313
  evolution of, 58–62
  in human malignancies, activation of, by chromosomal translocation, 38–40
  in nonviral cancer and cancer induced by replication-competent retroviruses, 18–19
  normal genes and, 26,28
  ras, mechanisms of activation of, 42–45
  in retinoblastoma, 406–407
  and reversibility of malignancy, 29–31
  viral, and proto-oncogenes, 35–38
Oncogenesis, genetic mechanisms of, 15–20
Oncogenic cell transformation, in C3H 1OT1/2 CL8 cells, inhibition and enhancement of, 225–233
Organ differences, and species differences, in metabolic activation of chemical carcinogens, 135–145
Organs, tissues, and cells, 382–383
Origin and reversibility of malignancy, 23–31
Ouabain resistance, chemically induced mutation to, in C3H 1OT1/2 cells
  chemical transformation and, 218–220
  development of assay for, 211–218
Oxidant defenses, 240

Pathological predisposition, survival and, as function of stem cell radiosensitivity, 376–377
Pathological response patterns, survival and, in chronic radiation leukemogenesis, 365–367
Patients, adult, with acute leukemia, karyotype and occupation in, 414–416

Perfect Z-DNA conformations, and distorted, of AAF-poly(dG-dC).poly(dG-dC), 475–477
Permissive factors, and protective, 306–310
Pharmacokinetics, and DNA dosimetry, in relating *in vitro* to *in vivo* actions of N-nitroso compounds, 109–119
Phenotype, promotion-resistant, 238–239
Phenotypic surface characteristics, of R1B6 cells, 294–295,297–298
Phorbol ester analogs, endogenous
  diacylglycerols as, 255–259
  receptors and, 249–260
Phorbol ester receptor, association of kinase activity with, 252–254
Phorbol ester recognition, and response, heterogeneity in, 250–253
Phorbol esters, induction of differentiation by, HL-60 variant reversibly resistant to, 287–300
Phorbol 12-myristate 13-acetate, and 1,25-$(OH)_2D_3$, 263–270
Phospholipid, in properties of phospholipid apo-receptor complex, 254–255
Physical agents, chemical and, combined with radiation, 394–397
PMA, *see* Phorbol 12-myristate 13-acetate
Poly(ADP-ribose), 309
Polycyclic (aromatic) hydrocarbon carcinogenesis
  intercalation in, 449–462
  mammalian cell mutation and, 433–444
Poly(dG-dC).poly(dG-dC), AAF-modified, *see* N-2-Acetylaminofluorene-poly(dG-dC).-poly(dG-dC)
Potencies, of chemicals that produce genetic damage, estimation of, 495–507
Preclinical responses of aplasia-prone and leukemia-prone subgroups, 369–373
Predisposition, pathological, survival and, as function of stem cell radiosensitivity, 376–377
Probes, for predicting carcinogen metabolism in different individuals, drugs as, 149–152
Progenitor clonotypes, radioresistant, 375–376
Progression, 70
  skin tumor, critical events in, 194–197
Prolonged life-span, emergence of clones with, in human diploid cells, 348–351
Promoters
  chemical carcinogens and, 177–185
  decreased G_T synthesis in response to, 239
  phorbol ester, receptors and endogenous analogs for, 249–260
  synergy between activated oncogene and, 180,182
Promoting agents, control of cell differentiation in cultured human cells after treatment with, 264–265
Promotion, 69,71
  initiation and, skin tumor, 189–192
  in mouse skin, inhibition of, 282–284
  multistage, 192–194
  regulation of, by $1\alpha25(OH)_2D_3$, 279–284
Promotion-resistant phenotype, 238–239
Promotion sensitivity genes, 240–244
Protective factors, permissive and, 306–310
Protein kinase C, activation of, by tumor promoters, 178–181
Proto-oncogenes
  and chromosome alterations, in carcinogenesis, 422–428
  ras, of human cells, 40
  viral oncogenes and, functions of, 35–38

Radiation
  chemical carcinogens and, transformation of normal into malignant cells in culture by, 23–25
  low-level, carcinogenic risks of, biological basis for assessing, 381–397
  ultraviolet, 395–396
Radiation carcinogenesis, 382–385
Radiation-induced neoplastic transformation, repair and misrepair in, 317–334
Radiation leukemogenesis, chronic, 363–378
Radiogenic transformation, of rodent and human cells, oncogenes and cellular controls in, 303–313
Radioresistance, acquired, of hematopoietic elements, as primary mechanism of recovery and accommodation, 373–375
Radioresistant progenitor clonotypes, 375–376
Radiosensitivity, stem cell, survival and pathological predisposition as function of, 376–377
ras genes, detection of, as human transforming genes, 41–42
ras oncogenes, 42–45
ras proto-oncogenes, of human cells, 40
RB gene, *see* Retinoblastoma gene
Receptor(s), phorbol ester
  association of kinase activity with, 252–254
  and endogenous analogs, 249–260
Recessive (regulatory?) susceptibility gene, human cancer, 403–407
Recognition
  of AAF-poly(dG-dC).poly(dG-dC), by nuclease S_1 and anti-cytidine antibodies, 468–471
  and response, phorbol ester, 250–253
Recovery, and accommodation, acquired radioresistance of hematopoietic elements as primary mechanism of, 373–375
Regulation
  of cell differentiation, by $1\alpha,25(OH)_2D_3$, 277–280
  of tumor promotion, by $1\alpha,25(OH)_2D_3$, 279–284
Regulatory (?) susceptibility gene, human cancer recessive, 403–407
Repair
  and misrepair

in radiation-induced neoplastic transformation, 317–334
in transformation due to fission-spectrum neutrons, 327–331,333
of sublethal and subeffective transformation damage by γ-rays, 326–328
Replication-competent retroviruses, 18–19
Response(s)
cellular
to carcinogen-induced DNA damage, 177–179
in chronic radiation leukemogenesis, 363–378
growth, differential, of radioresistant progenitor clonotypes, 375–376
preclinical, of aplasia-prone and leukemia-prone subgroups, 369–373
recognition and, phorbol ester, 250–253
to tumor promoters, decreased G_T synthesis in, 239
Response patterns, pathological, survival and, in chronic radiation leukemogenesis, 365–367
Retinoblastoma, oncogenes in, 406–407
Retinoblastoma gene, 403–407
Retroviruses
life cycle of, 15–18
replication-competent, cancer induced by, nonviral cancer and, oncogenes in, 18–19
Reversibility, of malignancy
oncogenes and, 29–31
origin and, 23–31
Reversibly resistant HL-60 variant, and induction of differentiation by phorbol esters, 287–300
Revertants, of mutant B19, HAT-resistant, 441–444
Risks, carcinogenic, of low-level radiation, biological basis for assessing, 381–397
Rodent cells, human and
analogous patterns of benzo(a)pyrene metabolism in, 123–131
radiogenic transformation of, oncogenes and cellular controls in, 303–313
R1B6 cells, 294–295,297–298

Salmonella, cell-mediated mutagenesis of, 141–144
Scavengers, free radical, and antioxidants, protection by, 308–310
Sex, and susceptibility to radiation carcinogenesis, 383–384
Short-term tests, for carcinogens, 496–497
Signals, membrane, genes and, involved in neoplastic transformation, 235–245
Single-hit concept of hepatocarcinogenesis, 83–91
Skin, mouse, *see* Mouse skin *entries*
Skin tumor initiation and promotion, critical targets and events in, 189–192
Skin tumor progression, 194–197
Skin tumorigenesis, multistage, 189–197
Smoking, cigarette, stimulatory effect of, on biotransformation of xenobiotics and steroid hormones, 153–156
Species differences
in data from liver cell-mediated assays, 145
organ differences and, in metabolic activation of chemical carcinogens, *in vitro* systems to study, 135–145
Spectra, mutational, from human cells, gradient denaturing gel electrophoresis in obtaining, 511–527
Stabilization, of Z-DNA conformation, by chemical carcinogenesis, 465–477
Stem cell radiosensitivity, 376–377
Steroid hormones, xenobiotics and, biotransformation of
effect of cigarette smoking on, 153–156
effects of diet on, 156–161
Stimulation
of melanin synthesis of B16 melanoma cells, 279
of terminal differentiation of epidermal keratinocytes, 278
Sublethal and subeffective transformation damage by γ-rays, repair of, 326–328
Sulfuric acid esters, as ultimate electrophilic and carcinogenic metabolites of some alkenylbenzenes and aromatic amines in mouse liver, 93–105
Surface characteristics, phenotypic, of R1B6 cells, 294–295,297–298
Survival
and pathological predisposition, as function of stem cell radiosensitivity, 376–377
and pathological response patterns, in chronic radiation leukemogenesis, 365–367
transformation and, 324–326
Susceptibility, to radiation carcinogenesis, factors affecting, 382–385
Susceptibility gene, human cancer recessive (regulatory?), 403–407
Synergy, between activated oncogene and tumor promoters, 180,182
Synthesis(es)
of 1-alkylbenzo(a)pyrenes, 452–454
decreased G_T, in response to tumor promoters, 239
melanin, of B16 melanoma cells, 279

Targets, critical, and events, in skin tumor initiation and promotion, 189–192
Terminal differentiation of, epidermal keratinocytes, stimulation of, 278
Tests, short-term, for carcinogens, 496–497
12-O-Tetradecanoylphorbol-13-acetate-resistant variants, isolation of, 290–292
Therapy, cytotoxic, acute leukemia secondary to, 409–415
Three-tier screen, 502–507
Tissues, and cells, organs and, 382–383

TPA, *see* 12-O-Tetradecanoylphorbol-13-acetate
Transcription, gene, in tumor cells, 182–184
Transformation
  chemically induced
    of BALB 3T3 cells, enhancement of, 279–281
    in hamster embryo cells, by 1,25-$(OH)_2D_3$, enhancement of, 266–270
  differentiation and, control of, *in vitro*, by PMA and 1,25-$(OH)_2D_3$, 263–270
  due to fission-spectrum neutrons, repair/misrepair in, 327–331,333
  malignant, of human diploid cells, mechanisms of, 337–352
  morphologic, of human diploid cells, 338–342
  neoplastic, *see* Neoplastic transformation
  of normal into malignant cells in culture, 23–25
  oncogenic cell, in C3H 1OT1/2 CL8 cells, inhibition and enhancement of, 225–233
  permissive and protective factors in, 306–310
  radiogenic, of rodent and human cells, oncogenes and cellular controls in, 303–313
  and survival, 324–326
Transformation assay system, 318–319
Transformation damage, sublethal and subeffective, by γ-rays, repair of, 326–328
Transformation dependence, on seeding density, 322–324
Transforming genes
  human
    and chromosome changes, 420–422
    detection of *ras* genes as, 41–42
  of human malignancies, 35–45
  in JB6 system, 243–244
Translocation, chromosomal, activation of oncogenes in human malignancies by, 38–40
Trisialoganglioside synthesis, 239
Tumor, skin, *see* Skin tumor *entries*
Tumor cells, 182–184
Tumor promoters, *see* Promoters
Tumor promotion, *see* Promotion
Tumorigenesis, skin, multistage, 189–197
Tumorigenicity experiments, with 1-alkylbenzo(a)pyrenes, 455,458

Ultimate electrophilic and carcinogenic metabolites, of some alkenylbenzenes and aromatic amines in mouse liver, sulfuric acid esters as, 93–105
Ultraviolet radiation, 395–396
Uncoupling, of normal controls, growth factors, differentiation factors and, 25–27
Urinary excretion of adducts, 173–174

Viral oncogenes, and proto-oncogenes, functions of, 35–38
Vitamin D_3, biologically active metabolite of, induction of macrophage cell differentiation in HL-60 cells by, 265–268
V79 cells, 136–141

Xenobiotic metabolism
  factors influencing, 152–153
  interindividual differences in, 147–161
  in migrant populations, 160
Xenobiotics, and steroid hormones, biotransformation of
  effect of cigarette smoking on, 153–156
  effects of diet on, 156–161
X-ray-induced chromosomal rearrangements in long-term cultures, 341–345